I0044706

# Biofuels: Production and Technologies

# Biofuels: Production and Technologies

Edited by Damian Price

New York

Published by Syrawood Publishing House,
750 Third Avenue, 9th Floor,
New York, NY 10017, USA
www.syrawoodpublishinghouse.com

**Biofuels: Production and Technologies**
Edited by Damian Price



International Standard Book Number: 978-1-68286-609-2 (Hardback)



**Cataloging-in-Publication Data**

Biofuels : production and technologies / edited by Damian Price.
p. cm.
Includes bibliographical references and index.
ISBN 978-1-68286-609-2
1. Biomass energy. I. Price, Damian.
TP339 .B56 2018
333.953 9--dc23

# TABLE OF CONTENTS

# PREFACE

This book was inspired by the evolution of our times; to answer the curiosity of inquisitive minds. Many developments have occurred across the globe in the recent past which has transformed the progress in the field.

Biofuel is the alternative form of energy, which is derived from biological processes like anaerobic digestion, etc. It is also generated directly from plants or plant based activities like agriculture and domestic waste. Biomass is also a main provider of biofuel in liquid, gas, and solid form. The book presents researches and studies performed by experts across the globe on biofuels and their generation. The topics covered herein offer new insights in the field of biofuel energy. This book will help new researchers by foregrounding their knowledge in this branch.

This book was developed from a mere concept to drafts to chapters and finally compiled together as a complete text to benefit the readers across all nations. To ensure the quality of the content we instilled two significant steps in our procedure. The first was to appoint an editorial team that would verify the data and statistics provided in the book and also select the most appropriate and valuable contributions from the plentiful contributions we received from authors worldwide. The next step was to appoint an expert of the topic as the Editor-in-Chief, who would head the project and finally make the necessary amendments and modifications to make the text reader-friendly. I was then commissioned to examine all the material to present the topics in the most comprehensible and productive format.

I would like to take this opportunity to thank all the contributing authors who were supportive enough to contribute their time and knowledge to this project. I also wish to convey my regards to my family who have been extremely supportive during the entire project.

**Editor**

# 1

# A review on the effect of proton exchange membranes in microbial fuel cells

Mostafa Rahimnejad *[1], Gholamreza Bakeri[1], Ghasem Najafpour[1], Mostafa Ghasemi[2], Sang-Eun Oh[3]

[1]*Biotechnology Research Lab., Faculty of Chemical Engineering, Babol University of Technology, Babol, Iran*

[2]*Fuel Cell Institute, Universiti Kebangsaan Malaysia, 43600 UKM Bangi, Selangor Darul Ehsan, Malaysia*

[3]*Department of Biological Environment, Kangwon National University, Chuncheon, Kangwon-do, Republic of Korea*

## HIGHLIGHTS

- MFC is a novel knowledge that can be used to obtain bioenergy in the form bioelectricity.
- Transfer of produced electrons to anode is one of the main parts in MFCs.
- Some MFCs needs artificial electron shuttle in their anaerobic anode compartment.
- The important goal of MFCs is to reach a suitable power for application in small electrical devices.

## ABSTRACT

Microorganisms in microbial fuel cells (MFC) liberate electrons while the electron donors are consumed. In the anaerobic anode compartment, substrates such as carbohydrates are utilized and as a result bioelectricity is produced in the MFC. MFCs may be utilized as electricity generators in small devices such as biosensors. MFCs still face practical barriers such as low generated power and current density. Recently, a great deal of attention has been given to MFCs due to their ability to operate at mild conditions and using different biodegradable substrates as fuel. The MFC consists of anode and cathode compartments. Active microorganisms are actively catabolized to carbon sources, therefore generating bioelectricity. The produced electron is transmitted to the anode surface but the generated protons must pass through the proton exchange membrane (PEM) in order to reach the cathode compartment. PEM as a key factor affecting electricity generation in MFCs has been investigated here and its importance fully discussed.

**Keywords:**
Microbial fuel cell (MFC)
Bioelectricity
Proton exchange membrane

## 1. Introduction

In a near future, fossil fuels will be depleted. Furthermore, traditional sources of energy have many a lot of disadvantages such as greenhouse gas production. In fact the emissions produced from these energy sources through human activities has proven to be the main cause of global warming and climate change (Barat et al., 2008; Najafpour et al., 2011). Several countries in the world, however have responded to the threats of energy security and global warming by diversifying their fuel sources to include renewable and alternative energy and developing clean energy technologies to replace the conventional ones (Daud et al., 2011).

Generating energy from renewable sources, such as biomass, is not only reliable and sustainable but also helps reduce global carbon dioxide emissions (Jung and Regan, 2007; Greenman et al., 2009; Kim and Chang, 2009; Oh et al., 2009).

A key method for generating renewable and alternative energy is through use of fuel cell technology. However, most fuel cell technologies require hydrogen, which is derived from fossil fuels (Jafary et al., 2013). The use of fossil fuels may lead to global warming, environmental pollution and climate change (Gunkel, 2009). Therefore, generating hydrogen from fossil fuels may not be a suitable alternative for replacing an energy resource (Stoica et al.,

* Corresponding author
E-mail address: Rahimnejad@nit.ac.ir & Rahimnejad_mostafa@yahoo.com (M. Rahimnejad).

2009). Microbial fuel cells (MFC) are one type of fuel cells, which are reliable for limited power production (Oh and Logan, 2005).

MFC is a novel knowledge that can be used to obtain bioenergy in the form of hydrogen and/or electricity, directly from different organic and inorganic compounds, while simultaneously treating biodegradable contaminants in wastewaters (Oh and Logan, 2005; Rahimnejad et al., 2012a). In MFCs, the electrons are provided from chemical bonds with the aid of active microorganisms such as enzymes or bacteria. Then, the generated electrons are transported to the anode electrode and the produced protons are moved through a proton exchange membrane toward the cathode compartment (Wen et al., 2009). The electron flows through an electrical external circuit while the anode is connected to the cathode (Fatemi et al.,2012; Rahimnejad et al., 2011b; Rahimnejad et al., 2012b). The flow of electron creates a current (I) and a power (P). The reduction of organic substances in the anode was catalyzed by living organism in the anode chamber (Chen et al., 2008; Rahimnejad et al., 2009).

## 2. Effective parameter on MFCs

The performance of MFC may be enhanced through several important process parameters which are critical to its operation such as: (i) cell metabolism, (ii) microbial electron transfer, (iii) proton exchange membrane transfer, (iv) external and internal resistances and (v) cathode oxidation. These process parameters have great influence on the transfer of the electron and power generation (Rabaey et al., 2003; Jafary et al., 2012).

The basic part in MFCs technology is the active biocatalyst. Active electro genic bacterial strains can transfer the produced electrons via metabolism across the cell membrane to an external electrode directly without adding any artificial components (Tardast et al.,2014). This mechanism plays an important role in harvesting bioelectricity in the MFC bioreactor. Even though the mechanism of extracellular electron transfer has not yet been fully understood, some possible pathways have been suggested, including direct outer membrane c-type cytochrome/anode coupling, through either redox electron shuttles or electrically conductive pili (see Fig. 1). Fig. 1 shows the principles of two-chambers of an MFC (Qian and Morse, 2011). Generally, oxygen is used as the final electron acceptor in the cathode. Eventually the combination of electrons and protons with oxygen forms water and ends this transfer cycle. Oxidized mediators can further accelerate the water formation process in the cathode chamber (Heitner-Wirguin, 1996).

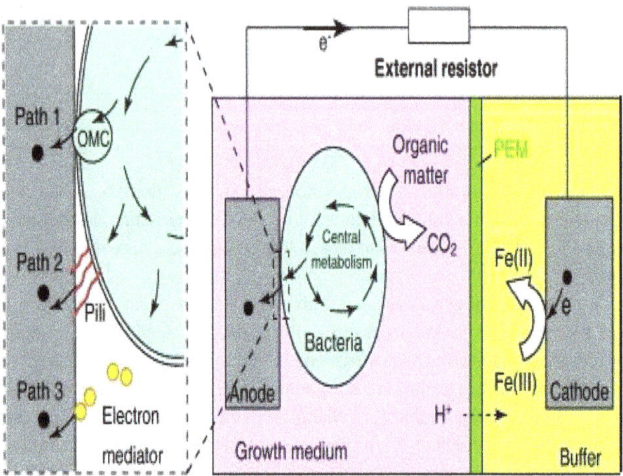


**Fig.1.** Principles of a two-chamber MFC.

Traditional MFCs consist of a cathode and anode compartment that is separated by a salt bridge or proton exchange membrane (PEM) (Fatemi et al., 2010). The active microorganisms are inculcated in the anaerobic anode chamber, where this biocatalyst used substrate and generate bioelectricity via their central metabolism (Jafary et al.,2013; Rahimnejad et al.,2011b; Mokhtarian et al., 2012a). Some of the produced electrons can transfer to the anode surface and then form an external circuit, move to the cathode and react with oxidants in the cathode electrode surface (Rahimnejad et al., 2010). To preserve the neutrality of the electro chemicals, the generated protons in the anode compartment are passed through the PEM to the cathode compartment. The important factors that need to be taken into account when investigating the MFCs performance include Columbic efficiency, power and current density, biological oxygen demand removal efficiency and sustainability (Oh et al., 2009; Qian and Morse, 2011).

### *2.1. Electron Transfer in MFCs*

In an anaerobic anode compartment, electro genic active microorganisms catalyze the oxidation of organic matter and generate electrons (represented by black circles in Fig.1) in their central metabolism. Some of these produced electrons are extracellularly transferred to an anode electrode via distinct pathways, path 1: including through direct outer membrane protein/anode coupling, path 2: conductive pili, and/or path 3: via self-secreted electron shuttles (Qian and Morse, 2011). Microorganisms as biocatalyst in the MFC consumed different substrates (such as glucose) as their source of carbon in the anode chamber and the produced electrons and protons. In the case of glucose being used as fuel for the MFC, the anodic and catholic reactions have been presented in equations 1 and 2.

$$C_6H_{12}O_6 + 6H_2O \longrightarrow 6CO_2 + 24\ e- + 24H^+ \qquad \text{(Eq. 1)}$$

$$6O_2 + 24\ e- + 24H^+ \longrightarrow 12H_2O \qquad \text{(Eq. 2)}$$

The performance of the MFCs can be improved through the addition of artificial electron mediators (Rahimnejad et al., 2009b; Mokhtarian et al., 2012b; Rahimnejad et al., 2012b). Electron mediators are used to shuttle electrons from the broth to the anode electrode surface. Mediators are artificial compounds or produced by the microorganism itself. Some active microorganisms produce nanowires to transfer the produced electrons directly without using any mediator but other organisms need to add artificial electron shuttles into the anode chamber (Rahimnejad et al., 2011). Park and Zeikus (2000) investigated the interactions between bacterial cultures and electron mediators. The effect of thionine and neutral red as mediators for the oxidation and reduction of energy carriers such as nicotinamide adenine dinucleotide ($NAD^+$) was investigated. The biomolecules, $NAD^+$ and NADH are in the oxidized and reduced forms, respectively. Several types of mediators were used in MFCs to enhance the electron transfer (Bennetto et al., 1985).

The soluble redox mediators have been added to the anode chamber to improve electron transfer. Several researchers have developed advanced anode materials by impregnating the anode with chemical catalysts (Park and Zeikus, 2000; Choi et al., 2004).

Thionine is one of the potential mediators for transferring the produced electron to the anode surface in MFCs. Thionine as mediator is not involved in any biochemical reaction. It has been reported that thionine may not be necessary for short incubation time while for long durations, thionine enhances electron transfer (Choi et al., 2007; Rahimnejad et al., 2012b). Table 1 shows some of the MFCs that were examined with mediators and different components as substrate.

Recently, MFCs as a new renewable source of energy have been extensively reviewed by different researchers. Their investigation includes studies on the different substrates used in MFCs (Pant et al., 2010), the different Nano-composite materials used as electrode material for MFCs (Ghasemi et al., 2013d; Ghasemi et al., 2013e). The development of MFCs and their applications (Franks and Nevin, 2010), decreasing the size of MFCs

**Table 1**
Maximum generated power and current of MFC with different types of mediators and microorganisms

| Reference | Current density | Power density | Microorganism | Substrate | Mediator |
|---|---|---|---|---|---|
| (Najafpour et al. 2011) | 1600 mA $m^{-2}$ | 190 mW $m^{-2}$ | *Saccharomyces cerevisiae* | Glucose | Neutral Red |
| (Ringeisen et al. 2006) | 44.4 mA $m^{-2}$ | 22.2 mW $m^{-2}$ | *Shewanellaoneidensis* | Lactate | Anthraquinone-2,6-disulfonate (AQDS) |
| (Thygesen et al. 2009) | 85 mA $m^{-2}$ | 28 mW $m^{-2}$ | Domestic waste water | Glucose | Humic acid |
| (Vega , Fernández 1987) | | _ | *Streptococcus lactis* | Glucose | Ferric chelate complex |
| (Thygesen et al. 2009) | 589 mA $m^{-2}$ | 123 mW $m^{-2}$ | Domestic waste water | Acetate | Humic acid |
| (Rabaey, Korneel et al. 2005) | _ | 479m W $m^{-3}$ | Mixed consortium | Glucose, Sucrose | Mediator-less |
| (Thygesen et al. 2009) | 145 mA $m^{-2}$ | 32 mW $m^{-2}$ | Domestic waste water | Xylose | Humic acid |

(Wang et al., 2011a), the introduction of several terminologies and in MFCs (Logan et al., 2006), the mechanisms used for current generation (Logan, 2009), state of the art information on MFCs and recent progresses in MFC technologies (Du et al. 2007) comparison of MFCs with conventional anaerobic digestion (Pham et al. 2006), cathodic limitations in MFCs (Rismani-Yazdi et al., 2008) and electrode material in MFC (Zhou et al., 2011). But, a comprehensive review on the effect of proton exchange membrane (PEM) is still lacking. PEM is important for any MFCs as it acts to transfer the produced protons from the anode to the cathode compartment. The efficiency and economic viability of MFCs depend strongly on the performance of PEM. The aim of this paper is to highlight the PEMs materials that have been used in MFCs, their improvement and also their effect on the performance of MFCs. (Tardast et al., 2012; Tardast et al., 2014).

*2.2. Resistances in MFCs*

Several parameters affect the performance of MFCs and the generated bioelectricity, namely microbial inoculums, chemical substrates, mass transfer areas, absence or existence of proton exchange materials, mechanism of electron transfer to the anode surface, the internal and external resistance of cells, solution ionic strength, electrode materials and the electrode spacing (Park and Zeikus, 2000; Gil et al., 2003; Liu and Logan, 2004; Li et al., 2011).

Mass transfer is also one key parameter in MFCs. Fig. 2 shows all different resistances in MFCs and also biological fuels. There are three kinds of over potentials such as over potentials for activation, ohmic losses and concentration polarization. For MFCs, the activation over potential appears to be the major limiting factor. Furthermore, membrane resistance has an important role on MFC performances because the produced protons must be transferred from the anaerobic anode to the cathode compartment (Rabaey et al., 2005b). In addition to this, there is another important parameter related to PEM. The cathode of MFCs works in aerobic conditions while the anode would be working in anaerobic conditions. This means the oxygen that exists in this chamber should diffuse from the cathode to the anode chamber through PEM that reduces the performance of MFCs. These two parameters affect the power generated by MFCs and must be taken into consideration. If the oxygen diffuses through PEM and makes the anode aerobic, the MFC cannot produce power and can only be applied for COD removal and wastewater treatment (Ghasemi et al., 2012).

Quantitative investigation of the electrochemical dynamics and resistances at different parts of MFCs can be conducted by a potentiostat or electrochemical station. Exploration of different MFC configurations, materials for the anode and cathode electrodes, bacterial strains, substrates and kinds of PEMs are the major focus of current MFC researches.

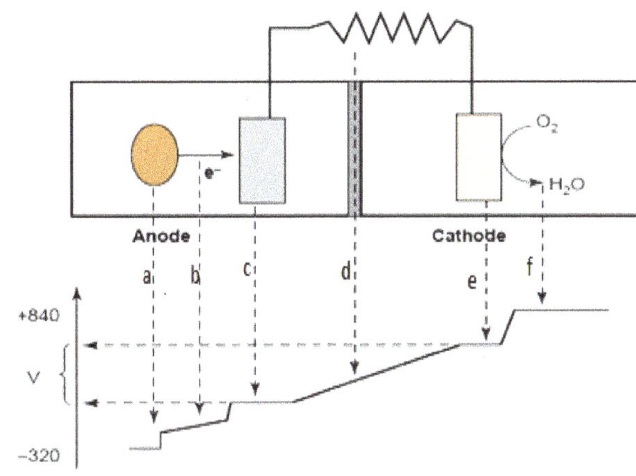


**Fig.2.** Potential losses during electron transfer in a MFC.
**a**: loss owing to bacterial electron transfer, **b**: losses owing to electrolyte resistance, **c**: losses at the anode, **d**: losses at the MFC resistance (useful potential difference) and membrane resistance losses, **e**: losses at the cathode, **f**: losses owing to electron acceptor reduction (Rabaey and Verstraete, 2005).

*2.3. Proton Exchange Membranes*

Most of the MFCs consist of two separate parts. In a two-chamber design, the anode and the cathode compartments are separated by an ion-selective membrane, allowing proton transfer from the anode to the cathode and preventing oxygen diffusion in the anode chamber from the cathode compartment. The membrane in the MFCs plays an important role on MFC performance. The membrane must have good capability for exchanging protons (Watanabe, 2008). Generally, there are two types for PEM; porous proton exchange membranes and nonporous membranes called dense membranes (Mayahi et al., 2013). In fuel cells, the main duty of dense membranes is to separate the anode and the cathode and to prevent the migration of the anode electrolyte to the cathode compartment as well as preventing the air from moving, which was purged in the cathode compartment, to the anode compartment (Leong et al., 2013). Fig. 3 shows the micrograph for the cross section of a porous membrane. It should be noted that porous or nonporous membranes are distinguished by their cross section. The micrograph for the cross section of a nonporous membrane is shown in Fig. 4.

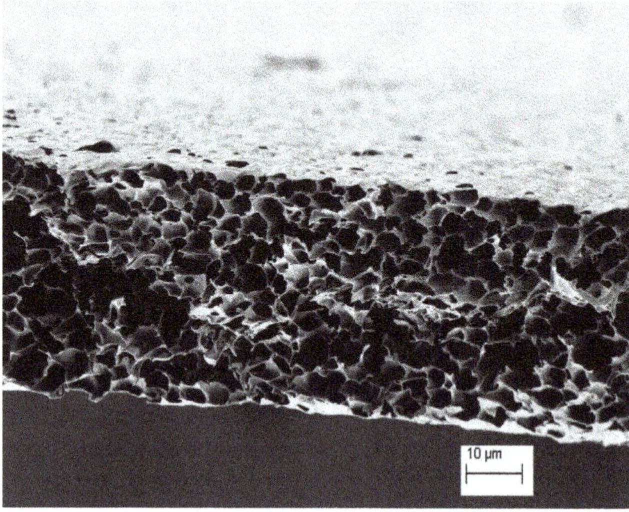


**Fig. 3.** Cross section SEM image of porous membrane.

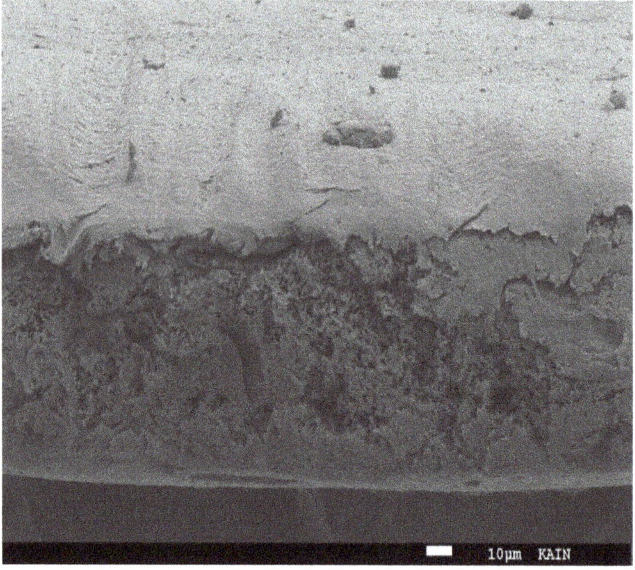


**Fig. 4.** Cross section SEM image of nonporous membrane.

From the SEM images, it is obvious that the porous membrane has a lot of pores along its cross section while there is no pore for the dense membrane. The AFM images of a dense membrane have been shown in Fig. 5 (a-b) revealing a dense layer on top of the membrane without any pores.

(a)

(b)

**Fig. 5.** a) 3D AFM image of a dense membrane, b) 2D AFM image of a dense membrane.

### 2.3.1. *Nafion as traditional PEM*

In the MFC, the Nafion membrane equilibrates with the cation species present in the anolyte and catholyte (Rahimnejad et al., 2010). This equilibration quickly changes the membrane from its proton form to a form in which mainly other cation species occupy the negatively charged sulfonate groups. More than 99.999% of the sulfonate groups are occupied with non-proton cations, because the sulfonate groups of Nafion have a higher affinity for most other cation species (Okada et al., 1997; Okada et al., 1998; Kelly et al., 2005). Subsequently, these cations combined with the sulfonate groups of Nafion stop the movement of protons that are produced at the time of substrate degradation.

In addition, other cation species have a higher concentration in the anolyte than protons which make proton transport slightly minor compared to the transport of other cations, causing a decrease in MFC performance. The diffusion coefficient of protons in the Nafion is relatively higher than other cations.

Chae et al. (2007) also, considered cation transport in an uninoculated MFC and reported an increase in the concentration of the cation species in the catholyte. The cation transport rates were slower than the reported ones in previous studies using an inoculated MFC (Rozendal et al., 2006).

Currently, the most available PEM for MFCs is Nafion from Dupont but this cannot operate efficiently at temperatures higher than 90 °C due to thermal instability (Rowe and Li, 2001; Ghassemi et al., 2006).The fuel cell electrochemical processes research group at the Fuel Cell Institute however, have successfully developed a new high temperature composite called Nafion-silicon oxide ($SiO^2$)- acid (PWA) a composite membrane with lower resistance, higher proton conductivity, higher current density and better thermal stability at 90 °C than the Nafion membrane from Dupont (Daud et al.,. 2004; Mahreni et al., 2009) and the Aciplex membrane from Asahi (Wang et al.,2011a). Rozendal et al. (2006) examined the effects of cation transport through Nafion 117 membrane on the cathode PH and MFC performance. In a two-compartment MFC, the number of cations other than the protons ($K^+$, $Na^+$, $Mg^{2+}$, $Ca^{2+}$) transported from the anode compartment to the cathode compartment were found to be the same as the number of electrons transferred through the circuit. An analysis of the membrane of the MFC showed that $K^+$ and $Na^+$ occupied about 74% of the sulfonate residues of the membrane. The cation transport was not driven by the concentration gradient, but was an electro dialysis process. This means that virtually no proton was transported in the MFC and that electroneutrality was sustained mainly through the transport of cations and not through proton transport. This phenomenon causes a number of electrochemical and microbiological problems for the efficient operation of MFCs. The anode compartment is acidified, raising the anode potential and producing adverse conditions for the microorganisms catalyzing the anode reaction, while the cathode compartment is alkalized, which lowers the cathode potential (Gil et al., 2003; Liu and Logan, 2004). This preferential transport of cations rather than protons may be avoided by either removing the membrane (Jang et al., 2004; Liu and Logan, 2004) or by using an electrolyte containing a low cation concentration.

Ghasemi et al. (2012) has developed a new Nano-composite membrane and compared it with Nafion 117. Their creation operates by activating the carbon Nano fiber/Nafion PEM and applying it in the MFC. Their data shows that this Nano- composite membrane can produce about 1.5 times more power than the Nafion 117. Also the CNF without activation and Nafion produced 27% more power than Nafion 117. They concluded that CNF and ACNF increase the conductivity of the membrane as well as the porosity, so the capability of membranes for proton exchange increases.

Furthermore, in another research, this group evaluated the effect of pretreatment on membrane performance. Nafion 117 was the PEM that was evaluated before and after the treatment. The effect of biofouling on membrane performance was also investigated. The results show that the minimum amount of power generation belongs to the biofouled Nafion 117 which is 20.9 mW $m-^2$. This means biofouling has a negative effect on membrane performance. The untreated membrane produced about 52.8 mW $m-^2$ of power whereas the treated membrane produced about 2 times more power equivalent to 103 mW $m-^2$. Results showed that pretreatment is highly effective in membrane performance and biofouling can have a very unfavorable effect on the membrane (Ghasemi et al., 2013f).

MFC is a device for simultaneous wastewater treatment and energy production but one of the obstacles for the commercialization of MFC is the high price of PEM. Ghasemi et al. (2013e) compared the economic investigation of MFCs that are working with Nafion 117 and solfonated poly ether ether ketone (SPEEK). They found that the MFC working with Nafion 117 can produce 106.7 mW $m^{-2}$ of power that is much more than the power produced by SPEEK (77.3 mW $m^{-2}$). But the COD removal of the system working with SPEEK was 88% which is higher than the system working with Nafion 117 (76%). They compared the economic issues of both systems and found that the power per cost of the MFC with SPEEK as PEM is 2 times higher than the MFC with Nafion 117. This means that the system with SPEEK is two times more economical than the system with Nafion 117.

Due to the high price of Nafion 117, scientists have always been interested in replacing Nafion 117 with a less expensive PEM. Mokhtarian et al. (2013) used Nafion 112 and four different Nafion 112 /Pani in the MFC as a separator. They prepared these separators by pretreating the Nafion 112. Then, they immersed the Nafion 112 in an aniline solution that was dissolved in HCl for 1, 2, 3 and 4 hours and called the end product Nafion/Pani1, Nafion/Pani2, Nafion/Pani3 and Nafion/Pani4. These four membranes and Nafion 112 were then applied to the MFC to see whether or not they could be used as PEMs. Results showed that among the membranes, Nafion/Pani3 produced the highest power density of (124 mW $m^{-2}$) which is 93% of the power produced by Nafion 117. Results The results also showed that Nafion 112 could be modified for application in the MFC.

*2.4. Effect of mass transfer area on MFC performances*

Nafion mass transfer area affects the production of power in MFCs as was shown in Fig. 6. Three different mass transfer areas (3.14 $cm^2$, 9 $cm^2$ and 16 $cm^2$) were tested and the results were presented in Fig. 6(a,b). The MFC memberane allows the generated hydrogen ions in the anode chamber to be transferred to the cathode chamber (Rabaey et al., 2005a; Cheng et al., 2006; Venkata et al., 2007; Aelterman et al., 2008). The results show that the maximum current and power occur when the area of Nafion is 16 $cm^2$. The maximum power and current generated were 152 mW $m^{-2}$ and 772 mA $m^{-2}$, respectively.

**Table 2**
Effect of mass transfer area on performance of MFC.

| PEM surface area ($cm^2$) | maximum voltage (mV) | maximum power density (mW $m^{-2}$) | maximum current density (mA $m^{-2}$) |
|---|---|---|---|
| 3.14 | 848 | 105.8 | 595 |
| 9 | 851 | 126 | 710 |
| 16 | 850 | 152 | 772 |

Furthermore, the thickness of PEM has an important role in MFCs' performances. In one study, *Saccharomyces cerevisia* was used as a biocatalyst for power generation in a dual chamber MFC. Nafion 112 and 117 were selected as membranes to transport the produced proton from the anode to the cathode compartment at ambient temperatures and pressures. Initial glucose concentrations were 30 g/l. The maximum obtained voltage and the, current and power density for Nafion 117 were 668 mV, 60.28 mA $m^{-2}$ and 9.95 mW $m^{-2}$ respectively. For Nafion 112 these figures were 670 mV, 150.6 mA $m^{-2}$ and 31.32 mW $m^{-2}$ respectively (Rahimnejad et al., 2010).

## 3. Problems associated with commercial PEM in MFCs

While Nafion is the common PEM used these days, there are several problems associated with Nafion *membranes*. These problems include oxygen leakage from the cathode to the anode chamber, high costs, substrate losses, cation transport and accumulation rather than that of protons and biofouling (Chae et al., 2008). Because of these disadvantages, researchers are working to fabricate a new kind of PEM which does not have these negative features and performs better than the Nafion membrane (Liu and Logan, 2004). Nowadays, due to the many existing and potential applications of polymer/inorganic nanoparticle membranes in energy, environment and biomedical materials, more attention is being paid to membrane science and technology. Nanoparticles improve separation performance by generating preferential permeation paths while they prevent the permeation of undesired species as well as increasing thermal and mechanical properties (Taurozzi et al., 2008; Jadav, 2009; Pan et al., 2010). This means the distribution of nanoparticles through the polymer matrix modifies chemical and physical properties of polymeric membranes (Mahreni et al., 2009). Recently, due to the unique and promising properties of $Fe_3O_4$ nanoparticles (magnetic, conductive, easy synthesis, eco-friendly and catalytic characteristic), intensive attention has been paid to them (Iida et al., 2007; Chen et al., 2009).

However until now, there have been different problems in making MFCs economical. MFC performance like other fuel cells is dependent on power, current density and the rate of fuel oxidation. Several important factors can influence the rate of fuel oxidation including the catalytic activity of the anode, fuel diffusion and the diffusion and consumption of electrons and protons.

One of the important challenges is the cost of the catalyst (such as platinum) that is used for accelerating the rate of oxygen-reduction reactions which account for more than half of the cost associated with MFCs (Lefebvre et al. 2009). Many researches have been done to replace or decrease the consumption of platinum or use less expensive and stable non-noble metals as cathodic catalyst to make it more practical (Kerzenmacher et al., 2008). Nowadays, nanostructured carbon-based materials specially carbon nanotubes are becoming popular catalysts or are used as a support for the catalyst due to their high surface area (Baughman et al., 2002), high mechanical strength (Meincke et al., 2004), high electrical conductivity (Berber et al., 2000) and catalytic activity (Gong et al., 2009). The higher catalytic activity of CNT-based materials may be due to the high surface area that cause better dispersion of materials as well as creating more space for functionalization and bonds (Ghasemi et al., 2011). Wang et al. (Wang et al., 2011c) used carbon nanotubes for modification of the air cathodeès single chamber MFC. They concluded that the power produced by CNT was more than double that of traditional carbon cloth cathodes.

In addition, Ghasemi et al. (2013e) compared the effect of carbon nanotubes on increasing the power generation of MFCs. They compared an MFC that is working with Pt as cathode catalyst and another one which is working by CNT/Pt composite cathode catalyst and concluded that the power generation of the MFC which is working by CNT/Pt is 1.3 times higher than neat Pt. Furthermore, they tested Pani/Vanadium as an alternative cathode catalyst in the MFC. Due to the nature of Nano-compsoite conducting polymers, Pani/$V_2O_5$ has high catalytic activity. Compared to the MFC with Pt used as its cathode catalyst, the nanostructure Pani/$V_2O_5$ generated 79.3 mW $m^{-2}$ of power, 10% more than the Pt which generated 72.1 mW $m^{-2}$ power.

Ghasemi et al. (2013d) also studied the effect of using macrocyclic compounds as an alternative to Pt as cathode catalyst for the MFC. They applied phthalocyanine (Pc), copper phthalocyanine (CuPc) and nickel nanoparticles as macrocyclic in the cathode catalyst of the MFC and compared their performance with that of the Pt as the most common cathode catalyst used in MFCs. Their results showed that CuPc produced 118 mW $m^{-2}$ power which is very close to that produced by the Pt (120.8 mW $m^{-2}$). Nickel nanoparticles also produced 94.4 mW $m^{-2}$ power which shows they can be used as cathode catalyst in MFCs. Although the produced bioelectricity from MFCs has improved considerably and researchers are also working on obtaining better results, the generated power is related to small lab-scale systems and the MFCs' scale-up is still a big challenge. Moreover, the high cost of PEM, the potential for biofouling and related high internal resistance restrains bioelectricity production and limits the practical application of MFCs (Hu, 2008).

There are some practical ways for overcoming the existing limitations in regards to MFCs. It is agreed that most MFCs generate too little power for any envisioned applications. Besides, the high cost of metal catalysts such as platinum which is usually needed on a cathode is also a big hindrance for the scale-up of these systems. The open air biocathodes proposed by Clauwaert et al. (2007) might be a possible solution in the future.

The cathode is the most challenging aspect of the MFC design due to its need for a three-phase interface: air (oxygen), water (protons), and solid (electricity). So far, the cathode is more likely to limit power generation than the anode. Most of the MFCs use Pt as a catalyst in the cathode electrode but this is too expensive and one of the challenges facing MFCs. The replacement of platinized cathodes with non-platinized ones with a similar efficiency is a major improvement in this area. The use of manganese dioxide as an alternative cathode catalyst in MFCs and stainless steel and nickel alloys in MECs has also been suggested (Pant et al., 2010).

## 4. Important applications of MFCs

Our petroleum resources will be depleted in about 200 years and after that, vehicles will no longer be equipped with petrol tanks. Researchers in the world are working to find an alternative for this. One very good alternatives which is less wasteful and cleaner, is producing bioelectricity for vehicles directly from different substrates such as carbohydrate sources using MFCs. Complete oxidation of a monosaccharide such as glucose or disaccharide like sucrose to water and carbon dioxide can generate $16\times10^6$ J/Kg energy. This amounts to about 5 kWh of generated electrical energy. The most important goal of MFCs is to reach a suitable power generation level for application in small electrical devices. Rahimnejad et al. used MFC stacks as a power source and succeeded in turning on one digital clock and ten LED lamps. These small consumer devices managed to operate successfully for the duration of 2 days (Rahimnejad et al., 2012a). Another application of MFCs for waste water treatment (Izadi and Rahimnejad, 2013). Active microorganisms present in the anode compartment can discharge the dual duty of degrading effluent and bioelectricity production. An active biocatalyst can oxidize organic compounds presented in waste materials and produce electricity. But the produced power in these systems is still too little. If the generated power level increases in future, MFCs can decrease operating costs in waste water treatment plants (Najafpour et al., 2010; Rahimnejad et al., 2013). Different kinds of MFC reactor designs based on chemical engineering principles such as packed bed reactors, fluidized bed reactors, dual chamber reactors, single chamber reactors etc. are under investigation to reach this important aim. Scientists have reported that to remove nitrogen and organic matters from leachate, biological treatment is prevalently used as a credible and highly cost-effective method (Gotvajn et al., 2009; Mehmood et al., 2009). Rabaey et al. demonstrated that MFCs by using specific microbes can remove sulfides from wastewater (Rabaey et al., 2006). Up to 90% of the COD can be removed in some cases (Puig et al.,2011 ; Wang et al.,2012) and a columbic efficiency as high as 80% has been reported (Kim et al., 2005).

The application of MFCs as biosensors for pollutant analysis and process monitoring are another application of this technology (Chang et al., 2005). Batteries have restricted lifetime and must be changed or recharged, thus MFCs are suitable for powering electrochemical sensors and are small telemetry systems that can transmit obtained signals to remote receivers (Ieropoulos et al., 2005; Greenman et al., 2009).

## 5. Conclusion

The idea of generating electricity in biological fuel cells theoretically exists, but as a practical method for energy production, it is quite new. MFC is a new technology for bioelectricity production from sustainable materials. This new source of energy can produce bioelectricity by using microorganisms or enzymes as an active biocatalyst. The present study has revealed that MFCs have a good ability for production of low voltage electricity and PEM has an important role on two chamber MFCs performances. MFCs produce current through the action of bacteria that can pass electrons to an anode, the negative electrode of a fuel cell. The electrons flow from the anode through a wire to a cathode. Some MFCs don't need any artificial mediators to pass the produced electrons but some others need mediators in the anode compartment. In MFCs, however, operating with wastewater at neutral pH conditions, the concentrations of other cation species (e.g. $Na^+$, $K^+$, $Ca^{2+}$ and $Mg^{2+}$) are typically 105 times higher than that for protons. Though Nafion is known as a PEM, but parallel to all other commercial cation exchange membranes other cations can pass through it.

**References**


Aelterman, P., Versichele M., Marzorati M., Boon N., Verstraete W., 2008. Loading rate and external resistance control the electricity generation of microbial fuel cells with different three-dimensional anodes. Bioresour. Technol. 99(18), 8895-8902.

Barat, A., Crane M., Ruskin H. J., 2008. Quantitative multi-agent models for simulating protein release from PLGA bioerodible nano-and microspheres. J. Pharm. Biomed. Anal. 48(2), 361-36

Baughman, R. H., Zakhidov A. A. , De Heer W. A., 2002. Carbon nanotubes--the route toward applications. Science. 297(5582), 787-785.

Bennetto, H., Delaney G., Mason J., Roller S., Stirling J., Thurston C., 1985. The sucrose fuel cell: efficient biomass conversion using a microbial catalyst. Biotechnol. Lett. 7(10), 699-704.

Berber, S., Kwon Y. K., Tománek D., 2000. Unusually high thermal conductivity of carbon nanotubes. Phys. Rev. Lett. 84(20), 4613-4616.

Chae, K., Choi M., Ajayi F., Park W., Chang I., Kim I., 2008. Mass Transport through a Proton Exchange Membrane (Nafion) in Microbial Fuel Cells. Energy Fuels. 22(1), 169-176.

Chae, K. J., Choi M., Ajayi F. F., Park W., Chang I. S. , Kim I. S., 2007. Mass Transport through a Proton Exchange Membrane (Nafion) in Microbial Fuel Cells†. Energy Fuels. 22(1), 169-176.

Chang, I. S., Moon H., Jang J. K., Kim B. H., 2005. Improvement of a microbial fuel cell performance as a BOD sensor using respiratory inhibitors. Biosens. Bioelectron. 20(9), 1856-1859.

Chen, G.-W., Choi S.-J., Lee T.-H., Lee G.-Y., Cha J.-H. , Kim C.-W., 2008. Application of biocathode in microbial fuel cells: cell performance and microbial community. Appl. Microbiol. Biotechnol. 79(3), 379-388.

Chen, J., Wang F., Huang K., Liu Y. , Liu S., 2009. Preparation of Fe3O4 nanoparticles with adjustable morphology. Int. J. Energy Res. 475(1-2), 898-902.

Cheng, S., Liu H., Logan B. E., 2006. Increased power generation in a continuous flow MFC with advective flow through the porous anode and reduced electrode spacing. Environ. Sci. Technol. 40(7), 2426-2432.

Choi, Y., Jung E., Park H., Jung S., Kim S., 2007. Effect of initial carbon sources on the performance of a microbial fuel cell containing environmental microorganism micrococcus luteus. Bulltin . Korean. Chem. Society. 28(9), 1591-1598.

Choi, Y., Jung E., Park H., Paik S. R., Jung S. , Kim S., 2004. Construction of microbial fuel cells using thermophilic microorganisms, Bacillus licheniformis and Bacillus thermoglucosidasius. Bulltin. Korean. Chem. Society. 25(6), 813-818.

Clauwaert, P., Van der Ha D., Boon N., Verbeken K., Verhaege M., Rabaey K. , Verstraete W., 2007. Open air biocathode enables effective electricity generation with microbial fuel cells. Environ. Sci. Technol. 41(21), 7564-7569.

Daud, W. R. W., Mohamad A. B., Kadhum A. A. H., Chebbi R. , Iyuke S. E., 2004. Performance optimisation of PEM fuel cell during MEA fabrication. Energy Convers. Manage. 45(20), 3239-3249.

Daud, W. R. W., Najafpour G. , Rahimnejad M., 2011. Clean Energy for Tomorrow: Towards Zero Emission and Carbon Free Future: A Review. Iranica J. Energy Environ. 2(3), 262-273.

Du, Z., Li H., Gu T., 2007. A state of the art review on microbial fuel cells: a promising technology for wastewater treatment and bioenergy. Biotechnol. Adv. 25(5), 464-482.

Fatemi, S., Ghoreyshi A., Najafpour G. , Rahimnejad M.,2012. Bioelectricity Generation in Mediator-Less Microbial Fuel Cell: Application of Pure and Mixed Cultures. Iranica J. Energy Environ. 3, 2079-2115.

Franks, A. E., Nevin K., 2010. Microbial fuel cells, a current rev 7 Energies 3, 899-919.

Ghasemi, M., Daud W. R. W., Hassan S. H., Oh S.-E., Ismail M., Rahimnejad M., Jahim J. M., 2013a. Nano-structured Carbon as Electrode Material in Microbial Fuel Cells: A Comprehensive Review. Int. J. Energy Res. 580, 245-255.

Ghasemi, M., Daud W. R. W., Ismail A. F., Jafari Y., Ismail M., Mayahi A. , Othman J., 2013b. Simultaneous wastewater treatment and electricity generation by microbial fuel cell: Performance comparison and cost investigation of using Nafion 117 and SPEEK as separators. Desalination 325(0), 1-6.

Ghasemi, M., Daud W. R. W., Rahimnejad M., Rezayi M., Fatemi A., Jafari Y., Somalu M. , Manzour A., 2013c. Copper-phthalocyanine and nickel nanoparticles as novel cathode catalysts in microbial fuel cells. Int. J. Hydrogen Energy.23, 9533-9540.

Ghasemi, M., Daud W. R. W., Rahimnejad M., Rezayi M., Fatemi A., Jafari Y., Somalu M. R. , Manzour A., 2013d. Copper-phthalocyanine and nickel nanoparticles as novel cathode catalysts in microbial fuel cells. Int. J. Hydrogen Energy .38(22), 9533-9540.

Ghasemi, M., Ismail M., Kamarudin S. K., Saeedfar K., Daud W. R. W., Hassan S. H. A., Heng L. Y., Alam J. , Oh S.-E., 2013e. Carbon nanotube as an alternative cathode support and catalyst for microbial fuel cells. Appl. Energy. 102(0), 1050-1056.

Ghasemi, M., Shahgaldi S., Ismail M., Kim B. H., Yaakob Z. , Wan Daud W. R., 2011. Activated carbon nanofibers as an alternative cathode catalyst to platinum in a two-chamber microbial fuel cell. Int. J. Hydrogen Energy. 36(21), 13746-13752.

Ghasemi, M., Shahgaldi S., Ismail M., Yaakob Z., Daud W. R. W., 2012a. New generation of carbon nanocomposite proton exchange membranes in microbial fuel cell systems. Chem. Eng. J. 184, 82-89.

Ghasemi, M., Wan Daud W. R., Ismail M., Rahimnejad M., Ismail A. F., Leong J. X., Miskan M. , Ben Liew K., 2013f. Effect of pre-treatment and biofouling of proton exchange membrane on microbial fuel cell performance. Int. J. Hydrogen Energy 38(13), 4845-485.

Ghassemi, H., McGrath J. E. , Zawodzinski T. A., 2006. Multiblock sulfonated-fluorinated poly (arylene ether) s for a proton exchange membrane fuel cell. Polymer 47(11), 4132-4139.

Ghoreishi, K. B., Ghasemi M., Rahimnejad M., Yarmo M. A., Daud W. R. W., Asim N. , Ismail M., 2014. Development and application of vanadium oxide/polyaniline composite as a novel cathode catalyst in microbial fuel cell. Int. J. Energy Res. 38 (1), 70-77.

Gil, G.-C., Chang I.-S., Kim B. H., Kim M., Jang J.-K., Park H. S. , Kim H. J., 2003. Operational parameters affecting the performannce of a mediator-less microbial fuel cell. Biosens. Bioelectron. 18(4), 327-334.

Gong, K., Du F., Xia Z., Durstock M. , Dai L., 2009. Nitrogen-doped carbon nanotube arrays with high electrocatalytic activity for oxygen reduction. Science 323(5915), 760.

Gotvajn, A., Tisler T., Zagorc-Koncan J., 2009. Comparison of different treatment strategies for industrial landfill leachate. J. Hazard. Mater. 162(2), 1446-1456.

Greenman, J., Gálvez A., Giusti L., Ieropoulos I., 2009. Electricity from landfill leachate using microbial fuel cells: comparison with a biological aerated filter. Enzyme Microbial Technol. 44(2), 112-119.

Gunkel, G., 2009. Hydropower–A green energy? Tropical reservoirs and greenhouse gas emissions. CLEAN–Soil, Air, Water 37(9), 726-734.

Heitner-Wirguin, C., 1996. Recent advances in perfluorinated ionomer membranes: structure, properties and applications. J. Membr. Sci. 120(1), 1-33.

Hu, Z., 2008. Electricity generation by a baffle-chamber membraneless microbial fuel cell. J. Power Sources 179(1), 27-33.

Ieropoulos, I., Greenman J., Melhuish C., Hart J., 2005. Energy accumulation and improved performance in microbial fuel cells. J. Power Sources 145(2), 253-256.

Iida, H., Takayanagi K., Nakanishi T. , Osaka T., 2007. Synthesis of Fe3O4 nanoparticles with various sizes and magnetic properties by controlled hydrolysis. J. Colloid Interf. Sci. 314(1), 274-280.

Izadi, P., Rahimnejad M., 2013. Simultaneous electricity generation and sulfide removal via a dual chamber microbial fuel cell. Biofuel Res. J. 1(1), 34-38.

Jadav, G. L., Singh P. S., 2009. Synthesis of novel silica-polyamide nanocomposite membrane with enhanced properties. J. Membr. Sci. 328(1-2), 257-267.

Jafary, T., Ghoreyshi A. A., Najafpour G. D., Fatemi S. , Rahimnejad M., 2012. Investigation on performance of microbial fuel cells based on carbon sources and kinetic models. Int. J. Energy Res. 37 (12), 1539-1549.

Jafary, T., Najafpour G., Ghoreyshi A., Haghparast F., Rahimnejad M. , Zare H., Bioelectricity power generation from organic substrate in a Microbial fuel cell using Saccharomyces cerevisiae as biocatalysts. Fuel Cells, 4, 1182.

Jafary, T., Rahimnejad M., Ghoreyshi A. A., Najafpour G., Hghparast F. , Daud W. R. W., 2013. Assessment of bioelectricity production in microbial fuel cells through series and parallel connections. Energy Convers. Manage. 75, 256-262.

Jung, S., Regan J. M., 2007. Comparison of anode bacterial communities and performance in microbial fuel cells with different electron donors. Appl. Microbiol. Biotechnol. 77(2), 393-402.

Kelly, M. J., Fafilek G., Besenhard J. O., Kronberger H. , Nauer G. E., 2005. Contaminant absorption and conductivity in polymer electrolyte membranes. J. Power Sources. 145(2), 249-252.

Kerzenmacher, S., Ducrée J., Zengerle R., Von Stetten F., 2008. Energy harvesting by implantable abiotically catalyzed glucose fuel cells. J. Power Sources. 182(1), 1-17.

Kim, D., Chang I. S., 2009. Electricity generation from synthesis gas by microbial processes: CO fermentation and microbial fuel cell technology. Bioresour. Technol. 100(19), 4527-4530.

Kim, J. R., Min B., Logan B. E., 2005. Evaluation of procedures to acclimate a microbial fuel cell for electricity production. Appl. Microbiol. Biotechnol. 68(1), 23-30.

Lefebvre, O., Ooi W. K., Tang Z., Abdullah-Al-Mamun M., Chua D. H. C. , Ng H. Y., 2009. Optimization of a Pt-free cathode suitable for practical applications of microbial fuel cells. Bioresour. Technol. 100(20), 4907-4910.

Leong, J. X., Daud W. R. W., Ghasemi M., Liew K. B. , Ismail M., 2013. Ion exchange membranes as separators in microbial fuel cells for bioenergy conversion: A comprehensive review. Renew. Sust. Energy Rev. 28(0), 575-587.

Li, W.-W., Sheng G.-P., Liu X.-W. , Yu H.-Q., 2011. Recent advances in the separators for microbial fuel cells. Bioresour. Technol. 102(1), 244-252..

Liu, H., Logan B., 2004. Electricity generation using an air-cathode single chamber microbial fuel cell in the presence and absence of a proton exchange membrane. Environ. Sci. Technol 38(14), 4040-4046.

Logan, B. E., 2009. Exoelectrogenic bacteria that power microbial fuel cells. Nat. Rev. Microbiol. 7(5), 375-381.

Logan, B. E., Hamelers B., Rozendal R., Schröder U., Keller J., Freguia S., Aelterman P., Verstraete W. , Rabaey K., 2006. Microbial fuel cells: methodology and technology. Environ. Sci. Technol. 40(17), 5181-5192.

Mahreni, A., Mohamad A., Kadhum A., Daud W., Iyuke S., 2009. Nafion/silicon oxide/phosphotungstic acid nanocomposite membrane with enhanced proton conductivity. J. Membr. Sci. 327(1-2), 32-40.

Mayahi, A., Ismail A. F., Ilbeygi H., Othman M. H. D., Ghasemi M., Norddin M. N. A. M. , Matsuura T., 2013. Effect of operating temperature on the behavior of promising SPEEK/cSMM electrolyte membrane for DMFCs. Sep. Purif. Technol. 106, 72-81.

Mehmood, M., Adetutu E., Nedwell D., Ball A., 2009. In situ microbial treatment of landfill leachate using aerated lagoons. Bioresour. Technol. 100(10), 2741-2744.

Meincke, O., Kaempfer D., Weickmann H., Friedrich C., Vathauer M., Warth H., 2004. Mechanical properties and electrical conductivity of carbon-nanotube filled polyamide-6 and its blends with acrylonitrile/butadiene/styrene. Polymer 45(3), 739-748.

Mokhtarian, N., Ghasemi M., Daud W. R. W., Ismail M., Najafpour G., Alam J., 2013. Improvement of Microbial Fuel Cell Performance by Using Nafion Polyaniline Composite Membranes as a Separator. J. Fuel Cell Sci. Technol. 10(4), 041008-041008-041006.

Mokhtarian, N., Rahimnejad M., Najafpour G., Daud W. R. W. , Ghoreyshi A., 2012a. Effect of different substrate on performance of microbial fuel cell. Afr. J. Biotechnol. 11(14), 3363-3369.

Mokhtarian, N., Wan Ramli W., Rahimnejad M. , Najafpour G., 2012b. Bioelectricity generation in biological fuel cell with and without mediators. World Appl. Sci. J. 18(4), 559-567.

Najafpour, G., Rahimnejad M., Ghoreshi A., 2011. The Enhancement of a Microbial Fuel Cell for Electrical Output Using Mediators and Oxidizing Agents. Energ. Source. Part A 33(24), 2239-2248.

Najafpour, G., Rahimnejad M., Mokhtarian N., Daud W. R. W. , Ghore A., 2010. Bioconversion of whey to electrical energy in a biofuel cell u Saccharomyces cerevisiae. World Appl. Sci. J. 8, 1-5.

Oh, S., Kim J., Joo J., Logan B., 2009. Effects of applied voltages and dissolved oxygen on sustained power generation by microbial fuel cells. Water Sci. Technol. 60(5), 1311.

Oh, S. , Logan B. E., 2005. Hydrogen and electricity production from a food processing wastewater using fermentation and microbial fuel cell technologies. Water Res. 39(19), 4673-4682.

Okada, T., Møller-Holst S., Gorseth O., Kjelstrup S., 1998. Transport and equilibrium properties of Nafion® membranes with $H^+$ and $Na^+$ ions. J. Electroanal. Chemi. 442(1), 137-145.

Okada, T., Nakamura N., Yuasa M. , Sekine I., 1997. Ion and water transport characteristics in membranes for polymer electrolyte fuel cells containin $H^+$ and $Ca^{2+}$ cations. J. Electrochem. Society 144(8), 2744-2750.

Pan, F., Cheng Q., Jia H., Jiang Z., 2010. Facile approach to polymer-inorganic nanocomposite membrane through a biomineralization-inspired process. J. Membr. Sci. 357(1-2), 171-177.

Pant, D., Van Bogaert G., Diels L., Vanbroekhoven K., 2010. A review of the substrates used in microbial fuel cells for sustainable energy production. Bioresour. Technol.101 (6), 1533-1543.

Park, D. H. , Zeikus J. G., 2000. Electricity generation in microbial fuel cells using neutral red as an electronophore. Appl. Environ. Microbiol. 66(4), 1292-1297.

Pham, T., Rabaey K., Aelterman P., Clauwaert P., De Schamphelaire L., Boon N. , Verstraete W., 2006. Microbial fuel cells in relation to conventional anaerobic digestion technology. Eng. Life Sci. 6(3), 285-292.

Puig, S., Serra M., Coma M., Cabr© M., Dolors Balaguer M. , Colprim J.,2011. Microbial fuel cell application in landfill leachate treatment. J. Hazard. Mater. 185(2), 763-767.

Qian, F., Morse D. E., 2011. Miniaturizing microbial fuel cells. Trends Biotechnol. 29(2), 62-69.

Rabaey, K., Boon N., Hofte M. , Verstraete W., 2005a. Microbial phenazine production enhances electron transfer in biofuel cells. Environ. Sci. Technol 39(9), 3401-3408.

Rabaey, K., Lissens G., Siciliano S. , Verstraete W., 2003. A microbial fuel cell capable of converting glucose to electricity at high rate and efficiency. Biotechnol. Lett. 25(18), 1531-1535.

Rabaey, K., Ossieur W., Verhaege M. , Verstraete W., 2005b. Continuous microbial fuel cells convert carbohydrates to electricity. Water Sci. Technol. 52(1), 515-523.

Rabaey, K., Van de Sompel K., Maignien L., Boon N., Aelterman P., Clauwaert P., De Schamphelaire L., Hai The Pham, Vermeulen J. , Verhaege M., 2006. Microbial fuel cells for sulfide removal. Environ. Sci. Technol. 40(17), 5218-5224.

Rabaey, K. , Verstraete W., 2005. Microbial fuel cells: novel biotechnology for energy generation. Trends Biotechnol. 23(6), 291-298.

Rahimnejad, M., Ghasemi M., Najafpour G., Ghoreyshi A., Bakeri G., Nejad S. K. H. , Talebnia F., 2013. Aceton removal and bioelectricity generation in dual chamber microbial fuel. Am. J. Biochem. Biotechnol. 8(4), 304.

Rahimnejad, M., Ghoreyshi A., Najafpour G., Younesi H. , Shakeri M., 2012a. A novel microbial fuel cell stack for continuous production of clean energy. Int. J. Hydrogen Energy. 37(7), 5992-6000.

Rahimnejad, M., Ghoreyshi A. A., Najafpour G., Jafary T., 2011a. Power generation from organic substrate in batch and continuous flow microbial fuel cell operations. Appl. Energy. 88(11), 3999-4004.

Rahimnejad, M., Jafari T., Haghparast F., Najafpour G. D. , Goreyshi A., 2010. Nafion as a nanoproton conductor in microbial fuel cells. Turk. J. Eng. Environ. Sci. 34, 289-292.

Rahimnejad, M., Mokhtarian N., Najafpour G., Daud W., Ghoreyshi A., 2009a. Low Voltage Power Generation in aBiofuel Cell Using Anaerobic Cultures. World Appl. Sci. J. 6(11), 1585-1588.

Rahimnejad, M., Mokhtarian N., Najafpour G., Ghoreyshi A. A. , Dahud W. R. W. 2009b. Effective parameters on performance of microbial fuel cell. Environmental and Computer Science. ICECS'09. Second International Conference on, IEEE.

Rahimnejad, M., Najafpour G., Ghoreyshi A., Shakeri M. , Zare H., 2011b. Methylene blue as electron promoters in microbial fuel cell. Int. J. Hydrogen Energy. 36(20), 13335-13341.

Rahimnejad, M., Najafpour G. D., Ghoreyshi A. A., Talebnia F., Premier G. C., Bakeri G., Kim J. R. , Oh S.-E., 2012b. Thionine increases electricity generation from microbial fuel cell using Saccharomyces cerevisiae and exoelectrogenic mixed culture. J. Microbiol. 50(4), 575-580.

Ringeisen, B. R., Henderson E., Wu P. K., Pietron J., Ray R., Little B., Biffinger J. C. , Jones-Meehan J. M., 2006. High power density from a miniature microbial fuel cell using Shewanella oneidensis DSP10. Environ. Sci. Technol. 40(8), 2629-2634.

Rismani-Yazdi, H., Carver S. M., Christy A. D. , Tuovinen O. H., 2008. Cathodic limitations in microbial fuel cells: an overview. J. Power Source. .180(2), 683-694.

Rowe, A., Li X., 2001. Mathematical modeling of proton exchange membrane fuel cells. J. Power Source. 102(1-2), 82-96.

Rozendal, R. A., Hamelers H. V. , Buisman C. J., 2006. Effects of membrane cation transport on pH and microbial fuel cell performance. Environ. Sci. Technol. 40(17), 5206-5211.

Stoica, A., Sandberg M. , Holby O., 2009. Energy use and recovery strategies within wastewater treatment and sludge handling at pulp and paper mills. Bioresour. Technol.100(14), 3497-3505..

Tardast, A., Najafpour G., Rahimnejad M. , Amiri A., 2012. Bioelectrical power generation in a membrane less microbial fuel cell. World Applied Sciences J. 16(2), 179-182.

Tardast, A., Rahimnejad M., Najafpour G., Ghoreyshi A., Premier G. C., Bakeri G. , Oh S.-E., 2014. Use of artificial neural network for the prediction of bioelectricity production in a membrane less microbial fuel cell. Fuel. 117, 697-703.

Tardast, A., Rahimnejad M., Najafpour G. D., Ghoreyshi A. A. , Zare H., 2012. Fabrication and Operation of a Novel Membrane-less Microbial Fuel Cell as a Bioelectricity Generator. Iranica J. Energy Environ. 3, 1-5.

Taurozzi, J. S., Arul H., Bosak V. Z., Burban A. F., Voice T. C., Bruening M. L., Tarabara V. V., 2008. Effect of filler incorporation route on the properties of polysulfone-silver nanocomposite membranes of different porosities. J. Membr. Sci. 25 (1), 58-68.

Thygesen, A., Poulsen F. W., Min B., Angelidaki I. , Thomsen A. B., 2009. The effect of different substrates and humic acid on power generation in microbial fuel cell operation. Bioresour. Technol. 100(3), 1186-1191.

Vega, C. A. , Fernández I., 1987. Mediating effect of ferric chelate compounds in microbial fuel cells with *Lactobacillus plantarum*, *Streptococcus lactis*, and *Erwinia dissolvens*. Bioelectrochem. Bioenerg. 17(2), 217-222.

Venkata Mohan, S., Veer Raghavulu S., Srikanth S. , Sarma P. N., 2007. Bioelectricity production by mediatorless microbial fuel cell under acidophilic condition using wastewater as substrate: Influence of substrate loading rate. Curr. Sci. 92(12), 1720-1726.

Wang, H.-Y., Bernarda A., Huang C.-Y., Lee D.-J., Chang J.-S., 2011a. Micro-sized microbial fuel cell: a mini-review. Bioresour. Technol. 102(1), 235-243.

Wang, H., Wu Z., Plaseied A., Jenkins P., Simpson L., Engtrakul C. , Ren Z., 2011b. Carbon nanotube modified air-cathodes for electricity production in microbial fuel cells. J. Power Source. 196 (18), 7465–7469.

Wang, S., Dong F., Li Z., Jin L., 2012. Preparation and properties of sulfonated poly (phthalazinone ether sulfone ketone)/tungsten oxide composite membranes. Asia Pacific J. Chem. Eng. 7 (4), 528–533.

Wang, Y. P., Liu X. W., Li W. W., Li F., Wang Y. K., Sheng G. P., Zeng R. J. , Yu H. Q.,2011c. A microbial fuel cell membrane bioreactor integrated system for cost-effective wastewater treatment. Appl. Energy. 98, 230–235.

Watanabe, K., 2008. Recent developments in microbial fuel cell technologies for sustainable bioenergy. J. Biosci. Bioeng. 106(6), 528-536.

Wen, Q., Wu Y., Cao D., Zhao L. , Sun Q., 2009. Electricity generation and modeling of microbial fuel cell from continuous beer brewery wastewater. Bioresour. Technol.100(18), 4171-4175.

Zhou, M., Chi M., Luo J., He H. , Jin T., 2011. An overview of electrode materials in microbial fuel cells. J. Power Source. 196(10), 4427- 4435.

# 2

# Recent trends in biodiesel production

Meisam Tabatabaei[1,2,*], Keikhosro Karimi[3], Ilona Sárvári Horváth[4], Rajeev Kumar[5]

[1] *Microbial Biotechnology and Biosafety Department, Agricultural Biotechnology Research Institute of Iran (ABRII), AREEO, Karaj, Iran.*

[2] *Biofuel Research Team (BRTeam), Karaj, Iran.*

[3] *Department of Chemical Engineering, Isfahan University of Technology, 84156-83111 Isfahan, Iran.*

[4] *Swedish Centre for Resource Recovery, University of Borås, 501 90 Borås, Sweden.*

[5] *Center for Environmental Research and Technology (CE-CERT), Bourns College of Engineering, University of California, Riverside, California, USA.*

## HIGHLIGHTS

- *Recent trends and innovations in biodiesel production were comprehensively reviewed.*
- *Upstream, mainstream, and downstream strategies for economizing biodiesel production were elaborated.*
- *Integrated strategiesfor enhancing sustainability of biodiesel production processes were discussed.*

## GRAPHICAL ABSTRACT

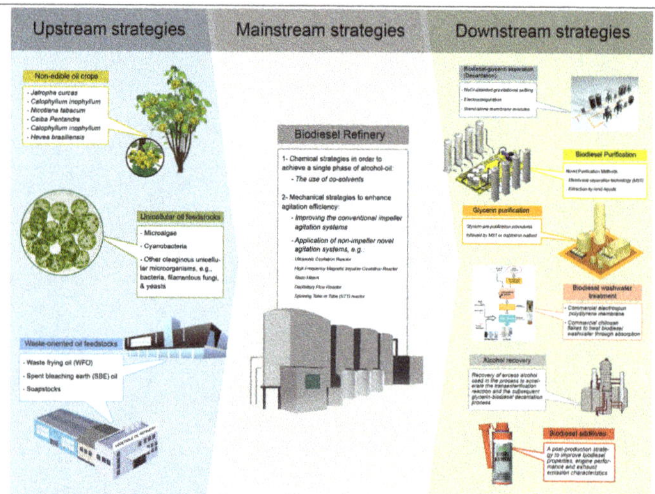


## ABSTRACT

This article fully discusses the recent trends in the production of one the most attractive types of biofuels, i.e., biodiesel.with a focus on the existing obstacles for its large scale production. Moreover, recent innovations/improvements under three categories of upstream, mainstream, and downstream processes are also presented. Upstream strategies are mainly focused on seeking more sustainable oil feedstocks and/or enhancing the quality of waste-oriented ones. The mainstream strategies section highlights the numerous attempts made to enhance agitation efficiency including chemical and/or mechanical strategies. Finally, the innovative downstream strategies basically dealing with 1) separation of biodiesel and glycerin, 2) purification of biodiesel and glycerin, and 3) improving the characteristics of the produced fuel, are comprehensively reviewed.

**Keywords:**
Biofuels Production
Recent Trends
Biodiesel
Economizing
Upstream strategies
Downstream strategies

* Corresponding author
E-mail address: meisam_tab@yahoo.com

**Contents**



## 1. Introduction

Biodiesel is attracting an increasing deal of attention worldwide for it is currently the only renewable energy carrier which could directly replace diesel fuel in compression ignition engines. Moreover, life cycle analysis (LCA) data for biodiesel suggests positive cumulative energy values when compared to petroleum-derived diesel (Vonortas and Papayannakos, 2014). Besides, biodiesel chemically known as monoalkyl esters of long chain fatty acids (FAME) (Singh and Taggar, 2014), also sustains many other advantageous features. Those include reduction of most exhaust emissions in comparison with petrodiesel, higher cetane number, biodegradability, lack of sulphur, inherent lubricity, positive energy balance, higher flash point, compatibility with the existing fuel distribution infrastructure, renewability, and domestic origin (Hajjari et al., 2014). The last feature not only could potentially secure a continuous, consistent, and economic feedstock supply, but also could provide opportunities for indigenous development of especially rural and isolated regions (Pinzi et al., 2014).

Among the many challenges yet to be overcome before one could portray biodiesel as a sustainable alternative for decades to come is an economic feedstock supply. In fact, it is now well-documented that the price of feedstock could account for 70-88% of the total biodiesel production cost and hence is considered as the most significant factor affecting the economic viability of the biodiesel market (Hasheminejad et al., 2011). The most common feedstocks currently used in biodiesel production are vegetable oils derived from edible oil crops, such as rapeseed, palm, soybean, and sunflower. However, biodiesel from edible oils is controversial to the extent that even in a number of occasions has been called "Crime Against Humanity" (Ferrett, 2007), by some social movements or individuals accusing biofuels in general and bioethanol/biodiesel in particular of being the main reason behind increased global food market prices and diminishing supply.

So, it is needless to mention that although biodiesel seems currently economically viable/competitive according to the latest Clean Cities Alternative Fuel Price Report (2015), released by the United States Department of Energy (US-DOE)(diesel *vs.* biodiesel (B20); 2.88 to 2.92 USD/diesel gallon equivalent); however, this scenario is doomed to change dramatically over time if edible oil crops are no longer an option. Thus, to provide non-competitive biodiesel (with food) in a sustainable and reasonably-priced manner, utilization of non-edible oleaginous plants as well as low-cost feedstocks such as waste frying oil (WFO) must be considered as a crucial factor. On the other hand, it is worth noting that the controversial 'food *vs.* fuel' competition cannot be completely avoided if valuable agricultural land is going to be used for the production of non-edible oil crops. Therefore, the only options left to sustain this business economically while avoiding its controversial features would be to benefit waste-oriented and/or non-edible oil feedstocks grown on marginal or non-agricultural lands. As for the WFO of either plant or animal origin, the main challenging issue is its high free fatty acid (FFA) content. In fact, it has been reported that biodiesel yield could drop down to 6% when the FFA content increases just above 5% wt. (Moser, 2011).

## 2. Strategies for economizing biodiesel production

Therefore, given the fact that the main interest is increasingly shifting towards larger scale biodiesel production, and also due to the above-mentioned existing obstacles, there is a strong market drive for more innovative, integrated, and efficient processes. Such innovations/improvements fall under three categories of upstream, mainstream, and downstream processes while integration of these processes is of great interest for it could enhance the economic viability of the whole biodiesel production process.

### *2.1. Upstream strategies*

Such solutions mainly deal with seeking more sustainable oil feedstocks and/or enhancing the quality of waste-oriented ones.

#### *2.1.1. Waste-oriented oil feedstocks*

Apart from WFO, which has been a focus of numerous studies, other waste oil resources of no commercial value such as spent bleaching earth (SBE) oil (Sahafi et al., 2015) and soapstocks ( Azocar et al., 2010) have also been proposed. It also should be noted that animal fats/lipids such as chicken fat, tallow, lard, waste fish oil (Sharma et al., 2014) and so on are not truly of no commercial value for they are in high demand in many parts of the globe as food/feed ingredients and hence, are not considered as ideal waste oil feedstocks for sustainable biodiesel production practices.

Annual global production of WFO is estimated at around 17 million tones (Gui et al., 2008) which if all collected, could have well supplied 50% of the total oil demand for the approximately 30 million tons of biodiesel produced in the year 2014. However, despite the growing efforts in increasing the WFO share in the market, this valuable waste oil feedstock is still under-utilized. For instance, in the United States, only less than 20% of the 4.5 million tons biodiesel generated in the year 2014 was of WFO origin (US-EIA, 2014).

SBE is an industrial waste generated in the vegetable oil refining industry after bleaching process of the crude oil (Canackci and Gerpen, 2003) and could contain 20 - 40 %wt. oil. SBE may present a fire hazard if not properly stored or disposed of at landfills (Sahafi et al., 2015). Huang and Chang (2010) estimated that on average 0.5 million tons of SBE oil is generated annually in the world. Sahafi et al. (2015) argued that based on their preliminary financial analysis, the economic feasibility and availability of SBE residual oil would make it a better option for biodiesel production than using the refined vegetable oil or WFO. Moreover, a number of studies reported significant improvements in the properties and performance of biodiesel-diesel fuel blends by dissolving waste polymers such as expanded polystyrene (EPS) in biodiesel produced from WFO and SBE oil (Mohammadi et al., 2012, 2013, 2014). Such a strategy not only offers an innovative way of recovering energy contained in waste polymers but also, given the achieved improvements reported, could enhance the economic features of waste-oriented biodiesel.

Soapstock is a by-product from vegetable oil refinement (Pinzi et al., 2014). In 2009, the worldwide generation of soybean soapstock alone stood at 2.16 million tons, 6% of the soybean oil produced in the same year (Azocar et al., 2010). Despite of its considerable global production quantity, when compared to WFO and SBE oil, soapstock does not stand a chance in terms of quality. This is ascribed to soapstock`s high water content of about 50% wt. which is also difficult to be removed because it is emulsified with lipids (Pinzi et al., 2014). Waste oil recovered from edible oil mills effluents through the application of novel technologies such as electrospun nanofibrous filters (Sundaramurthy et al., 2014) can also be considered as a low-cost, widely available, emerging, and interesting source for biodiesel production (Shirazi et al., 2014a).

The main disadvantageous features shared by all these waste oil resources are their high FFA content (which can reach values up to 12%) and low oxidative stability. The former has to be dealt with prior to the reaction as it negatively affects the common alkaline transesterification process through saponification reaction reducing the biodiesel yield, preventing the separation of biodiesel-glycerin phases, and finally increasing the washwater generated (El Sabagh et al., 2011). A number of methods have been proposed so far in order to pre-treat WFO to ensure FFA content of below 0.5% wt. (Zhang et al., 2003). These include steam injection (Lertsathapornsuk et al., 2005), column chromatography (Ki-Teak and Foglia, 2002), ion-exchange resins (Shibasaki-Kitakawa et al., 2007), neutralization by film vacuum evaporation (Cvengros and Cvengrosova, 2004), vacuum filtration (Dias et al., 2008), and finally the most usual procedure of esterification of FFA with homogenous or heterogeneous acid catalysts (**Table 1**) (Ghadge and Raheman, 2005; Chung et al., 2008; Zhang and Jiang, 2008; Lu et al., 2009; Bojan and Durairaj, 2010; Hayyan et al., 2010; Montefrio et al., 2010; Kombe et al., 2012; Ouachab and Tsoutsos, 2012).

### *2.1.2. Non-edible oil crops*

Non-edible oil crops such as *Jatropha curcas*, *Calophyllum inophyllum*, *Nicotiana tabacum*, *Ceiba Pentandra*, *Calophyllum inophyllum*, and *Hevea brasiliensis* are considered highly sustainable feedstocks for biodiesel production provided that they result in no competition over water or land used for conventional agricultural practices. In better words, these crops should be cultivated on marginal lands where no irrigation is performed. Laying such hard conditions would make it difficult for non-edible oil crops to catch up with the currently profitable biodiesel market mostly (>80%) relying on edible oil crops. Moreover, non-edible oil feedstocks are also known for their high FFA content requiring pretreatment (Sivakumar et al., 2013).

On the other hand, it is worth mentioning that the properties of biodiesel produced from oil feedstocks including non-edible oil crops strongly depend on fatty acid profile of the oil (Talebi et al., 2013). Talebi et al. (2014) recently introduced the BiodieselAnalyzer© software capable of predicting properties of a prospective biodiesel solely based on the fatty acid methyl ester (FAME) profile of the oil feedstock used. The bioprospecting software could cover a wide range of biodiesel quality parameters, i.e., unsaturation level (including saturated and unsaturated fatty acids, degree of unsaturation), cetane number, cold flow properties (including cloud point and cold filter plugging point), oxidation stability (including allylic and bis-allylic position equivalents and oxidation stability), higher heating value, kinematic viscosity, and density. This could lead to an insight into the quality of the produced product and would eliminate the need for lengthy and costly production and analysis steps. The software is available on the public domain at http://www.brteam.ir/biodieselanalyzer.

Silitonga et al. (2013) compared the quality parameters of biodiesel produced from a number of certain oil feedstocks used in their investigation and the values reported for the same crops in the literature. They confirmed significant variations in some parameters resulting from different fatty acid profiles of the oils used (**Table 2**) (Ramadhas et al., 2005; Sarin et a., 2007; Sulistyo et al., 2008; Devan and Mahalakshmi, 2009; Ong et al., 2011; Mofijur et al., 2012; Silitonga et al., 2013).

### *2.1.3. Unicellular oil feedstocks*

Microalgae is a feedstock that may help meet the growing global demands for biodiesel while posing potentially fewer environmental/food security threats than conventional oil feedstocks of either edible or non-edible nature. This is ascribed to the fact that not only these autotrophic

**Table 1.**
Comparative analysis for homogenous and heterogeneous acid catalyst esterification as pretreatment step for scale-up application.

| Catalyst type | | Catalyst conc. (% Wt. Oil) | Methanol:Oil molar ratio | T (°C) | Time | Initial FFA amount (mg KOH/g) | Post-treatment FFA amount (mg KOH/g) [1] | Conversion (%) | Operating cost (USD/ton oil) [2] | References |
|---|---|---|---|---|---|---|---|---|---|---|
| Sulfuric acid | | 2% | 24:1 | 60 | 80 min | 45.51 | 1.2 | 97 | 718 | Zhang and Jiang, 2008 |
| | | 20% | 35:1 | 40 | 1 h | 53 | 1.2 | 97 | 1152 | Ouachab and Tsoutsos, 2012 |
| | | 2% | 12:1 | 60 | 1 h | 38 | 4.8 | 87 | 366 | Ghadge and Raheman, 2005 |
| | | 1% | 9:1 | 60 | 1 h | 13.7 | 2.3 | 83 | 272 | Bojan and Durairaj, 2010 |
| | | 2% | 16:1 | 60 | 2 h | 11 | 0.6 | 95 | 484 | Kombe et al., 2012 |
| Solid acid | Mordenite | 1% | 30:1 | 60 | 3 h | 1.25 | 0.2 | 80.9 | 935 | Chung et al., 2008 |
| | Metatitanic acid | 4% | 20:1 | 90 | 2 h | 14 | 0.4 | 97 | 1190 | Lu et al., 2009 |
| | PTSA | 0.75% | 10:1 | 60 | 1 h | 43 | 3.9 | 90 | 305 | Hayyan et al., 2010 |
| | Carbonized vegetable oil asphalt | 0.2% | 16.8:1 | 220 | 4.5 h | NA[3] | NA | 80.5 | 525 | Montefrio et al., 2010 |

[1] Post-treatment FFA amount (mg KOH/g) of ≤1 has been reported as suitable for proceeding with alkali-catalyzed transesterification reaction (Montefrio et al., 2010).

[2] In order to calculate the overall cost of each procedure, the following assumptions were made:

- For economic assessment, average prices were considered as follows: industrial methanol at 800 USD/ton; industrial grade sulfuric acid at 600 USD/ton; PTSA (95%-98% resin catalyst) at 1400 USD/ton; carbonized vegetable oil asphalt at 12000 USD/ton (100 ml sulfuric acid is required for production of 5 g carbonized vegetable oil asphalt); metatitanic acid at 15000 USD /ton; mordenite molecular sieve at 5000 USD /ton.
- All calculations were made for pretreating 1 ton oil feedstock and the cost of equipment was not taken into consideration.

[3] NA: not available.

**Table 2.**
Comparison of the properties of methyl esters obtained from various vegetable oil.

| Vegetable oil methyl esters | Kinematic viscosity at 40 °C ($mm^2/s$) | Density at 15 °C ($kg/m^3$) | Calorific value (MJ/kg) | Flash point (°C) | Pour point (°C) | Cloud point (°C) | Oxidation stability (h, 110 °C) | References |
|---|---|---|---|---|---|---|---|---|
| *Jatropha curcas* | 4.48 | 864.0 | 40.224 | 160.5 | 3.0 | 5.8 | 9.41 | Silitonga et al., 2013 |
| *J. curcas* | 4.84 | 879.0 | – | 191.0 | – | – | – | Mofijur et al., 2012 |
| *Sterculia foetida* | 4.92 | 873.0 | 40.167 | 160.5 | −3.0 | 1.2 | 3.44 | Silitonga et al., 2013 |
| *S. foetida* | 6.00 | 875.0 | 40.211 | 162.0 | 1.0 | – | – | Devan and Mahalakshmi, 2009 |
| *Calophyllum inophyllum* | 4.57 | 872.5 | 40.204 | 158.5 | 6.0 | 6.0 | 13.08 | Silitonga et al., 2013 |
| *C. inophyllum* | 4.00 | 869.0 | 41.397 | 140.0 | 4.3 | 13.2 | – | Ong et al., 2011 |
| *Alureitas moluccana* | 3.84 | 869.0 | 40.127 | 165.5 | 8.0 | 8.0 | 5.31 | Silitonga et al., 2013 |
| *A. molucanna* | 4.12 | 886.9* | – | – | – | – | – | Sulistyo et al., 2008 |
| *Hevea brasiliensis* | 4.93 | 886.8 | 39.605 | 166.5 | 3.0 | 0.0 | 8.61 | Silitonga et al., 2013 |
| *H. brasiliensis* | 5.81 | 874.0 | 36.500 | 130.0 | −8.0 | 4.0 | – | Ramadhas et al., 2005 |
| Palm oil | 4.45 | 857.0 | 40.511 | 156.5 | 10.5 | 10.5 | 7.50 | Silitonga et al., 2013 |
| Palm oil | 4.50 | – | – | 135.0 | – | 16 | 13.37 | Sarin et al., 2007 |
| Diesel fuel | 2.91 | 839.0 | 45.825 | 71.50 | 1.0 | 2.0 | 23.70 | – |

*At 20 °C

microorganisms can be cultivated and harvested continuously throughout the year (Chen et al., 2011), but some also grow extremely fast doubling their biomass within a day (Chisti, 2008). Moreover, despite sharing the same basic photosynthetic machinery as the C3 land plants (Chisti, 2013a), microalgal species are more efficient in converting sunlight to biochemical energy; 5-8.3% *vs.* 2.4-4.6%, respectively (Zhu et al., 2008; Stephenson et al., 2011; Chisti, 2013b). Such high theoretical productivities and oil accumulation exhibited by certain algal species (**Table 3**) have made biodiesel production by algal oil transesterification an ultimate choice for numerous research and development attempts (Griffiths et al., 2012; Beetul et al., 2014). Regardless of the growth systems applied (i.e., phototrophic or heterotrophic), the following steps are involved in processing algal biomass including, microalgae growth, harvest, dewatering, and drying (Daroch et al., 2013).

**Table 3.**
Oil content of some microalgal species (Chisti, 2007; Tabatabaei et al., 2011).

| Microalgal species | Oil content (% dry wt.) |
|---|---|
| *Amphora sp. (Persian Gulf)* | 24 |
| *Ankistrodesmus sp.* | 17.5 |
| *Botryococcus braunii* | 25–75 |
| *Chlorella emersonii* | 18.5 |
| *Chlorella protothecoides* | 18 |
| *Crypthecodinium cohnii* | 20 |
| *Chlorella vulgaris* | 17 |
| *Cylindrotheca* sp. | 16–37 |
| *Dunaliella primolecta* | 23 |
| *D. salina (UTEX)* | 24 |
| *Isochrysis* sp. | 25–33 |
| *Monallanthus salina* | > 20 |
| *Nannochloris* sp. | 20–35 |
| *Nannochloropsis* sp. | 31–68 |
| *Neochloris oleoabundans* | 35–54 |
| *Nitzschia* sp. | 45–47 |
| *Phaeodactylum tricornutum* | 20–30 |
| *Schizochytrium* sp. | 50–77 |
| *Scenedesmus sp.* | 16 |
| *Tetraselmis sueica* | 15–23 |

Harvesting and dewatering of microalgal biomass (increasing biomass concentration from 0.02–0.5% (Brennan and Owende, 2010) to 15–20% (Heasman et al., 2000)) are major bottlenecks to commercialize algal biodiesel as energy requirements for their production exceed the energy which could be potentially obtained from algal biomass (Chisti, 2007). This is attributed to the small size of algal cells and their low concentration in the culture media (Bilad et al., 2014). Such obstacles have had a significant contribution to the sheer reality that algal fuels including algal biodiesel, despite of their unique attributes, have not yet been produced at commercial scale. This could be clearly comprehended through the latest monthly biodiesel production report released in May 2014 by the US Energy Information Administration (US-EIA, 2014). A recent economic viability analysis argued that the estimated production cost for a barrel of algal oil stands at USD 456.12–559.44 (Sun et al., 2011; Haase et al., 2013). This is still way higher that the current average price of crude oil. Nevertheless, there have been some large - scale demonstrations plants (ABO, 2015), which have generated less than 500 tons of algal biodiesel during the years 2013 and 2014 (US-EIA, 2014).

As a result, most existing large-scale microalgal plants are currently aimed at producing high-value products such as nutrition supplement and cosmetics instead of biofuels since for such compounds using energy-intensive centrifugation-based harvesting system is economical (Brennan and Owende, 2010; Lundquist et al., 2010). Therefore, for algal biodiesel to be economically attractive, these bottlenecks should be removed. Membrane-based processes have come to the center of attention in various applications including algal biomass processes during the last decade due to their several advantages, e.g., high efficiency under mild operational conditions, compact equipment, low operational time/energy, ease of integration with other processes, and high process scale-up capacity (Shirazi et al., 2013a; Bilad et al., 2014; Shirazi et al., 2014b, c; Shirazi and Tabatabaei, 2014; Shirazi et al., 2015). However, membranes also suffer from several drawbacks restricting their large-scale application in algal fuel production. These include concentration polarization, membrane fouling, low membrane life-span, low selectivity, and permeance (Bilad et al., 2014). Mostly commercially-available organic polymer-based membranes, due to their wide availability, high chemical compatibility, variety of designs and reasonable cost (Strathmann, 2011), have been investigated for microalgae harvesting so far (Gerardo et al., 2014).

In a recent attempt, Shuman et al. (2014) successfully tested an ultra-low energy method for rapid pre-concentrating microalgae using electro-coagulation–flocculation method. They managed to achieve rapid separation of >90% of microalgal cells within 120 min while >90% of the cells were still alive after processing. The minimum energy density input required for effective separation in their study was 0.03 $kWh/m^3$. This was significantly lower that the required energy input (ranging from 0.3 to 2.23 $kWh/m^3$ (Bhave et al., 2012; Buckwalter et al., 2013; Gerardo et al., 2013) for the membrane-based separation processes reported in the literature.

Finally, two methods of algal oil transformation could be performed 1) a two-step method, i.e., oil extraction followed by oil transesterification and 2) single step *in situ* transesterification of algal oils to biodiesel (Daroch et al., 2013). Having said all, algal biodiesel is not destined to reach its economically viable commercialization stage till the cost of producing algal biomass is reduced significantly. In an estimate, Chisti (2013a) argued that algal biomass with an oil content of 40% wt. has to be produced at a cost of ≲$0.25/kg, if algal oil is to compete with petroleum given its current price of $100/barrel. He concluded that the actual cost of producing the biomass at present appears is at least 10-fold greater. Chisti (2013a) also insisted that widespread use of algal fuels is unlikely in the short run but specific applications such as in aviation may be likely in the medium term. Finally, genetic and metabolic engineering of microalgae to enhance oil production and to ease its recovery seem indispensable parts of algal biodiesel commercialization scenario in the long run (Tabatabaei et al., 2011; Chisti, 2013a; Talebi et al., 2014).

Cyanobacteria, oxygenic photosynthetic bacteria, play a significant role in global biological carbon sequestration, oxygen production and the nitrogen cycle (Parmar et al., 2011). Similar to microalgae, cyanobacteria as oil-producing unicellular organisms also offer unique features including fast cell growth, simple nutrient requirements, i.e., water, sunlight, and $CO_2$ (Ruffing, 2011). Equally important, they are naturally transformable and as a result could be potentially improved by genetic engineering (Machado and Atsumi, 2012). Due to their natural diversity, the capacity of cyanobacteria to grow in a variety of locations, even those unfit for agriculture, could be exploited for biofuel production. Karatay and Dönmez (2011) investigated a number of thermophile cyanobacteria for biodiesel production. They reported lipid contents of 42.8% for *Synechococcus* sp., 45.0% for *Cyanobacterium aponinum*, and 38.2% for *Phormidium* sp. under optimum conditions. Liu et al. (2010a) genetically engineered the cyanobacterium *Synechococcus elongates* PCC7942 in order to increase FFA content and achieved a production efficiency of up to 133 ± 12 mg/L per day at a cell density of 0.23 g of dry weight per liter. This was almost 3 folds higher than the lipid content achieved in a genetically engineered *E. coli* strain (Liu et al., 2010b). Such findings further mark cyanobacteria as promising feedstock for biodiesel production.

Besides microalgae and cyanobacteria, other oleaginous unicellular microorganisms including bacteria, filamentous fungi, and yeasts have also been utilized for biodiesel production under the same brand of "third generation biodiesel". Yeasts not only sustain the unique features of microalgae such as high oil accumulation of up to 70% dry wt., and that they can be genetically modified to enhance production (Liang and Jiang, 2013), but also compared to microalgae offer other advantageous features, i.e., potentially faster growth rates, higher density growth, less susceptibility to viral infection, and bacterial contamination (Sitepu et al., 2014). Among the yeast species investigated, *Rhodotorula* sp., *Cryptococcus* sp., *Lipomyces* sp., and *Candida* sp. are known to have the highest capability to accumulate oil (Beopoulos et al., 2011).

### 2.2. Mainstream strategies

Conventionally, biodiesel is produced through the agitation of the reagents, i.e., oil, alcohol (mainly methanol), and catalyst at about 60 °C (just below the boiling point of methanol i.e. 64.7 °C) for around 1 h. Currently, the majority of industrial biodiesel production practices worldwide are batch or continuous processes with mechanical agitation (Noipin and Kumar, 2014). However, since oil and alcohol are not well miscible, mixing efficiency is therefore the main challenge faced. The most efficient mixing is achieved when the alcohol–oil interfacial area is maximized by decreasing the droplet size of the reactants i.e. alcohol and oil as much as possible. Theoretically, this could be as low as the sizes of the molecules involved in the reaction. Therefore, both the agitation and temperature are indispensible elements required to accomplish a successful transesterification reaction. Numerous attempts have been made to enhance agitation efficiency including chemical and/or mechanical strategies.

Chemical strategies involve the use of a co-solvent in order to achieve a single phase of alcohol-oil (Boocock, 2004). The co-solvents used should 1) be completely miscible in both the alcohol and oil and 2) have a boiling point close to that of the alcohol used (e.g., methanol), so that they could be easily co-distilled and recovered/recycled upon the termination of the reaction. Cyclic ethers such as tetrahydrofuran (THF), 1,4-dioxane, diethyl ether, methyl tertiary butyl ether, and diisopropyl ether (Boocock, 2004), owing to their hydrophilic oxygen atom capable of forming hydrogen bonds with alcohols, and their hydrophobic hydrocarbon portion capable of solubilizing oils, meet the first condition required for an ideal co-solvent. Having included the second condition, THF (boiling point: 66 °C) is regarded as the most ideal co-solvent especially if methanol is used in the transesterification reaction.

Mechanical strategies used to enhance agitation efficiency fall into three different categories:

1) Improving the conventional impeller agitation systems (Hosseini t al., 2012). For instance, Hosseini et al. reported a reactor equipped with a helical ribbon-like agitator using which at 900 rpm stirring speed and after 20 min residence time, 97.3% conversion of triglycerides to methyl esters was achieved.

2) Application of non-impeller novel agitation systems in which highly efficient mechanical energy is provided for mixing and initiating the transesterification reaction. These include ultrasound-based agitation systems, e.g., ultrasonic cavitation reactor (Singh et al., 2007), high frequency magnetic impulse cavitation reactor (Oh et al., 2012), static mixers (Hompson and He, 2007), oscillatory flow reactors (Harvey et al., 2003), and spinning tube in tube (STT) reactors developed by Four Rivers BioEnergy Company, Inc. (Qiu et al., 2010).

And finally, 3) application of novel systems in which no agitation is applied but conditions required for a successful transesterification are provided. These include microwave reactors which utilize microwave irradiation to transfer energy directly into reactants and consequently accelerate the rate of reaction (Barnard et al., 2007) and membrane reactors (Atadashi et al., 2011). In fact, the latter integrates reaction and membrane-based separation into a single process and increase the rate of equilibrium-limited transesterification reaction by constantly removing the products, i.e, biodiesel from the reactants stream via membranes (Qiu et al., 2010).

It is worth quoting that the final characteristics of biodiesel could be influenced by the procedure through which the fuel has been produced. For instance, Sajjadi et al. (2015) investigated the influence of sonoluminescence transesterification on biodiesel physicochemical properties and compared the results to those of traditional mechanical stirring. They argued that based on the experimental results, the transesterification with ultrasound irradiation could change the biodiesel density by about 0.3 kg/m$^3$; the viscosity by 0.12 mm$^2$/s; the pour point by about 1-2 °C, and the flash point by 5 °C compared to the traditional method (Sajjadi et al., 2015).

### 2.3. Downstream strategies

Innovative downstream strategies basically deal with 1) separation of biodiesel and glycerin, 2) purification of biodiesel and glycerin, and also 3) improving the characteristics of the produced fuel.

#### 2.3.1. Biodiesel-glycerin separation (decantation)

Separation of biodiesel and glycerin, the main by-product of the transesterification reaction, is a slow process and is usually achieved by gravitational settling. This could lead to longer operating times, bigger equipment and larger amount of steel and consequently increased production cost (Shirazi et al., 2013b). Therefore, acceleration of glycerol/biodiesel decantation could play an important role in the overall biodiesel refinery process. In a recent study, Shirazi et al. (2013b) reported the application of NaCl-assisted gravitational settling as an economizing strategy. They argued that the addition of 1 g conventional NaCl salt/100 ml glycerol–biodiesel mixture decreased the glycerol settling time significantly by 100% while maintaining the methyl ester purity as high as the control (0 g NaCl). In a different study, Noureddin et al. (2014) investigated the interactive effects of prominent parameters, i.e., temperature (25-65 °C), NaCl addition (0-2 g/100ml), and methanol concentrations (10–30 vol.%) on decantation behavior of glycerol/biodiesel mixture. They reported that decantation time was significantly decreased by 200% (3 min) under the optimum conditions, i.e., 45 °C, 1 g NaCl addition and 20% excess methanol (Noureddin et al., 2014).

Electrocoagulation at high voltages could also significantly accelerate the decantation rate (Tabatabaei and Khatamifar, 2009). As mentioned earlier, membrane reactors could integrate both reaction and separation stages, eliminating the downstream biodiesel/glycerin separation stage. Stand-alone membrane modules could also be used in order to separate glycerin from biodiesel after the termination of the reaction. In a study, nanocomposite solvent resistant polyimide (PI) membranes with a variety of functionalized multiwall carbon nanotubes (MWCO) in the range of ultra to nanofiltration were synthesized by different MWCNTs loadings via phase inversion method. These membranes were then used to remove the glycerol dispersed in crude biodiesel (Peyravi et al., 2015). The authors claimed that the synthesized solvent resistant nanofiltration (NCSR) membranes achieved excellent glycerol removal up to 100% glycerol rejection without significant decline in flux permeation. Moreover, the presence of MWCNTs in the PI membrane structure resulted in enhanced chemical and thermal resistance as a result of polymer chain mobility limitation. They also stated that the antifouling properties of the modified membranes were improved compared to the neat PI membrane (Peyravi et al., 2015).

### 2.3.2. *Biodiesel purification*

Biodiesel should undergo a purification step in order to meet the ASTM D6751 or EN 14214 quality standards for biodiesel (B100). Purification of biodiesel is conventionally carried out by wet and/or dry washing. These processes along with the novel purification methods such as membrane separation technology (MST) and extraction by ionic liquids have been well reviewed very recently by Stojković et al. (2014). In their comprehensive review, they highlighted the existing controversial observances in the literature on the purification efficiency (refining yield, fuel properties, and the fulfillment of prescribed standard limits) of wet and dry washing and suggested that various purification methods need to be evaluated for their real purification efficiency under identical conditions (Stojković et al., 2014).

### 2.3.3. *Biodiesel washwater treatment*

Nevertheless, since currently water washing is the most widely used process in industrial-scale biodiesel refineries; therefore, treating the huge amount of highly polluting washwater generated on a daily basis (3–10 L water/L biodiesel; COD, 35000 mg/L; BOD, 29000 mg/L) is of critical importance. Recently, Shirazi et al. (2013c) applied commercial electrospun polystyrene membrane for treatment of biodiesel washwater and achieved promising reduction rates of 75%, 55%, 92%, 96%, and 30% for COD, BOD, TS, TDS, and TSS, respectively. Using a different approach, Pitakpoolsil and Hunsom (2014) used commercial chitosan flakes to treat biodiesel washwater through absorption. They reported that by a single adsorption within 3 h in the presence of 3.5 g chitosan/L at a mixing rate of 300 rpm, BOD, COD, and oil & grease were reduced by 59.3%, 87.9%, and 66.2%, respectively. They also indicated that repetitive adsorption for four times using fresh flakes further enhanced the removal of BOD, COD, and oil & grease up to 93.6%, 97.6%, and 95.8%, respectively. Their findings could have been more promising if the commercial chitosan used could be regenerated. However, the NaOH-regenerated chitosan sustained only 40% of its adsorption capacity (Pitakpoolsil and Hunsom, 2014). Therefore, commercial electrospun polystyrene membranes seem to comparatively hold greater promises for industrial-scale applications given their availability and relatively low cost.

### 2.3.4. *Glycerin purification*

Glycerin could also be purified and sold as a strategy to economize the whole production process (Hasheminejad et al., 2011). Among the methods currently used for glycerol purification are MST and distillation (Javani et al., 2012). MST is more cost-effective than distillation provided that crude glycerin undergoes a pre- purification in order to reduce salts and organic nonglycerol matter (ONGM, such as methyl ester) (Manosak et al., 2011). A number of attempts have been made so far in order to develop an efficient pre-purification procedure for crude glycerin involving acidification and neutralization steps. For instance, Hajek and Skopal (2010) purified crude glycerin to a final purity of 86% while they also obtained high quality FFAs with a purity of 99.5 wt.% by including a saponification step. $KH_2PO_4$ was also produced through the acidification step which could potentially be used as fertilizer. Kongjao et al. (2010) produced glycerin with a purity of around 93.34% by extracting crude glycerin with ethanol. In a similar study, Manosak et al. (2011) used a better precipitant, i.e., isopropanol (IPA), and increased the purity of the glycerin to 95.74 wt.%. In an innovative investigation, Javani et al. (2012) through a step-by-step approach further increased the purity of crude glycerin to as high as 96.08 wt.%. They also managed to generate high quality potassium phosphate salts, i.e., $KH_2PO_4$ and $K_2HPO_4$ as well as FFAs with a purity of 98%, 98.05%, and 99.58%, respectively (Javani et al., 2012).

### 2.3.5. *Alcohol recovery (recycling)*

One of the downstream strategies usually neglected is the recovery/recycling of the excess alcohol (mostly methanol) fraction from both biodiesel and glycerin. In fact, due to the reversible nature of the transesterification reaction, excess alcohol is mostly used to drive the reaction towards the final product, i.e., biodiesel. Under conventional transesterification conditions, i.e., methanol to oil molar ratio of 6:1, the recoverable methanol from biodiesel and glycerol are around 2% and 25%, respectively (Mythili et al., 2014). Therefore, inclusion of this downstream strategy could significantly influence the economic viability of the process.

On the other hand, it should be noted that the presence of excess methanol and its progressive evaporation generally affects the decantation time negatively through the formation of miniemulsions (Noureddin et al., 2014). More specifically, methanol is adsorbed at the interface between the glycerin and biodiesel phases and consequently reduces the interfacial tension (IFT). In better words, the adsorbed molecules of methanol would form a mechanically strong and elastic interfacial film that would act as a barrier against aggregation and coalescence (Mythili et al., 2014). However, if the excess methanol exceeds a certain level, the surplus obviously enters the glycerin phase more than the biodiesel phase due to the fact that there are extensive possibilities for hydrogen bonding between the glycerin molecules and methanol. The contained methanol in the glycerol phase would result in decreased viscosity of glycerin which consequently facilitate its speedy decantation (Noureddin et al., 2014). The exact amount of the excess alcohol to be used in the process in order to accelerate the glycerin-biodiesel decantation process needs to be determined according to a specific biodiesel production practice. Such strategy would be logical only if downstream recovery of methanol from biodiesel and glycerin has been implemented and that the production system is well contained to prevent leakage of the methanol vapor.

### 2.3.6. *Biodiesel additives*

A post-production strategy which could potentially enhance the economic viability of the whole production cycle through value addition to biodiesel or its blends is to improve biodiesel properties, engine performance, and exhaust emission characteristics. Various types of additives such as oxygenated additives (Ribeiro et al., 2007), antioxidants (Hajjari et al., 2014), cetane number improvers (Venkateswarlu, 2015), lubricity improvers (Anastopoulos et al., 2001), cold flow improvers

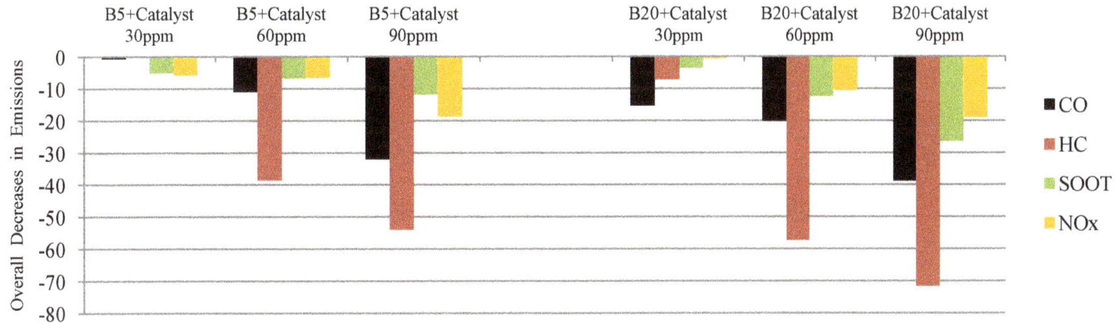


**Fig.1.** Overall decrease in emissions (i.e., CO, HC, Soot and $NO_x$) achieved by addition of the $CeO_2$-MWCNTs nanocatalyst at different concentrations (30, 60, and 90 ppm) compared to catalyst free B5 and B20. (Mirzajanzadeh et al., 2015), Copyright (2015), with permission from Elsevier.

(Mohammadi et al., 2014), and combustion improvers (Sadhik Basha and Anand, 2012; Kannan et al., 2011; Mirzajanzadeh et al., 2015) are used in biodiesel to meet specification limits and to enhance quality. In a very recent study, Mirzajanzadeh et al. (2015) introduced a novel soluble hybrid nanocatalyst as a promising combustion improver. The hybrid nanocatalyst containing cerium oxide on amide-functionalized MWCNT ($CeO_2$-MWCNTs), owing to the high surface area of the soluble nano-sized catalyst particles and their proper distribution along with catalytic oxidation reaction, resulted in significant overall improvements in the combustion reaction specially in B20 containing 90 ppm of the catalyst $B20_{(90\ ppm)}$. More specifically, all pollutants, i.e., $NO_x$, CO, HC, and soot were reduced by up to 18.9%, 38.8%, 71.4%, and 26.3%, respectively, in $B20_{(90\ ppm)}$ compared to neat B20. The innovated fuel blend also increased engine performance parameters, i.e., power and torque by up to 7.81%, 4.91%, respectively, and decreased fuel consumption by 4.50% (**Fig. 1**) (Mirzajanzadeh et al., 2015). The authors concluded that the unique oxygen donation/absorption properties of $CeO_2$ resulted in CO oxidation reaction. They also stated that $CeO_2$ nano particles owing to their decreasing impact on peak temperature in the combustion chamber led to decreased production of nitrogen oxides (NOx) (Mirzajanzadeh et al., 2015).

## 4. Concluding remarks

Despite decades-long research conducted on various aspects of biodiesel production in order to improve the economic viability of this unique renewable energy carriers, its future sustainability remains uncertain. This is mainly ascribed to the insufficient oil feedstock available to meet the growing demands for biodiesel, and on the other hand, the controversial food *vs.* fuel crisis. Moreover, given the recent falling oil prices, maintaining biodiesel's market price competitive to that of petroleum-derived diesel will be a challenge as well. In line with these points, enhancing the economic aspects of biodiesel production through integrated strategies targeting different stages (i.e. upstream, mainstream, and downstream) is vital (**Fig. 2**).

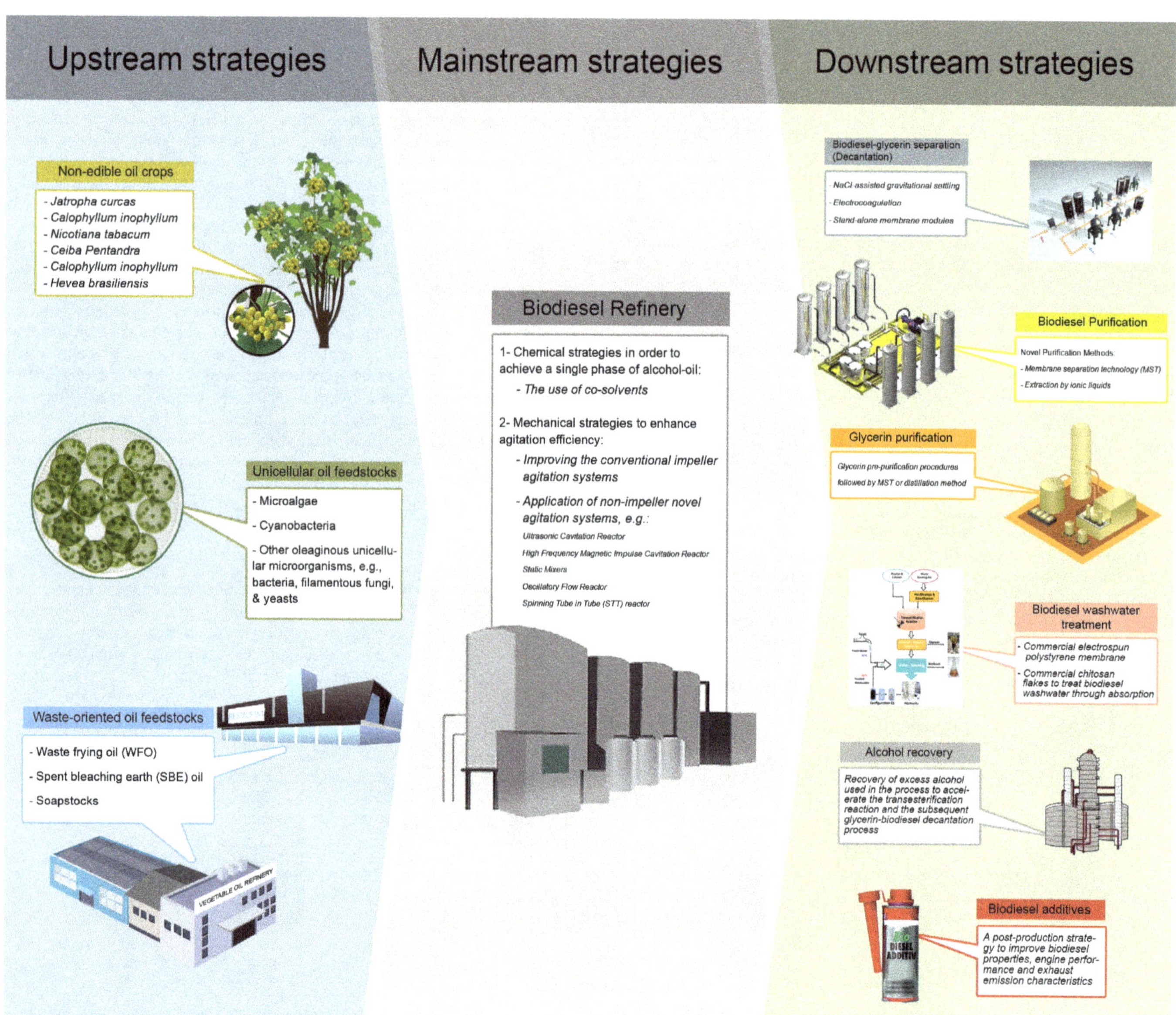


**Fig.2.** Enhancing the economic aspects of biodiesel production through integrated strategies targeting different stages (i.e. upstream, mainstream, and downstream).

**References**


Algal Biomass Organization (ABO), Available at http://allaboutalgae.com (accessed on 10 August 2015).

Anastopoulos, G., Lois, E., Serdari, A., Zanikos, F., Stournas, S., Kalligeros, S., 2001. Lubrication properties of low-sulfur diesel fuels in the presence of specific types of fatty acid derivatives. Energy Fuels. 15(1), 106-112.

Atadashi, I.M., Aroua, M.K., Abdul Aziz, A.R., Sulaiman, N.M.N., 2011. Membrane biodiesel production and refining technology: a critical review. Renew. Sustain. Energy Rev.15(9), 5051-5062.

Azocar, L., Ciudad, G., Heipieper, H.J., Navia, R., 2010. Biotechnological processes for biodiesel production using alternative oils. Appl. Microbiol. Biotechnol. 88, 621-636.

Barnard, T.M., Leadbeater, N.E., Boucher, M.B., Stencel, L.M., Wilhite, B.A., 2007. Continuous-flow preparation of biodiesel using microwave heating. Energy Fuels. 21(3), 1777-1781.

Beetul, K., Sadally, S.B., Taleb-Hossenkhan, Bhagooli, R., Puchooa, D., 2014. An investigation of biodiesel production from microalgae found in Mauritian waters. Biofuel Res. J. 1(2), 58-64.

Beopoulos, A., Nicaud, J.M., Gaillardin, C., 2011. An overview of lipid metabolism in yeasts and its impact on biotechnological processes. Appl. Microbiol. Biotechnol. 90, 1193-1206.

Bhave, R., Kuritz, T., Powell, L., Adcock, D., 2012. Membrane-based energy efficient dewatering of microalgae in biofuels production and recovery of value added co-products. Environ. Sci. Technol. 46, 5599-5606.

Bilad, M.R., Arafat, H.A., Vankelecom, I.F.J., 2014. Membrane technology in microalgae cultivation and harvesting: a review. Biotechnol. Adv. *32*(7), 1283-1300.

Bojan, S.G., Durairaj, S.K., 2010. Producing biodiesel from high free fatty acid *Jatropha curcas* oil by a two step method - an Indian case study. J. Sustain. Energy Environ. 3, 63-66.

Boocock, D.G.B., 2004. Process for production of fatty acid methyl esters from fatty acid triglycerides. US Patent 6712867.

Brennan, L., Owende, P., 2010. Biofuels from microalgae - a review of technologies for production, processing, and extractions of biofuels and co-products. Renew. Sustain. Energy Rev. 14(2), 557-577.

Buckwalter, P., Embaye, T., Gormly, S., Trent, J.D., 2013. Dewatering microalgae by forward osmosis. Desalination. 312, 19-22.

Canackci, M., Gerpen, J.V., 2003. A pilot plant to produce biodiesel from high free fatty acid feedstocks. Trans. ASAE. 46(4), 945-954.

Chen, C.Y., Yeh, K.L., Aisyah, R., Lee, D.J., Chang, J.S., 2011. Cultivation, photobioreactor design and harvesting of microalgae for biodiesel production: a critical review. Bioresour. Technol. 102(1), 71-81.

Chisti, Y., 2007. Biodiesel from microalgae. Biotechnol. Adv. 25, 294-306.

Chisti, Y., 2008. Biodiesel from microalgae beats bioethanol. Trends Biotechnol. 26(3), 126-131.

Chisti, Y., 2013a. Constraints to commercialization of algal fuels. J. Biotechnol. 167(3), 201-214.

Chisti, Y., 2013b. Raceways-based production of algal crude oil. Green. 39(3-4), 195-216.

Chung, K.H., Chang, D.R., Park, B.G., 2008. Removal of free fatty acid in waste frying oil by esterification with methanol on zeolite catalysts. Bioresour. Technol. 99, 7438-7443.

Clean cities alternative fuel price report April 2015. U.S. Department of Energy. Available at http://www.afdc.energy.gov/uploads/publication/alternative_fuel_price_report_april_2015.pdf

Cvengros, J., Cvengrosova, Z., 2004. Used frying oils and fats and their utilization in the production of methyl esters of high fatty acids. Biomass Bioenergy. 27, 173-181.

Daroch, M., Geng, S., Wang, G., 2013. Recent advances in liquid biofuel production from algal feedstocks. Appl. Energy. 102, 1371-1381.

Devan, P.K., Mahalakshmi, N.V., 2009. Study of the performance, emission and combustion characteristics of a diesel engine using poon oil-based fuels. Fuel process. Technol. 90(4), 513-519.

Dias, J.M., Alvim-Ferraz, M.C.M., Almeida, M.F., 2008. Comparison of different homogeneous alkali catalysts during transesterification of waste and virgin oils and evaluation of biodiesel quality. Fuel. 87, 3572-3578.

El Sabagh, S.M., Keera S.K., Taman, A.R., 2011. The characterization of biodiesel fuel from waste frying oil. Energy Sources Part A. 33, 401-409.

Ferrett, G., 2007. Biofuels 'crime against humanity'. BBC News. 27.

Gerardo, M.L., Oatley-Radcliffe, D.L., Lovitt, R.W., 2013. Minimizing the energy requirement of dewatering *Scenedesmus* sp. by microfiltration: performance, costs, and feasibility. Environ. Sci. Technol. 48(1), 845-853.

Gerardo, M.L., Oatley-Radcliffe, D.L., Lovitt, R.W., 2014. Integration of membrane technology in microalgae biorefineries. J. Membr. Sci. 464, 86-99.

Ghadge, S.V. Raheman, H., 2005. Biodiesel production from Mahua (*Madhuca indica*) oil having high free fatty acids. Biomass Bioenergy. 28, 601-605.

Griffiths, M.J., van Hille, R.P., Harrison, S.T.L., 2012. Lipid productivity, settling potential and fatty acid profile of 11 microalgal species grown under nitrogen replete and limited conditions. J. Appl. Phycol. 24(5), 989-1001.

Gui, M.M., Lee, K.T., Bhatia, S., 2008. Feasibility of edible oil vs. non-edible oil vs. waste edible oil as biodiesel feedstock. Energy. 33, 1646-1653.

Haase, R., Bielicki, J., Kuzma, J., 2013. Innovation in emerging energy technologies: A case study analysis to inform the path forward for algal biofuels. Energy Policy. 61, 1595-1607.

Hájek, M., Skopal, F., 2010. Treatment of glycerol phase formed by biodiesel production. Bioresour. Technol. 101(9), 3242-3245.

Hajjari, M., Ardjmand, M., Tabatabaei, M., 2014. Experimental investigation of the effect of cerium oxide nanoparticles as a combustion-improving additive on biodiesel oxidative stability: mechanism. RSC Adv. 4(28), 14352-14356.

Harvey, A.P., Mackley, M.R., Seliger, T., 2003. Process intensification of biodiesel production using a continuous oscillatory flow reactor. J. Chem. Technol. Biotechnol. 78(2-3), 338-341.

Hasheminejad, M., Tabatabaei, M., Mansourpanah, Y., Khatami far, M., Javani, A., 2011. Upstream and downstream strategies to economize biodiesel production. Bioresour. Technol. 102(2), 461-468, 2011.

Hayyan, A., Alam, Md.Z., Mirghani, M.E.S., Kabbashi, N.A., Hakimi, N.I.N.M., Siran, Y.M., Tahiruddin, S., 2010. Sludge palm oil as a renewable raw material for biodiesel production by two-step processes. Bioresour. Technol. 101, 7804-7811.

Heasman, M., Diemar, J., O'connor, W., Sushames, T., Foulkes, L., 2000. Development of extended shelf-life microalgae concentrate diets harvested by centrifugation for bivalve mollusks - a summary. Aquacult. Res. 31(8-9), 637-659.

Hompson, J.C., He, B.B., 2007. Biodiesel production using static mixers. Trans. ASABE. 50(1), 161-165.

Hosseini, M., Nikbakht, A.M., Tabatabaei, M., 2012. Biodiesel production in batch tank reactor equipped to helical ribbon-like agitator. Mod. Appl. Sci. 6(3), 40-45.

Huang, Y.P., Chang, J.I., 2010. Biodiesel production from residual oils recovered from spent bleaching earth. Renew. Energy. 35, 269-274.

Javani, A., Hasheminejad, M., Tahvildari, K., Tabatabaei, M., 2012. High quality potassium phosphate production through step-by-step glycerol purification: A strategy to economize biodiesel production. Bioresour. Technol. 104, 788-790.

Kannan, G.R., Karvembu, R., Anand, R., 2011. Effect of metal based additive on performance emission and combustion characteristics of diesel engine fuelled with biodiesel. Appl. Energy. 88(11), 3694-3703.

Karatay, S.E., Dönmez, G., 2011. Microbial oil production from thermophile cyanobacteria for biodiesel production. Appl. Energy. 88(11), 3632-3635.

Ki-Teak, L., Foglia, T.A., 2002. Production of alkyl ester as biodiesel from fractionated lard and restaurant grease. J. Am. Oil Chem. Soc. 79(2), 191-195.

Kombe, G.G., Temu, A.K., Rajabu, H.M., Mrema, G.D., 2012. High free fatty acid (FFA) feedstock pre-treatment method for biodiesel production. Proc. 2nd Int. Conf. Adv. Eng. Technol. 176-182.

Kongjao, S., Damronglerd, S., Hunsom, M., 2010. Purification of crude glycerol derived from waste used-oil methyl ester plant. Korean J. Chem. Eng. 27(3), 944-949.

Lertsathapornsuk V., Pairintra R., Ruangying P., Krisnangkur K., 2005. Continuous transethylation of vegetable oils by microwave irradiation, In: Proc. 1st Conf. Energy Network. Thailand, RE11-RE14.

Liang, M.H., Jiang, J.G., 2013. Advancing oleaginous microorganisms to produce lipid via metabolic engineering technology. Prog. Lipid Res. 52(4), 395-408.

Liu, X., Brune, D., Vermaas, W., Curtiss, R., 2010a. Production and secretion of fatty acids in genetically engineered cyanobacteria. Proc. Natl. Acad. Sci. U.S.A.

Liu, T., Vora, H., Khosla, C., 2010b. Quantitative analysis and engineering of fatty acid biosynthesis in *E. coli*. Metab. Eng. 12(4), 378-386.

Lu, H., Liu, Y., Zhou, H., Yang, Y., Chen, M., Liang, B., 2009. Production of biodiesel from *Jatropha curcas* L. oil. Comput. Chem. Eng. 33, 1091-1096.

Lundquist, T.J., Woertz, I.C., Quinn, N.W.T., Benemann, J.R., 2010. A realistic technology and engineering assessment of algae biofuel production. Report of Energy Biosciences Institute, University of California, USA.

Machado, I.M., Atsumi, S., 2012. Cyanobacterial biofuel production. J. Biotechnol. 162(1), 50-56.

Manosak, R., Limpattayanate, S., Hunsom, H., 2011. Sequential-refining of crude glycerol derived from waste used-oil methyl ester plant via a combined process of chemical and adsorption. Fuel Process. Technol. 92(1), 92-99.

Mirzajanzadeh, M., Tabatabaei, M., Ardjmand, M., Rashidi, A., Ghobadian, B., Barkhi, M., Pazouki, M., 2015. A novel soluble nano-catalysts in diesel–biodiesel fuel blends to improve diesel engines performance and reduce exhaust emissions. Fuel. 139, 374- 382.

Mofijur, M., Masjuki, H.H., Kalam, M.A., Hazrat, M.A., Liaquat, A.M., Shahabuddin, M., Varman, M., 2012. Prospects of biodiesel from Jatropha in Malaysia. Renew. Sustain. Energy Rev. 16(7), 5007-5020.

Mohammadi, P., Nikbakht, A.M., Tabatabaei, M., Farhadi, K., Mohebbi, A., Khatami far, M., 2012. Experimental investigation of performance and emission characteristics of DI diesel engine fueled with polymer waste dissolved in biodiesel-blended diesel fuel. Energy. 46(1), 596-605.

Mohammadi, P., Tabatabaei, M., Nikbakht, A.M., Farhadi, K., Castaldi, M.J., 2013. Simultaneous energy recovery from waste polymers in biodiesel and improving fuel properties. Waste Biomass Valorization. 4(1), 105-116.

Mohammadi, P., Tabatabaei, M., Nikbakht, A.M., Esmaeili, Z., 2014. Improvement of the cold flow characteristics of biodiesel containing dissolved polymer wastes using acetone. Biofuel Res. J. 1(1), 26-29.

Montefrio, M.J., Xinwen, T., Obbard, J.P., Montefrio, M.J., Xinwen, T., Obbard, J.P., 2010. Recovery and pre-treatment of fats, oil and grease from grease interceptors for biodiesel production. Appl. Energy. 87, 3155-3161.

Moser, B.R., 2011. Biodiesel production, properties, and feedstocks. Biofuels, Springer New York, 285-347.

Mythili, R., Venkatachalam, P., Subramanian, P., Uma, D., 2014. Recovery of side streams in biodiesel production process. Fuel.117, 103-108.

Noipin, K.N., Kumar, S., 2014. Optimization of ethyl ester production assisted by ultrasonic irradiation. Ultrason. Sonochem. 22, 548-558.

Noureddin, A., Shirazi, M.M.A., Tofeily, J., Kazemi, P., Motaee, E., Kargari, A., Mostafaei, M., Akia, M., Hamieh, T., Tabatabaei, M., 2014. Accelerated decantation of biodiesel–glycerol mixtures: Optimization of a critical stage in biodiesel biorefinery. Sep. Purif. Technol. 132, 272-280.

Oh, P.P., Lau, H.L.N., Chen, J., Chong, M.F., Choo, Y.M., 2012. A review on conventional technologies and emerging process intensification (PI) methods for biodiesel production. Renew. Sustain. Energy Rev. 16(7), 5131-5145.

Ong, H.C., Mahlia, T.M.I., Masjuki, H.H., Norhasyima, R.S., 2011. Comparison of palm oil, *Jatropha curcas* and *Calophyllum inophyllum* for biodiesel: a review. Renew. Sustain. Energy Rev. 15(8), 3501-3515.

Ouachab, N., Tsoutsos, T., 2012. Study of the acid pretreatment and biodiesel production from olive pomace oil. J. Chem. Technol. Biotechnol. 88, 1175-1181.

Parmar, A., Singh, N.K., Pandey, A., Gnansounou, E., Madamwar, D., 2011. Cyanobacteria and microalgae: a positive prospect for biofuels. Bioresour. Technol. 102(2), 10163-10172.

Peyravi, M., Rahimpour, A., Jahanshahi, M., 2015. Developing nanocomposite PI membranes: morphology and performance to glycerol removal at the downstream processing of biodiesel production. J. Membr. Sci. 473, 72-84.

Pinzi, S., Leiva, D., López-García, I., Redel-Macías, M.D., Dorado, M.P., 2014. Latest trends in feedstocks for biodiesel production. Biofuels, Bioprod. Biorefin. 8(1), 126-143.

Pitakpoolsil, W., Hunsom, M., 2014. Treatment of biodiesel wastewater by adsorption with commercial chitosan flakes: parameter optimization and process kinetics. J. Environ. Manage. 133, 284-292.

Qiu, Z., Zhao, L., Weatherley, L., 2010. Process intensification technologies in continuous biodiesel production. Chem. Eng. Process. Process Intensif. 49(4), 323-330.

Ramadhas, A.S., Jayaraj, S., Muraleedharan, C., 2005. Biodiesel production from high FFA rubber seed oil. Fuel. 84(4), 335-340.

Ribeiro, N.M., Pinto, A.C., Quintella, C.M., da Rocha, G.O., Teixeira, L.S.G., Guarieiro, L.L.N., do Carmo Rangel, M., et al., 2007. The role of additives for diesel and diesel blended (ethanol or biodiesel) fuels: a review. Energy Fuels. 21(4), 2433-2445.

Ruffing, A.M., 2011. Engineered cyanobacteria: teaching an old bug new tricks. Bioeng. Bugs. 3, 136-149.

Sadhik Basha, J., Anand, R.B., 2012. Effects of nanoparticle additive in the water-diesel emulsion fuel on the performance, emission and combustion characteristics of a diesel engine. Int. J. Veh. Des. 59(2-3), 164-181.

Sahafi, S.M., Goli, S.A.H., Tabatabaei, M., Nikbakht, A.M., Pourvosoghi, N., 2015. The reuse of waste cooking oil and spent bleaching earth to produce biodiesel. Energy Sources Part A. In press.

Sajjadi, B., Abdul Aziz, A.R., Ibrahim, S., 2015. Mechanistic analysis of cavitation assisted transesterification on biodiesel characteristics. Ultrason. Sonochem. 473, 72-84.

Sarin, R., Sharma, M., Sinharay, S., Malhotra, R.K., 2007. Jatropha–palm biodiesel blends: an optimum mix for Asia. Fuel. 86(10), 1365-1371.

Sharma, Y.C., Singh, B., Madhu, D., Liu, Y., Yaakob, Z., 2014. Fast synthesis of high quality biodiesel from 'waste fish oil' by single step transestrification. Biofuel Res. J. 1(3), 78-80.

Shibasaki-Kitakawa, N., Honda, H., Kuribayashi, H., Toda, T., Fukumura, T., Yonemoto, T., 2007. Biodiesel production using anionic ion-exchange resin as heterogeneous catalyst. Bioresour. Technol. 98, 416-421.

Shirazi, M.M.A., Bastani, D., Kargari, A., Tabatabaei, M., 2013a. Characterization of polymeric membranes for membrane distillation using atomic force microscopy. Desalin. Water Treat. 51(31-33), 6003-6008.

Shirazi, M.M.A., Kargari, A., Tabatabaei, M., Mostafaeid, B., Akia, M., Barkhi, M., Shirazi, M.J.A., 2013b. Acceleration of biodiesel–glycerol decantation through NaCl-assisted gravitational settling: a strategy to economize biodiesel production. Bioresour. Technol. 134, 401-406.

Shirazi, M.M.A., Kargari, A., Bazgir, S., Tabatabaei, M., Shirazi, M.J.A., Abdullah, M.S., Matsuura, T., Ismail, A.F., 2013c. Characterization of electrospun polystyrene membrane for treatment of biodiesel's water-washing effluent using atomic force microscopy. Desalination. 329 , 1-8.

Shirazi, M.J.A., Bazgir, S., Shirazi, M.M.A., 2014a. Edible oil mill effluent; a low-cost source for economizing biodiesel production: Electrospun nanofibrous coalescing filtration approach. Biofuel Res. J. 1(1), 39-42.

Shirazi, M.M.A., Kargari, A., Tabatabaei, M., 2014b. Evaluation of commercial PTFE membranes in desalination by direct contact membrane distillation. Chem. Eng. Process. Process Intensif. 76, 16-25.

Shirazi, M.M.A., Kargari, A., Tabatabaei, M., Ismail, A.F., Matsuura, T., 2014c. Assessment of atomic force microscopy for characterization of PTFE membranes for membrane distillation (MD) process. Desalin. Water Treat. 1-10.

Shirazi, M.M.A., Kargari, A., Tabatabaei, M., 2015. Sweeping gas membrane distillation (SGMD) as an alternative for integration of bioethanol processing: study on a commercial membrane and operating parameters. Chem. Eng. Commun. 202(4), 457-466.

Shirazi, M.M.A., Tabatabaei, M., 2014. Biofuels, In: Sharma, U.C., Kumar, S., Prasad, R. (Eds.), Energy Science and Technology (Vol. 6), Studium Press LLC, USA.

Shuman, T.R., Mason, G., Marsolek, M.D., Lin, Y., Reeve, D., Schacht, A., 2014. An ultra-low energy method for rapidly pre-concentrating microalgae. Bioresour. Technol. 158, 217-224.

Silitonga, A.S., Masjuki, H.H., Mahlia, T.M.I., Ong, H.C., Chong, W.T., Boosroh, M.H., 2013. Overview properties of biodiesel diesel blends from edible and non-edible feedstock. Renew. Sustain. Energy Rev. 22, 346-360.

Singh, A.K., Fernando, S.D., Hernandez, R., 2007. Base-catalyzed fast transesterification of soybean oil using ultrasonication. Energy Fuels. 21(2), 1161-1164.

Singh, I., Taggar, M.S., 2014. Recent Trends in Biodiesel Production—An Overview. Int. J. Appl. Eng. Res. 9(10), 1151-1158.

Sitepu, I.R., Garay, L.A., Sestric, R., Levin, D., Block, D.E., Bruce German, J., Boundy-Mills, K.L., 2014. Oleaginous yeasts for biodiesel: current and future trends in biology and production. Biotechnol. Adv. 32(7), 1336-1360.

Sivakumar, P., Sindhanaiselvan, S., Gandhi, N.N., Devi, S.S., Renganathan, S., 2013. Optimization and kinetic studies on biodiesel production from underutilized *Ceiba Pentandra* oil. Fuel. 103, 693-698.

Stephenson, P.G., Moore, C.M., Terry, M.J., Zubkov, M.V., Bibby, T.S., 2011. Improving photosynthesis for algal biofuels: toward a green revolution. Trends Biotechnol. 29(12), 615-623.

Stojković, I.J., Stamenković, O.S., Povrenović, D.S., Veljković, V.B., 2014. Purification technologies for crude biodiesel obtained by alkali-catalyzed Transestrification. Renew. Sustain. Energy Rev. 32, 1-15.

Strathmann, H., 2011. Introduction to membrane science and technology. Wiley-VCH, Weinheim.

Sulistyo, H., Rahayu, S.S., Winoto, G., Suardjaja, I.M., 2008. Biodiesel production from high iodine number candlenut oil. World Acad. Sci. Eng. Technol. 48, 485-488.

Sun, A., Davis, R., Starbuck, M., Ben-Amotz, A., Pate, R., Pienkos, P., 2011. Comparative cost analysis of algal oil production for biofuels. Energy. 36, 5169-5179.

Sundaramurthy, J., Li, N., Kumar, P.S., Ramakrishna, S., 2014. Perspective of electrospun nanofibers in energy and environment. Biofuel Res. J. 1(2), 44-54.

Tabatabaei, M., Khatami far, M., 2009. Pro spinning tube in tube reactor (Pro-STT) equipped with high voltage system for the production of biodiesel and glycerin. IR Patent 62533.

Tabatabaei, M., Tohidfar, M., Salehi Jouzani, G., Safarnejad, M.R., Pazouki, M., 2011. Biodiesel production from genetically engineered microalgae: Future of bioenergy in Iran. Renew. Sustain. Energy Rev. 15(4), 1918-1927.

Talebi, A.F., Mohtashami, S.K., Tabatabaei, M., Tohidfar, M., Bagheri, A., Zeinalabedini, M., Hadavand Mirzaei, H., Mirzajanzadeh, M., Malekzadeh Shafaroudi, S., Bakhtiari, S., 2013. Fatty acids profiling: a selective criterion for screening microalgae strains for biodiesel production. Algal Res. 2(3), 258-267.

Talebi, A.F., Tabatabaei, M., Chisti, Y., 2014. BiodieselAnalyzer©: a user-friendly software for predicting the properties of prospective biodiesel. Biofuel Res. J. 2, 55-57.

Talebi, A.F., Tohidfar, M., Bagheri, A., Lyon, S.R., Salehi-Ashtiani, K., Tabatabaei, M., 2014. Manipulation of carbon flux into fatty acid biosynthesis pathway in *Dunaliella salina* using *AccD* and *ME* genes to enhance lipid content and to improve produced biodiesel quality. Biofuel Res. J. 1(3), 91-97.

US Energy Information Administration (US-EIA), 2014. Monthly biodiesel production report.

Venkateswarlu, K., Murthy, B.S.R., Subbarao, V.V., 2015. An experimental investigation to study the effect of fuel additives and exhaust gas recirculation on combustion and emissions of diesel–biodiesel blends. J. Braz. Soc. Mech. Sci. Eng. 1-10.

Vonortas, A., Papayannakos, N., 2014. Comparative analysis of biodiesel versus green diesel. Wiley Interdiscip. Rev. Energy Environ. 3(1), 3-23.

Zhang, Y., Dube, M.A., McLean, D.D., Kates, M., 2003. Biodiesel production from waste cooking oil: 1. Process design and technological assessment. Bioresour. Technol. 89, 1-16.

Zhang, J., Jiang, L., 2008. Acid-catalyzed esterification of *Zanthoxylum bungeanum* seed oil with high free fatty acids for biodiesel production. Bioresour. Technol. 99, 8995-8998.

Zhu, X.G., Long, S.P., Ort, D.R., 2008. What is the maximum efficiency with which photosynthesis can convert solar energy into biomass?. Curr. Opin. Biotechnol. 19, 153-159.

# Pyrolysis of *Parinari polyandra* Benth fruit shell for bio-oil production

Temitope E. Odetoye[1,2]*, Kolawole R. Onifade[3], Muhammad S. AbuBakar[2], James O. Titiloye[2,4]

[1]*Department of Chemical Engineering, University of Ilorin, PMB1515, Ilorin, Nigeria*

[2]*European Bioenergy Research Institute, CEAC, Aston University, Birmingham, United Kingdom*

[3]*Department of Chemical Engineering, LadokeAkintola University of Technology, Ogbomoso, Nigeria*

[4]*College of Engineering, Swansea University, Swansea, SA2 8PP, United Kingdom*

## HIGHLIGHTS

- Bio-oil production from non-conventional *Parinari polyandra* B. fruit shell is an intermediate pyrolysis process.
- Bio-oils were obtained using a fixed bed reactor within a temperature range of 375 - 550°C.
- The presence of valuable compounds suggests high potentials for industrial applications.
- The presence of acetic acids in bio-oil suggests the need to upgrade the oil before utilization as a fuel.

## GRAPHICAL ABSTRACT

## ABSTRACT

Non-conventional agricultural residues such as *Parinari polyandra* Benth fruit shell (PPBFS) are potential sources of biomass feedstock that have not been investigated for bio oil production. In this study, PPBFS was pyrolyzed *via* an intermediate pyrolysis process for the production of bio oil. The bio oils were obtained using a fixed bed reactor within a temperature range of 375–550 °C and were characterized to determine their physicochemical properties. The most abundant organic compounds present were acetic acid, toluene, 2-cyclopenten-1-one, 2-furanmethanol, phenol, guaiacol and 2,6-dimethoxyphenol. The bio-oil produced at 550 °C possessed a higher quantity of desirable compounds than those produced at lower temperatures. The presence of acetic acids in the bio-oil suggested the need to upgrade the bio-oil before utilization as a fuel source.

**Keywords:**
Biomass
Biofuel
Bio-oil
Pyrolysis
*Parinari polyandra* Benth
Agricultural residue

* Corresponding author
E-mail address: todetoye@yahoo.com

## 1. Introduction

There is a mounting interest in harnessing a variety of renewable resources for energy supply (IEA, 2011; Akia et al, 2014). This is the consequence of an increasing awareness on the need to find alternative forms of energy to address the negative effects of the use of fossil fuels on the environment. The growing energy needs of the world, the hike in global fuel prices as well as the depletion of the non-renewable fossil fuel reserves are also among the main reasons directing more research towards renewable fuels (E4tech, 2008; Akia et al, 2014; Kumar, 2014).

Biomass which is regarded as the third world largest primary energy resource has been a research focus for production of various types of alternative fuels including bio-oil (Bradely, 2006). This research focus has in turn necessitated the investigation of more biomass sources as feedstock. On the other hand, recently there has been an increase in research interest on biomass conversion via pyrolysis (IEA, 2011). Pyrolysis is one of the thermochemical conversion processes involving thermal decomposition of biomass usually above 400 °C in the absence of oxygen (Czernik and Bridgwater, 2004).

A vast number of plant materials including energy crops and various agricultural wastes are being considered as bio-oil feedstock (Abnisia et al., 2011, Omar et al., 2011; Garcia et al.,2012; Greenhalf et al., 2012; Park et al, 2012; Balan et al., 2014; Beetul et al, 2014.; Kumar, 2014)..The quality of bio-oil obtained depends on the properties of the biomass as well as operating conditions (Park et al., 2012). Despite the fact that there have been various research works conducted on a number of energy crops and agricultural wastes for bio-oil production such as rice husk (Garcia et al., 2012), cassava rhizomes (Pattiya et al., 2008), miscanthus (Hodgson et al,2010) corn stover and bagasse among others, there is still a growing need to investigate non-conventional biomass feedstock particularly those of no other applications such as *Parinari polyandra* Benth.

*Parinari polyandra* Benth is a savannah plant found in West Africa extending from Mali to Sudan. It is also known as *Maranthes polyandra* Benth, belonging to the family *Rosasceae*. The tree is about 8 m high with glossy leaves that are elliptical and are usually rounded at both ends. The fruits are smooth and about 2.5 cm long having a deep red or blackish purple color depending on the variety (Fig. 1). The endosperm has a yellow white appearance with a thick seed coat containing the oily mass. The fresh seed kernel contains between 31-60% oil depending on the variety and season of harvest (Olatunji et al., 1996).

**Fig.1.** *Parinari polyandra* Benth fruits: (a) fresh (b) dried (c) without seed, and (d) milled.

*Parinari polyandra* Benth is currently very much under-utilized partly due to its non-edibility and also owing to lack of extensive research on the fruit and seeds. The main utilization has been that of the plants parts for trado-medicinal purposes (Iweala & Oludare, 2011). Although, the seed oil is considered not edible, it has been found to be suitable for preparation of alkyd resins of desirable properties (Odetoye et al., 2013a, Odetoye et al., 2014). The fruit shell, being a major waste generated from the seed oil utilization, has been previously characterized (Odetoye et al., 2013b; Titiloye et al., 2013). As reported in our previous work, the thermochemical properties obtained for parinari fruit shell determined using ASTM standard methods including ultimate, inorganic, and proximate analyses as well as the higher heating value of 20.5 MJ/kg were positive indicators for bio-oil production as shown in Table 1. (Odetoye et al., 2013b).

**Table1.**
Proximate analysis, ultimate analysis, inorganic analysis and higher heating values of *Parinari polyandra* Benth fruit shell (Odetoye et al., 2013 b).

| Ultimate analysis | % (dry basis) | Inorganic analysis | ppm (dry basis) |
|---|---|---|---|
| C | 48.04 | Al | 37 |
| H | 5.76 | B | 13 |
| N | 2.13 | Ba | 24 |
| S | 0.10 | Ca | 3600 |
| Cl | 0.44 | Cd | ND* |
| O[a] | 43.53 | Cr | 2 |
| | | Cu | 8 |
| | | Fe | 92 |
| | | Hg | 1 |
| **Proximate analysis** | **%** | K | 11200 |
| Moisture content | 2.7 | Mg | 1700 |
| Ash content | 4.7 | Mn | 17 |
| Volatile content | 78.2 | Na | 54 |
| Fixed carbon | 17.1 | Ni | 3 |
| | | P | 1200 |
| | | Pb | ND |
| | | Rb | 18 |
| | | Si | 164 |
| **Higher heating value** | **MJ/kg** | Sr | 14 |
| Bomb calorimeter | 20.50 | Ti | 4 |
| Calculated HHV[1] | 18.96 | V | 1 |
| Calculated HHV[2] | 18.90 | Zn | 10 |

*ND: not detected
[a] Oxygen by difference
[1] (Channiwala& Parikh, 2002)
[2] (Wai ,2005)

In this study, we report for the first time the pyrolysis of *Parinari polyandra* Benth fruit shells in a fixed bed reactor and investigate its suitability as a new feedstock for bio-oil production *via* intermediate pyrolysis. To the best of our knowledge no pyrolysis work has been reported on *Parinari polyandra* Benth before.

## 2. Materials and methods

### *2.1. Materials*

*Parinari polyandra* Benth fruits were collected from trees at the University of Ilorin, Ilorin, Nigeria (8°30'N 4 ° 33'E) during the month of November. The fruits were cut into halves to remove the seeds from the woody endocarp and were subsequently sun-dried for five days. The dried fruits were ground using a Retsh SM 200 heavy-duty cutting mill fitted with a 2 mm particle size screen. The samples were characterized and compared to other similar biomasses.

### *2.2. Methods*

#### *2.2.1. Pyrolysis procedure*

The intermediate pyrolysis experiments were carried out on a bench scale fixed-bed reactor at reaction temperatures of 550, 450 and 375 °C. The reactor was made of quartz glass tube with an internal diameter of 30 mm and a height of 390 mm. It was connected to a primary condenser used for the collection of condensable bio-oil. The non-condensable gases released were scrubbed with isopropanol before sending a stream of the gases to the GC-TCD HP Series II and the remaining gases vented through the fume cupboard.

The experimental set up is as shown in Figure 2. Nitrogen gas flow into the reactor was maintained at a flow rate of 50 $cm^3$/min. The heating rate of the reactor was 25°C/min. The bio-oils were obtained from the oil pots, condenser and the connecting tube after each experimental run. The bio-oil produced was initially in a gaseous phase when leaving the reactor at a relatively fast rate due to the high temperatures used. On reaching the bottle-like condensers filled with dry ice, the condensable gases turned into liquid in the condenser. The secondary condenser implemented in the system ensured that the adequate condensation was achieved. The bio-oil samples used for the subsequent analyses were collected mainly from the oil pot and the condensers without the use of a solvent.

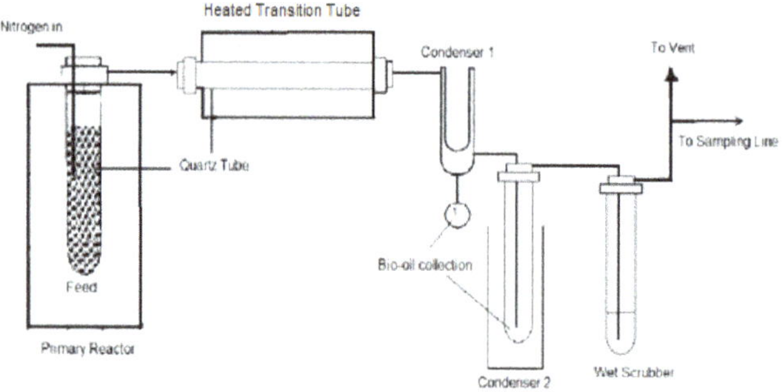


**Fig.2.** Experimental set up for intermediate pyrolysis experiment

The bio-oil yield was calculated considering the oil entrapped in the glassware condensers, by weighing each part of the glassware apparatus before and after each pyrolysis experiment. The biochar yield was also measured by weighing the reactor tube and the reactor head before and after the reaction since the char (solid) remained mainly in the reactor tube and head after the reaction. The mass of the non-condensable gases was obtained by difference. The effects of reaction temperatures (375, 450 and 550 °C) on products yield and properties were also investigated

*2.3. Bio-oil characterization*

The bio-oil products obtained were characterized to determine their quality and composition.

*2.3.1. Density, pH and water content determinations*

The pH values of the bio-oils prepared were determined using a Sartorius pH meter model PB-11. Mettler Toledo Portable Lab Densimeter was used to determine the density of the bio-oil samples. Water content was determined using the Karl Fischer volumetric titration method based on the ASTM D1744 standard

*2.3.2. Heating value and elemental analysis*

The elemental analysis of the parinari shell bio-oil was carried out by MEDAC Ltd. Surrey, U.K. A unified correlation approach was employed to calculate the higher heating values (HHV) using an empirical equation suggested by Saidur*et al.(*2011):

$$HHV = 0.3516 (C)+1.16225 (H)-0.1109 (O)+0.0628 (N)+0.10465 (S) \quad \text{(Eq. 1)}$$

*2.3.3. Gas chromatograph-mass spectrometer (GC-MS) analysis*

The composition of the bio-oil produced was determined using the Hewlett Packard 5890 Series II Gas Chromatograph equipped with a Hewlett Packard 5972 mass selective detector. Helium was used as the carrier gas with a DB 1706 non-polar capillary column. The initial oven temperature was 40 °C and ramped up to 290 °C at a rate of 3 °C/min. The injection temperature was held at 310 °C with a volume of 5µl. The dilution solvent used was ethanol and the dilution rate was 1:5. Identification of compounds in the spectral and chromatograph data was done with the aid of NIST mass spectra database.

**3. Results and discussion**

*3.1. Effect of temperature on yield*

As shown in Figure 3, relatively higher liquid yield was obtained when the pyrolysis experiment was run at 550 °C compared to 375 and 450 °C suggesting that the liquid product yield increased by increasing the temperature. The 550 °C run was also characterized with relatively lower char content of 35.6%, while the highest char content of the product was obtained at 375 °C.

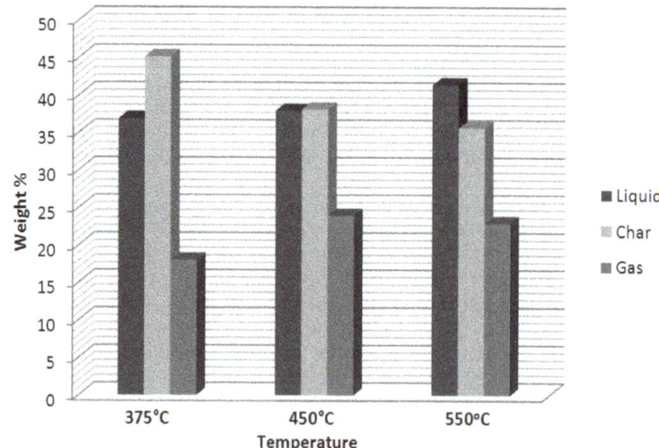


**Fig.3.** Product yield distribution of Parinari fruit shell obtained through pyrolysis at different temperatures

The liquid and char products obtained at 450 °C were of comparable values. The liquid product yield tended to increase with the rise in process temperature. This trend observed in the product yield distribution is similar to the pyrolysis results obtained by Volli and Singh for sesame, mustard and neem de-oiled cakes (Volli&Singh, 2012). The relatively higher temperature required for the parinari biomass to decompose into favorable liquid product may be due to the high lignin concentration of 30 wt. % yielding phenols and phenolic compounds (Kim et al., 2012; Titiloye et al., 2013).

The biochar yield at 450 and 550 °C are relatively close in percentage value. This observation is actually in line with the thermogravimetric analysis result reported in our earlier work (Odetoye et al., 2013b). In fact, the decomposition of cellulose and hemicellulose which are the main components occurred before 450 °C while lignin decomposition occurred relatively slower at around 400 °C, hence, little difference was observed in the biochar yield after 400 °C.

**Fig.4.** *Parinari polyandra* fruit shell bio-oil samples

### *3.2. Bio-oil characterization*

The bio-oil samples obtained were dark brown in colour (Fig.4) and were composed of aqueous and organic phases which were relatively semi-homogenous. The liquids were observed to separate completely into 2 phases when stored. The parinari bio-oil exhibited homogenous characteristics and free flowing properties. The pH of the bio-oil tended to increase with increasing processing temperature and the sample prepared at 550 °C had the highest pH of 4.56 (Table 2).

**Table 2.**
Some of the physical and chemical properties of the bio-oil prepared (organic phase)

| Properties | Unit | 375° C | 450° C | 550° C |
|---|---|---|---|---|
| pH | | 4.12 | 4.32 | 4.56 |
| Density | g/cm$^3$ | 1.043 | 1.027 | 1.021 |
| Water content | wt % | 13.995 | 16.17 | 14.78 |
| Elemental composition | wt % | | | |
| C | | 44.74 | 66.43 | 69.31 |
| H | | 9.15 | 8.28 | 8.97 |
| N | | 1.27 | 2.73 | 2.89 |
| O* | | 44.27 | 21.38 | 18.79 |
| S | | 0.52 | 0.79 | 0.73 |
| Cl | | <0.05 | 0.42 | <0.05 |
| HHV | MJ/kg | 21.85 | 30.76 | 32.84 |
| Empirical Formula | | $CH_{2.45}O_{0.74}N_{0.02}$ | $CH_{1.5}O_{0.24}N_{0.04}$ | $CH_{1.55}O_{0.20}N_{0.04}$ |
| H/C molar ratio | | 2.5 | 1.5 | 1.6 |
| O/C molar ratio | | 0.74 | 0.24 | 0.2 |

The calculated heating value of 30.76 MJ/kg for the bio-oil obtained through pyrolysis at 450 °C was comparable to those of some other biomasses reported in the literature. More specifically, it is slightly higher than 26.22 MJ/kg obtained for corncob (Demiralet al., 2012) and comparable to the 30 MJ/kg of neem seed cake bio-oil (Volli & Singh, 2012). The elemental analysis of the parinari bio-oil showed that the oxygen content decreased as the pyrolysis temperature increased, suggesting the deoxygenation reaction taking place. The chlorine content was relatively higher while the sulphur content was desirably low in parinari bio-oil. Hence, there is a lower risk of sulphur dioxide formation.

### *3.3. Bio-oil chemical composition*

The bio-oils consisted of a complex mixture of various compounds. The GC-MS chromatogram for a typical parinari fruit shell sample pyrolysis oil is shown in Figure 5. The most prominent peaks identified, corresponding chemical names, retention times, chemical formula, molecular weights and peak areas measured are summarised in Table 3 and Figures 6, 7, 8 and 9. The identified compounds included alkenes, phenols, carboxylic acids and their derivatives reported for the bio-oils obtained from some other biomasses (Volli & Singh, 2012). As presented in Figure5, the main constituents in parinari bio-oil were acetic acid, toluene, 2-cyclopenten-1-one, 2-furanmethanol, phenol, guaiacol and 2,6-dimethoxyphenolas. Phenolic compounds such as guaiacol and 2,6-dimethoxyphenol are among the valuable components that can be obtained from bio-oil (Demiral et al, 2012)

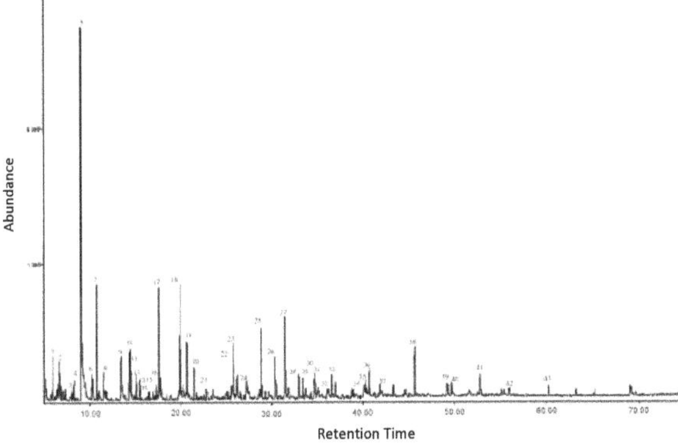


**Fig.5**. A typical GC-MS chromatogram for *Parinari polyandra* bio-oil

**Table 3.**
Most prominent identified compounds of Parinari bio-oil

| Peak ID | RT (min) | Compound Name | Formulae | RMM* | Area % 375°C | Area % 450°C | Area % 550°C |
|---|---|---|---|---|---|---|---|
| 1 | 6.048 | 2-methylfuran | $C_5H_6O$ | 82.10 | 0.51 | 0.96 | 1.25 |
| 2 | 6.772 | 2-butanone | $C_4H_8O$ | 72.11 | 0.71 | 1.45 | 1.38 |
| 3 | 8.083 | 2-methylbutanal | $C_5H_{10}O$ | 86.13 | 0.25 | 0.38 | 0.29 |
| 4 | 8.336 | 2,5-Dimethylfuran | $C_6H_8O$ | 96.13 | 0.15 | 0.53 | 0.49 |
| 5 | 9.083 | Acetic Acid | $C_2H_4O_2$ | 60.05 | 18.96 | 18.2 | 13.4 |
| 6 | 10.267 | Hydroxyacetone | $C_3H_6O_2$ | 74.08 | 1.18 | 0.8 | 0.66 |
| 7 | 10.784 | Toluene | $C_7H_8$ | 92.14 | 0.65 | 3.67 | 5.28 |
| 8 | 11.509 | Pyridine | $C_5H_5N$ | 79.10 | 0.73 | 1.29 | 0.77 |
| 9 | 13.429 | Propanoic acid | $C_3H_6O_2$ | 74.08 | 1.6 | 2 | 1.21 |
| 10 | 14.372 | Cyclopentanone | $C_5H_8O$ | 84.12 | 0.62 | 1.7 | 1.43 |
| 11 | 14.51 | 1-hydroxy-2-butanone | $C_4H_8O_2$ | 88.11 | 1.1 | 0.76 | - |
| 12 | 15.05 | Ethylbenzene | $C_8H_{10}$ | 106.17 | - | 0.82 | 1.32 |
| 13 | 15.441 | p-Xylene | $C_8H_{10}$ | 106.17 | - | 0.7 | 1.54 |
| 14 | 16.464 | Cyclohexanone | $C_6H_{10}O$ | 98.14 | - | 0.28 | 0.33 |
| 15 | 16.901 | m-Xylene | $C_8H_{10}$ | 106.17 | 0.21 | 0.28 | 0.6 |
| 16 | 17.269 | Styrene | $C_8H_8$ | 104.15 | - | 0.57 | 0.75 |
| 17 | 17.603 | 2-Cyclopenten-1-one | $C_5H_6O$ | 82.10 | 4.03 | 4.51 | 3.84 |
| 18 | 19.936 | 2-Furanmethanol | $C_5H_6O_2$ | 98.10 | 5.98 | 4.66 | 3.76 |
| 19 | 20.661 | 2-Cyclopenten-1-one, 2-methyl- | $C_6H_8O$ | 96.13 | 1.31 | 2.09 | 2.23 |
| 20 | 21.431 | 2-Furyl Methyl Ketone | $C_6H_6O_2$ | 110.11 | 1.27 | 1.33 | 1.22 |
| 21 | 23.535 | 2-hydroxy-2-cyclopenten-1-one | $C_5H_6O_2$ | 98.00 | 1.09 | 0.72 | - |
| 22 | 25.731 | 3-Methyl-2-Cyclopentenone | $C_6H_8O$ | 96.13 | 1.84 | 2.67 | 2.55 |
| 23 | 26.226 | Tetrahydro-2-furanmethanol | $C_5H_{10}O_2$ | 102.13 | 2 | 1.12 | 1.13 |
| 24 | 27.249 | 2,4-dimethyl-2-oxazoline-4-methanol | $C_6H_{11}NO_2$ | 129.16 | 0.23 | 1.38 | 0.82 |
| 25 | 28.813 | Maple lactone / 2-hydroxy-3-methyl-2-cyclopenten-1-one | $C_6H_8O_2$ | 112.13 | 4.28 | 3.62 | 3.48 |
| 26 | 30.353 | Phenol | $C_6H_6O$ | 94.11 | 1.11 | 2.06 | 4.19 |
| 27 | 31.377 | Guaiacol | $C_7H_8O_2$ | 124.14 | 5.39 | 3.86 | 3.49 |
| 28 | 32.883 | 2-Methylphenol | $C_7H_8O$ | 108.14 | 0.36 | 1.04 | 2.36 |
| 29 | 33.4 | 3-Ethyl-2-hydroxy-2-cyclopenten-1-one | $C_7H_{10}O_2$ | 126.15 | 1.45 | 0.91 | 0.81 |
| 30 | 34.619 | p-cresol | $C_7H_8O$ | 108.14 | 0.55 | 0.83 | 1.51 |
| 31 | 34.711 | m-Cresol | $C_7H_8O$ | 108.14 | 0.37 | 1.33 | 3.58 |
| 32 | 36.585 | Isocreosol / 5-methylguaiacol | $C_8H_{10}O_2$ | 138.17 | 1.38 | 0.91 | 0.8 |
| 33 | 37.01 | 2,4-Dimethylphenol | $C_8H_{10}O$ | 122.16 | 0.29 | 0.65 | 1.6 |
| 34 | 39.045 | 4-Ethylphenol | $C_8H_{10}O$ | 122.16 | - | 0.2 | 0.5 |
| 35 | 40.161 | Dianhydromannitol | $C_6H_{10}O_4$ | 146.14 | 1.06 | 0.95 | 0.76 |
| 36 | 40.666 | 4-Ethylguaiacol | $C_9H_{12}O_2$ | 152.19 | 1.64 | 1.29 | 1.33 |
| 37 | 41.862 | 1,4:3,6-dianhydro-a-d-glucopyranose | $C_6H_8O_4$ | 144.13 | 1.02 | 0.68 | 0.65 |
| 38 | 45.576 | 2,6-Dimethoxyphenol | $C_8H_{10}O_3$ | 154.17 | 5.51 | 2.44 | 2.23 |
| 39 | 49.175 | Isoeugenol | $C_{10}H_{12}O_2$ | 164.20 | 1.36 | 0.55 | 0.9 |
| 40 | 49.623 | 4-methoxy-3-(methoxymethyl)-phenol | $C_9H_{12}O_3$ | 168.19 | 2.29 | 0.98 | 1.24 |
| 41 | 52.762 | 1,2,3-trimethoxy-5-methylbenzene | $C_{10}H_{14}O_3$ | 182.22 | 2.05 | 1.12 | 1.23 |
| 42 | 55.924 | 2,4-hexadienedioic acid, 3,4-diethyl-, dimethyl ester, (EZ)- | $C_{12}H_{18}O_4$ | 226.00 | - | 0.47 | 0.52 |
| 43 | 60.155 | 4-Allyl-2,6-dimethoxyphenol | $C_{11}H_{14}O_3$ | 194.23 | 1.34 | 0.57 | 0.92 |
| | | TOTAL | | | 75.87 | 77.33 | 78.35 |

*Relative molecular mass

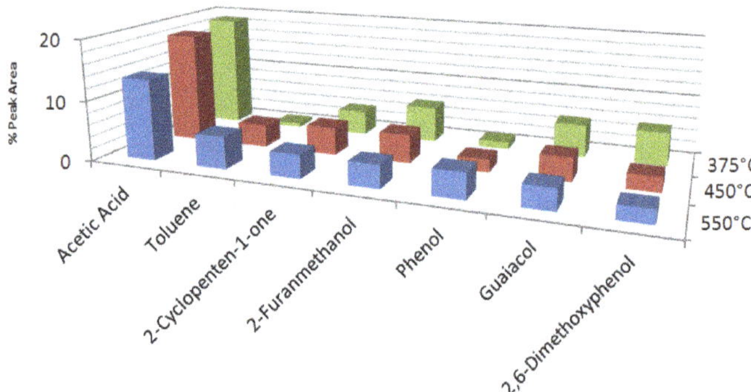


**Fig.6**. Main chemical constituents of Parinari fruit shell bio-oil.

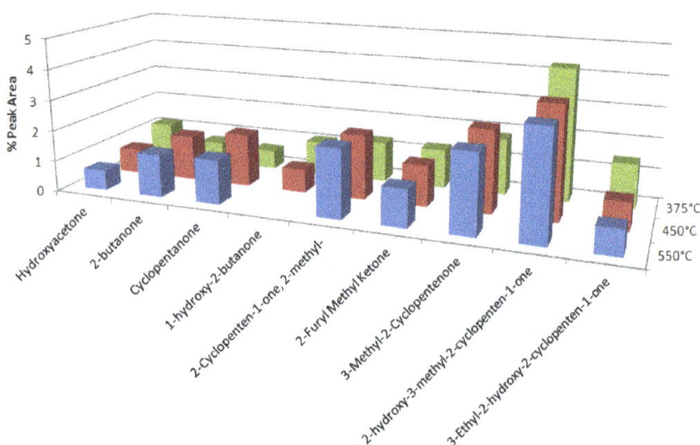


**Fig.7**. Main ketone constituents in Parinari fruit shell bio-oil.

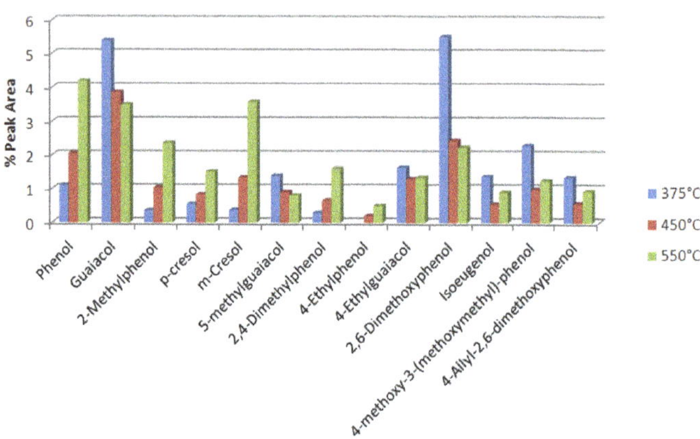


**Fig.8**. Phenol and its derivatives in Parinari fruit shell bio-oil.

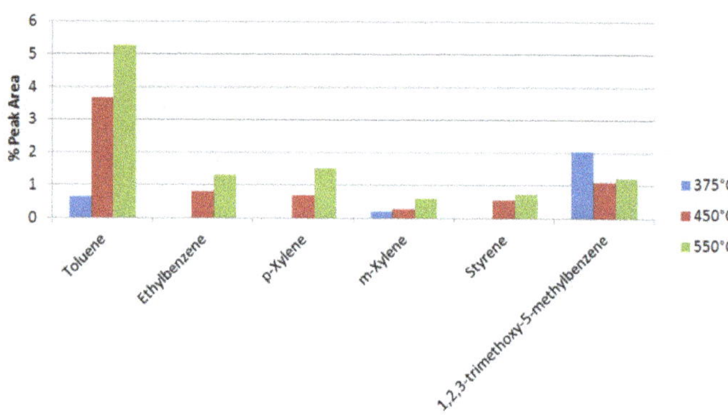


**Fig.9**. Main BTX aromatic hydrocarbons in Parinari fruit shell bio-oil.

The highest temperature of 550 °C favored the production of phenolic compounds and their derivatives while the oxygenated phenolic derivatives were produced at lowest temperature of 375 °C. The significant presence of phenolics and aromatics in the bio-oil can be attributed to lignin which is a phenolic bio-polymer (Park et al., 2012). Acetic acid is the main organic acid contributing up to 19% peak area. The presence of these organic acids is associated with low pH values measured for the bio-oil and are generally known to be undesirable in bio-oils. The increase in pH achieved by increasing the pyrolysis temperature correlated with the decrease in the acetic acid peak area as detected by GC.

This suggests that higher temperature of 550 °C was more favourable for the preparation of bio-oil resulting in less acidity. Furthermore, the relatively high content of acetic acid suggests that the bio-oil may need to be upgraded before it can be used as fuel in engines. BTX and related xylene derivatives compounds (Fig. 9) were found to increase with increasing temperature with the exception of 1,2,3 trimethoxy-5-methylbenzene which showed an increase at 375 °C.

## 4. Conclusion

Bio-oil has been successfully produced from *Parinari polyandra* Benth fruit shell using an intermediate pyrolysis process. The properties of the bio-oil obtained herein was favorably comparable with those of other bio-oils reported in the literature. The findings of the present study revealed that the pyrolysis temperature had a significant effect on the properties and quality of bio-oil obtained from Parinari seed shell. More specifically, the bio-oil produced at 550 °C had more desirable properties than those produced at 375 and 450 °C. However, the PPBFS bio-oil needs to be upgraded before it can be utilized as a fuel substitute particularly in engines as it consisted of various complex organic compounds such as acetic acid. Finally, due to the availability of PPBFS as a biomass feedstock, the parinari bio-oil is strongly suggested to be used as an alternative biofuel. Moreover, the presence of valuable compounds such as phenolic compounds in the produced bio-oil indicates potentials industrial applications as well.

## 5. Acknowledgments

The authors express their gratitude for the financial support provided by the School of Engineering and Applied Science, Aston University, Birmingham, U.K.

## References

Abnisia, F, WanDaud, W.M.A., Shu, J.N., 2011. Optimization and characterization studies on bio-oil production from palm shell by pyrolysis using response surface technology. Biomass Bioenerg. 35, 3604-3616.

Akia, M., Yazdani, F., Motaee, E., Han, D., Arandiyan, H., 2014. A review on conversion of biomass to biofuel by nanocatalysts. Biofuel Res. J. 1, 16-25.

ASTM Standard D1744, Standard Test Method for Determination of Water in Liquid Petroleum Products by Karl Fisher Reagent, ASTM International, United States.

Balan, V., Chiaramonti, D., Kumar, S., 2013. Review of US and EU initiatives toward development, demonstration, and commercialization of lignocellulosic biofuels. Biofuels, Bioprod. Biorefin. 7(6),733-760.

Beetul, K., Sadally, S.B., Hossenkhan, N., Puchooa, D., 2014. An investigation of biodiesel production from microalgae found in Mauritian waters. Biofuel Res. J. 2,58-64.

Bradley, D., 2006. IEA Bioenergy Task 40- Climate change solutions European market study for bio-oil (pyrolysis oil). IEA Bioenergy Task 40-Bio-trade report. p1-85.

Czernik, S., Bridgwater, A.V., 2004. Overview of applications of biomass fast pyrolysis oil. Energ. Fuel. 18, 590-598.

Demiral, I., Eryazıcı, A., Sensoz, A., 2012. 2012Bio-oil production from pyrolysis of corn-cob (*Zea mays* L.). Biomass Bioenerg. 36,43-49.

E4tech Report, 2008. Biofuel Review: Advanced technologies overview. p1-12.

Garcia, R., Pizarro, C., Lavin, A.G., Bueno, J.L., 2012. Characterization of Spanish biomass waste for energy use. Bioresour. Technol. 103,249- 258.

Greenhalf, C.E., Nowakowski, D.J., Bridgwater, A.V., Titiloye, J., Yates, N., Richie, A., Shield, I., 2012. Thermochemical Characterisation of Straws and High Yielding Perennial Grasses. Ind. Crop. Prod. 36, 449-459.

Greenhalf, C.E., Nowakowski, D.J., Harms, A.B., Titiloye, J.O., Bridgwater, A.V., 2012. Sequential pyrolysis of willow SRC at low and high heating rates – Implications for selective pyrolysis. Fuel. 93, 692-702.

Hodgson, E.M., Lister, S.J., Bridgewater, A.V., Clifton-Brown, J., Donnison, I.S., 2010. Genotypic and environmentally derived variation in the cell wall composition of *Miscanthus*in relation to its use as a biomass feedstock. Biomass Bioenerg. 34, 652 - 660.

International Energy Agency (IEA) Report, 2011. Biofuel for transport technology roadmap. p1-56.

Iweala, E.E., Oludare, F.D., 2011. Hypoglycemic effects,biochemical and histological changes of *Spondiasmombin* Linn. and*Parinari polyandra* Benth Seeds ethanolicextractinalloxan-induced diabetic rats. J. Pharm. Toxicology. 6, 101-112.

Kim, S., Jung, S., Kim J., 2010. Fast pyrolysis of palm kernel shells: Influence of operation parameters on the bio-oil yield and the yield of phenol and phenolic compounds. Bioresour. Technol. 01,9294-9300.

Kumar, S., 2014. Hydrothermal Processing of Biomass for Biofuels. Biofuel Res. J. 1(2), 43.

Odetoye, T.E., Onifade, K.R., Abu Bakar, M.S., Titiloye, J.O., 2013. Thermochemical characterization of *Parinari polyandra* Benth fruit shell for bio-oil production. Ind. Crop. Prod., 44, 62-66.

Odetoye, T.E., Ogunniyi, D.S. Olatunji, G.A., 2013.Studies on the preparation of *Parinari polyandra* Benthseed oil alkyd resins. J. Appl. Polym. Sci. 1276, 4610-4616.

Odetoye, T.E., Ogunniyi, D.S., Olatunji, G.A., 2014. Refining and Characterization of *Parinari polyandra* Benth seed oil for Industrial Utilization. Nig. J. Pure Appl. Sci., 27 (in press).

Olatunji, G.A., Ogunleye, A.J., Lawani, S.A.,1996. Studies on the seed oil of *Parinari polyandra* Benth; Proximate Chemical composition.Nig. J. Pure & Appl. Sci., 6, 177-179.

Omar, R., Idris, A., Yunus, R., Khalid, K., Isma, M.I., 2011. Characterization of empty fruit bunch for microwave –assisted pyrolysis. Fuel 90, 1536-1544.

Park, Y., Yoo, M.L., Lee, H.W., Park, S.H., Jung, S., Park, S., Kim, S., 2012. Effect of operation conditions on pyrolysis characteristics of agricultural residues. Renew. Energ. 42,125-130.

Pattiya, A., Titiloye, J.O., Bridgwater, A.V., 2008. Fast pyrolysis of cassava rhizome in the presence of catalysts. J. Anal. Appl. Pyrol. 81,72-79.

Saidur, R., Abdelaziz, E.A., Demirbas, A., Hossain, M.S., Mekhilef, S., 2011. A review on biomass as a fuel for boilers, Renew. Sustain. Energ. Rev. 15(5), 2262-2289.

Titiloye, J.O., Abu Bakar, M.S., Odetoye, T.E.,2013. Thermochemical Characterisation of Agricultural Wastes from West Africa. Ind. Crop. Prod. 47, 199- 203.

Volli, V., Singh, R.K., 2012. Production of bio-oil from de-oiled cakes by thermal pyrolysis. Fuel. 96, 579-585.

White, J.E., Catallo, W.J., Legendre, B.L., 2011. Biomass pyrolysis kinetics: A comparative critical review with relevant agricultural residue case studies. J. Anal. Appl. Pyrol. 91,1–33.

# 4

# Simultaneous Saccharification and Fermentation of Cassava Waste for Ethanol Production

C. Pothiraj[1], A. Arun[2], M. Eyini[3,*]

[1] Department of Botany, Alagappa Government Arts College, Karaikudi 630003, Tamilnadu, India.

[2] Department of Energy Science, Alagappa University, Karaikudi 630004, Tamilnadu, India.

[3] Research Centre in Botany, Thiagarajar College, Madurai 625009, Tamilnadu, India.

## HIGHLIGHTS

- *SSF showed significantly higher ethanol yield than the separate enzymatic saccharification and fermentation.*
- *Microbial saccharification and fermentation of cassava waste using S. diastaticus and Z. mobilis led to improved ethanol production.*
- *Ethanol yield of 20.4 g $L^{-1}$ in 36 h (91.3% of the theoretical yield) was achieved.*

## GRAPHICAL ABSTRACT

SEM Picture of *S. diastaticus and Z. mobilis*

Cassava Waste

Ethanol yield of 20.4 g $L^{-1}$ in 36 h (91.3% of the theoretical yield) was achieved

## ABSTRACT

The efficiency of enzymatic and microbial saccharification of cassava waste for ethanol production was investigated and the effective parameters were optimized. The mixture of amylase and amyloglucosidase (AMG) resulted in a significantly higher rate of saccharification (79.6%) than the amylase alone (68.7%). Simultaneous saccharification and fermentation (SSF) yielded 6.2 g $L^{-1}$ ethanol representing 64.5% of the theoretical yield. Saccharification and fermentation using pure and co-cultures of fungal isolates including *Rhizopus stolonifer*, *Aspergillus terreus*, *Saccharomyces diastaticus* and *Zymomonas mobilis* revealed that the co-culture system involving *S. diastaticus* and *Z. mobilis* was highly suitable for the bio-conversion of cassava waste into ethanol, resulting in 20.4 g $L^{-1}$ in 36 h (91.3% of the theoretical yield).

**Keywords:**
Cassava waste
Ethanol
Enzymatic and microbial saccharification
Simultaneous saccharification and fermentation (SSF)

* Corresponding author
E-mail address: jeyini2005@yahoo.co.in

## 1. Introduction

The conventional energy resources could hardly meet the increasing energy demands (Zhu et al., 2005), and hence, biofuels such bioethanol have turned into promising alternatives to the fossil fuels used in the transportation sector. Ethanol has tremendous applications in chemical, pharmaceutical, and food industries in the form of raw material, solvent, and fuel (Alvira et al., 2010; Ohgren et al., 2006). An important issue regarding the bioethanol production is the process economy. Research efforts have been focused on designing economically viable processes capable of sustainably producing high amounts of fuel bioethanol. The cost effectiveness of bioethanol production through hydrolysis of starchy substrates by using enzymatic/microbial processes has been proved commercially viable (Baras et al., 2002, Kim and Dale, 2002).

In India, cassava (*Manihot esculenta* Crantz) is grown largely over 3.9 million hectares producing $60\times10^6$ million tones of tubers annually. Cassava is an industrial crop for the production of sago, vermicelli, and starch, whereas each ton of cassava tuber processed for sago and starch, yields half a ton of fibrous residue as waste. Cassava waste contains starch (50% zdry weight), cellulose, hemicellulose, and ashes in the extractable materials. Using cassava waste for ethanol production offers huge opportunities owing to the enormous availability of this inexpensive raw material in the cassava growing countries (Hermiati et. al., 2012). Lots of emphasis has been given to screen feasible bioprocess methodologies for efficient conversion of cassava waste to fuel ethanol. Since, starch derived from any plant sources is a complex molecule, therefore, it require various hydrolytic enzymes to be converted into simple fermentable sugars. A number of strategies have been adopted for the construction of starch-utilizing systems, which include the addition of amylolytic enzymes in culture broth, as well as mixed-culture fermentation.

Conversion of both cellulosic and starchy materials in a single process can be achieved by co-culturing two or more compatible microorganisms with the ability to utilize the materials. Fungal co-culturing offers a means to improve hydrolysis of residues, and also to enhance biomass utilization which would minimize the need for additional enzymes in the bioconversion process. Co-cultures have many advantages compared to their monocultures, including improved productivity, adaptability, and substrate utilization. Moreover, fermentation technologies utilizing strains of *Zymomonas mobilis* instead of the traditional yeast have been proposed by a number of authors, as these strains have been shown to ferment under fully anaerobic conditions with faster specific rates of glucose uptake and ethanol production close to the theoretical yield. The aim of the present work was to evaluate the efficiency of enzymatic and microbial liquefaction and saccharification of cassava waste for cost-effective production of ethanol using *Z. mobilis*.

## 2. Materials and methods

### 2.1. Materials and chemicals

Raw cassava waste (also known as thippi) was obtained from the Varalaxmi Sago Industry, Namagiri, Salem, Tamilnadu, India. It was sun dried, coarsely ground to uniform size (5 mm), stored in gunny bags, and used within 1 month after procurement. Enzymes including α-amylase (E.C3.2.1.1), amyloglucosidase (AMG)(E.C. 3.2.1.3), and cellulase (E.C.3.2.1.4) used in the experiments were kindly gifted by Novo Nordisk (India).

### 2.2. Fungal isolates from cassava waste

The primary isolation of fungi from cassava waste dispersed soil was done by serial dilution technique using potato dextrose agar medium at pH 6.5-7.0. The plates were maintained under aerobic conditions at 30 °C for 48 h. The isolates were identified based on their morphological and microscopical characteristics (Alexopolous et al., 1996). The fungal isolates were pre-cultured on 2% Potato Dextrose Agar (PDA) medium for 14 d. *Saccharomyces diastaticus* (NCIM 3314) and *Z. mobilis* (NCIM B806) were obtained from the National Chemical Lab, Pune, India. The rich medium used for *Z. mobilis* consisted of 2% glucose, 0.2% $KH_2PO_4$, 1% yeast extract, and 2% agar at pH 5.6-6.4. MGYP broth medium was used for the cultivation of *Saccharomyces diastaticus.*

### 2.3. Analysis of cassava waste composition

The composition of cassava waste was analyzed for the various parameters including starch (Arditti and Dunn, 1969), cellulose (Updegraff, 1969), hemicellulose (Deschatelets and Yu, 1986), lignin (Chesson, 1978), reducing sugars (Miller, 1959), and protein (Lowry et al., 1951).

### 2.4. Enzyme liquefaction and saccharification

#### 2.4.1. Liquefaction using α-amylase

Optimization of the enzyme concentration for liquefaction of cassava waste was done following the method of Amutha and Gunasekaran (1994) using various concentrations of α-amylase (1-15 mg $g^{-1}$ biomass). The optimal values of biomass concentration (50-150 g) and incubation time (12, 24, 36, 48, and 60 h) were obtained by step-wise experiments where the specified parameters were changed by keeping all other parameters constant. The pH of the reaction mixture in all the optimization experiments was kept constant at 6.0. The amount of the reducing sugars released was analyzed by DNS method (Miller, 1959).

#### 2.4.2. Saccharification using AMG

The liquefied slurry obtained through the liquefaction was subjected to saccharification with various concentration of AMG (1-5 mg $g^{-1}$ biomass) at 55°C and pH 4.5. The optimal values of various parameters such as biomass concentration (50 – 150g) and incubation time (12, 24, 36, 48, and 60 h) were obtained as mentioned above. The amount of reducing sugars released was also analyzed as described earlier.

#### 2.4.3. Saccharification using enzyme mixture

Cellulase (0.15 g $g^{-1}$ of cassava waste) was added along with AMG, and the saccharification was carried out under the same conditions presented earlier and the liberated reducing sugar was estimated.

### 2.5. Enzymatic saccharification and microbial fermentation

The liquefied slurry obtained was subjected to saccharification with optimum concentration of AMG (2 mg $g^{-1}$ biomass) while *Z. mobilis* was also inoculated to the liquefied cassava waste at the time of AMG addition. Simultaneous saccharification and fermentation (SSF) was carried out at room temperature (30±2 °C) as described by Amutha and Gunasekaran (1994).

Batch fermentations involving free cells of *Z. mobilis* were also performed using the production medium developed from separate enzymatic liquefaction and saccharification. Inoculum was prepared in the rich medium and after 24 h of incubation, 1 ml of the culture was added to 50 ml of the production medium. The flasks were incubated at 30°C with 120 rpm for 48 h. Aliquots were taken every 12 h and analyzed.

### 2.6. Microbial saccharification of cassava waste and ethanol fermentation

Microbial saccharification and fermentation were carried out using pure and mixed cultures of *S. diastaticus* or the fungal isolates (isolated from cassava waste disposed soil) as described by Zabala et al., (1994). Erlenmeyer flasks (250 ml) containing 5 g cassava waste in 50 ml distilled water were autoclaved at 121°C for 15 min. Spore suspensions were prepared by repeated washings of 7 d old cultures of the fungal isolates in PDA plates with distilled water. Late-log phase culture of *S. diastaticus* in MGYP broth or the spore suspensions were used as inoculum (10% v/v) at a constant OD of 0.5 for the yeast and 1.3 for the fungi. The flasks were incubated at room temperature (30±2 °C). The mixed fermentation was started with *S. diastaticus* or the fungal isolates as pure cultures, and then late-log phase culture of *Z. mobilis* (10%v/v) was inoculated either simultaneously or at 12 h intervals. Aliquots of the fermented substrates were taken at every 12 h and analyzed. All experiments were carried out in triplicate, and the data are the mean values with standard deviations of <3%.

*2.7. Analytical Methods*

Biomass composition was determined according to Doelle and Greenfield (1985). The supernatant obtained by centrifugation of the culture broth at 6,000 ×g was used for reducing sugars (Miller, 1959) and ethanol (Caputi et al., 1968) analyses. The kinetic parameters of ethanol fermentation were calculated according to Abate et al. (1996).

## 3. Results and discussion

*3.1. Strain isolation and identification*

Two phenotypically different fungal colonies were obtained from cassava waste. Based on the morphology and microscopical characters, the two isolates were identified as *Rhizopus stolonifer* and *Aspergillus terreus* (Pothiraj et al., 2006 and 2007).

*3.2. Cassava waste composition*

Cassava waste was found to contain 52% starch, 13.4% cellulose, 9.38% hemicellulose, 11% lignin, 1.14% free reducing sugars, and 2.88% protein (dry weight basis).

*3.3. Optimization of liquefaction and saccharification of cassava waste*

The degree of liquefaction and saccharification of cassava waste depends on various factors such as enzyme concentration, incubation time, and substrate concentration. For optimizing the conditions of cassava waste liquefaction, different concentrations of commercial α-amylase (1-15 mg $g^{-1}$ cassava waste) were added to cassava waste (10% w/v) and incubated for 1.5 h at 75°C. The maximum concentration of the reducing sugars obtained was at 18.70 g/l using 11.25 mg/g enzyme (**Fig. 1**).

Further increase in the concentration of the enzyme did not result in a corresponding increase in the concentration of the reducing sugars in the liquefied slurry. Hence, the enzyme concentration of 11.25 mg/g cassava waste was considered as optimal for liquefaction.

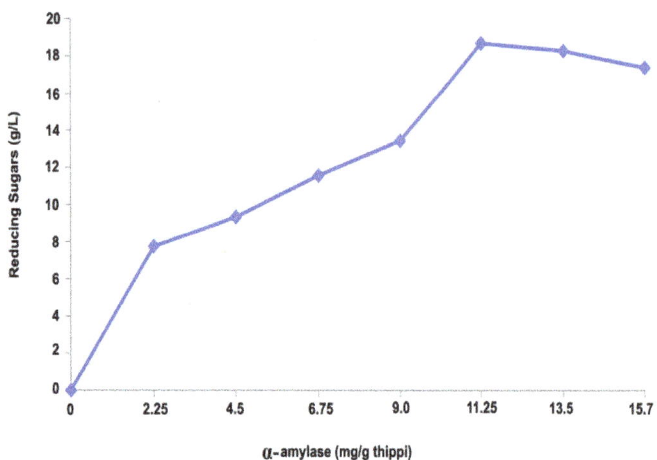


**Fig.1.** Influence of varying concentrations of α-amylase on liquefaction of cassava waste.

Optimization of the liquefaction time was carried out using the optimum concentration of the enzyme (11.25 mg/g cassava waste). The results showed that the release of the reducing sugars was highest (18.9 g/l) during the first 90 min; thus, this was chosen as the optimum time required for liquefaction (**Fig. 2**).

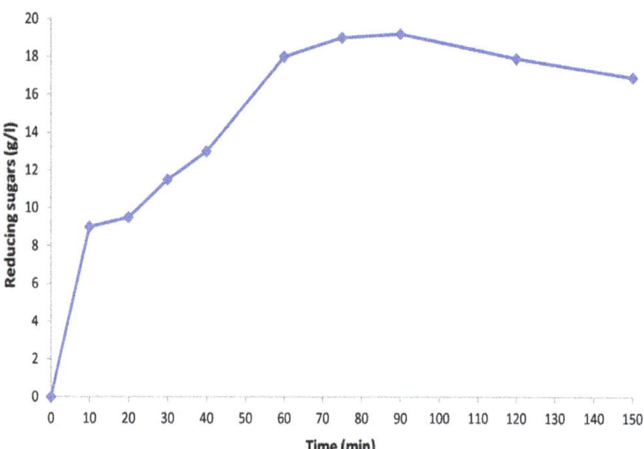


**Fig.2.** Influence of increase in time on liquefaction of cassava waste by α-amylase.

In order to optimize the concentration of AMG for saccharification, various AMG concentrations (1-10 mg $g^{-1}$ cassava waste) were used with the liquefied slurry.

This experiment was carried under optimum conditions for AMG activity i.e. pH 4.5 and 55°C. It was found out that during a constant saccharification time (22 h), the maximum concentration of reducing sugars (39.63 g/l) could be obtained at an enzyme concentration of 5 mg/g cassava waste (**Fig. 3**), and beyond which no appreciable increase in the reducing sugars was obtained. Therefore, this AMG concentration was chosen as the optimum concentration for further experiments.

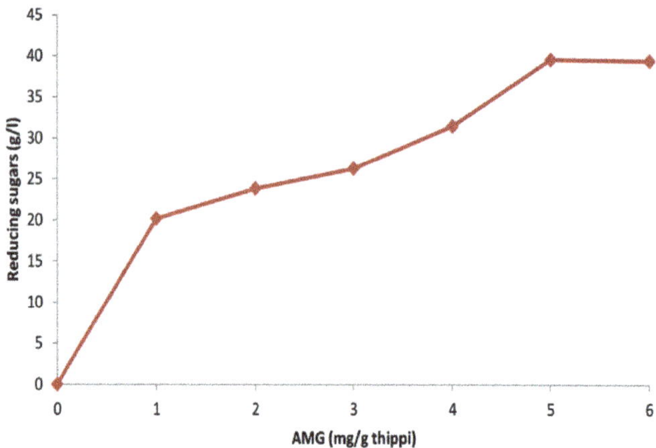


**Fig.3.** Influence of varying concentrations of AMG on saccharification of cassava waste.

The saccharification time of the liquefied slurry was optimized by using the obtained optimum concentration of AMG (5 mg/g cassava waste). The results obtained revealed that maximum saccharification (39.56 g/l reducing sugars) could be obtained within 22 h (**Fig. 4**). There was no significant increase in the reducing sugar concentration beyond 22 h of saccharification. Hence, this period was considered optimal for saccharification.

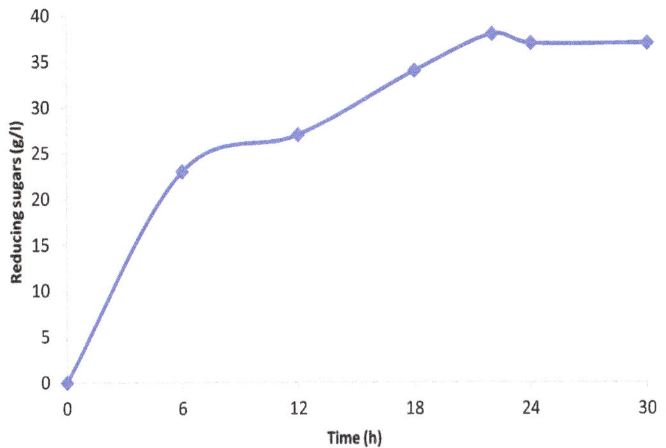


**Fig.4.** Influence of increase in time on saccharificationof cassava waste by amyloglucosidase.

### *3.4. Enzymatic saccharification and fermentation*

The reducing sugars yield after the liquefaction and saccharification was measured at 39.5 g/l from 100 g/l substrate concentration, representing 76% saccharification (**Fig. 5**). While by using the combination of cellulase and AMG, a higher yield of reducing sugars (46 g/l), representing 79.6% saccharification was achieved. Arasaratnam and Balasubramaniam (1993) obtained a similar glucose yield on corn flour by using a two enzyme hydrolysis method. Simillarly, Mojovic et al. (2006) also reported a high conversion rate (dextrose equivalent 92.1%) of corn meal by using the same method.

The maximum ethanol concentration achieved by the enzymatic saccharification and fermentation was observed after 36 h with a productivity of 9.3 g/l (45.6 % of the theoretical yield). While the maximum ethanol concentration of 13.6 g/l, representing 57.7 % of the theoretical yield, was observed after 36 h of fermentation with *Z. mobilis* when cellulase was used additionally with AMG for the saccharification of cassava waste (**Table 1**). These results are in agreement with those of Dabas et al. (1997) who obtained 6-6.4% (v/v) ethanol within 36 h at 30°C from 25% wheat mash saccharified by a combination of α-amylase and amyloglucosidase and fermented by *S. cerevisiae*. In a different study, Mojovic et al. (2006) achieved 78.5 % of
The theoretical yield of ethanol and the highest volumetric productivity of 1.6 g/l/h utilizing 17.5 % corn meal in 48 h using *S. cerevisiae*. The lower product yield obtained in the current study may be due to the different experimental conditions applied and the fermenting organism used.

### *3.5. Simultaneous saccharification and fermentation*

To reduce the complete process time and achieve beneficial energy savings as suggested by Mojovic et al. (2006), the fermenting organism *Z. mobilis* was inoculated along with AMG and both saccharification and fermentation were carried out at room temperature (30°C). Volumetric ethanol productivity by *Z. mobilis* was higher in AMG enzyme saccharification (0.38g $l^{-1}$ $h^{-1}$) while specific ethanol productivity was higher in amylase enzyme saccharification (0.171 g $l^{-1}$ $h^{-1}$). Simultaneous enzyme saccharification and fermentation performed better in terms of ethanol conversion and theoretical yield than the two separate enzymatic saccharification and fermentation. Though the ethanol yield was relatively low (6.2 g/l), the highest ethanol conversion of 33.2%, representing 64.5% of the theoretical yield was obtained in this SSF process. It was also efficient in saving time by reducing the time of the complete process by 10 h as the highest product yield was obtained after 49.5 h (**Table 1**).

**Table 1.**
Effect of the type of enzyme saccharification on the alcoholic fermentation of cassava waste by *Z. mobilis.*

| Organisms | Sugar concentration after saccharificaton or liquefaction (g/l) | Saccharification (%) | Fermentation time (h) | Ethanol (g/l) | Biomass (g/l) | Volumetric ethanol productivity (g $l^{-1}h^{-1}$) | Specific ethanol productivity (g $l^{-1}h^{-1}$) | Ethanol conversion (%) | Percent of the theoretical yield (%) |
|---|---|---|---|---|---|---|---|---|---|
| **Enzymatic saccharification using AMG** | | | | | | | | | |
| *Z. mobilis* | 39.73 | 68.76 | 36 | 9.26 | 1.52 | 0.26 | 0.171 | 23.3 | 45.6 |
| **Enzymatic saccharification using cellulose and AMG** | | | | | | | | | |
| *Z. mobilis* | 46 | 79.61 | 36 | 13.57 | 2.36 | 0.38 | 0.161 | 29.5 | 57.7 |
| **Simultaneous saccharification and fermentation** | | | | | | | | | |
| *Z. mobilis* | 18.7 | 32.36 | 48 | 6.2 | 2.1 | 0.13 | 0.061 | 33.2 | 64.5 |

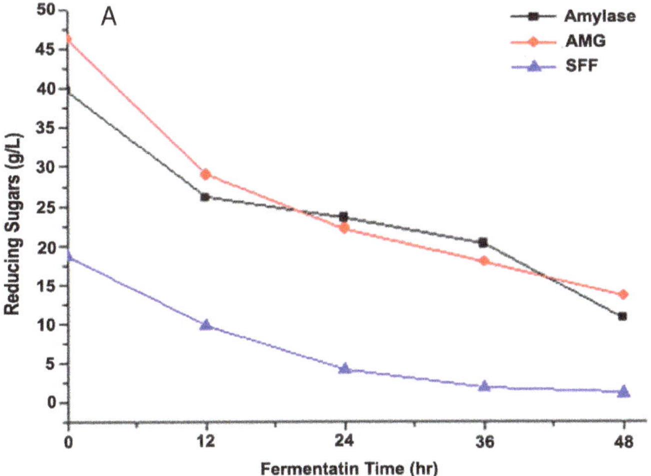


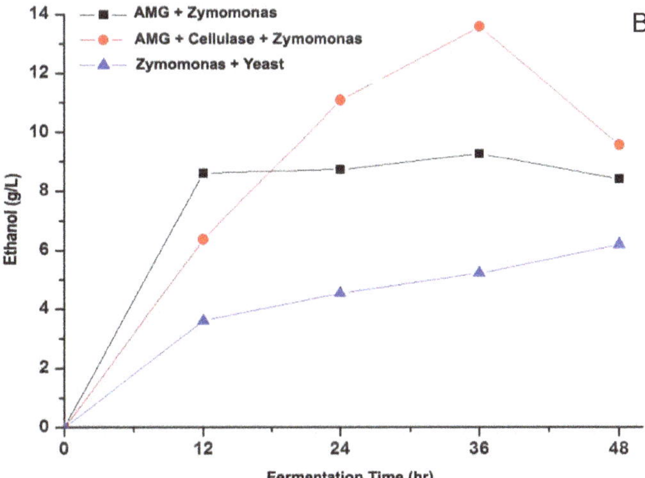


Fig. 5. A: Utilization of reducing sugar during enzymatic saccharification (amylase; AMG and SSF). B: Ethanol production during enzymatic saccharification (amylase + AMG) and simultaneous saccharification (amylase + AMG + Zymomonas mobilis).

The reducing sugars yield (18.7 g/l) from 100 g/l cassava waste in simultaneous saccharification experiment was lower compared to the yield in enzymatic liquefaction followed by saccharification (39.7 and 46 g/l, respectively). This was expected because in SSF, the reducing sugars were assessed after liquefaction with α-amylase and since the organisms were inoculated along with AMG, the actual yield of the reducing sugars due to saccharification could not be found as both saccharification and fermentation proceeded simultaneously. Similar results were observed by Mojovic et al., (2006) in their SSF experiments on corn meal using *S. cerevisiae* (**Table 1**).

*3.6. Microbial saccharification and fermentation*

Direct fermentation of cassava waste was carried out in submerged state employing the pure cultures of the yeast *S. diastaticus* or the spore suspensions of the fungi *R. stolonifer* or *A. terreus* in separate experiments.

The mixed fermentations were started as a single process with a pure culture of *S. diastatiicus, R. stolonifer* or *A. terreus. Z. mobilis* was added at different times of fermentation (including simultaneous inoculation) in order to determine the appropriate time to initiate the mixed fermentation.

*3.6.1. Fermentation using S. diastaticus and Z. mobilis*

The biggest increment in the reducing sugars yield was observed after 24 and 36 h of fermentation using pure culture of *S. diastaticus* and the maximum reducing sugars content of 43.8 g/l was detected at 36 h (**Table 2**). In simultaneously-inoculated mixed culture fermentation, the variations of glucose concentration during fermentation time indicated a synergism between the saccharifying and the fermenting organisms.

**Table 2.**
Submerged fermentation of cassava waste (100 g/l) by pure and mixed culture of *S. diastaticus* and *Z. mobilis*.

| Culture | Period of fermentation (h) | | | | | |
|---|---|---|---|---|---|---|
| | 0 | 12 | 24 | 36 | 48 | 60 |
| | **Reducing sugars (g/l)** | | | | | |
| a | 1.13 | 11.78 | 15.56 | 43.76 | 41.60 | 40.40 |
| b | - | 15.32 | 13.68 | 11.48 | 9.68 | 8.16 |
| c | - | - | 12.58 | 12.08 | 11.38 | 11.02 |
| d | - | - | - | 15.96 | 15.38 | 15.08 |
| e | - | - | - | - | 29.02 | 16.7 |
| | **Biomass (g/l)** | | | | | |
| a | - | 1.92 | 2.18 | 2.67 | 2.82 | 2.91 |
| b | - | 2.28 | 2.67 | 3.12 | 3.24 | 3.38 |
| c | - | - | 2.30 | 2.87 | 3.34 | 3.27 |
| d | - | - | - | 3.34 | 3.41 | 3.41 |
| e | - | - | - | - | 3.45 | 3.49 |
| | **Ethanol (g/l)** | | | | | |
| a | - | 8.02 | 10.5 | 17.92 | 17.78 | 17.65 |
| b | - | 11.26 | 14.66 | 20.42 | 20.70 | 20.56 |
| c | - | - | 13.88 | 18.04 | 17.96 | 17.88 |
| d | - | - | - | 14.68 | 18.34 | 18.16 |
| e | - | - | - | - | 18.7 | 18.48 |

Pure culture a: *S. diastaticus*
Mixed cultures b: Simultaneous inoculations of *S. diastaticus* and *Z. mobilis*.
c, d, e: *S. diastaticus* inoculated with *Z. mobilis* after 12, 24 and 36 h of incubation, respectively.

Direct fermentation of cassava waste by the pure culture of *S. diastaticus* produced the maximum amount of ethanol at 17.9 g/l at 36 h. Simultaneous co-culturing of *S. diastaticus and Z. mobilis* resulted in a higher final ethanol concentration (20.42 g/l) during the same period, and the ethanol yield remained constantly high till the end of fermentation. No significant difference in ethanol production was observed when the yeast was allowed to grow for 12 to 36 h before *Z. mobilis* was added. In the other mixed fermentations, the maximum ethanol concentrations detected ranged between these two values and were obtained either after 36 h or 48 h.

*3.6.2. Fermentation using R. stolonifer and Z. mobilis*

The release of reducing sugars by the pure culture of *R. stolonifer* increased slowly peaking at 42.76 g/l after 60 h fermentation. This was 24 h more to yield nearly the same reducing sugar concentration released by *S. diastaticus.*

**Table 3** shows the reducing sugar, biomass, and ethanol production results using pure cultures of *R. stolonifer* and mixed fermentations of *R. stolonifer* with *Z. mobilis*. Reducing sugar levels obtained by mixed co-culture fermentation dropped continuously reaching nearly half the values obtained by using the pure culture *R. stolonifer* fermentation at 48 h. All the mixed processes showed higher values of microbial biomass than the single pure culture fermentation processes. Maximum 14% increase in biomass was obtained at 36 h fermentation.

**Table 3.**
Submerged fermentation of cassava waste (100 g/l) by pure and mixed cultures of *Rhizopus stolonifer* and *Z. mobilis*.

| Culture | Reducing sugars (g/l) | | | | |
|---|---|---|---|---|---|
| a | 7.90 | 20.34 | 22.88 | 25.14 | 42.76 |
| b | 38.14 | 34.23 | 29.42 | 19.68 | 17.43 |
| c | - | 16.84 | 15.74 | 14.52 | 12.44 |
| d | - | - | 21.92 | 18.42 | 16.54 |
| e | - | - | - | 24.62 | 22.46 |
| | **Biomass (g/l)** | | | | |
| a | 1.61 | 1.89 | 2.21 | 2.76 | 2.98 |
| b | 1.84 | 2.21 | 2.51 | 2.98 | 3.18 |
| c | - | 2.05 | 2.45 | 3.10 | 3.15 |
| d | - | - | 2.94 | 3.04 | 3.1 |
| e | - | - | - | 3.08 | 3.4 |
| | **Alcohol (g/l)** | | | | |
| a | 7.56 | 8.44 | 12.4 | 13.02 | 13.62 |
| b | 9.01± | 10.38 | 15.16 | 17.48 | 18.67 |
| c | - | 12.28 | 13.02 | 14.22 | 14.82 |
| d | - | - | 14.96 | 15.14 | 14.66 |
| e | - | - | - | 16.92 | 17.22 |

Pure culture a: *R. stolonifer.*
Mixed cultures b: Simultaneous inoculations of *R. stolonifer* and *Z. mobilis*.
c, d, e: *R. stolonifer* inoculated with Z. mobilis after 12, 24 and 36 h of incubation, respectively.

The maximum ethanol concentration achieved by direct fermentation of cassava waste using the pure culture of *R. stolonifer* was recorded at 13.6 g/l after 60h fermentation. Simultaneous co-culturing of *R. stolonifer* and *Z. mobilis* resulted in a higher final ethanol concentration (18.7 g/l) at the same time. The maximum final ethanol concentrations in the other mixed fermentations were obtained at either 48 or 60 h fermentation; however, none of these processes were better than the simultaneously-inoculated mixed culture system of *R. stolonifer* and *Z. mobilis.*

*3.6.3. Fermentation using A. terreus and Z. mobilis*

*A. terreus* directly fermented cassava waste into ethanol and the highest increment in the reducing sugar concentration was obtained 24 h earlier compared to the *R. stolonifer* fermentation and the reducing sugar level was maintained till the end of the fermentation period (**Table 4**). The reducing sugars left in the substrate at the end of the fermentation period were least in this simultaneous co-culture system. The residual sugars content progressively increased by time of the inoculation of *Z. mobilis* in to the fungal culture. Inoculation of *Z. mobilis* after 60 h growing of *A. terreus* resulted in the highest increase (18%) in biomass compared to the pure culture.

**Table 4.**
Submerged fermentation of cassava waste (100 g/l) by pure and mixed cultures of *A. terreus* and *Z. mobilis*.

| Culture | Period of fermentation (h) | | | | |
|---|---|---|---|---|---|
| | 12 | 24 | 36 | 48 | 60 |
| | | | **Reducing sugars (g/l)** | | |
| a | 5.84 | 18.88 | 40.58 | 42.46 | 42.60 |
| b | 41.74 | 36.51 | 19.40 | 14.14 | 12.40 |
| c | - | 21.26 | 20.06 | 17.64 | 14.82 |
| d | - | - | 24.52 | 22.92 | 21.66 |
| e | - | - | - | 26.58 | 20.10 |
| | | | **Biomass (g/l)** | | |
| a | 1.45 | 1.79 | 2.18 | 2.59 | 2.71 |
| b | 1.73 | 2.01 | 2.23 | 2.78 | 2.99 |
| c | - | 1.98 | 2.28 | 2.61 | 2.97 |
| d | - | - | 2.48 | 2.81 | 3.15 |
| e | - | - | - | 2.85 | 3.19 |
| | | | **Alcohol (g/l)** | | |
| a | | 7.80 | 12.60 | 12.68 | 12.82 |
| b | 10.10 | 16.68 | 18.58 | 18.32 | 18.18 |
| c | - | 10.10 | 10.68 | 12.58 | 14.32 |
| d | - | - | 13.02 | 13.78 | 14.44 |
| e | - | - | - | 15.14 | 15.64 |

Pure culture a: *A. terreus*
Mixed cultures b: Simultaneous inoculations of *A. terreus* and *Z. mobilis*
c, d, e: *A. terreus* inoculated with *Z. mobilis* after12, 24 and 36 h of incubation, respectively.

these organisms were capable of direct conversion of cassava waste into ethanol.

It was observed that the saccharification efficiency of *R. stolonifer* and *A. terreus* was significantly higher in the presence of the fermenting organism *Z. mobilis* than that of their respective pure cultures (**Tables 3 and 4**). These data suggested that ethanol produced during fermentation had a stimulatory effect on enzyme action.

The highest percent of the theoretical ethanol yield observed in the present investigation was 91.3% in the mixed cultures of *S. diastaticus* and *Z. mobilis.* The mixed culture systems of *R. stolonifer* with *Z. mobilis* and *A. terreus* with *Z. mobilis* led to 85.4% and 89.5% of the theoretical ethanol yield, respectively. These results clearly demonstrated that starch saccharification potential of *S. diastaticus* was higher than that of *R. stolonifer* and *A. terreus.* Similar results have been reported by Abate et al. (1996).

The major limitation in starch or carbohydrate conversion into ethanol was the rate at which a saccharifying organism could hydrolyse the complex carbohydrates. The total time required for the complete process of starch conversion in to ethanol is also a crucial factor determining the overall economy of the process.

In this study, 6.2 g/l ethanol from 18.7 g/l reducing sugar resulted from the liquefied cassava waste was achieved after simultaneous inoculation of AMG and *Z. mobilis* at 48h and 30°C. Since the saccharification was effectively performed along with the fermentation at the room temperature (30°C), the energy required for a separate saccharification step by AMG (optimum temperature 5of 5°C) could be saved.

Direct fermentation of cassava waste by *R. stolonfier* and *A. terreus* showed a higher ethanol concentration (13.62 g/l, 12.8 g/l, respectively) but the production was delayed for 60 h.

Direct fermentation by *S. diastaticus* was superior to the fungal fermentations in terms of product yield and total time duration. Simultaneous mixed cultures of either of these fungi with *Z. mobilis* gave 18.6 g/l ethanol yield after 60 h and 48 h, respectively. In *A. terreus* and *Z. mobilis* mixed culture system, the ethanol production was attained in a shorter fermentation period (48 h).

**Table 5.**
Effect of microbial saccharification on the alcoholic fermentation of cassava waste by *Z. mobilis*.

| Organism | Fermentation time (h) | Maximum ethanol (g/l) | Maximum biomass (g/l) | Volumetric ethanol productivity ($g\ l^{-1} h^{-1}$) | Specific ethanol productivity ($g\ l^{-1} h^{-1}$) | Percent of the theoretical yield (%) |
|---|---|---|---|---|---|---|
| | | | **Microbial saccharification and fermentation (using AMG)** | | | |
| *S. diastaticus* | 36 | 17.92 | 2.67 | 0.497 | 0.187 | 80.1 |
| *S. diastaticus* + *Z. mobilis* | 36 | 20.42 | 3.12 | 0.567 | 0.18 | 91.3 |
| | | | **Microbial saccharification and fermentation (using cellulose and AMG)** | | | |
| *R. stolonifer* | 60z | 13.62 | 2.98 | 0.26 | 0.087 | 62.3 |
| *R. stolonifer* + *Z. mobilis* | 60 | 18.67 | 3.18 | 0.31 | 0.10 | 85.4 |
| | | | **Microbial saccharification and fermentation** | | | |
| *A. terreus* | 60 | 12.8 | 2.18 | 0.36 | 0.165 | 61.7 |
| *A. terreus* + *Z. mobilis* | 48 | 18.58 | 2.23 | 0.52 | 0.233 | 89.5 |

Direct fermentation of cassava waste by the pure culture of *A. terreus* produced the maximum ethanol content of 12.8 g/l after 60 h fermentation. Simultaneous co-culturing of *A. terreus* and *Z. mobilis* resulted in a higher ethanol concentration at earlier time (16.7 g/l at 24 h). Ethanol concentrations near this value were only obtained at 60 h in the other mixed culture fermentations.

The mixed culturing of *S. diastaticus* with *Z. mobilis* proved to be the best method for converting cassava waste carbohydrates into biomass protein and ethanol (**Tables 2-5**).

The production of ethanol in the submerged fermentation of cassava waste by pure cultures of *S. diastaticus* or *A. terreus* or *R. stolonifer* indicated that

There are a few reports in the literature on the use of fungi in pure or mixed cultures for bioconversion of starchy substrates into ethanol. For instance, Dabas et al. (1997) used the hydrolytic enzymes and *A. awamori* for ethanol production from cassava starch and obtained 10% ethanol in five d at 30°C.

The results of all the microbial saccharification and fermentation experiments showed that the simultaneous mixed cultures system involving *S. diastaticus* with *Z. mobilis* was best suited for the bioconversion of cassava waste into ethanol. This system produced the highest ethanol concentration of 20.4 g/l after 36 h fermentation from 100 g/l cassava waste representing 91.3% theoretical yield.

### 4. Conclusion

In the present study, simultaneous enzymatic saccharification and fermentation showed significantly higher ethanol yield than the separate enzymatic saccharification and fermentation. The production of ethanol in the submerged fermentation of cassava waste by pure cultures of *S. diastaticus* or *A. terreus* or *R. stolonifer* indicated that these organisms were capable of direct conversion of cassava waste into ethanol. The results of the present study indicated that the microbial saccharification and fermentation of cassava waste involving *S. diastaticus* with *Z. mobilis* can be used for improved ethanol production from cassava waste or other starch residues for reducing the ethanol production cost by saving energy, the cost of pure hydrolytic enzymes and by reducing the total time of the complete process.

### References

Abate, C., Callieri, D., Rodriguez, E., Garro, O., 1996. Ethanol production by a mixed culture of flocculent strains of *Zymomonas mobilis* and *Saccharomyces sp.* Appl. Microbiol. Biotechnol. 45, 580-583.

Alexopolous, C.J., Mims, C.W., Blackwell, M., 1996. Introductory Mycology. (4th ed.) John Wiley and Sons, Inc., New York.

Alvira, P., Tomás-Pejó, E., Ballesteros, M., Negro, M.J., 2010. Pretreatment technologies for an efficient bioethanol production process based on enzymatic hydrolysis: a review. Bioresour. Technol. 101, 4851-4861.

Amutha, R., Gunasekaran, P., 1994. Simultaneous saccharification and fermentation of cassava starch using *Zymomonas mobilis*. J. Microbiol. Biotechnol. 9, 22-34.

Arasaratnam, V., Balasubramaniam, K., 1993. Synergistic action of α-amylase and glucoamylase on raw corn. Starch-Stärke, 45(6), 231-233.

Arditti, J. and Dunn, A., 1969. Experimental plant physiology: experiments in cellular and plant physiology. Holt, Rinehart and Winston. Inc., New York, pp, 8.

Baras, J., Gacesa, S., Pejin, D., 2002. Ethanol is a strategic raw material. Chem. Ind. 56, 89-105.

Caputi, A., Veda, M., Brown, T., 1968. Spectrophotometric determination of ethanol in wine. American J. Enol. Vitic. 19, 160-165.

Chesson, A., 1978. The maceration of linen flax under anaerobic condition. J. Appl. Bacteriol. 45, 219-230.

Dabas, R., Verma, V.K., Chaudhary, K., 1997. Ethanol production from wheat starch. Indian J. Microbiol. 37, 49-50.

Deschatelets, L. and Yu, E.K.C., 1986. A simple pentose assay for biomass conversion. J. Appl. Microbiol. Biotechnol. 24, 379-385.

Doelle, H.W., Greenfield, P.F., 1985. The production of ethanol from sucrose using *Zymomonas mobilis*. Appl. Microbiol. Biotechnol. 22, 405-410.

Hermiati, E., Mangunwidjaja, D., Sunarti, T., Suparno, O., Prasetya, B., 2012. Potential utilization of cassava pulp for ethanol production in Indonesia. J. Sci. Res. Essays. 7, 100-106.

Kim, S., Dale, B.E., 2002. Allocation procedure in ethanol production system from corn grain I. System expansion. Int. J. Life Cycle Assess. 7, 237-243.

Lowry, O.H., Rosebrough, N.J., Farr, A.L., Randall, R.J., 1951. Protein measurement with the folin phenol reagent. J. Biol. chem. 193, 265-275.

Miller, G.L., 1959. Use of DNS reagent for the determination of reducing sugars. Anal. Chem. 31, 426-428.

Mojovic, L., Nikolic, S., Rakin, M., Vukasinovic, M., 2006. Production of bioethanol from corn meal hydrolyzates. Fuel. 85, 1750-1755.

Ohgren, K., Rudolf, A., Galbe, M., Zacchi, G., 2006. Fuel ethanol production from steam-pretreated corn stover using SSF at higher dry matter content. Biomass Bioenergy. 30, 863-869.

Pothiraj, C. Balaji, P., Eyini,M., 2006 .Raw Starch Degrading Amylase Production by Various Fungal Cultures Grown on Cassava Waste. Mycobiology 34, 128-130.

Pothiraj,C., Eyini, M., 2007. Enzyme Activities and Substrate Degradation by Fungal Isolates on CassavaWaste During Solid State Fermentation. Mycobiology 35, 196-204.

Updegraff, D.M., 1969. Semi micro determination of cellulose biological materials. Anal. Biochem. 32, 420-424.

Zabala, I., Ferrer, A., Ledesma, A., Alello, C., 1994. Microbial protein production by submerged fermentation of mixed cellulolytic cultures. Adv. Bioproc. Eng. Springer Netherlands. pp. 455-460.

Zhu, S., Wu, Y., Zinniu,Y., Xuan, Z., Cunwen, W., Faquan, Y., Siwei, J., Yufeng, Z., Shaoyong, T., Yongping, X., 2005. Simultanious saccharification and fermentation of microwave / alkali pre-treated rice straw to ethanol. Biosys. Eng. 92(2), 229-235.

# 5

# Photoelectrochemical cells based on photosynthetic systems

Roman A. Voloshin[1,*], Vladimir D. Kreslavski[1,2], Sergey K. Zharmukhamedov[2], Vladimir S. Bedbenov[1], Seeram Ramakrishna[3], Suleyman I. Allakhverdiev[1,2,4,*]

[1] *Controlled Photobiosynthesis Laboratory, Institute of Plant Physiology, Russian Academy of Sciences, Botanicheskaya Street 35, Moscow 127276, Russia.*

[2] *Institute of Basic Biological Problems, Russian Academy of Sciences, Pushchino, Moscow Region 142290, Russia.*

[3] *Department Center for Nanofibers and Nanotechnology, Department of Mechanical Engineering, National University of Singapore, 117576, Singapore.*

[4] *Department of Plant Physiology, Faculty of Biology, M.V. Lomonosov Moscow State University, Leninskie Gory 1-12, Moscow 119991, Russia.*

## HIGHLIGHTS

- *Photobioelectrochemical photoconverters based on photosynthetic systems are discussed*
- *Strategies used to improve the efficiency of photobioelectrochemical cells were presented*
- *Advantages and disadvantages of photobioelectrochemical cells were highlighted*

## GRAPHICAL ABSTRACT

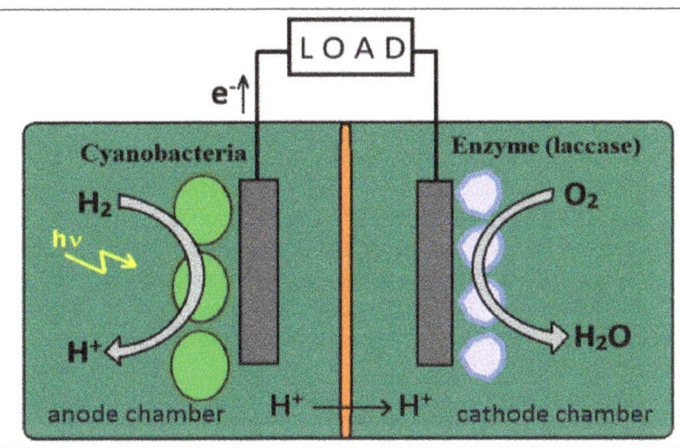


## ABSTRACT

Photosynthesis is a process which converts light energy into energy contained in the chemical bonds of organic compounds by photosynthetic pigments such as chlorophyll (Chl *a*, *b*, *c*, *d*, *f*) or bacteriochlorophyll. It occurs in phototrophic organisms, which include higher plants and many types of photosynthetic bacteria, including cyanobacteria. In the case of the oxygenic photosynthesis, water is a donor of both electrons and protons, and solar radiation serves as inexhaustible source of energy. Efficiency of energy conversion in the primary processes of photosynthesis is close to 100%. Therefore, for many years photosynthesis has attracted the attention of researchers and designers looking for alternative energy systems as one of the most efficient and eco-friendly pathways of energy conversion. The latest advances in the design of optimal solar cells include the creation of converters based on thylakoid membranes, photosystems, and whole cells of cyanobacteria immobilized on nanostructured electrode (gold nanoparticles, carbon nanotubes, nanoparticles of ZnO and $TiO_2$). The mode of solar energy conversion in photosynthesis has a great potential as a source of renewable energy while it is sustainable and environmentally safety as well. Application of pigments such as Chl *f* and Chl *d* (unlike Chl *a* and Chl *b*), by absorbing the far-red and near infrared region of the spectrum (in the range 700-750 nm), will allow to increase the efficiency of such light transforming systems. This review article presents the last achievements in the field of energy photoconverters based on photosynthetic systems.

**Keywords:**
Photobioelectrochemical cell
Self-assembling layer
Thylakoids
Photosystem 1
Photosystem 2
Nanostructures

* Corresponding authors
E-mail address: voloshinra@gmail.com (R.A. Voloshin); suleyman.allakhverdiev@gmail.com (S.I. Allakhverdiev)

**Abbreviations**

| | |
|---|---|
| Asc | Ascorbic acid |
| BOD | Bilirubin oxidase |
| BQ | 1,4-benzoquinone |
| Cyt $b_6f$ | Cytochrome $b_6f$ complex |
| Chl | Chlorophyll |
| CNT | Carbon nanotube |
| DCMU | 3-(3,4-dichlorophenyl)-1,1-dimethylurea |
| DCIP | 2,6-dichlorphenolindophenol |
| ETC | Electron transport chain |
| Fd | Ferredoxin |
| FNR | Ferredoxin:NADP-oxidoreductase |
| FTO | Fluorine tin oxide |
| GNP | Gold nanoparticle |
| HNQ | 2-hydroxy-1,4-naftahinon |
| ITO | Indium tin oxide |
| MWCNT | Multi-walled carbon nanotube |
| OEC | Oxygen-evolving complex |
| PS1 | Photosystem I |
| PS2 | Photosystem II |

## 1. Introduction

The energy crisis and environmental problems are among the most important challenges that humanity must solve in the XXI century. Many of the current investigations are focused on the development of energy sources which must be renewable, sustainable and eco-friendly. Now, the available sources of renewable energy, including solar, wind, rain, waves and geothermal heat, could generate only approximately 16% of the energy used (Sekar and Ramasamy, 2015). Sunlight is the most accessible and reliable among these renewable energy sources. One of the main pathways of solar energy conversion is photosynthesis. Higher plants, microalgae and some bacteria implement photosynthesis. In the course of photosynthesis, water or another electron donor, carbon dioxide and light are used to produce carbohydrates and other organic compounds.

On the other hand, photovoltaic semiconductor devices have also been developed that could generate electric power by converting sunlight directly into electricity. The coefficient of efficiency of the light energy conversion into the electric current by commercial silicon photovoltaic cells is typically less than 20% (Blankenship et al., 2011). Unfortunately, materials and components used in photovoltaic systems are exhaustible and cannot be fully recycled . As photosynthetic organisms operate with a quantum yield close to 100%, therefore, it is reasonable to use this natural process for energy conversion applications. Recently, experts in the field of artificial photosynthesis reviewed and critically analyzed the photosynthetic and photovoltaic energy conversion mechanisms and concluded that it is difficult to compare the conversion efficiency of the current photovoltaic cells with the living photosynthesizing cells because they are completely different systems (Blankenship et al., 2011). The efficiency of photovoltaic cells can be measured based on the output power divided by the total solar radiation spectrum. However, this method does not take into account the storage and transportation of energy. Photovoltaic energy is generally stored in the batteries, which increases their production cost while this also increases the expenses required for the maintenance of such systems. On the contrary, photosynthesis stores solar energy in the form of energy of chemical bonds, which can further be converted into electrical energy (Sekar and Ramasamy, 2015).

## 2. Solar cells

Solar cells are used to convert energy of solar radiations into electrical energy. Since searching for alternative energy sources is very serious, the development of effective and inexpensive solar cells is of particular interest. Currently, there are many different types of solar energy converters. The so-called solar cells or photoelements are devices serving to convert solar energy into usable energy such as an electrical one and can be divided into two types, though maybe not very accurately. They are termed as regenerative cells and photosynthetic cells (Grätzel, 2001). In the regenerative cells, sun light is converted into electricity. This process is not accompanied with any subsequent chemical reactions while the photosynthetic cells generate hydrogen controlled by light (Grätzel, 2001).

Major general steps can be identified for all types of solar cells (Das, 2004):

A) *Absorption of light by photoactive component*

Photoactive component is the substance inside the cell, which absorbs photons. A semiconductor acts as a photoactive component in conventional photovoltaic solar cells while an organic pigment (photosensitizer molecule) serves as a photoactive component in dye-sensitized solar cells. Absorption of a photon leads to certain changes in the energy of the photosensitizer molecule, which is necessary for the further generation of current or activation of chemical reactions leading to the synthesis of molecular hydrogen. In photoelements based on biological systems, a photosensitizer molecule is excited by light. Excited photosensitizer molecule has a very low redox potential, thus, further charge separation is possible.

B) *The charge separation.*

In photoelements using plant or bacterial photosystems, which include pigments, such as Chl *a*, *b*, *d*, *f* and a set of electron carriers, charge separation occurs due to a series of redox reactions. A light-excited pigment molecule quickly transfers electrons to the primary electron acceptor and is then reduced by the primary electron donor. As a result of the charge separation, some voltage is generated in the photoelement

*C) The transfer of electrons to an external circuit for biofuel generation*

In the case of the elements that act as a photoelectric converter - regenerative cells - this step implies an electron transfer to the electrode, and further to an external circuit. For photosynthetic cells, charge separation leads to the activation of the chain of redox pairs, resulting in the formation of molecular hydrogen.

Photoelectric cells, in which organic photoactive material act as photoactive elements, have several advantages over traditional silicon solar cells. The cost of production of such cells is less due to less strict requirements for their production, and the field of their application is wider (Das, 2004). The main disadvantage is that they do not reach the efficiency of the inorganic solar cells (Grätzel, 2001; Das, 2004). With the use of biological molecules and systems in solar cells, a great increase in the efficiency of solar energy absorption is expected.

Currently, solar cells containing inorganic semiconductors, such as mono- and polycrystalline silicon, are used for commercial applications including small devices, such as solar panels on roofs, pocket calculators, and water pumps. These traditional solar batteries can use less than 20% of the incident solar light (Blankenship et al., 2011). Production of silicon solar cells requires energy-intensive processes, high temperatures (400-1400°C) and clean vacuumed conditions, all contributing to their high cost (Das, 2004). On the contrary, production of solar cells based on biological photoactive components does not impose much cost.

## 3. Photosynthesis

Photosynthesis is the process by which the sunlight energy is converted into chemical energy of various organic compounds and is carried out by photosynthesizing organisms. Photosynthesis serves as the ultimate source of energy for all kinds of life on Earth. Photosynthetic organisms use the sunlight energy for synthesis of glucose and other organic compounds which in turn are sources of energy and essential metabolites for heterotrophic organisms (Blankenship, 2002; Blankenship, 2010). This process takes place in two stages: the light one is the light absorption by photosynthetic pigments and the formation of ATP and NADPH; and the dark one is the biosynthesis of carbohydrates. During the dark stage carbon dioxide ($CO_2$) acts as a carbon substrate, reduced form of the NADP (NADPH) molecule as a proton source, and ATP molecule as a source of energy. The electron transport chain (ETC) is an essential element of the light stage of photosynthesis. Through ETC an electron is

$NADP^+$ molecule and reduces it. An external source of electrons is then required to reduce the oxidized pigment molecule. In the case when water acts as an electron source, such type of photosynthesis is named "oxygenic", because molecular oxygen is evolved as a result of the water decomposition (Blankenship and Hartman, 1998). The key source of oxygen in the atmosphere is the oxygenic photosynthesis which manifests itself in all plants, microalgae, and cyanobacteria.

The oxygenic photosynthesis could be summarized through the following general equation (**Eq. 1**):

$$CO_2 + H_2O + h\nu \rightarrow [CH_2O] + O_2 \quad \text{(Eq. 1)}$$

Where $h\nu$ stands for light quantum energy. Light stage processes of the oxygenic photosynthesis take place in membrane structures called thylakoids. In eukaryotic cells of green plants, thylakoids are localized in specific photosynthetic organelles - chloroplasts. Space limited by chloroplast membrane is designated stroma, and the space inside the thylakoid is designated lumen. Thus, one side of the thylakoid membrane faces the stroma, and the other side faces the lumen. In prokaryotic cells (cyanobacteria), thylakoids are located directly in the cytoplasm (Bryant, 1994). Reactions involved in the light stage of the photosynthesis occur in thylakoids.

Light energy is not immediately converted into ATP energy. In fact, it is initially stored in the form of a transmembrane electrochemical potential, which is formed due to the proton transfer by lipophilic transporters through the thylakoid membrane from the stroma into the lumen. As a result, the lumen becomes acidic and the stroma is alkalized. At the expense of the energy of the created potentials ($\Delta\mu H^+$) difference, the enzyme ATP synthase embedded in the thylakoid membrane starts functioning (Andralojc and Harris, 1992).

Light stage of photosynthesis is, in fact, a chain of enzymatic reactions. In higher plants, four transmembrane protein enzymes (**Fig. 1**) catalyze these reactions: photosystem I (PS1), photosystem II (PS2), cytochrome $b_6f$ complex (Cyt $b_6f$) and ATP synthase (Nelson and Yocum, 2006). PS2 catalyzes the electron transfer reaction from the water molecule to plastoquinone. The Cyt $b_6f$ takes part in the oxidation of plastoquinole and reduction of plastocyanin which mediates the transport of electrons from PS2 to PS1 as well as proton transfer from stroma into lumen (Andralojc and Harris, 1992). PS1 catalyzes the oxidation of lipophilic electron carrier, plastocyanine, and the ferredoxin reduction. The enzyme, ferredoxin:NADP-oxidoreductase (FNR) catalyses the $NADP^+$ reduction at the expense of the electrons from the reduced ferredoxin.

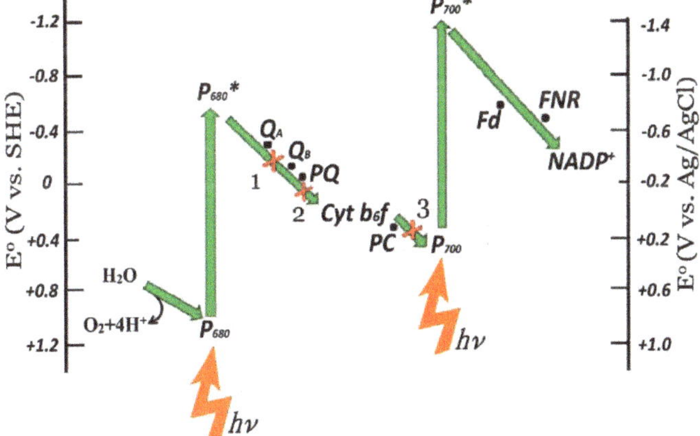


**Fig.1.** The scheme of the non-cyclic electron transport pathway in thylakoids of higher plants and the redox potentials of the components of electron transport chain. Redox potentials are measured *vs.* both the standard hydrogen electrode or SHE (left axis) and silver chloride electrode Ag/AgCl (right axis). $P_{680}$: primary electron donor in photosystem II; $P_{680}^*$: singlet exited state of $P_{680}$; $P_{700}$: primary electron donor in photosystem I; $P_{700}^*$: singlet exited state of $P_{700}$; $Q_A$ and $Q_B$ are primary and secondary quinone acceptors, respectively; PQ: mobile plastoquinone molecule which transfers electrons to the cytochrome $b_6f$ complex (Cyt $b_6f$); PC: plastocyanin; Fd: ferredoxin; FNR: ferredoxin-$NADP^+$ reductase; and $NADP^+$: Nicotinamide adenine dinucleotide phosphate. Red crosses represent reactions that can be inhibited by a) 3-(3,4-dichlorophenyl)-1,1-dimethylurea (DCMU); b) Dibromothymoquinone (DBMIB); c) Potassium cyanide (KCN) (adapted from Sekar et al., 2014).

The primary charge separation, which involves photosynthetic pigments, occurs in the photosynthetic RC. In the RC, the primary electron donor is at the inner lumenal side of thylakoid membrane whereas the primary electron acceptor is closer to the outer stromal side of the membrane. Thus, an electron captured from the molecules of the primary electron donor moves onto the opposite side of the thylakoid membrane (Andralojc and Harris, 1992). As a result of the charge separationm an electric charge on the membrane is generated.

The electron transport chain is activated by light. The photon excites the primary electron donor - a special pair. Chlorophyll is the pigment molecule that can be excited by light of a certain wavelength (**Fig. 2**). The basis of the chlorophyll structure is a heterocyclic ring consisting of four pyrrole rings, connected by methine bridges (Scheer, 2006). Four nitrogen atoms within the chlorine ring are associated with magnesium ion ($Mg^{2+}$). A long hydrophobic phytol tail is attached to the fourth pyrrole ring, while a pigment molecule is attached to the membrane and is correctly oriented. There are several forms of chlorophyll with Chl *a* and Chl *b* as two main forms which are the most widespread in nature.

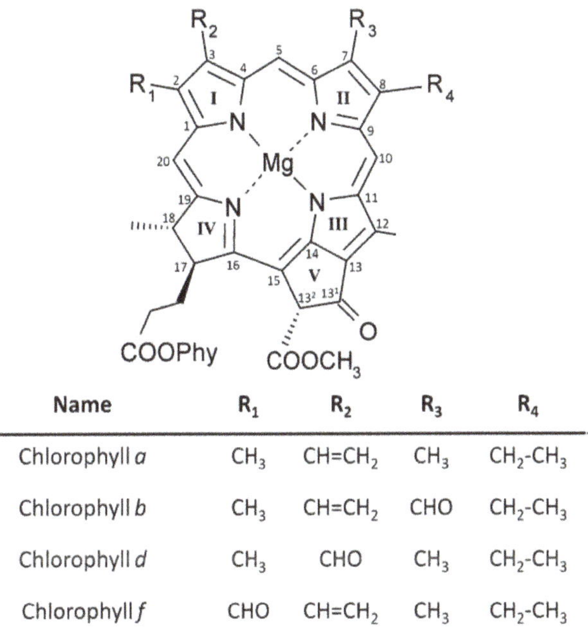


| Name | $R_1$ | $R_2$ | $R_3$ | $R_4$ |
|---|---|---|---|---|
| Chlorophyll *a* | $CH_3$ | $CH{=}CH_2$ | $CH_3$ | $CH_2$-$CH_3$ |
| Chlorophyll *b* | $CH_3$ | $CH{=}CH_2$ | CHO | $CH_2$-$CH_3$ |
| Chlorophyll *d* | $CH_3$ | CHO | $CH_3$ | $CH_2$-$CH_3$ |
| Chlorophyll *f* | CHO | $CH{=}CH_2$ | $CH_3$ | $CH_2$-$CH_3$ |

**Fig.2.** The structural formula of chlorophylls: Chl *a*, Chl *b*, Chl *d*, and Chl *f* (adapted from Loughlin et al., 2013).

Chl *a* serves as the primary electron donor in the RC, and Chl *b* is the accessory pigment of the antenna complexes. A free Chl *a* molecule absorbs light preferably in the wavelength ranges of 400-500 nm and 600-700 nm. Due to the use of other pigments, such as carotenoids, the absorption spectrum of the photosystems is much broader (Sandman, 2009). In addition to Chl *a* and Chl *b*, other forms of chlorophyll, Chl *d* and Chl *f* could also be found in antenna complexes of phototrophic organisms, such as cyanobacteria. Chl *d* is also present in the photosynthetic RC (Tomo et al., 2008, Tomo et al., 2014). The chemical difference between the Chl *b*, Chl *d*, Chl *f* and the Chl *a* is that methyl or vinyl group are substituted by formyl one (**Fig. 2**). The absorption spectra of the chlorophylls also differ from each other. More specifically, the long-wavelength maximum in the absorption spectrum of Chl *d* and Chl *f* markedly shifts towards longer wavelengths compared to that of the Chl *a* (shift up to 40 nm).The energy region (i.e. 380-710 nm) consists of photosynthetically-active radiations which constitutes about 40% of the total solar radiation reaching the Earth's surface (Blankenship and Chen, 2013). However, further expansion in the region ranging from 700 to 750 nm, leads to an increase in the overall energy intensity by about 19%, the creation of solar cells based on the use of photosynthetic components looks very attractive.

## 4. The use of components/systems of the photosynthetic apparatus to generate electricity

### *4.1. Thylakoids as photobiocatalysts*

Thylakoid membranes can be isolated from plants and immobilized on the electrode surface to generate a photocurrent. A team of researchers led by Robert Carpenter (1999) was the first to begin using thylakoid membranes isolated from spinach leaves as a photosensitizer. In their work, a platinum electrode was used as a final acceptor. Studies were carried out in the light and in the dark, in the presence and in the absence of potassium ferrocyanide as a mediator. Native thylakoids generated a photocurrent reaching 6-9 μA without a mediator, and by 4 times more current in the presence of potassium ferrocyanide. It was concluded that the photocurrent generation without any mediators is associated with direct transfer of electrons from the membrane proteins to the electrode surface or through the presence of molecules in the electrolyte that can function as mediators. Oxygen, capable of producing the superoxide radical, may be viewed as a mediator. In 2011, Bedford et al., immobilized thylakoids on conductive nanofibers. They used the electrospinning technique, important for the stabilization of the thylakoid membranes, for immobilizing the thylakoids on the conductive nanofibers. Upon illumination by red light with a wavelength of 625 nm, the maximum electric power generated by the cell surface (1 $cm^2$) was 24 mW.

It is possible to create a stable solar cell by combining the photosynthetic anode and biocatalytic cathode. The idea is to use photosynthetic organisms/organelle/photosystems for the water oxidation at the anode and the oxygen reduction to water at the cathode.

Calkins and colleagues (2013) created photobioelectrochemical cells using thylakoids isolated from spinach. Thylakoids were immobilized on the anode modified with multi-walled carbon nanotubes (MWCNT). Glass electrode modified by laccase/MWCNT system was used as the cathode (**Fig. 3a**). The work demonstrated maximum current density of 68 $mA/cm^2$ and the maximum power density of 5.3 $mW/cm^2$ (**Fig. 3b**). Composite electrode based on thylakoid-MWCNT produced a current density of 38 mA with/$cm^2$, which was by two orders higher than the predicted density.

The main advantage of the use of membrane thylakoids for photocurrent is that during the isolation process, the integral membrane protein complexes remain in their native state. This leads to greater stability and greater power output as compared to the results that may be achieved by using isolated protein complexes or RCs.

### *4.2. Photosystem I as photobiocatalyst*

Beside the thylakoid membrane preparations, some researchers have conducted studies to generate photocurrent by cells based on isolated photosystems. There are two major benefits of using photosystems as a photosensitizers compared to thylakoids as follows:

A) There is less influence exerted by the other redox systems on the electron transfer in the photosystem chain.

B) RCs are closer to the electrode which facilitates direct electron transfer to the electrode.

Formond et al. (2007) developed a photobioelectrochemical system with PS1 as the main photocatalytic subunit, cytochrome $C_6$ and ferredoxin as electron carriers and ferredoxin:NADP-oxidoreductase (FNR) as an electron acceptor (**Fig. 1**). In that experiment, a gold electrode was used. In an earlier investigation, Frolov et al. (2005) created a photobioelectrochemical cell that could generate a voltage of 0.498 ± 0.02 V. They used the PS1 preparations isolated from the cyanobacteria *Synechocystis sp*. PCC 6803. These systems are more stable than plant systems because in the preparations, the antenna pigment molecules (chlorophylls and carotenoids) are integrated into the nuclear subunits. More specifically, unlike in plant systems, the antenna pigments are associated with chlorophyll-protein complex only, which the latter is coupled to nuclear subunits. To stabilize such PS1, surfactant peptides, which are necessary for the stabilization of other plant and bacterial RCs, were not required (Das et al., 2004). Another important factor in their work was the mutation-based replacement of specific amino acids of the PS1 by cysteines. Stable, properly oriented monolayer of PS1 was formed through the formation of Au-S bonds between the thiol group of cysteine and purified hydrophilic gold surface. The procedure for creating the corresponding gold electrode included thermal treatment at 350°C.

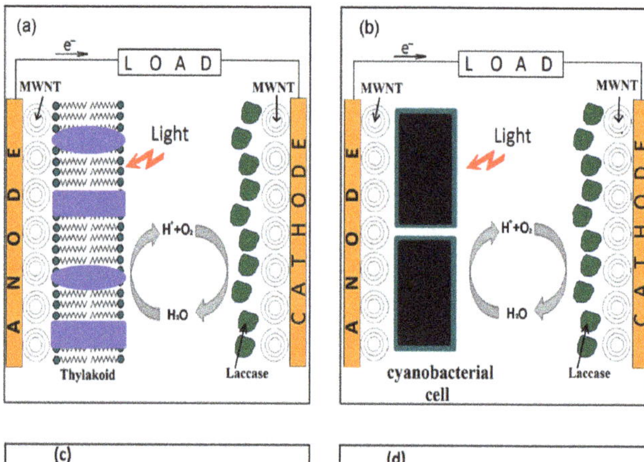


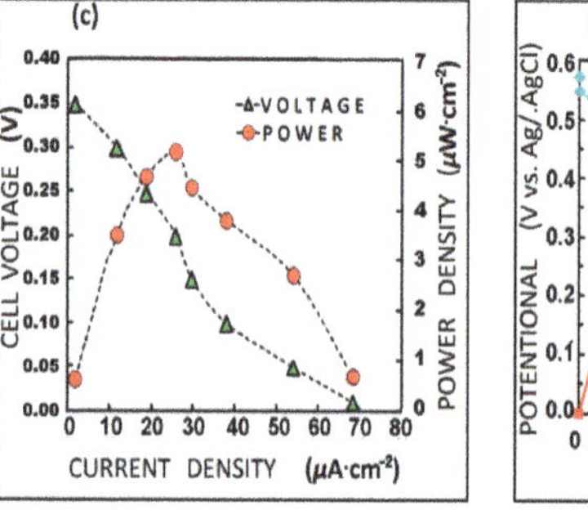


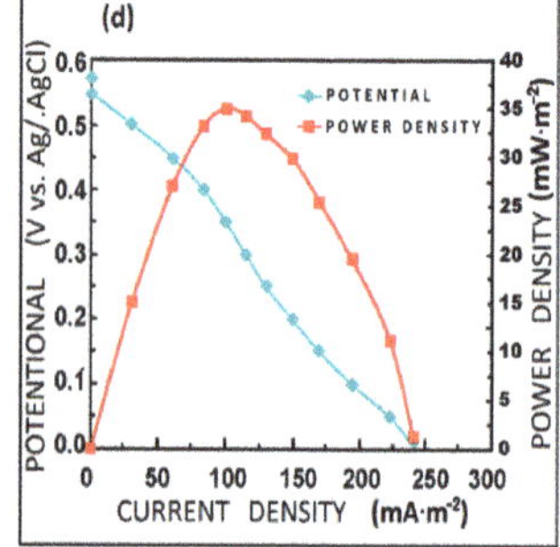


**Fig.3.** Schematic representation of the functioning photobioelectrochemical cells based on, a) the thylakoid-multi-walled carbon nanotubes (MWCNT) and, b) cyanobacteria *Nostoc sp.*-MWCNT. The graphs c) and d) present the dependences of the voltage and the flux density of the received energy on the current density for each of the cells shown (adapted from Calkins et al., 2013; Sekar et al., 2014).

In a different study, Faulkner et al. (2008) reported a fast way for creating a dense monolayer of PS1, isolated from spinach leaves, on a gold electrode. This method of the monolayer creation requiring vacuum conditions was 80 times faster than monolayer creation by method of photosystem precipitation from a solution. More specifically, PS1 was immobilized on the electrode modified with gold nanoparticles (GNP). In the presence of suitable mediators, the cell generated a photocurrent of 100 $nA/cm^2$ (**Fig. 4**).

However, photobioelectrochemical elements based on the PS1 monolayer were not sufficiently effective in cases when a large cross-sectional area of light absorption was required. In the same year, a photobioelectrochemical cell based on multilayer structures of PS1 was created (Frolov et al., 2008). To form the multilayer structures, PS1 complexes were platinized. The platinum ions facilitated the binding of the lumenal side of PS1 and the platinized stromal side resulting in the electrically connected multilayer. The first PS1 monolayer was attached to the gold surface through the bonds between the cysteine`s thiol groups in the mutant PS1 and the gold atoms. The next layer was then formed through the connection between the photosystem donor side of the next layer and the platinum atoms (**Fig. 4c**). The devices developed based on the two and three layers generated photovoltage outputs of 0.330 and 0.386 V, respectively (Frolov et al., 2008). Hereafter, work on the development of solar cells based on multilayer structures of PS1 was continued (Ciesielski et al., 2010; Mershin et al., 2012). The suggested method did not require the use of photosystems isolated from mutated cyanobacteria, nor the use of a high vacuum, making it more economical and less time-consuming. In this work, a plate of gold (thickness of about 125 nm) immobilized on a silicon substrate and a working surface of transparent plastic plate coated with lead oxide doped with indium served as cathode and anode of the photoelement, respectively. A cavity remaining between them, was half-filled with an electrolyte composed of 5 mM 2,6-dichlorphenolindophenol (DCIP), 100 mM ascorbic acid (Asc), and 100 mM NaCl in 5 mM phosphate buffer, pH 7.0. The other half was filled with a buffer solution containing PS1 complexes (about 9 μM), Triton-X100 (0.05% w/v), 0.14 M in 0.2 M $NaH_2PO_4$ at pH 7.0. The PS1 complexes were precipitated on a gold electrode for 7 d. As a result, a multi-layered structure of the PS1 complexes with a thickness of 1-2 μm was obtained. The obtained solar cell generated a photocurrent at a density of about 2 mA/cm$^2$ under illumination by a standard light intensity (clear sky at noon). The device demonstrated a considerable stability and retained activity under ambient conditions for at least 280 d (Ciesielski et al., 2010).

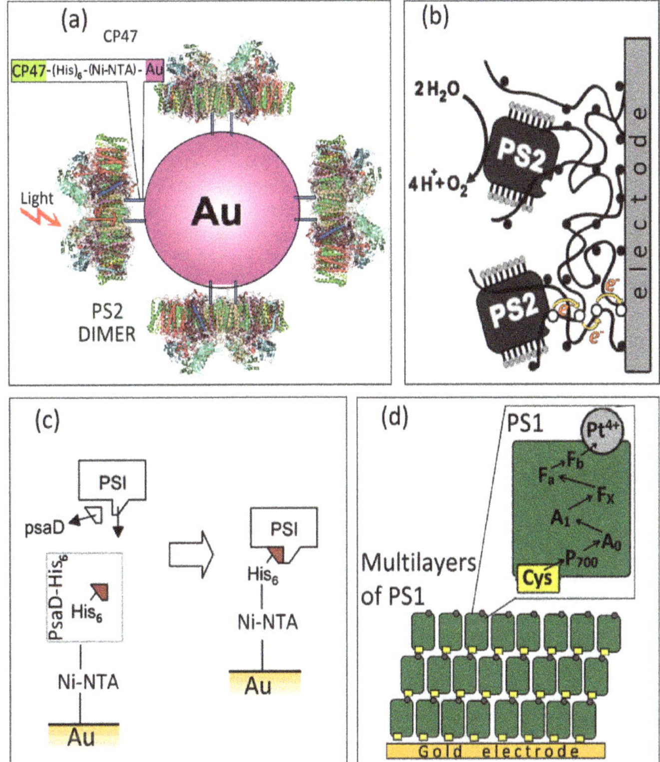


**Fig.4.** Models used for immobilizing photosystems on the electrode. a) connection of PS2 through gold nanoparticle (GNP; Au) histidine tag with Ni-nitrilotriacetic acid (Ni-NTA) attached to the C-terminus of the CP47 protein (adapted from Noji et al., 2011). b) PS2 associated with an osmium redox polymer containing a mediator network (adapted from Badura et al., 2008). Osmium complexes are represented by gray circles. Yellow arrows depict the electron transfer pathway by a hopping mechanism. c) Native PsaD subunit of PS1 replaced by PsaD-His which clings to the histidine tag and the entire structure is associated with the gold electrode through a Ni-NTA (adapted from Das et al., 2004), and d) the scheme of the cysteine mutants of the PS1 with Pt ion and the multilayer structure of such PS1 on a gold substrate (adapted from Frolov et al., 2008).

Sekar and Ramasamy (2015) recently reviewed the recent advances in photosynthetic energy conversion and stated that to date the highest current density of 362 mA/cm$^2$ and the energy flux density of 81 mkW/cm$^2$ using PS1 were obtained by Mershina et al. (2012). In their work, measurements were carried out under normal sunlight and the efficiency of solar cells with different semiconductor substrates i.e. nanocrystalline titanium dioxide $TiO_2$ and nanowires of zinc oxide (ZnO) were compared (**Fig. 5** ). The PS1 complexes were adsorbed on each of these two substrates. Stability of the isolated PS1 complexes was increased by the treatment with surface-active peptides. Peptide Ac-AAAAAAK-$NH_2$ - a sequence of six alanines and one lysine was used as the surfactant. This also promoted the selective adsorption of the PS1 on the substrates and increased the absorption of light. Such approach improved the photovoltaic performance. In this artificial system, cobalt electrolyte Z813 performed the role of plastocyanine, and ferredoxin was replaced by nanocrystalline $TiO_2$, or nanofiber ZnO (Mershin et al., 2012).

Overall, PS1 is a good photobiocatalyst, but has several disadvantages as a photosensitizer. Firstly, the process of the complex isolation is more laborious compared to the isolation of thylakoid membranes. Secondly, the isolated PS1 complex is less stable. Thirdly, for getting a continuous electron transfer, the RC of the PS1, $P_{700}$, requires an external electron donor with a redox potential approximately equal to the redox potential of plastocyanin and thus, the photosystem depends on other electron sources.

### *4.3. Photosystem II as photobiocatalyst*

The main advantage of PS2 *vs.* PS1 is the fact that the electron source, which is required to activate the electron transfer, is water, which is abundant in the environment (Sekar and Ramasamy, 2015). Unlike the PS1, which requires an electron donor, PS2 has an internal oxygen-evolving complex (OEC), also known as water-splitting complex, and depends on the availability of water and light. Here, electrons from $P_{680}$ are transferred to pheophytin, then to plastoquinone and further to the other ETC components (Nelson and Yocum, 2006). Reduction of plastoquinone in $Q_B$-site of PS2 by two electrons from plastoquinone in $Q_A$-site of PS2 and diffusion of the double reduced quinone ($PQH_2$) inside membrane are the major rate-limiting steps (Nelson and Yocum, 2006). Therefore, it is assumed that the water oxidation in PS2 should be accelerated if electrons from $Q_A^-$ could be efficiently transferred to an external electron acceptor (Ulas and Brudvig, 2011). Thus, in order for the electrons from PS2 to be transferred onto the electrode,the complex should

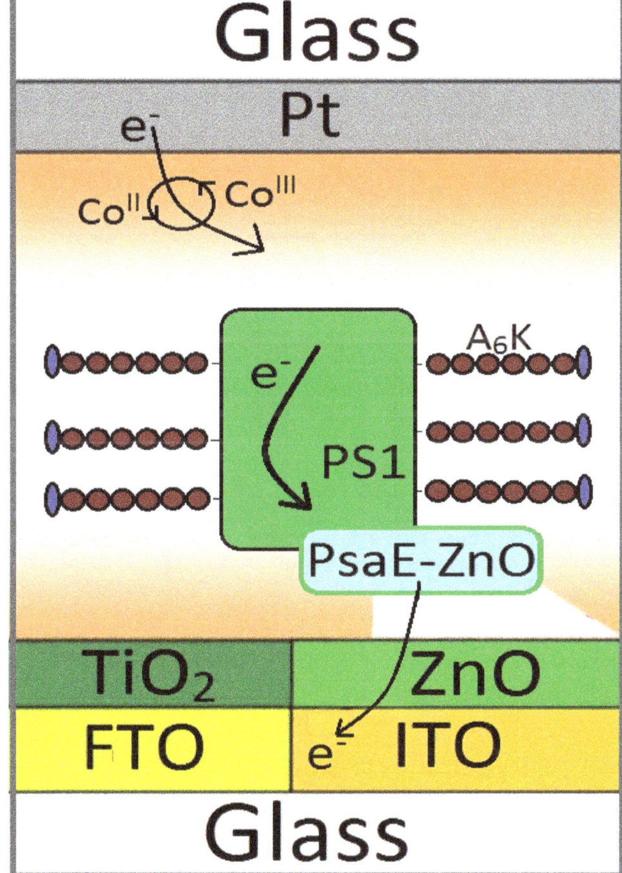


**Fig.5.** Schematic presentation of two Mershina's cells with zinc oxide and titanium dioxide. FTO - a layer of fluorine doped with tin oxide and, ITO - a layer of indium doped with tin oxide (adapted from Mershin et al., 2012). $A_6K$ - designed peptide detergents, acetyl-AAAAAAK.

come in contact with the surface of the electrode, or a mediator should carry out the electron transfer. In fact, it is difficult to achieve direct electron transfer from the PS2 to the electrode because the site of Pheo-PQ is localized deeply inside the PS2 (Badura et al., 2008).

For creation of efficient converters based on PS2, it is important to improve its stability and increase electron transport efficiency. To achieve that, Vittadello et al. (2010) reported the application of histidine-tagged protein complex of PS2 from *S. elongatus* covalently bound to a gold electrode treated with Ni(2+)-nitrilotriacetic acid (Ni-NTA). The current density of the resulting photobioelectrochemical cell reached 43 mkA/cm$^2$. On the other hand, while the photochemical energy conversion efficiency of the freshly isolated PS2 was 0.7, the same parameter for the PS2 immobilized on Au was 0.53. This clearly indicated that the PS2 complexes were photochemically stable even after immobilization, but long-term stability was not discussed by the authors (Vittadello et al., 2010).

Utilization of osmium-containing redox polymer based on poly-1-vinylimidazole is also an effective immobilization method, which could help maintain the stability as well as enhance the coating degree of the electrode by the PS2 complexes (**Fig. 4b**) (Badura et al., 2008). The polymer works as an immobilization matrix and a mediator. Such a system could facilitate the electron transfer from the PS2 complex to the electrode. The correct orientation of the immobilized complex could also facilitate the electron transfer. Recently, Noji and colleagues developed a nanodevice for the artificial water decomposition controlled by light, using a conjugates of PS2-GNP (Noji et al., 2011). The core of the PS2 complex, comprising a histidine tag on the C-terminal of CP47 protein, was immobilized on a GNP by Ni-NTA (**Fig. 4a**). GNPs used in this work were about 20 nm in diameter, and were able to bind four to five PS2 complexes. The efficiency of oxygen evolution by the developed PS2-GNP was comparable to that of the unbound PS2 (Noji et al., 2011).

Israeli scientists developed the photocell which was based on bacterial PS2 complexes isolated from the thermophilic bacterium *Mastigocladus laminosus*. The photoanode consisting of a matrix of 2-mercapto-1,4-benzoquinone (MBQ) was electro-polymerized on the gold surface. Then, PS2 complexes were immobilized on the surface. The anode was electrically connected to the cathode by a composite based on bilirubin oxidase/carbon nanotube (BOD/CNT). They claimed that photo-induced quinone-mediated electron transfer led to the generation of photocurrent with an output power of 0.1 W (Yehezkeli et al., 2012).

Both photosystems (PS1 and PS2) are highly sensitive to photoinhibition and are effectively protected by protective compounds present inside the chloroplast (Sekar and Ramasamy, 2015). It is evident that the stability of isolated photosystems will be impaired after their isolation from native environments. It should be noted that isolated PS2 is more unstable compared to PS1 and photocurrents of high density could be achieved in cells using PS1 complexes. Thus, the use of thylakoids as a photosensitizer should be more preferable as compared to isolated photosystems.

### *4.4. The bacterial cell as photobiocatalyst*

Photocells in which isolated photosynthetic structures such as thylakoids, PS1 and PS2 are used suffer from significant disadvantages. The components of these cells are quite expensive, have a short running time, require laboratory procedures such as isolation/purification and are relatively unstable. These limitations could be overcome if whole cells of photosynthetic microorganisms would be used as a biocatalyst or/and sensitizer.

In the last few years, some studies have been conducted to construct a photosynthetic microbial fuel cell (PMFC) based on whole cells of photosynthetic organisms such as cyanobacteria (Sekar and Ramasamy, 2015). In the anode chamber of PMFCs, photosynthetic organisms are used which could oxidize water using light. Only sunlight and water are required for a PMFC to function, whereas traditional MFCs based on bacteria such as *Gejbacter*, *Shewanella*, etc. require organic carbon sources such as glucose/lactate, and they produce $CO_2$ as final product. The general scheme of the combined cell is shown in **Figure 6**, which demonstrates (*a*) a hydrogen fuel cell with a platinum catalyst on the anode and the cathode, and (*b*) photobioelectrochemical photoelements based on cyanobacterial cell at the anode and the laccase enzyme at the cathode.

Various cyanobacteria were used in most effective PMFCs reported (Yagishita et al., 1993; Torimura et al., 2001; Pisciotta et al., 2010). In particular, the ability of the cyanobacteria *Nostoc sp*, in generating a photocurrent was investigated using various electrochemical methods. In a recent study, using the site-specific photosynthetic inhibitors (**Fig. 1**), the mechanism of direct electron transfer from *Nostoc sp* ETC on electrode was studied (Sekar et al., 2014). It was shown that the photobioelectrochemical cell with *Nostoc sp.* immobilized on the MWCNT-modified carbon electrode as an anode and laccase/MWCNT-modified cathode (**Fig. 3c**) generated a current with a maximum density of 25 mA/cm$^2$, while the maximum energy flux density achieved without mediators was only at 3.5 mW/cm$^2$ (**Fig. 3d**). While the cell based on thylakoids (**Fig. 3a**) generated a current with a maximum density of 10 mA/cm$^2$, and the maximum energy flux density achieved without mediators was recorded at 5 mW/cm$^2$ (**Fig. 3b**). Overall, the maximum current density by the photobioelectrochemical element based on the native photosynthetic cells was higher than that of the photoelement based on thylakoids.

One of the main advantages of cyanobacteria *vs.* individual components of the photosynthetic apparatus is that they are considerably less susceptible to dehydration. Currently, the power which could be generated by PMFCs is less than those achieved by the biofuel cell (Sekar and Ramasamy, 2015). However, their many advantages such as simplicity of operation, the utilization of available substrates e.g. water, as well as stress- resistance of PMFCs in comparison with the other biofuel cells mark them as promising options for the future.

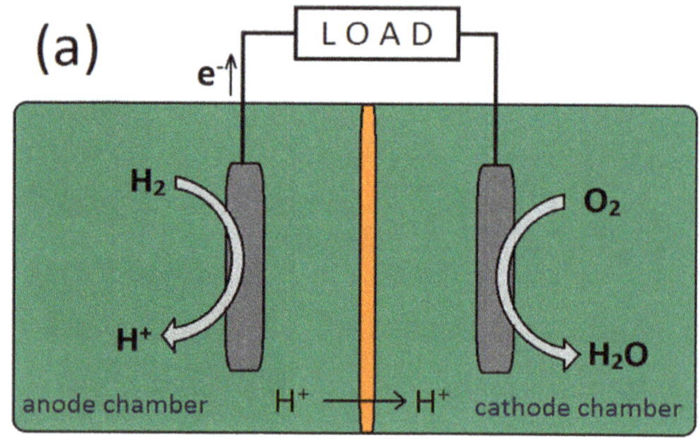


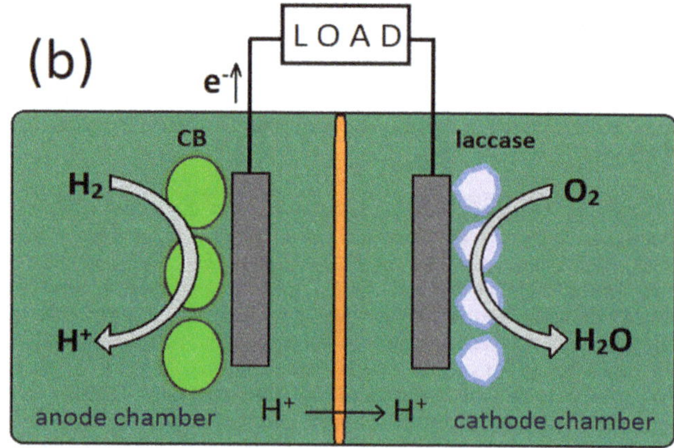


**Fig.6.** Schematic representation of the different forms of fuel cells: a) hydrogen fuel cell with a platinum catalyst on the anode and the cathode, and b) photobioelectrochemical elements based on cyanobacterial cell (CB) on the anode and laccase enzyme on the cathode (adapted from Sekar and Ramasamy, 2015).

## 5. Improving the efficiency of solar cells

### *5.1. The redox-active components: changing the direction of the electron flow*

Various redox-active components accept electrons from specific sites of ETC in accordance with their redox potential. Redox active sites of metalloproteins are usually hidden inside the PS2 complex (Badura et al., 2008), therefore, the electron transfer from immobilized photosystems onto the electrode may be limited. This limitation can be overcome by redirecting the electrons from the inner part of the protein to the surface (Sekar and Ramasamy, 2015). For instance, Larom et al. (2010) successfully used an artificial mediator to redirect electrons from $Q_A^-$ to an artificial acceptor at a distance of about 1.3 nm from the stromal side of the membrane. This change in the direction of electron flow along with additional blocking of $Q_B$ led to a reduction of oxidative damages at the expense of reducing the time of the intermediate electron transfer at the stage of $Q_A/Q_B$. In another study, Secar et al. (2014) used whole cells of cyanobacteria as photobiocatalysts in a photobioelectrochemical cell. They achieved a power density of about 10 μW/cm$^2$ by adding 1,4-benzoquinone (BQ) as a mediator which was three time higher vs. transferring electrons without a mediator using the systems *Nostoc*/MWCNT and laccase/MWCNT. Since the structure of the BQ is similar to that of PQ, the addition facilitated electron transfer from the PS2 to the MWCNT (Sekar et al., 2014). Previously, mediators such as BQ, 2,6-dimethyl-1-benzoquinone (DMBIB), and 2-hydroxy-1,4-naftahinon (HNQ) were also used for accepting the electrons from the cyanobacterial photosynthetic ETC (Yagishita et al., 1993; Torimura et al., 2001).

### *5.2. Bioengineering of photosynthetic RCs*

Primary processes of photosynthesis have high quantum yield reaching almost 100%, but the energy storage efficiency can reach as high as about 27% under ideal conditions, and much less under non-ideal ones (Fultz and Durst, 1982). This value is comparable to the efficiency of the modern silicon-based solar panels operating with an efficiency of approximately less than 20%. Note that some laboratory models demonstrated an effectiveness of 40% (Blankenship and Hartman, 1998; Grätzel, 2001; Sekar and Ramasamy, 2015). Moreover, the photosynthetic pigments usually absorb only light from the visible region of the spectrum (from 400 to 700 nm) (Blankenship and Hartman, 1998) unlike photovoltaic cells capable of absorbing light from the UV and near infrared regions as well. Thus, photosynthetic organisms utilize only about half of the incident solar energy. Nevertheless, photosynthetic efficiency can be improved by expanding the region of photosynthetic absorption by using bioengineering techniques. Since photosynthesis uses two photosystems which absorb light under the same conditions, the variation of band gaps could increase the efficiency. This approach could be feasible by using photoelements on a base of RC containing far-red and infra-red absorbing pigments similar to bacteriochlorophyll that absorbs light in the region up to 900 nm (Hanna and Nozik, 2006) or Chl *d* or Chl *f*, absorbing light in the region 400-750 nm (Blankenship and Chen, 2013). As a result, the absorption region may be significantly increased.

### *5.3. Biomimetics*

Biomimetic approach is aimed at constructing artificial systems mimicing the natural photosynthesis for the production of electricity or hydrogen. Synthetic sensitizers and catalysts are considered as a suitable alternative to unstable native systems. As a first-line strategy, porphyrins, phthalocyanines and their metal complexes which absorb light in the same optical region as the native molecules of chlorophyll are considered as such synthetic RCs. Covalently-linked cyclic porphyrins are more durable, but they are difficult to synthesize (Sanders, 2000). Non-covalently associated porphyrins easily degrade due to their sensitivity to changing environmental conditions (Iengo et al., 2003). The advantages of utilizing the porphyrin structures include stability of the RCs and accessibility compared to synthetic products while their disadvantage is the short lifetime in their excited state. Polypyridines containing transition metals have a longer lifetime in their high energy excited states (Krassen et al., 2011). However, they generally require more expensive metals (Pisciotta et al., 2010). It is worth noting that the biomimetic-based semiconductor materials mimicking the PS2`s OEC were designed to create energy devices during the earlly 1970`s (Fujishima and Honda, 1972).

## 6. Improving the efficiency of solar cells

### *6.1. Methods of immobilization and orientation of biocatalysts*

For optimum functioning of a photobioelectrochemical cell, it is necessary to immobilize photoactive molecules on a conductive substrate. Most photobioelectrochemical cells require peptides for immobilization of pigments on the electrode surface. Another important question is the correct orientation of pigment molecules. The studies conducted at Stanford (Goldsmith and Boxer, 1996) were focused on the orientation of photosynthetic RCs with respect to the electrode surface. The essence of the construction was that a poly-histidine tag was attached to the C-terminus of the M-subunit of the *Rhodobacter sphaeroides*` RC. With help of the tag, the construction was immobilized on a gold electrode containing self-assembling layer of alkanethiols with a head part as Ni-NTA. It was shown experimentally that the proximity of RCs to the electrode is important for the cell effective operation (Kincaid et al., 2006).

Many techniques were used for immobilization of photosynthetic complexes, including in particular, bioelectrocatalystic self-assembling monolayers (bio-SAMs) (Nakamura et al., 2000; Kincaid et al., 2006; Frolov, et al., 2008; Mershin et al., 2012); Ni-NTA was attached to poly-histidine tagged PS1 complexes (**Fig. 4d**) (Das, 2004; Sekar et al., 2014); the redox-active hydrogels (Badura et al., 2008); and fixation on CNTs by means of molecular binding reagents (Das et al., 2004). Each of these techniques provides various beneficial properties, including increasing the electrode surface area, increasing the rate of electron transfer between the electrode and photobiocatalyst and/or orientation of specific enzymes on the electrode. Unfortunately, enzyme immobilization reduces their activity compared to their native state. Therefore, the correct methods of immobilization should retain enzyme activity for a long time. Meunier, et al (2009) conducted a study in which thylakoids were adsorbed on a silicon matrix, thereby the stability of the native thylakoid suspension increased and it remained active for 30 d (Meunier et al., 2009). Immobilization should provide an optimum rate of electron transfer from the protein to the electrode, with minimal resistance. This can be achieved by correct orientation of proteins *vs.* the electrode surface or by use of intermediate carriers. Many studies showed that the correct orientation of a photosystem on the electrode results in an improved electron transport (Frolov, et al., 2008). In many studies, the correct orientation of the photosystems on a gold electrode provides specific binding of a histidine-tagged protein complex with Ni-NTA molecule anchored to a gold electrode (Noji et al., 2011; Das et al., 2004). Badura et al. (2008) used the osmium-containing polymer of polyvinyl-imidazole as modified electrode which worked as an immobilization device and a mediator (**Fig. 4b**). The immobilized PS2 in these modified systems were capable of generating a current density of 45 mA/cm$^2$ at light intensity of 2.65 mW/cm$^2$ (maximum wavelength at 675 nm) (Badura et al., 2008).

### *6.2. The stability of the isolated proteins*

The main problem associated with the use of isolated proteins as photosensitizers in photovoltaic cells is their very low stability. The main reason for protein destruction is photoinhibition of photosystems, especially in case of the PS 2. Photoinhibition caused by an excessive amount of radiation may damage the photosynthetic apparatus and hence destroy the chloroplast proteins. Photosystems are provided with some protective mechanisms *in vivo* which include antioxidants (Sekar and Ramasamy, 2015). However, once the proteins are isolated, natural self-healing mechanisms do not work. Thus, isolated proteins are more susceptible to damage and have a short lifetime. One of the methods to stabilize the photosynthetic complexes is through the simulation of the natural states of proteins. Surfactant peptides can be used to imitate the lipid membrane naturally stabilizing photosynthetic complexes. Such surfactants are designed as molecular nanomaterial to study the membrane proteins stability (Mershin et al., 2012) and consist of hydrophilic amino acids (aspartate or lysine) as the head of the chain and hydrophobic amino acids (alanine) as the tail. Das and colleagues (2004) used the peptides A6K (AAAAAAK) and V6D (VVVVVVD) as cationic and anionic surfactant peptides, respectively, for the stabilization of the photosynthetic complex during construction of a solid electrical device (**Fig. 4d**) and showed a short-circuit current density of 0.12 mA/cm$^2$ at the excitation light intensity of 10 W/cm$^2$ with a wavelength of 808 nm. Presumably, this direction is promising.

### *6.3. Increase of surface area*

Increase of the electrode surface area is a traditionally used method for improving the efficiency of functioning photobioelectrochemical cells. In many cases, the electrode itself is initially flat, and its geometry cannot be changed without destroying it. However, the electrode can be modified *via* nanomaterials, which could increase the real surface area due to the formation of nanostructures on the electrode surface with no planar topology. In this case, the working electrode area is larger than the area of the original flat surface, and it may adsorb more pigment molecules. These approaches include the use of GNPs (Noji et al., 2011), nanoporous gold electrodes (Ciesielskiet al., 2010), and redox hydrogels (Badura et al., 2008). In a study carried out by Mershin et al. (2012), two different forms of electrode modification by using nanocrystals of titanium dioxide and zinc oxide nanowires were compared (**Fig. 5**). Compared to the flat electrodes of the same size, the electrodes with the $TiO_2$ and ZnO had 200 and 30 times larger active areas, respectively. In fact, the samples based on ZnO were less effective due to the smaller coefficient of roughness. On the other hand, ZnO was found to be more conductive and less expensive compared to the $TiO_2$ (Mershin et al., 2012).

### *6.4. Direct or mediated transfer of electrons*

Another way to achieve the maximum current density in the cells on the basis of photosynthetic photosensitizers is to create a system that uses direct electron transfer from photosystem to electrode without using a mediator. As mentioned earlier, the mediators have lower redox potential, required for the efficient transfer of electrons, compared to the real electron source. The electrons lose some of their energy if they are transferred to the mediator, in contrast to transfer from the real source. The distance between redox site and electrode should also be minimized in order to ensure efficient transfer of electrons. The main disadvantage of direct electron transfer is the difficulty in ensuring continuous contact between the electron source and the electrode.

Furukawa and colleagues (2006) developed a photosynthetic biofuel cell using polyaniline as an electronic catalyst instead of mediators. Polyaniline has a good electrical conductivity and is compatible with the photosystem. Polyaniline also increases the surface area due to its nanostructure. During their experiment, a good efficiency was achieved for the developed cell: peak current density was about 150 mA/cm$^2$ and power density was measured at 5.3 mkW/cm$^2$. In another study conducted by Sekar et al. (2014), MWCNTs were successfully used for direct transfer of electrons, both in isolated spinach thylakoids and cyanobacteria *Nostoc sp.* (**Fig. 6b**).

### *6.5. Extension of the spectral range of the light absorption by photosystems*

Such extension is possible using Chl *d* or *f* (Allakhverdiev et al., 2010, ; Chen et al., 2010; Chen and Blankenship, 2011; Chen and Scheer, 2013; Loughlin et al., 2013; Tomo et al., 2011, 2014 ). However, the creation of artificial solar cells based on these chlorophylls is still in its early stages of development. Overall, designing solar cells using these chlorophylls seems to hold great promises.

## 7. Conclusion

Researchers in the field of artificial photosynthesis have focused on the development of systems that could effectively produce a sustainable energy from sunlight without requiring external fuels. These systems must have a high quantum yield and generate energy fluxes of high density in order to meet the energy demands. The more we learn about the nature, the closer we come to the creation of efficient energy solar cells using the photosynthetic apparatus and its components. Using systems mimicking the photosynthetic apparatus and the elements of photosynthetic systems in current energy generators and fuel cells is a quite promising direction (Marshall, 2014). However, lot of changes and improvements are needed before such photovoltaics could be widely used.

## 5. Acknowledgments

The work was supported by the Russian Scientific Foundation №14-14-00039.

## References

Allakhverdiev, S.I., Tomo, T., Shimada, Y., Kindo, H., Nagao, R., Klimov, V.V., Mimuro, M., 2010. Redox potential of pheophytin ***a*** in photosystem II of two cyanobacteria having the different special pair chlorophylls. Proc. Natl. Acad Sci. USA. 107(8), 3924-3929.

Andralojc, J., Harris, D.A., 1992. The chloroplast ATP-synthase — a light regulated enzyme. Biochem. Educ. 20(1), 44-48.

Badura, A., Guschin, D., Esper, B., Kothe, T., Neugebauer, S., Schuhmann, W., Rogner, M., 2008. Photo-induced electron transfer between photosystem 2 via crosslinked redox hydrogels. Electroanalysis. 20, 1043-1047.

Bedford, N.M., Winget, G.D., Punnamaraju, S., Steckl, A.J., 2011. Immobilization of stable thylakoid vesicles in conductive nanofibers by electrospinning. Biomacromolecules. 12(3), 778-784.

Blankenship, R.E., 2002. Molecular Mechanisms of Photosynthesis. Blackwell Science, Oxford.

Blankenship, R.E., 2010. Early Evolution of Photosynthesis. Plant Physiol.154, 434-438.

Blankenship, R.E., Chen, M., 2013. Spectral expansion and antenna reduction can enhance photosynthesis for energy production. Current Opinion in Chem. Biol. 17, 457-461.

Blankenship, R.E., Hartman, H., 1998. The origin and evolution of oxygenic photosynthesis. Trends Biochem. Sci. 23, 94-97.

Blankenship, R.E., Tiede, D.M., Barber, J., Brudvig, G.W., Fleming, G., Ghirardi, M.R., Gunner, M., Junge, W., Kramer, D.M., Melis, A., Moore, T.A., Moser, C.C., Nocera, D.G., Nozik, A.J., Ort, D.R., Parson, W.W., Prince, R.C., Sayre, R.T., 2011. Comparing photosynthetic and photovoltaic efficiencies and recognizing the potential for improvement, Science. 332, 805-809.

Bryant, D. B. (Ed.) 1994. The Molecular Biology of Cyanobacteria. Kluwer Academic Publishers, Dordrecht, Boston.

Calkins, J.O., Umasankar, Y., O'Neill, H., Ramasamy, R.P., 2013. High photoelectrochemical activity of thylakoid-carbon nanotube composites for photosynthetic energy conversion. Energy Environ. Sci. 6, 1891-1900.

Carpentier, R., Lemieux, S., Mimeault, M., Purcell, M., Goetze, D.C., 1999. A photoelectrochemical cell using immobilized photosynthetic membranes. Bioelectrochem. Bioenerg. 22, 391-401.

Chen, M., Blankenship, R.E., 2011. Expanding the solar spectrum used by photosynthesis. Trends Plant Sci. 16, 427-431.

Chen, M., Scheer, H., 2013. Extending the limit of natural photosynthesis and implications of technical light harvesting. J. Porphyrins Phthalocyanines. 17, 1-15.

Chen, M., Schliep, M., Willows, R.D., Cai, Z.-L., Neilan, B.A., Scheer, H., 2010. A red-shifted chlorophyll. Science. 329, 1318-1319.

Ciesielski, P.N., Hijazi, F.M., Scott, A.M., Faulkner, C.J., Beard, L., Emmett, K., Rosenthal, S.J., Cliffel, D., Jennings, G.K., 2010. Photosystem I – Based biohybrid photoelectrochemical cells. Bioresour. Technol. 101, 3047-3053.

Das, R., 2004. Photovoltaic devices using photosynthetic protein complexes. Doctoral dissertation. Ph.D. thesis, Massachusetts Institute of Technology, Boston.

Das, R., Kiley, P.J., Segal, M., Norville, J., Yu, A.A., Wang, L.Y., Trammell, S.A., Reddick, L.E., Kumar, R., Stellacci, F., Lebedev, N., Schnur, J., Bruce, B.D., Zhang, S.G., Baldo, M., 2004. Integration of Photosynthetic Protein Molecular Complexes in Solid-State Electronic Devices. Nano Lett. 4, 1079-1083.

Faulkner, C.J., Lees, S., Ciesielski, P.N., Cliffel, D.E., Jennings, G.K., 2008. Rapid assembly of photosystem I monolayers on gold electrodes. Langmuir. 24, 8409-8412.

Fourmond, V., Lagoutte, B., Setif, P., Leibl, W., Demaille, C., 2007. Electrochemical study of a reconstituted photosynthetic electron-transfer chain. J. Am. Chem. Soc. 129, 9201-9209.

Frolov, L., Wilner, O., Carmeli, C., Carmeli, I., 2008. Fabrication of oriented multilayers of photosystem I proteins on solid surfaces by auto-metallization. Adv. Mater. 20, 263-266.

Fujishima, A., Honda, K., 1972. Electrochemical photolysis of water at a semiconductor electrode. Nature. 238, 37-38.

Fultz, M.L., Durst, R.A., 1982. Mediator compounds for the electrochemical study of biological redox systems – a compilation. Anal. Chim. Acta. 140, 1-18.

Furukawa,Y., Moriuchi, T., Morishima, K., 2006.Design principle and prototyping of a direct photosynthetic/metabolic biofuel cell (DPMFC). J. Micromech. Microeng. 16, 220-225.

Grätzel, M., 2001. Photoelectrochemical cells. Nature. 414, 338-344.

Grätzel, M., 2007. Photovoltaic and photoelectrochemical conversion of solar energy. Phil. Trans. R. Soc. 365, 993-1005.

Goldsmith, J.O., Boxer, S.G., 1996. Rapid isolation of bacterial photosynthetic reaction centers with an engineered poly-histidine tag. Biochim. Biophys. Acta, Bioenerg. 1276, 171-175.

Hanna, M.C., Nozik, A.J., 2006. Solar conversion efficiency of photovoltaic and photoelectrolysis cells with carrier multiplication absorbers. J. Appl. Phys. 100, 1-8.

Iengo, E., Zangrando, E., Alessio, E., 2003. Discrete supramolecular assemblies of porphyrins mediated by coordination compounds. Eur. J. Inorg. Chem. 2003, 2371-2384.

Iwuchukwu, I.J., Vaughn, M., Myers, N., O'Neill, H., Frymier, P., Bruce, B.D., 2010. Selforganized photosynthetic nanoparticle for cell-free hydrogen prod. Nat. Nanotechnol. 5, 73-79.

Kincaid, H.A., Niedringhaus, T., Ciobanu, M., Cliffel, D.E., Jennings, G.K., 2006. Entrapment of photosystem I within self-assembled films. Langmuir. 22, 8114-8120.

Krassen, H., Ott, S., Heberle, J., 2011. In vitro hydrogen production – using energy fromthe sun. Phys. Chem. Chem. Phys. 13, 47-57.

Larom, S., Salama, F., Schuster, G., Adir, N., 2010. Engineering of an alternative electron transfer path in photosystem II. Proc. Natl. Acad. Sci. U.S.A. 107, 9650-9655.

Loughlin, P., Lin, Y., Chen, M., 2013. Chlorophyll *d* and *Acaryochloris marina*: current status. Photosynth. Res. 116(2-3), 277-293.

Marshall, J., 2014. Solar energy: Springtime for the artificial leaf. Nature. 510, 22-24.

Mershin, A., Matsumoto, K., Kaiser, L., Yu, D.Y., Vaughn, M., Nazeeruddin, M.K., Bruce, B.D., Graetzel, M., Zhang, S.G., 2012. Self-assembled photosystem-I biophotovoltaics on nanostructured TiO2 and ZnO. Sci. Rep. 2, 1-7.

Meunier, C.F., Van Cutsem, P., Kwon, Y.U., Su, B.L., 2009. Thylakoids entrapped within porous silica gel: towards living matter able to convert energy. J. Mater. Chem. 19, 1535-1542.

Nakamura, C., Hasegawa, M., Yasuda, Y., Miyake, J., 2000. Self-assembling photosynthetic reaction enters on electrodes for current generation. Appl. Biochem. Biotechnol. 84(6), 401-408.

Nelson, N., Yocum, C.F., 2006. Structure and function of photosystem I and II. Annu. Rev. Plant Biol. 57, 521-565.

Noji, T., Suzuki, H., Gotoh, T., Iwai, M., Ikeuchi, M., Tomo, T., Noguchi, T., 2011. Photosystem II-gold nanoparticle conjugate as a nanodevice for the development of artificial light-driven water-splitting systems. J. Phys. Chem. Lett. 2, 2448-2452.

Pisciotta,J.M., Zou, Y., Baskakov, I.V., 2010. Light-dependent electrogenic activity of cyanobacteria. PLoS ONE. 5(5), 1-10.

Sanders, J.K.M., 2000. Porphyrin Handbook. Academic Press, New York, USA.

Sandman, G., 2009. Evolution of carotenoid desaturation: the complication of a simple pathway. Arch. Biochem Biophys. 483,169-174.

Scheer, H., 2006. An overview of chlorophylls and bacteriochlorophylls: biochemistry, biophysics, functions and applications, in: Grimm, B., Porra, R.J., Rüdiger, W., Scheer, H. (Eds.) Chlorophylls and Bacteriochlorophylls: Biochemistry, Biophysics, Functions and Applications, Springer, Dordrecht, pp. 4-11.

Sekar, N., Ramasamy, R.P., 2015. Recent advances in photosynthetic energy conversion. J. Photochem. Photobiol., C. 22, 19-33.

Sekar, N., Umasankar, Y., Ramasamy, R.P., 2014. Photocurrent generation by immobilized cyanobacteria via direct electron transport in photo-bioelectrochemical cells. Phys. Chem. Chem. Phys.16(17), 7862-7871.

Tomo, T., Akimoto, S., Tsuchiya, T., Fukuya, M., Tanaka, K., Mimuro, M., 2008. Isolation and spectral characterization of Photosystem II reaction center from *Synechocystis* sp. PCC 6803. Photosynth. Res. 98, 293-302.

Tomo, T., Shinoda, T., Chen, M., Allakhverdiev, S.I., Akimoto, S., 2014. Energy transfer processes in chlorophyll *f*-containing cyanobacteria using time-resolved fluorescence spectroscopy on intact cells. Biochim. Biophys. Acta.1837, 1484-1489.

Torimura, M., Miki, A., Wadano, A., Kano, K., Ikeda, T., 2001. Electrochemical investigation of cyanobacteria Synechococcus sp. PCC7942-catalyzed photoreduction of exogenous quinones and photoelectrochemical oxidation of water. J. Electroanal. Chem. 496, 21-28.

Vittadello, M., Gorbunov, M.Y., Mastrogiovanni, D.T., Wielunski, L.S., Garfunkel, E.L., Guerrero, F., Kirilovsky, D., Sugiura, M., Rutherford, A.W., Safari, A., Falkowski, P.G., 2010. Photoelectron generation by photosystem II core complexes tethered to gold surfaces. ChemSusChem. 3, 471-475.

Ulas, G., Brudvig, G.W., 2011. Redirecting electron transfer in photosystem II from water to redox-active metal complexes. J. Am. Chem. Soc. 133, 13260-13263.

Whitney, S.M., Houtz, R.L., Alonso, H., 2011. Advancing our understanding and capacity to engineer nature's CO2-sequestering enzyme, Rubisco. Plant Physiol. 155, 27-35.

Yagishita, T., Horigome, T., Tanaka, K., 1993. Effects of light, CO2 and inhibitors on the current output of biofuel cells containing the photosynthetic organism Synechococcus sp. J. Chem. Technol. Biotechnol. 56, 393-399.

Yehezkeli, O., Tel-Vered, R., Wasserman, J., Trifonov, A., Michaeli, D., Nechushtai, R., Willner, I., 2012. Integrated photosystem II-based photo-bioelectrochemical cells. Nat. Commun. 3,742, 1-7.

# 6

# An investigation of biodiesel production from microalgae found in Mauritian waters

Keshini Beetul[1], Shamimtaz Bibi Sadally[2], Nawsheen Taleb-Hossenkhan[2], Ranjeet Bhagooli[2] & Daneshwar Puchooa[1*]

[1]*Faculty of Agriculture, University of Mauritius, Réduit, Mauritius.*

[2]*Faculty of Science, University of Mauritius, Réduit, Mauritius.*

## HIGHLIGHTS

- Total micro-phytoplankton count amounted to 6.59±1.27x10$^5$ cells L$^{-1}$which was dominated by diatoms (95.2%), followed by dinoflagellates (2.9%) and cyanobacteria (1.9%).
- The cyanobacterial mats were identified as *Leptolyngbya* sp. and *Nodularia harveyana*, and the endosymbiotic dinoflagellates as the *Symbiodinium* clade C.
- There were also differences recorded in the % lipid of the different microalgae ($p<0.005$) - among all, *Symbiodinium* clade C had the highest with (38.39±6.58%).
- $^1$H and $^{13}$C NMR analyses indicated the presence of the acyl glycerols

## GRAPHICAL ABSTRACT

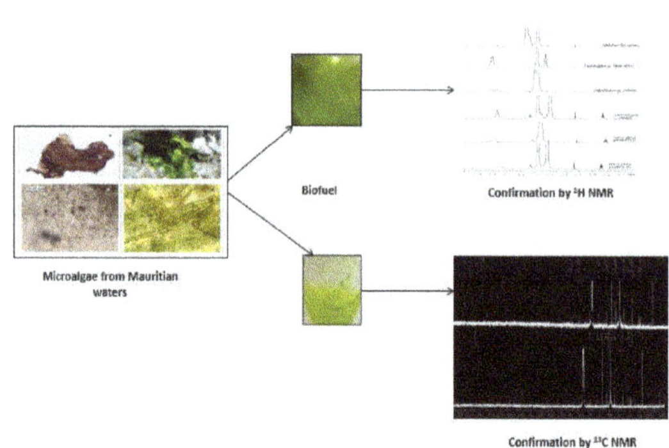


## ABSTRACT

The aim of this study was to assess the lipid content and the subsequent potential of different microalgae present in the Mauritian marine water to produce biodiesel. The share of micro-phytoplankton species in the water column was determined. The cyanobacterial mats and endosymbiotic dinoflagellates were characterised morphologically and genetically using RFLP. The samples were quantified gravimetrically and analysed using $^1$H &$^{13}$C NMR spectroscopy. Total micro-phytoplankton count amounted to 6.59±1.27x10$^5$ cells L$^{-1}$which was dominated by diatoms (95.2%), followed by dinoflagellates (2.9%) and cyanobacteria (1.9%). The cyanobacterial mats were identified as *Leptolyngbya* sp. and *Nodularia harveyana*, and the RFLP characterised the endosymbiotic dinoflagellates as the *Symbiodinium* clade C. The highest amount of lipid was recorded in the Symbiodinium clade C (38.39±6.58%). $^1$H and $^{13}$C NMR analyses indicated the presence of acyl glycerols. An attempt to synthesise biodiesel by alkaline trans-esterification reaction was also performed and the presence of biodiesel was detected using the Fourier Transform Infrared Spectroscopy. The Infrared analysis yielded peaks at around 1738cm$^{-1}$ and 1200cm$^{-1}$ characteristic of the carbonyl and ether groups respectively, indicating the presence of biodiesel.

**Keywords:**
Biodiesel
Lipid
Microalgae
Mauritian waters

## 1. Introduction

In the twenty-first century, the atmospheric carbon dioxide level is 30% higher than the pre-industrial era (NOAA Earth System Research Laboratory 2013; Bernstein et al. 2007; Hofmann et al. 2009) sharing a direct relationship with the hike in fossil fuel burning since 1950 (US Energy Information Administration, 2012). In 2011, the global transport sector had a 28% energy share and accounted for almost a quarter of the world's carbon dioxide emissions (US Energy Information Administration 2012). Carbon dioxide is the major greenhouse gas contributing to global warming and ocean acidification; hence, triggering concern worldwide (Bernstein et al., 2007). The situational irony is that mankind is exploiting a dwindling global oil reserve (OPEC, 2012) at the expense of the environment-causing energy insecurity and climate catastrophes. Biodiesel is the ultimate solution for replacing the petroleum-centered transports industry (Chisti, 2008), subsequently, reducing the sector's carbon dioxide emissions (Bernstein et al.

* Corresponding author
E-mail address: sudeshp@uom.ac.mu (D. Puchooa).

2007). In addition to its greater energy currency than bioethanol (Chisti, 2008), it also conforms to the current diesel engines (Wang et al. 2000). However, the biofuel industry is being subjected to controversies including food insecurity due to the divergence of staple crops which are being used for biofuel production (Tenenbaum, 2008). Choosing a proper feedstock is therefore crucial. Lipids obtained from non-edible feedstocks are popular because they do not compete with the food market. Also, the prohibitive cost of edibleoils prevents their use in Biodiesel preparation. This is while nonedible oils are affordable for Biodiesel production (Karmee & Chadha, 2005). Nonedible oils which have been used for biodiesel production include Jatropha and Pongamia (Karmee et al. 2004; 2006). Microalgae, being anonedible lipid source, is another potential candidate for biodiesel production as it does not compete with food commodities (Gouveia, 2011) and has a high lipid content which is usually between 20-50% (Chisti, 2007). Studies have focused mostly on eukaryotic species such as *Botryococcus braunii,* Chlorella sp., *Chlamydomonas reinhardtii* and *Nannocloropsis* sp. because of their relatively higher lipid content (Scott et al. 2010). Cyanobacteria – a microalgae prokaryote – is also gaining momentum in the biodiesel production arena with respect to its fast growth rate and lipid content (Quintana et al. 2011).

The aim of this study was to gain preliminary data on the potential of different microalgae - micro-phytoplankton, filamentous cyanobacteria and endosymbiotic dinoflagellates - found in the Mauritian waters in order to identify prospective biodiesel feedstocks. The focus was on the determination of the moisture content as well as lipid content. An attempt to synthesise biodiesel from the microalgae's lipids was also carried out.

## 2. Materials & methods

### 2.1. Collection and Identification of Samples

The microalgae which were collected from the Mauritian lagoons were cyanobacteria, micro-phytoplankton and endo-symbiotic dinoflagellates. The collection period was from September to December 2012 as illustrated in Table 1.

**Table 1**
Samples collection sites and the sampling periods.

| Sample | Collection Site | Sampling Period |
|---|---|---|
| Cyanobacterial Mats | Albion & Belle Mare | November & December |
| Micro-Phytoplankton | Flic en Flac | September |
| Endosymbiotic Dinoflagellates | Flic en Flac | September |

The cyanobacterial mats were collected live and stored in a cooler bag along with some seawater. The micro-phytoplankton was collected as described by Sadally and coworkers. (2011). As for the *Symbiodinium* sp., part of the coral *Fungia repanda* was collected and the endosymbiotic dinoflagellates were water-picked and centrifuged at 5000rpm for 5 minutes.

The morphology and microscopic structures observed under a light microscope were used to identify the cyanobacterial mats while the micro-phyotoplankton was counted and identified following the methods described by Sadally et al. (2011). The endosymbiotic dinoflagellates were identified with respect to their cnidarian host and the clade by performing a Restriction Fragment Polymorphism (RFLP) using the restriction enzymes Taq1 and *Sau3A*. The microalgae were stored below -20°C for further experimental procedures.

### 2.2. Determination of Moisture Content

The moisture content of the respective microalgae was determined gravimetrically. They were weighed, using the electronic mass balance - METTLER TOLEDO B303-S, before and after drying in a drying cabinet at 80°C. The dried cyanobacterial mats were ground and stored in opaque plastic bottles at 4°C for lipid extraction.

### 2.3. Lipid Extraction

Total lipids were extracted by a modified Folch method with chloroform: methanol (1:1, v/v) (Ryckebosch et al. 2011) and the volume of organic solvent was determined according to the principles used by Halim et al. (2011). The volume of organic solvent was calculated for each microalgae species using the following equations:

$$Vt = \frac{Sw \times 100}{4}$$

Where **Vt** is the total volume of solvent (ml) and **Sw** is the amount of dried sample (mg)

$$Vs = \frac{4 \times Vt}{5}$$

Where **Vs** is the volume of choloroform: methanol mixture (ml).

$$Vw = \frac{1 \times Vt}{5}$$

Where **Vw** is the volume of water (ml)

### 2.4. Analysis of Total Lipids

$^{13}C$ & $^{1}H$ NMR spectroscopy were performed to establish the lipid profile of the different samples. A few drops of the lipids that had been dissolved in the extracting solvent were added to an NMR capillary tube followed by deuterated chloroform using a Pasteur pipette. The NMR spectrum was calculated at 250 MHz for both $^{13}C$ & $^{1}H$ using the NMR machine 250 MHz Bruker.

### 2.5. Biodiesel Synthesis

Biodiesel was only synthesized with the cyanobacterial mats' lipid extracts through the trans-esterification reaction as described by Hossain et al. (2008). For a brief period, 18.5 mMol sodium methoxide in methanol was mixed with the lipid extracts in a quickfit flask which was then placed on the electric shaker MODEL G25 for 1.5hours at 300rpm.

### 2.6. Biodiesel Purification

The trans-esterified mixture was then transferred to a corning tube and left to settle for about 16 hours. After the settling period, the corning tube was centrifuged at 2000rpm for 10 minutes as described by Prommuak et al. (2011). An attempt was made to purify the upper layer through water-wash using 60°C distilled water as suggested by Nakpong & Wootthikanokkhan (2010) and dry-wash using silica gel. Both washes were unsuccessful.

### 2.7. Biodiesel Analysis

The presence of biodiesel was determined by using FT-IR spectroscopy. A few drops of the trans-esterified mixture were placed on the FT-IR - BRUKER ALPHA - crystal. Absorbant measurements were carried out using 32 scans.

### 2.8. Statistical Analysis

A one-way analysis of variance (ANOVA) and a post-hoc Tukey's test with $a$ = 0.05 were performed to statistically evaluate the results using MINITAB 15.

## 3. Results and discussions

### 3.1. Identification

Figure 1 illustrates the share of micro-phytoplanktons in the water column collected at Flic en Flac.

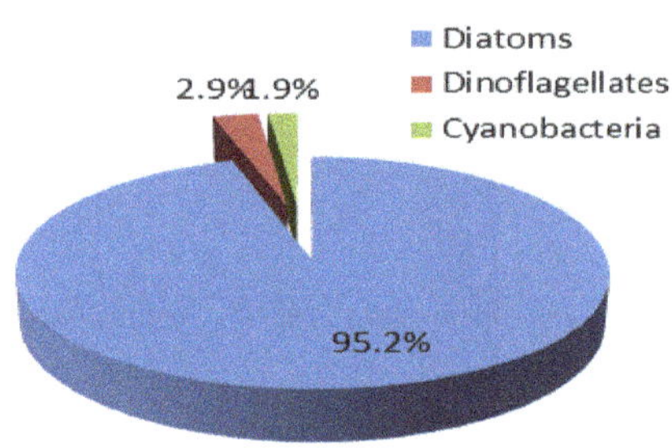


**Fig.1.** Percentage of Micro-phytoplankton

The total micro-phytoplankton density amounted to (6.59±1.27) × $10^5$ cells/L with the diatoms bearing the biggest share (95.2%) while the cyanobacteria had the smallest (1.9%). The RFLP analysis of the endosymbiotic dinoflagellates, as depicted in Figure 2, showed that the latter was the *Symbiodinium* clade C.

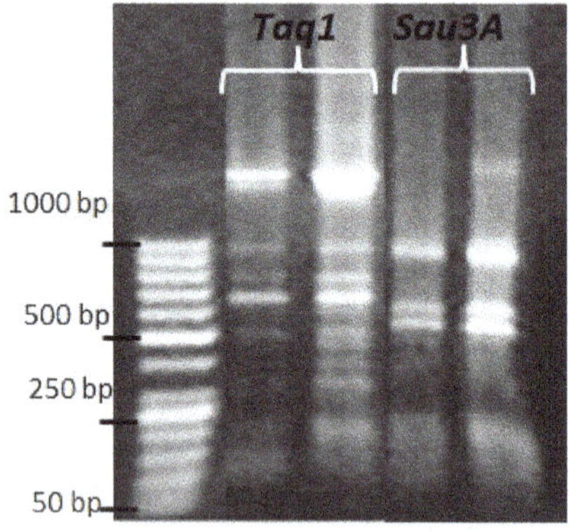


**Fig.2.** RFLP pattern of Clade C Symbiodinium from Fungia repanda. Lanes 1&2: restriction digests using Taq1; Lanes 3&4: restriction digests using *Sau3A*.

The morphology and microscopic structures of the cyanobacterial mats have been illustrated in Figure 3 and the latter were identified as *Leptolyngbya* sp. (Charpy et al., 2010) and *Nodularia harveyana* (Lyra et al., 2005) respectively.

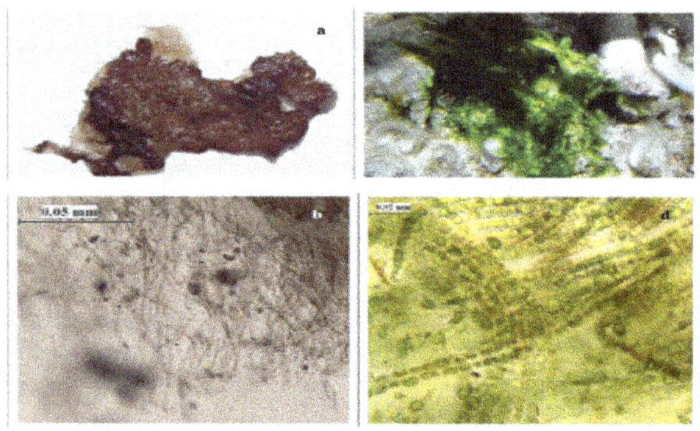


**Fig.3.** (a) Mat of *Leptolyngbya* sp. (b) Filaments of *Leptolyngbya* sp. with single trichomes (c) Mat of *Nodularia harveyana* (d) Filamentous structures of *Nodularia harveyana* – uniseriate and heterocytous.

### 3.2. Moisture Content

As depicted in Figure 4, the moisture content of the microalgae under study was highest in micro-phytoplankton and lowest in the *Symbiodinium* clade C (P<0.05).

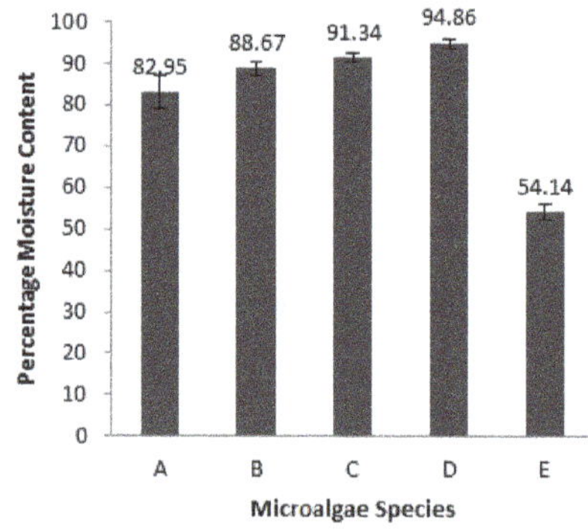


**Fig.4.** Moisture content of the microalgae species (mean± SD; n=3). A: *Leptolyngbya* sp. (Albion), B: *Leptolyngbya* sp. (Belle Mare), C: *Nodularia harveyana*, D: Micro-phytoplankton & E: *Symbiodinium* clade C.

The moisture content of the *Leptolyngbya* sp. at Albion was not significantly different from that of *Leptolyngbya* sp. collected at Belle Mare. Furthermore, no significant differences were observed between *Leptolyngbya* sp. (Belle Mare), *Nodularia harveyana* (Albion), *Nodularia harveyana* (Albion) and micro-phytoplankton. The moisture content tallied with that of other literature (Patil et al. 2008). This particular trait is attributed to the poikilohydric-characteristic of the microalgae (Campbell et al., 2004); consequently, high energy input is required to remove the water molecules. Nonetheless, the removal of water is mandatory as water molecules around the cells and within the cells act as barriers against the organic solvent (Mercer & Amerta, 2011). No additional pre-treatments were deemed to be necessary due to their negligible effects (Rittmann, 2008), preceding lipid extraction.

### 3.3. Lipid Extracts

The lipid extracts of the different microalgae exhibited different colors as illustrated in Figure 5.

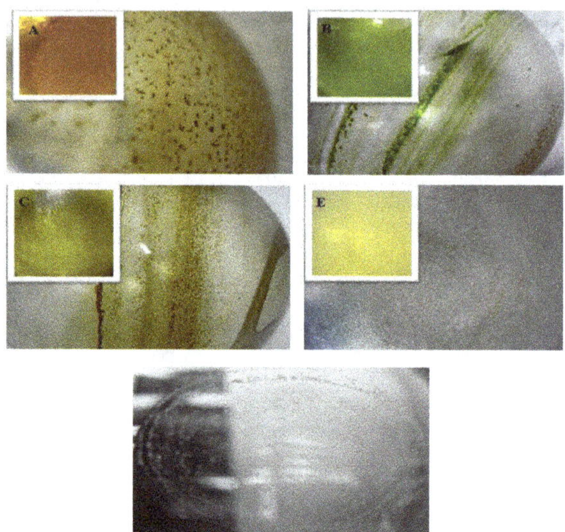


**Fig.5.** Pictures in the boxes depict the aqueous extracts' colours while the others illustrate the pigments' colours of the extracts of the different microalgae. *Aqueous extract for the micro-phytoplankton was colourless. A: *Leptolyngbya* sp. (Albion), B: *Leptolyngbya* sp. (Belle Mare), C: *Nodularia harveyana*, D: Micro-phytoplankton & **E**: *Symbiodinium* clade C.

The aqueous extract of the *Leptolyngbya* sp. collected at different locations varied significantly-reddish brown (Albion) and pale green (Belle Mare). However, the pigments observed following evaporation of the organic solvent showed that both have similar pigments but with different ratios. On the other hand, the aqueous extract of *Nodularia harveyana* was green containing brownish/green pigments. The micro-phytoplankton yielded a colorless aqueous extract with white pigments while the *Symbiodinium* clade C gave a yellow aqueous extract with greenish yellow pigments. It was observed that the lipid extracts were of different colours as illustrated in Figure 5. Algae have different classes of photosynthetic pigments namely chlorophyll, carotenoids comprising of carotene and xantophylls, and phycobilins while phycobilin pigments are the main light harvesting pigments in cyanobacteria. Among the different classes of photosynthetic pigments, the chlorophylls and carotenes are the fat soluble molecules which can be extracted from the thylakoid membranes using organic solvents such as methanol (Robertson, 2011). For the cyanobacteria, the pigments can be identified mainly as chlorophyll and beta carotene. However, it may be that some of the phycobilin pigments tagged along with the lipids during the extraction as in the case of *Nodularia harveyana*. Simis et al. (2012) showed that phycocyanin, which might be mixed with phycoerythrocyanin, is the most abundant photosynthetic accessory pigment. Apart from the chlorophyll pigments, *Leptolyngbya* spp. contain rhodopsin which can be observed in Figure 5 as a red coloration. Finally, Daigo et al. (2008) observed the presence of zeaxanthin in the *Symbiodinium* genus; hence, explaining the characteristic yellowish colour obtained along with chlorophyll pigments Daigo et al. (2008). It has been reported that both rhodopsin and zeaxanthin can be extracted using organic solvents (Darszon et al., 1978; Chen et al., 2005) which implies that the red pigments obtained in *Leptolyngbya* spp. and the yellow colour observed in the lipid extract of the *Symbiodinium* genus are due to rhodopsin and zeaxanthin respectively.

### 3.4. Total Lipid Content

The quantification of the total lipid content has been illustrated in Table 2. Symbiodinium clade C had relatively high total lipid content compared to the cyanobacterial mats ($P<0.05$).

**Table 2**
Percentage of Lipid (mean ± SD; n=5; except *Nodularia harveyana*: n=4 & *Symbiodinium* clade C: n=3) *ND: Not Determined

| Microalgae | Percentage of Lipids |
|---|---|
| *Leptolyngbya* sp. (Albion) | 2.75±0.21 |
| *Leptolyngbya* sp. (Belle Mare) | 19.09±0.26 |
| *Nodularia harveyana* (Albion) | 10.12±0.19 |
| Micro-phytoplankton (Flic en Flac) | ND |
| *Symbiodinium* clade C (Flic en Flac) | 38.39±6.58 |

Moreover, the total lipid content of the cyanobacterial mats collected at different locations was significantly different from each other. The lipid content of microalgae depends strongly on the storage structure of the organism (Rittmann, 2008), the enzymes involved (Wada & Sato, 2010) as well as the environmental conditions (Karatay & Donmez, 2011; Sakthivel et al., 2011). On the one hand, eukaryotic algae accumulate important amounts of lipids as reserves in the form of triacylglycerols and diacylglycerol for the membrane structures (Wada & Sato, 2010) while cyanobacteria accumulate lipids in their thylakoid membranes only (Rittmann, 2008), agreeing with the theory that cyanobacterial genomes lack the gene coding for the triacylglycerol synthesis (NREL 2012). On the other hand, the difference in the percentage of lipid among the different cyanobacterial mats may be explained by the enzymes present in cyanobacterial cells which differ from one species to another (Wada & Sato, 2010). The environmental factors such as the pH and nitrate content should not be overlooked as studies have shown that these two factors influence the accumulation of lipids in the cyanobacteria (Karatay & Donmez, 2011; Sakthivel et al., 2011).

*Symbiodinium* clade C produced a relatively huge lipid content due to its mutualistic relationship with corals for faster calcification (Davy et al., 2012). The main energy store of corals is lipids – metabolites synthesized from the primary products obtained from the photosynthetically-fixed carbon translocated from the dinoflagellates symbionts (Davy et al., 2012). Kellogg and Patton (1983) showed that primary metabolites are not the only ones which are transferred to the coral; the dinoflagellates also release fat droplets suggesting that lipids are also translocated to the corals. Such activities account for the high metabolic rate of dinoflagellates as they have to cater for their own energy supply as well as that of their hosts. Consequently, they have to accumulate a significant amount of lipids as well as other products during and following photosynthesis. Nevertheless, the moisture content of the dinoflagellates was relatively very low because those organisms have to provide metabolites for two bodies at once leaving less space for moisture accumulation in the cells. Yet, it should be noted that such characteristics may differ in the isolated dinoflagellates (Sutton & Hoegh-Guldberg, 1990).

### 3.5 $^{1}H$ & $^{13}C$ NMR Analysis

The proton NMR analysis yielded the results which have been illustrated in Figure 6.

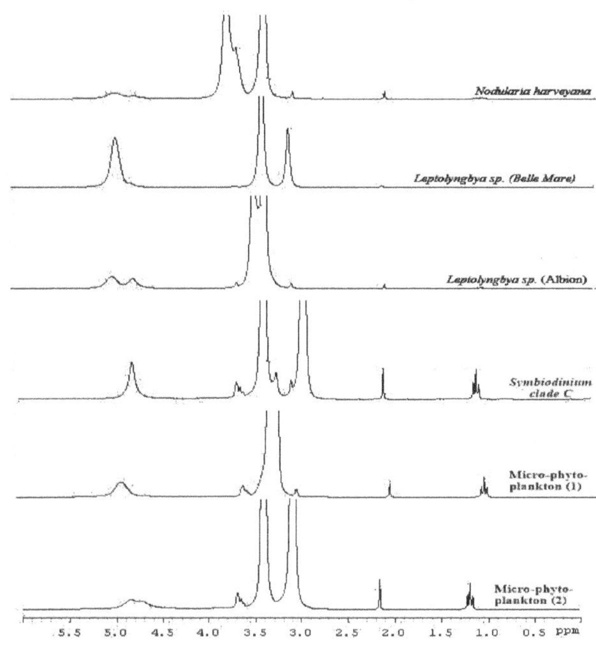


**Fig.6.** $^{1}H$ NMR analysis. The singlet between 2.0ppm ppm and 2.5ppm is for the methyl in the acetoxy group (COOR-$CH_3$). The two singlets overlapping peaks between 3.0ppm and 3.5ppm is characteristics of phospholipids. The peaks between 4.5ppm and 5.0ppm show the presence of glycerol methyl protons.

It was observed that the peaks were moved downfield due to the hydrogen bonds formed between the lipids and the methanol in which they were dissolved for storage. 4.3ppm is the benchmark for the presence of acyglycerols (Adosraku et al., 1994). However, due to the formation of inappropriate hydrogen bonds, that particular peak moved downfield – between 4.5ppm and 5ppm. The NMR spectra show that the *Leptolyngbya* sp. (Belle Mare) and Symbiodinium clade C have the highest amount of acyglycerols in the prokaryotes and eukaryotes arena respectively. Also, the lipid profile of the micro-phytotplankton is similar irrespective of the site of collection in the lagoon.

Figure 7 illustrates the $^{13}C$ Lipid profile of the eukaryotic (*Symbiodinium* clade C) and prokaryotic (*Leptolyngbya* sp.) microalgae. The resonance for ether (–C-O) at chemical shifts 57.8ppm, 62.3ppm and 70.6ppm may indicate the presence of sulfoquinovosyldiacylglycerol, phosphatidylgycerol and acylglycerols which are usually present in microalgae (Wada & Sato, 2010).

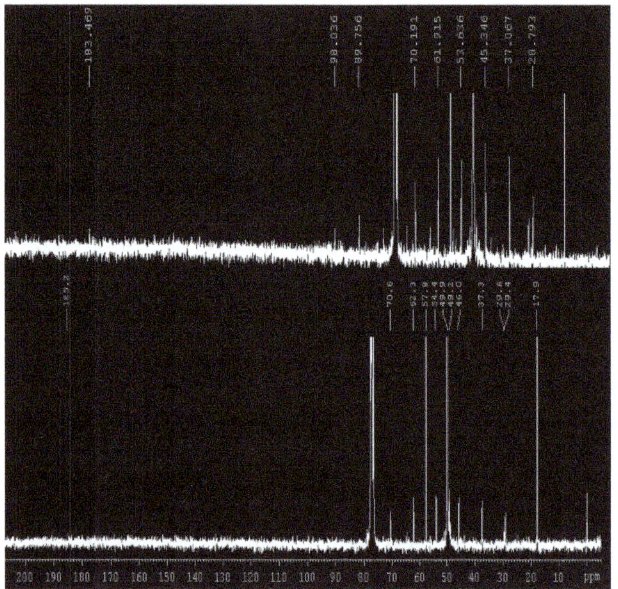

**Fig.7.** $^{13}C$ NMR spectrum of *Symbiodinium* clade C (Flic-en-Flac) and *Lepolyngbya* sp. (Belle Mare). The resonance for the methyl group and methylene group are between the range of 0ppm to 30ppm and 15ppm to 55ppm respectively. It should be noted that the peak at 49.2ppm was not counted as a carbon atom in the lipid sample as this peak codes for methanol. The resonance for ether (–C-O) is between 55ppm and 90ppm. The peaks 183.5ppm and 185.2 ppm indicates the presence of carbonyl group found in esters confirming the presence of lipids.

The high number of carbon is crucial in the biodiesel industry as the longer and more saturated are the fatty acid chain the higher will be the cetane number. The cetane number is a standard for the measurement of ignition delay time of an engine and a larger cetane number for a diesel fuel leads to a shorter ignition delay (Knothe, 2006).

### *3.6. Biodiesel Analysis*

As mentioned in the methodology section, both the water-wash and the dry-wash were unsuccessful; consequently, the biodiesel volume could not be quantified. However, during the washing attempt, a golden yellow solution, as illustrated in Figure 8, was obtained indicating the presence of biodiesel.

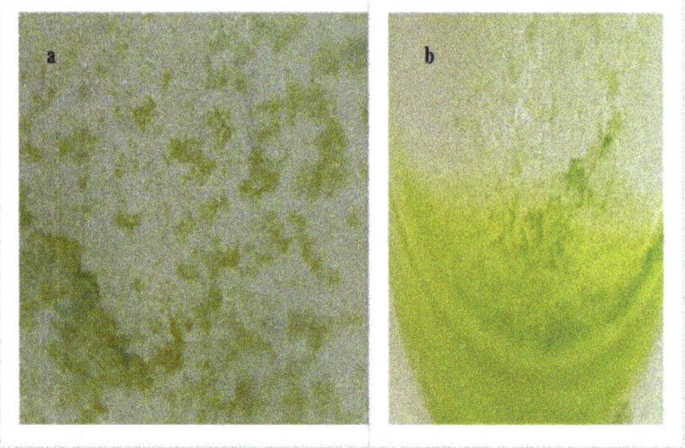


**Fig.8.** The brownish colour in (a) indicates the presence of the biodiesel synthesised during the trans-esterification reaction. Due to the minute amount produced, the latter appear as pigments but when methanol is added to it a golden yellow colour is obtained confirming the presence of biodiesel.

The FT IR analysis carried out indicated the presence of biodiesel for *Leptolyngbya* sp. (Belle Mare) and *Nodularia harveyana* (Albion) lipids only. As illustrated in Figure 9, peaks at 1738.30cm$^{-1}$ for C=O and 1171.49cm$^{-1}$ and 1249.31cm$^{-1}$ for –C-O were obtained for the Leptolynbya sp. (Belle Mare). As for the *Nodularia harveyana*, peaks at 1737.84cm$^{-1}$ for C=O and 1189.43cm$^{-1}$ for –C-O were obtained (Knothe, 2006).

The alkali-catalysed process was carried out for the trans-esterification step in this study. Following the settling time, an important amount of glycerine deposit was observed which was of a semi-solid composition. Gerpen et al. (2004) suggested that the amount of catalyst to be used should be 1% of the extracted lipids' mass but according to Hossain et al. (2008), the amount of catalyst used was sometimes 200 times higher than the expected amount generating an excessive number of water molecules which are toxic to triacylglycerols and diacylglycerols. The amount of lipid extracts being very small, the water molecules were enough to hydrolyse most of the acylglycerols in some cases, accounting for the semi-solid glycerine observed; but this may also be explained by the highly saturated lipids which might have been present (Gerpen et al., 2004). It should be noted that the amount of glycerine obtained was not considered since the qualitative data would have been deceptive comprising of the excessive sodium hydroxide salt.

Following glycerine and biodiesel separation, purification is mandatory to bring down the pH to 7. During the washing process, two phases must be observed in the separating funnel because biodiesel has a lower density than water – 0.93g/L (Sander & Murthy, 2010). Yet, during the water-wash no two phases were observed. This is because of the small amount of biodiesel obtained following the transesterification step. The relationship between oil and biodiesel is a 1:1 ratio provided that only triacylglycerol, diacylglycerol or monoacylglycerol are present. The average lipid content of the *Leptolyngbya* sp. collected at Albion was 0.056g. Assuming that the oil had no free fatty acids, the amount of biodiesel obtained for the latter should be around 58.16μl (Gerpen et al., 2004). However, due to the high hydrolysis of the acylglycerol, the volume of biodiesel was expected to be much lower ruling out the utility of the water-wash for this study. A dry wash was also attempted but to no avail. While drying out, all methanols evaporated leaving the pigment-like substance. The golden yellow pigments clearly indicated the presence of a small quantity of biodiesel. This is not surprising because Daroch et al. (2013) reported that many studies failed to produce biodiesel from microalgae as the alkali-catalysed reaction failed to convert the lipids into biodiesel efficiently.

Since, quantification was hard to achieve, detecting the presence of biodiesel was done by using FTIR following the evaporation of the excess methanol in the biodiesel mixture. The presence of fatty acid methyl esters were detected only for the *Leptolyngbya* sp. collected at Belle Mare and the *Nodularia harveyana.* According to the equation established by (Gerpen et al., 2004) the volumes of biodiesel which should have been obtained for these species were 170.57±1.74μl and 64.66±1.45μl respectively. The species' lipids were more resistant to hydrolysis than the one collected at Albion.

The biodiesel also known as fatty acid methyl esters have the ester bond which is linked to a $-CH_3$. The esters' bonds IR spectral region normally have peaks in the range of 1730-1750 cm$^{-1}$ and 1000$^{-1}$300 cm$^{-1}$ for C=O and –C-O respectively. However, it should be noted that acylglycerols have the same bonds as well. The IR spectral region at around 1200cm$^{-1}$ for $-O-CH_3$ distinguishes between the fatty acid methyl esters and acylglycerols – presence of peak at 1200 cm$^{-1}$ indicating the presence of fatty acid methyl esters (Knothe, 2006). The peaks at 1249.31 cm$^{-1}$ and 1738.30 cm$^{-1}$ indicated the presence of fatty acid methyl esters in the biodiesel mixture synthesized from the lipid extract of the *Leptolyngbya* sp. collected at Belle Mare while the peaks at 1737.84cm$^{-1}$ and 1189.43cm$^{-1}$ indicated the presence of fatty acid methyl esters found in that of *Nodularia harveyana.*

Among the cyanobacterial mats used in this study, the *Leptolyngbya* sp. collected at Belle Mare had the highest lipid content of 19.09±0.26% and the hypothetical biodiesel yield was 170.57±1.74μl. A Singh and Gu (2010) analysis showed that the biodiesel yield from the 30% oil-microalgae was 58, 700 l/ha while rapeseed and canola, jatropha, karanj, corn, soybeen, peanut, coconut and oil palm yielded 1190 l/ha, 1892 l/ha, 2590 l/ha, 1721 l/ha, 446 l/ha, 1059 l/ha, 2689 l/ha and 5959 l/ha respectively. If a simple proportion is applied to our study comparing the biodiesel yields, it can be speculated that since 30% oil-microalgae produces 58, 700 l/ha, a 19% oil-microalgae can produce 37, 177 l/ha which is still higher than the other oleaginous crops.

## 4. Conclusions

In this study, the potential of microalgae found in Mauritian waters for biodiesel production was investigated. The share of micro-phytoplankton species was determined. Total micro-phytoplankton count amounted to 6.59±1.27x105 cells $L^{-1}$ which was dominated by diatoms (95.2%), followed by dinoflagellates (2.9%) and cyanobacteria (1.9%). The cyanobacterial mats were identified as *Leptolyngbya* sp. and *Nodularia harveyana*, and the RFLP characterised the endosymbiotic dinoflagellates as the Symbiodinium clade C. The moisture content obtained was highest in micro-phytoplankton (94.86±0.99%) and lowest in the Symbiodinium clade C (54.14±1.97%). Differences in the % lipid of the different microalgae were also recorded ($p<0.005$) with Symbiodinium clade C having the highest (38.39±6.58%). $^{1}$H and $^{13}$C NMR analyses indicated the presence of the acyl glycerols and the Infrared analysis confirmed the presence of biodiesel. The preliminary data obtained in this study indicates the capacity of the different microalgae found in Mauritian waters for acting as a potential nonedible feedstock for the production of Biodiesel.

## Acknowledgements

The authors wish to thank the Faculties of Agriculture and Science at the University of Mauritius for supporting this work. The precious help provided by the technical staff of the Department of Agriculture and Food Science of the Faculty of Agriculture and the Departments of Biological Sciences and Chemistry of the Faculty of Science is gratefully acknowledged.

## References

Adosraku RK, Choi GTY, Constantinou-Kokotos V, Anderson MM, Gibbons WA. 1994. NMR lipid profiles of cells, tissues, and body fluids: proton NMR analysis of human erythrocyte lipids. Journal of lipid Research 35:1925 – 1931.

Bernstein L, Bosch P, Canziani 0, Chen Z, Christ R, Davidson O, Hare W, Huq S, Karoly D, Kattsov V, Kundzewicz Z, Liu J, Lohmann U, Manning M, Matsuno T, Menne B, Metz B, Mirza M, Nicholls N, Nurse L, Pachauri R, Palutikof J, Parry M, Qin D, Ravindranath N, Reisinger A, Ren J, Riahi K, Rosenzweig C, Rusticucci M, Schneider S, Sokona Y, Solomon S, Stott P, Stouffer R, Sugiyama T, Swart R, Tirpak D, Vogel C, Yohe G. 2007. Climate Change 2007: Synthesis Report – Summary for Policymakers. Intergovernmental Panel on Climate Change. http://www.ipcc.ch/pdf/assessment-report/ar4/syr/ar4_syr.pdf (accessed on 12th October 2013).

Campbell A, Cooke P, Cass K, Earl K. 2004. Plant and Animal Evolution. New Zealand, The University of Waikato. http://sci.waikato.ac.nz/evolution/aboutus.shtml (accessed on 2nd April 2013)

Charpy L, Palinska Ka, Casareto B, Langlade Mj, Suzuki Y, Abed Rmm, Golubic S. 2010. Dinitrogen-Fixing Cyanobacteria in Microbial Mats of Two Shallow Coral Reef Ecosystems. Microb Ecol 59:174-186.

Chen F, Li H, Wong R N, Jiang Y. 2005. Isolation and purification of the bioactive carotenoid zeaxanthin from the microalga Microcystis aeruginosa by high-speed counter-current chromatography. Journal of Chromatography A 1064(2):183-186

Chisti Y. 2007. Biodiesel from microalgae. Biotechnology Advances 25:294-306.

Chisti Y. 2008. Biodiesel from microalgae beats bioethanol. Trends in Biotechnology 26(3):126-131.

Daigo K, Nakano Y, Casareto Be, Suzuki Y, Shioi Y. 2008. High performance Liquid Chromatographic Analysis of Photosynthetic Pigments in Corals: An Existence of a Variety of Epizoic, Endozoic and Endolithic Algae. 11th International Coral reef Symposium, 7-11 July 2008 Fort Lauderdale, FL, USA; 123-127

Daroch M, Geng S, Wang G. 2013. Recent advances in liquid biofuel production from algal feedstocks. Applied Energy 102:1371-1381.

Darszon A, Philipp M, Zarco J, Montal M. 1978. Rhodopsin-phospholipi complexes in apolar solvents: Formation and properties. The Journal of Membrane Biology 43(1):71-90.

Davy SK, Allemand D, Weis VM. 2012. Cell Biology of Cnidarian Dinoflagellate Symbiosis. Microbiology and Molecular Biology Reviews 76(2):229-261.

Gerpen JV, Shanks B, Pruszko R, Clements D, Knothe G. 2004. Biodiesel Production Technology. USA: NREL (SR-510-36244). http://www.nrel.gov/docs/fy04osti/36244.pdf (accessed on 10th December 2012)

Gouveia L, 2011. Microalgae as Feedstock for Biofuels. In: Springer ed. Microalgae as Feedstock for Biofuels. New York: Springer Berlin Heidelberg, p 1-69.

Halim R, Gladman B, Danquah MK, Webley PA. 2011. Oil extraction from microalgae for biodiesel production. Bioresource Technology 102:178-185.

Hofmann D J, Butler H J, Tans PP. 2009. A new look at atmospheric carbon dioxide. Atmospheric Environment 43:2084-2086.

Hossain ABM, Shalleh A, Boyce AN, Naqiuddin M. 2008. Biodiesel fuel production from algae as renewable energy. American Journal of Biochemistry and Biotechnology 4(3):250-254.

Karatay SE, Donmez G. 2011. Microbial oil production from thermophile cyanobacteria for biodiesel production. Applied Energy 88(11):3632-3635.

Karmee SK, Mahesh P, Ravi R, Chadha A. 2004. Kinetic study of the base-catalyzed transesterification of monoglycerides from pongamia oil. Journal of the American Oil Chemists' Society 81(5): 425-430

Karmee SK, Chadha A. 2005. Preparation of biodiesel from crude oil of Pongamia piñata. Bioresource Technology. 96(13): 1425-1429

Karmee SK, Chandna D, Ravi R, Chadha A. 2006. Kinetics of base-catalysed transesterification of triglycerides from Pongama oil. Journal of the American Oil Chemists' Society. 83(10): 873-877.

Kellogg, RB., Patton, JS. 1983. Lipid droplets, medium of energy exchange in the symbiotic anemone Condylactis gigantea: a model coral polyp. Marine Biology 75(2-3); 137-149

Knothe G. 2006. Analyzing Biodiesel: Standards and Other Methods. JAOCS 83:823-833.

Lyra C, Laamanen M, Lehtimaki JM, Surakka A, Sivonen K. 2005. Benthic cyanobacteria of the genus Nodularia are non-toxic, with gas vacuoles, able to glide and genetically more diverse than planktonic Nodularia. International Journal of Systematic Microbiology 55:555-568.

Mercer P, Amerta RE. 2011. Development in oil extraction from microalgae. European Journal of Lipid Science and Technology 113(5):539-547.

Nakpong P, Wootthikanokkhan S. 2010. Roselle (Hibiscus sabdariffa L.) oil as an alternative feedstock for biodiesel production in Thailand. Fuel 89:1806-1811.

Noaa Earth System Research Laboratory. 2013. Atmospheric CO2. Mauna Loa. http://co2now.org/images/stories/data/co2-mlo-monthly-noaa-esrl.pdf (accessed on 12th October 2013)

NREL. 2012. NREL Creates New Pathways for Producing Biofuels and Acids from Cyanobacteria. USA, NREL. http://www.nrel.gov/docs/fy13osti/55974.pdf (accessed on 9th March 2013)

Organization of The Petroleum Exporting Countries. 2012. World Oil Outlook 2012. Austria: OPEC.

Patil V, Tran K, Giselrod HR. 2008. Towards Sustainable Production of Biofuels from Microalgae. International Journal of Molecular Sciences 9:1188-1195.

Prommuak C, Pavasant P, Quitain AT, Goto M, Shotipruk A, 2011. Microalgal Lipid Extraction and Evaluation of Single-Step Biodiesel Production. Engineering Journal 16(5):158-166.

Quintana N, Van Der Kooy F, Van De Rhee MD, Voshol G P, Verpoorte R. 2011. Renewable energy from Cyanobacteria : energy production optimization by metabolic pathway engineering. Appl Microbiol Biotechnol 91:471-490.

Rittmann BE. 2008. Opportunities for Renewable Bioenergy Using Microorganisms. Biotechnology and Bioengineering 100 (2):203-212.

Robertson D, 2011. Algal Pigments. USA, Clark University. http://www.clarku.edu/faculty/robertson/ (accessed on 30th March 2013).

Ryckebosch E, Muylaert K, Foubert I. 2011. Optimization of an Analytical Procedure for Extraction of Lipids from Microalgae. Journal of the American Oil Chemists' Society 89(2):189-198.

Sadally SB, Bhagooli R, Taleb-Hossenkhan N. 2011. Micro-Phytoplankton Distribution and Biomass at Two Lagoons around Mauritius Island. University of Mauritius Research Journal 18A:54-87.

Sander K, Murthy GS. 2010. Life cycle analysis of algae biodiesel. Int J Life Cyle Assess 15:704-714.

Sakthivel R, Elumalai S, Arif MM. 2011. Microalgae lipid research, past, present: A critical review for biodiesel production, in the future. Journal of Experimental Sciences 2(10):29-49.

Scott SA, Davey MP, Dennis JS, Horst I, Howe CJ, Lea-Smith D, Smith AG. 2010. Biodiesel from algae: challenges and prospects. Current Opinion in Biotechnology 21(3):277-286.

Simis SGH, Huot Y, Babin M, Seppala J, Metsamaa L. 2012. Optimization of variable fluorescence measurements of phytoplankton communities with cyanobacteria. Photosynthesis Research 112 (1):13-30.

Singh J, Gu S. 2010. Commercialization of potential of microalgae for biofuels production. Renewable and Sustainable Energy Reviews 14:2596-2610.

Sutton DC, Hoegh-Guldberg O. 1990. Host-Zooxanthella Interactions in Four Temperate Marine Invertebrate Symbioses: Assessment of Effect of Host Extracts on Symbionts. The Biological Bulletin 178:175-186.

Tenenbaum DJ. 2008. Food vs. Fuel. Environmental Health Perspectives 116(6):254-257.

US Energy Information Administration. 2012. Annual Energy Review 2011. United States: Energy Information Administration

Wada H, Sato N. 2010. Lipid Biosynthesis and its Regulation in Cyanobacteria. In: Wada H, Murata N, ed. Lipids in Photosynthesis. Netherlands: Springer Netherlands, p 157-177.

Wang WG, Lyons DW, Clark NN, Gautam M. 2000. Emissions from Nine Heavy Trucks Fueled by Diesel and Biodiesel Blend without Engine Modification. Environ. Sci. Technol. 34:933-939.

# 7

# A review of conversion processes for bioethanol production with a focus on syngas fermentation

Mamatha Devarapalli, Hasan K. Atiyeh*

*Department of Biosystems and Agricultural Engineering, Oklahoma State University, Stillwater, OK 74078, USA.*

## HIGHLIGHTS

- Summary of biological processes to produce ethanol from food based feedstocks.
- Overview of fermentation processes for ethanol production from biomass.
- Process development and reactor design are critical for feasible syngas fermentation.

## GRAPHICAL ABSTRACT

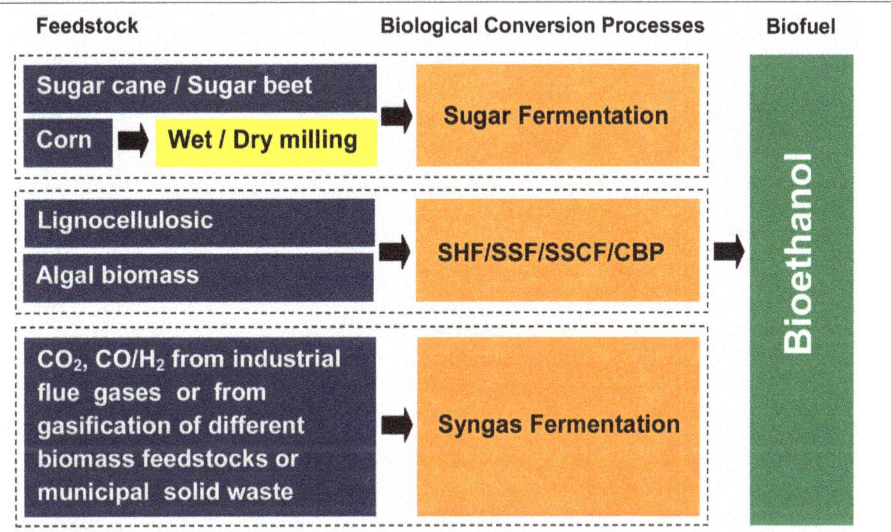


## ABSTRACT

Bioethanol production from corn is a well-established technology. However, emphasis on exploring non-food based feedstocks is intensified due to dispute over utilization of food based feedstocks to generate bioethanol. Chemical and biological conversion technologies for non-food based biomass feedstocks to biofuels have been developed. First generation bioethanol was produced from sugar based feedstocks such as corn and sugar cane. Availability of alternative feedstocks such as lignocellulosic and algal biomass and technology advancement led to the development of complex biological conversion processes, such as separate hydrolysis and fermentation (SHF), simultaneous saccharification and fermentation (SSF), simultaneous saccharification and co-fermentation (SSCF), consolidated bioprocessing (CBP), and syngas fermentation. SHF, SSF, SSCF, and CBP are direct fermentation processes in which biomass feedstocks are pretreated, hydrolyzed and then fermented into ethanol. Conversely, ethanol from syngas fermentation is an indirect fermentation that utilizes gaseous substrates (mixture of CO, $CO_2$ and $H_2$) made from industrial flue gases or gasification of biomass, coal or municipal solid waste. This review article provides an overview of the various biological processes for ethanol production from sugar, lignocellulosic, and algal biomass. This paper also provides a detailed insight on process development, bioreactor design, and advances and future directions in syngas fermentation.

**Keywords:**
Bioethanol
Conversion processes
Syngas fermentation

* Corresponding author
E-mail address: hasan.atiyeh@okstate.edu

**Contents**



## 1. Introduction

Renewable energy can be derived from sunlight, wind, water, geothermal, and biomass, which are considered sustainable and environmentally friendly. Conversely, non-renewable energy is derived from fossil fuels such as coal, oil and natural gas, which do not regenerate at sustainable rates (Twidell and Weir, 2003). Most of the world's energy demand is currently met using fossil fuels. The Energy Information Administration (EIA) reported that 70% of all oil consumed in the United States was used for transportation (EIA, 2015a). According to EIA's 2014 net imports data 27% of petroleum consumed in the U.S. was imported from foreign countries (EIA, 2015b). Factors such as high gas prices, rising concerns over national energy security and dependency on foreign oil imports, and environmental impacts of high oil usage have led to an increased focus on biofuel production (German et al., 2011).

Ethanol was the first biofuel produced from food-based feedstocks such as corn and sugarcane. The United States, being the largest producer of corn, have successfully commercialized corn ethanol production (Dien et al., 2002). However, the use of corn for biofuels raised debate over its potential interference with the food market. This gave rise to the use of non-food based feedstocks such as agricultural and forest residues, municipal wastes, lignocellulosic, and algal biomass for bioethanol production. Unlike crude oil, biomass feedstocks are diverse in their composition. Hence, different conversion processes have been developed to produce a variety of biofuels. This review article focuses on conversion processes pertinent to bioethanol production using different biomass feedstocks. Further, this article discusses the developments of syngas fermentation for ethanol production.

### *1.1. Bioethanol from sugar/starch*

First generation bioethanol is produced from corn and sugarcane using a well-established technology (Sims et al., 2008). The steps involved in production of ethanol from sugar and starch crops are shown in **Figure 1**.

Sugar crops such as sugar cane, sugar beet and sweet sorghum mostly consist of glucose, fructose, and sucrose as their major components (Bai et al., 2008). These fermentable sugars are extracted by grinding or crushing followed by fermentation to ethanol. Further, ethanol is separated from the products stream by distillation followed by dehydration.

Grains such as corn and wheat contain starch, which is a polysaccharide of glucose units linked by α (1-4) and α (1-6) glycosidic bonds (Pandey, 2010). Starch is not directly fermented by yeast. After milling the grains and extracting starch, starch is hydrolyzed into glucose using α-amylase and glucoamylase (Nigam and Singh, 1995). Glucose is then fermented to ethanol.

Production of ethanol from starch is performed by either dry grind or wet milling process (Bothast and Schlicher, 2005). The main difference between these two processes is the extraction method of glucose and co-products formed (Sims et al., 2008). In dry grind, whole corn is milled to produce ethanol along with high protein animal feed called dry distillers' grains with solubles (DDGS). In wet milling, steeping of corn is followed by separation of germ, fiber, and starch. Wet milling produces value added by-products such as corn sweeteners, oil, and corn gluten meal in addition to ethanol. Wet milling requires high capital cost and is less efficient in producing ethanol than dry grind process (Rausch and Belyea, 2006 ; Rodríguez et al., 2010). The high capital cost of wet milling process is due to separation of various corn components to co-produce value added by-products in addition to 2.5 gallons of ethanol per bushel of corn. However, whole corn is utilized in dry grind facilities maximizing capital return per gallon of ethanol. About 2.8 gallons ethanol are produced per bushel of corn via the dry grind process (Bothast and Schlicher, 2005). Most corn ethanol plants in the U.S. are dry grind facilities (USGC, 2012). One disadvantage of dry grind process is that the value of DDGS has decreased due to an increase in dry grind facilities. Thus, modified dry grind facilities have been proposed to recover germ and fiber from the corn grains and improve byproduct value (Rodríguez et al., 2010). The cost efficiency of ethanol production from food based feedstocks and impacts on change in land usage has been criticized (Rathmann et al., 2010). Such drawbacks of first generation bioethanol gave rise to the need for ethanol production from non-food based feedstocks such as biomass.

### *1.2. Bioethanol from cellulosic feedstocks*

The non-food based feedstocks used for production of second generation ethanol comprises of cellulosic biomass such as dedicated energy crops (e.g., switchgrass, miscanthus) and agricultural and wood residues (e.g., woodchips, cornstover, sugarcane bagasse, and sawdust) (Naik et al., 2010). Cellulosic biomass mainly consists of cellulose, hemicellulose, and lignin polymers interlinked in a heterogeneous matrix (Kitani and Hall, 1989). Cellulose is a linear polysaccharide consisting of several β(1-4) linked D-glucose units. Hemicellulose is a heteropolymer of xylose, mannose, galactose, rhamnose and arabinose. Lignin is a complex polymer of cross-linked aromatic compounds. Lignin acts as a protective barrier and hinders the depolymerization of cellulose and hemicellulose to fermentable sugars. Unlike first generation ethanol production, the process for conversion of cellulosic feedstocks to ethanol is complex (Stöcker, 2008; Szczodrak and Fiedurek, 1996). Cellulosic biomass is first pretreated either chemically or enzymatically to breakdown the polymeric units and increase the accessibility of C5-C6 sugars for microbial fermentation to produce ethanol. An overview of the biological conversion processes for ethanol production is discussed in sections 2 and 3.

Second generation bioethanol from cellulosic feedstocks was successfully demonstrated in pilot scale plant (Menetrez, 2014). Recently in 2014, 25 million gallons per year capacity commercial scale cellulosic ethanol plants were commissioned by POET-DSM and Abengoa Bioenergy (Lane, 2015; POET-DSM, 2014). Further, DuPont's 30 million gallon per year cellulosic ethanol plant is expected to start production in 2015. While commercialization of second generation ethanol plants looks promising, the

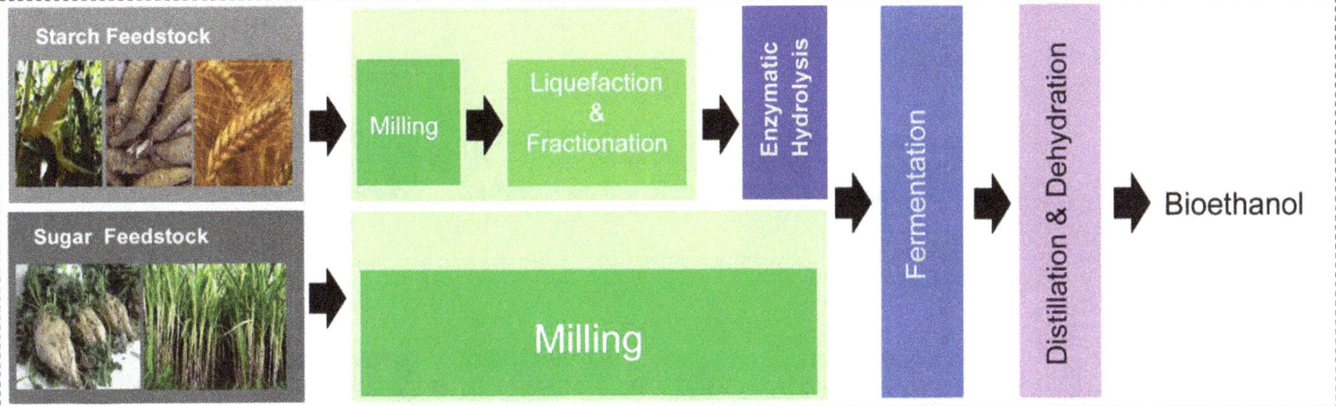


**Fig.1.** Bioethanol production from first generation biomass (Adapted from Sims et al., 2008).

sustainability of these plants will largely depend on the market availability of the feedstocks at reasonable prices. For the cellulosic ethanol industry to flourish, biomass feedstocks should be available at large scale and low cost. Most of the cellulosic feedstocks meet this requirement (Carriquiry et al., 2011). One of the main challenges of cellulosic ethanol commercialization is the impact of the change in land usage (Searchinger et al., 2008). The production of dedicated energy crops requires vast land area. However, land management practices are necessary to reduce any indirect carbon and nitrogen gas emissions that pose a threat to produce harmful greenhouse gases (GHGs) (Tilman et al., 2006). This disadvantage of cellulosic biomass gave rise to considering algal biomass as a potential feedstock for biofuels production.

### *1.3. Bioethanol from algal biomass*

Algal biomass can be used to produce a variety of biofuels such as hydrogen, diesel, isobutene, and ethanol (Cruz et al., 2014; Mussatto et al., 2010; Nayak et al., 2014; Posten and Schaub, 2009). Microalgae are unicellular plants that are either autotrophic or heterotrophic and can grow in diverse environment (Mata et al., 2010). Autotrophic algae harness sunlight and fix atmospheric $CO_2$ into carbohydrates such as starch and cellulose via photosynthesis. On the other hand, heterotrophic algae species can utilize small organic carbon compounds that are turned into lipids, protein, and oils (John et al., 2011). Conversely, macroalgae are large multicellular marine algae obtained from natural and cultivated resources. Harvested macroalgae (red, brown and green) are mainly used to produce hydrocolloids that constitute 10-40% of their biomass. Macroalgae has a low concentration of lipids and primarily contains 35-74% carbohydrates and 5-35% proteins (Ito and Hori, 1989). Conversely, most of the microalgae such as *Botryococcus braunii, Chlorella sp., Nannochloris sp., Nitzschia sp., Schizochytrium sp.* have at least 20-50% oil content (Chisti, 2007). Several studies have reported the production of bioethanol from both micro- and macro-algal biomass (Fasahati et al., 2015; Harun et al., 2010; Harun et al., 2014; John et al., 2011; Jung et al., 2013). Starch and cellulose are extracted from algae biomass using mechanical shear or by enzyme hydrolysis, after which they are utilized for bioethanol production (John et al., 2011). Enzymatic hydrolysis of cellulose from algae is simpler than from plant biomass due to negligible or no presence of lignin in algae. Various species of algae were reported to contain different starch and biomass content after oil extraction (John et al., 2011). Ethanol production from algal starch is similar to conversion processes of starch or sugars to ethanol discussed in section 1.1. The conversion technologies of algal and plant based cellulosic biomass to ethanol are similar, which are discussed in sections 2 and 3 of this review article.

Algae can grow on non-arable lands and do not change land usage. Further, $CO_2$ produced in industrial flue gases can be used to produce algal biomass (Brennan and Owende, 2010). Another main advantage of algal biomass is that it does not require fresh water for cultivation. Waste waterfrom industrial and domestic sewage can also be used for the cultivation of algal biomass (Mussatto et al., 2010).

The major obstacle for the commercialization of algal biofuels is process economics. Harvesting corresponds to 20-30% of total cultivation costs (Demirbas and Fatih Demirbas, 2011). Cultivation of microalgae through open ponds is economical but has inherent disadvantages such as low productivity, water loss, low $CO_2$ utilization, and high affinity to be contaminated by other algal strains (Chisti, 2007; John et al., 2011; Posten and Schaub, 2009). The disadvantages of open ponds led to development of closed photobioreactors, which facilitate higher productivity, less contamination, and less water loss. However, photobioreactors suffered from $CO_2$, $O_2$ and pH gradients, wall growth, fouling, hydrodynamic stress, and high scale up costs (John et al., 2011). While macroalgae has recently gained renewed interest as bioethanol feedstock; its process economics are not fully addressed. Nevertheless, a recent quantitative sustainability assessment on macroalgae reported it to have a potential as a sustainable bioethanol feedstock (Park et al., 2014).

Conversion of non-food based feedstocks to bioethanol and other products can be broadly classified into chemical and biological processes. Further, biological conversion of biomass can be through direct or indirect fermentation. Bioethanol can be produced through direct fermentation of the biomass via hydrolysis-fermentation and through indirect fermentation via syngas fermentation. In this article, ethanol production through hydrolysis-fermentation is briefly discussed followed by a detailed review of syngas fermentation process an indirect biomass conversion process to produce bioethanol. Discussion on thermochemical conversion processes can be found elsewhere and is out of scope of this review article (Dutta et al., 2011; Perales et al., 2011).

## 2. Hydrolysis fermentation

Biological conversion of lignocellulosic biomass to ethanol consists of three main steps namely pretreatment, hydrolysis and fermentation. Different pretreatment methods have been employed to disrupt the cell wall and expose the cellulose, hemicellulose fibers for further processing. Pretreatment methods are mainly divided into (i) physical (milling and grinding), (ii) physiochemical (steam pretreatment/auto hydrolysis, hydrothermolysis, and wet oxidation), (iii) chemical (alkali, dilute acid, oxidizing agents, and organic solvents), (iv) biological or a combination of these methods (Alvira et al., 2010; Mood et al., 2013). After biomass pretreatment, the cellulose and hemicellulose are broken down into monomers by acid or enzymatic hydrolysis (Sun and Cheng, 2002). Next, fermentation is carried out to convert these monomeric sugars into alcohols using yeast or bacteria (Liu et al., 2015a; Liu et al., 2015b; Pessani et al., 2011).

Four process configurations for ethanol production are possible based on the degree to which the above mentioned steps are consolidated as shown in **Figure 2**.

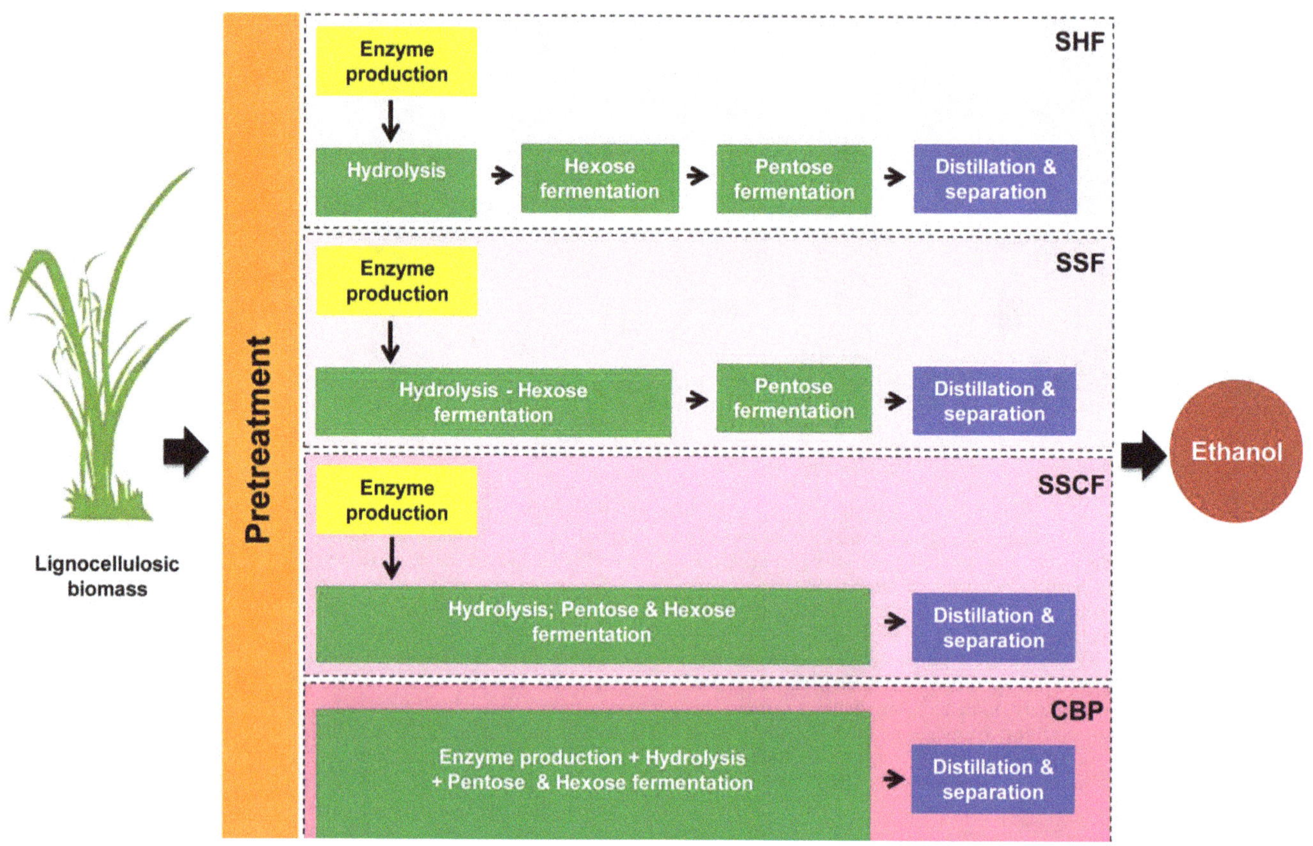


**Fig.2.** Bioethanol lignocellulosic biomass process configurations (i) Separate Hydrolysis & Fermentation (SHF) (ii) Simultaneous Saccharification & Fermentation (SSF) (iii) Simultaneous Saccharification & Co-Fermentation (SSCF) (iv) Consolidated Bioprocessing (CBP) (Adapted from Hamelinck et al., 2005).

Process integration reduces capital cost and makes the biofuel production process more efficient and economically viable (Cardona Alzate and Sánchez Toro, 2006; Hahn-Hägerdal et al., 2006; Hamelinck et al., 2005). In Separate Hydrolysis and Fermentation (SHF) configuration, the enzyme production, hydrolysis of biomass, hexose and pentose fermentation are carried out in separate reactors (Lynd et al., 2002). In SHF, hydrolysis and fermentation can occur at their optimum conditions. However, the accumulation of glucose and cellobiose during hydrolysis inhibit the cellulases and reduce their efficiencies (Margeot et al., 2009).

The disadvantages of SHF led to the development of Simultaneous Saccharification and Fermentation (SSF) process (Wright et al., 1988). In SSF, both cellulose hydrolysis and hexose fermentation occur in the same reactor. This results in relieving the end product inhibition on the cellulases as the sugars are immediately consumed by the fermenting microorganism (Hahn-Hägerdal et al., 2006). However, SSF process has some limitations. There is a trade-off between the cost of enzymes production and hydrolysis fermentation process (Lynd et al., 2002). In SSF, the rate of enzyme production limits the rate of alcohol production. In addition, cellulases used for hydrolysis and the fermenting microorganisms usually have different optimum pH and temperature conditions. It is important to have compatible conditions for both the enzyme and the microorganism. Another issue with SSF is that most microorganisms used for fermentation of glucose cannot utilize xylose, a hemicellulose hydrolysis product (Lin and Tanaka, 2006).

In Simultaneous Saccharification and Co-fermentation (SSCF) process, glucose and xylose are co-fermented in the same reactor. Strains of *Saccharomyces cerevisiae* and *Zymomonas mobilis* are genetically engineered to co-ferment both glucose and xylose (Dien et al., 2003; Hahn-Hägerdal et al., 2007; Öhgren et al., 2006; Zhang et al., 1995).

Another method of process integration is the Consolidated BioProcessing (CBP), in which one single microorganism is used for hydrolysis and fermentation steps. This potentially reduces the capital costs and increases process efficiency (Lynd et al., 2002). However, microorganisms which can both produce enzymes for hydrolysis of biomass and then ferment released sugars are still in the early development stage (Lynd et al., 2005).

The main advantage of biochemical conversion technologies is the high product selectivity of the biocatalyst (Foust et al., 2009). The enzymes that catalyze the biochemical reactions produce highly specific products. Hence, metabolic engineering and synthetic biology can be used to alter the metabolic pathway and regulate only specific enzymes to increase the desired product yields (Fischer et al., 2008; Percival Zhang et al., 2006). Another advantage of the biochemical processes is that they are usually operated at ambient temperature and pressure, unlike the chemical processes. However, lignin is not utilized in biochemical processes. Ethanol production from lignocellulosic feedstocks using biochemical processes is more difficult compared to corn ethanol production (Lynd et al., 2008). This is attributed to the high costs associated with pretreatment and enzymatic hydrolysis. Research areas that should be addressed to increase the economic feasibility of biochemical conversion processes include (i) improving effectiveness of biomass pretreatment, (ii) increasing enzymatic hydrolysis yields, (iii) decreasing enzyme cost, (iv) reuse of enzymes, (v) genetically modifying microorganisms for efficient fermentation of pentose and hexose sugars, and (vi) producing high value co-products to improve process economics.

## 3. Syngas fermentation

Syngas fermentation is an indirect conversion process for the production of alcohols, organic acids and other products. Unlike hydrolysis fermentation processes, syngas fermentation is referred to as an indirect fermentation because the feedstocks are not directly fed in the fermentor to form products. Feedstocks are first gasified into syngas, which is then cleaned and cooled before it is fed into the fermentor to make products. Non-food based feedstocks such as agricultural residue, municipal solid wastes, energy crops, coal, and petcoke can be gasified to produce syngas. Syngas is mainly a mixture of CO, $CO_2$, and $H_2$. However depending on the type of feedstock and gasification system used, small amounts of tars, $CH_4$, $C_2H_2$, $C_2H_4$ , $H_2S$, $NH_3$, carbonyl sulfide (COS), hydrogen cyanide (HCN) and nitric oxide are also detected in the syngas (Ahmed et al., 2006; Xu et al., 2011; Xu and Lewis, 2012). Tars can foul equipment and along with other contaminants such as nitric oxide, $H_2S$ and HCN can inhibit growth and enzyme activity. For example, presence of 150 ppm of nitric oxide in the biomass-derived syngas inhibited hydrogenase ($H_2$ase) activity of *C. carboxidivorans* P7 (Ahmed et al., 2006). However, the same study reported that *C. carboxidivorans* P7 adapted and grew in the presence of tars in the syngas. Although some contaminants such as $H_2S$ and $NH_3$ can be used as nutrients by syngas fermenting microorganisms, high levels of $NH_3$ can inhibit growth and enzyme activity. $NH_3$ in syngas is converted into ammonium ion ($NH_4^+$) in the fermentation medium and an increase in $NH_4^+$ in the medium to 0.7 M caused 50% inhibition of $H_2$ase activity of *Clostridium ragsdalei* (Xu and Lewis, 2012). A review of biomass derived syngas contaminants and suggested gas cleanup technologies are presented in Woolcock and Brown (2013). These include electrostatic separation, filtration, wet scrubbing, adsorption, thermal and catalytic cracking.

In addition to the syngas produced from gasification of biomass, industrial waste gas streams containing CO, $CO_2$ or $H_2$ can also be converted by acetogens to biofuels and chemicals. Under anaerobic conditions, acetogens such as *C. ljungdahlii, C. carboxidivorans, A. bacchi* and *C. ragsdalei* serve as biocatalysts (Liou et al., 2005; Liu et al., 2012; Phillips et al., 1994; Wilkins and Atiyeh, 2011).

In syngas fermentation, acetogens metabolize CO, $CO_2$, and $H_2$ to alcohols and organic acids. The overall biochemical reactions to convert syngas to ethanol and acetic acid are shown below (Klasson et al., 1990a; Vega et al., 1990).

$$6CO + 3H_2O \rightarrow C_2H_5OH + 4CO_2 \quad (1)$$
$$2CO_2 + 6H_2 \rightarrow C_2H_5OH + 3H_2O \quad (2)$$
$$4CO + 2H_2O \rightarrow CH_3COOH + 2CO_2 \quad (3)$$
$$2CO_2 + 4H_2 \rightarrow CH_3COOH + 2H_2O \quad (4)$$

CO and/or $H_2$ can supply the electrons used in the enzymatic reactions. However, CO and $CO_2$ are used as a carbon source. As per the stoichiometry, if only CO is used as the sole carbon and energy source then the carbon conversion efficiency to ethanol will only be 33%, while, 67% of the carbon are lost as $CO_2$ as per **Eq.1**. However, if both CO and $H_2$ are utilized then **Eq.1** and **Eq. 2** are combined into **Eq. 5**.

$$3CO + 3H_2 \rightarrow C_2H_5OH + CO_2 \quad (5)$$

When equimolar amounts of CO and $H_2$ are provided, the maximum carbon conversion efficiency to ethanol increases to 67%. On the other hand, when CO and $H_2$ are utilized solely to make acetic acid, then the carbon conversion efficiency to acetic acid is 100% as indicated in **Eq. 6**.

$$2CO + 2H_2 \rightarrow CH_3COOH \quad (6)$$

It is important to note that if only CO is utilized to produce acetic acid then only 50% carbon conversion efficiency can be achieved. The carbon conversion efficiency is high when electrons are supplied by $H_2$ and CO is utilized as the carbon source. However, $H_2$ utilization decreases because hydrogenase activity is inhibited by CO (Terrill et al., 2012; Ukpong et al., 2012). This results in CO utilization as both carbon and energy source decreasing the overall conversion efficiency of the process (Ahmed and Lewis, 2007). While the stoichiometry provides an estimate of the maximum theoretical yields of products from the substrates, the actual production rates and yields vary depending on the microorganism, gas mixture, medium components and fermentation conditions (Gao et al., 2013; Phillips et al., 2015; Zeikus, 1980).

### *3.1. Biocatalysts*

CO can be anaerobically metabolized by photosynthetic, acetogenic, carboxydotrophic, and methanogenic microorganisms to produce hydrogen, methane, acetate, butyrate, ethanol, and butanol as end products (Abrini et al., 1994). Among the different anaerobes, acetogens have been of prime interest due to their ability to grow chemolithotrophically (i.e., use inorganic reduced compounds as energy source) and produce ethanol and butanol along with acetate and butyrate from CO, $CO_2$, $H_2$, formate, and methanol (Mohammadi et al., 2011).

*Moorella thermoacetica* (formerly called *C. thermoaceticum*) is the most extensively studied acetogen (Fontaine et al., 1942). This microorganism was used to determine the acetyl-CoA pathway enzymology in the laboratories of Harland Goff Wood and Lars Gerhard Ljungdahl (Drake et al., 2008). To date, there are more than 100 acetogenic species isolated from a variety of habitats such as sediments, soils, sludge, and intestinal tracts of animals (Drake et al., 2008). Most of the microorganisms currently known to ferment syngas to ethanol are predominantly mesophilic with operating temperatures in the range of 30-40 °C (Munasinghe and Khanal, 2010a). The most widely studied mesophilic microorganisms are *C. aceticum, Acetobacterium woodii, C. ljungdahlii C. carboxidivorans, C. autoethanogenum* and *C. ragsdalei* (Abubackar et al., 2015; Phillips et al., 1993; Phillips et al., 2015; Ukpong et al., 2012; Younesi et al., 2005).

Acetogens metabolize single carbon source compounds via the acetyl-CoA pathway, also called the Wood-Ljungdahl pathway to (i) synthesize acetyl moiety of acetyl-CoA from $CO_2$, (ii) conserve energy, (iii) assimilate $CO_2$ to cell carbon (Ljungdhal, 1986; Wood et al., 1986). Acetyl-CoA is a major metabolic intermediate in acetogens and can be utilized to produce ethanol, butanol, hexanol, acetate, butyrate, hexanoate, and cell mass (Phillips et al., 2015). A list of selected syngas fermenting microorganisms, alcohol and organic acid concentrations, ethanol yield from CO and ethanol productivity are shown in **Table 1**.

The Wood-Ljungdahl pathway is a linear and reductive pathway unlike cyclic $CO_2$-fixing processes such as the Calvin and tricarboxylic acid cycles (Madigan et al., 2003). Acetogens cannot utilize the Calvin cycle that is employed by photosynthetic and chemosynthetic autotrophs because it lacks ribulose diphosphate carboxylase enzyme (Wood et al., 1986). The Wood-Ljungdahl pathway is considered to occur in both oxidation and reduction directions. Conversion of $CO_2$ to acetate is a reduction process. However, acetate can be converted back to $CO_2$ through oxidation (Ragsdale, 1997). Acetogens conserve energy by reduction of CO, and/or $CO_2$, and $H_2$ to acetate. In the Wood-Ljungdahl pathway, synthesis of the acetyl-CoA occurs through two branches, the methyl branch and carbonyl branch. Acetyl-CoA can then be converted to other products including acetate, ethanol, and cell mass (Drake and Daniel, 2004). The pathway for the conversion of acetyl-CoA to acetate is called acetogenesis and the conversion of acetyl-CoA to ethanol is called solventogenesis. The electrons necessary for the reduction reactions in the pathway come from oxidation of $H_2$ by hydrogenase and/or from oxidation of CO by carbon monoxide dehydrogenase (CODH) as shown in **Eqs.7** and **8**.

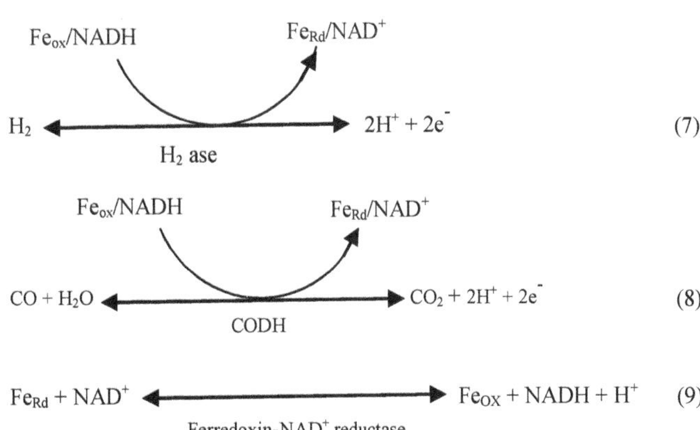


The reducing power donated by $H_2$ or CO are carried by electron carrier pairs NADH/NAD$^+$ , NADPH/NADP$^+$ or ferredoxin (Ljungdhal, 1986) as shown in **Eq. 7** through **Eq. 9**. While electrons are carried by the electron carrier pairs, adenosine triphosphate (ATP) transports the chemical energy within the cells for metabolism. The hydrolysis of the phosphate bonds releases energy and converts ATP to adenosine diphosphate (ADP).

**Table 1.**
Alcohol and organic acid concentrations, yields and productivities during syngas fermentation using various biocatalysts.

| Biocatalysts | Reactor/gas composition [a] | Products (g/L) | Yield from CO [b](%) | Productivity[c] (mg/L·h) | References |
|---|---|---|---|---|---|
| *Clostridium ljungdahlii* | CSTR with cell recycle (55% CO, 20% $H_2$, 10% $CO_2$ and 15% Ar) | Ethanol: 48<br>Acetate: 3.0 | 70.2 | 168.0 | (Phillips et al., 1993) |
| | CSTR without cell recycle (55% CO, 20% H2, 10% CO2 and 15% Ar) | Ethanol: 6.50<br>Acetate: 5.43 | 38.9 | 48.8 | (Mohammadi et al., 2012) |
| | Two stage CSTR & bubble column with cell recycle (60%CO, 35% $H_2$ and 5% $CO_2$) | Ethanol: 19.7<br>Acetate: 8.6 | 100 | 306.4 | (Richter et al., 2013) |
| *Clostridium carboxidivorans* | Bubble column reactor without cell recycle (25% CO, 15% $CO_2$, 60% $N_2$) | Ethanol: 1.6<br>Acetate: 0.4<br>Butanol: 0.6 | 39.6 | 42.7 | (Rajagopalan et al., 2002) |
| | HFR (20% CO, 15% $CO_2$, 5% $H_2$, 60% $N_2$) | Ethanol: 24.0<br>Acetate: 5.0 | 72.0 | 112.5 | (Shen et al., 2014a) |
| | Bubble column reactor (20% CO, 15% $CO_2$, 5% $H_2$ 60% $N_2$) | Ethanol: 3.2<br>Acetate: 2.35 | 51.0 | 64.1 | (Shen et al., 2014b) |
| | MBR (20% CO, 15% $CO_2$, 5% $H_2$, 60% $N_2$) | Ethanol: 4.9<br>Acetate: 3.1 | 51.0 | 97.9 | |
| | Serum bottles ( 70% CO, 20% $H_2$, 10% $CO_2$) | Ethanol: 3.0<br>Acetate: 0.5<br>Butanol: 1.0<br>Hexanol: 0.9 | ND[d] | 21.4 | (Phillips et al., 2015) |
| *Clostridium ragsdalei* | CSTR (20% CO, 15% $CO_2$, 5% $H_2$, 60% $N_2$) | Ethanol: 9.6<br>Acetate: 3.4 | 60.0 | 26.7 | (Maddipati et al., 2011) |
| Mixed culture *of Alkalibaculum bacchi & C. propionicum* | CSTR without cell recycle (28% CO, 60% $H_2$, 12% $N_2$) | Ethanol: 8.0<br>Acetate: 1.1<br>Propanol: 6.0<br>Butanol: 1.1 | 30.6 | 40.0 | (Liu et al., 2014a) |
| *Clostridium autoethanogenum* | CSTR without cell recycle (100% CO) | Ethanol: 0.9<br>Acetate: 0.9 | ND | 4.5 | (Abubackar et al., 2015) |

[a] CSTR: continuous stirred tank reactor; HFR: hollow fiber membrane reactor; MBR: monolithic biofilm reactor
[b] Ethanol yield = (mol EtOH consumed/mol CO consumed)*100%/(1 mol EtOH/6 mol CO)
[c] Ethanol productivity
[d] ND: not determined

The redox reactions involved in the Wood-Ljungdahl pathway to form ethanol, acetate, and cell mass are shown in **Figure 3**.

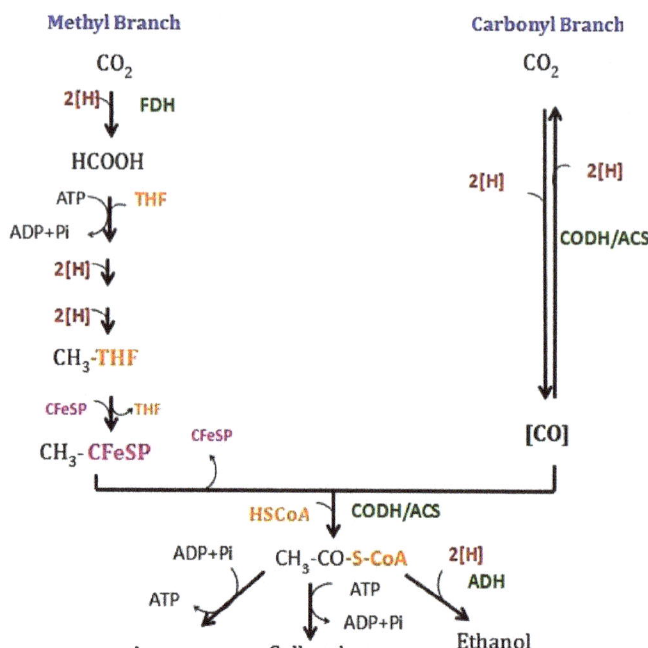


**Fig.3.** Overview of Wood-Ljungdahl pathway. Green text indicates enzymes, orange text indicates coenzymes and pink text indicates co-protein involved in the metabolic pathway. FDH: formate dehydrogenase; CODH/ACS: bifunctional carbon monoxide dehydrogenase/acetyl-CoA synthase; ADH: alcohol dehydrogenase; CFeSP: corrinoid iron(Fe)-sulfur(S) protein; THF: tetrahydrofolate (vitamin B9, folic acid derivative); HSCoA: thiol (SH) functional group Coenzyme A; $CH_3$-CO-S-CoA: acetyl-Coenzyme A intermediate (Adapted from Drake and Daniel, 2004).

### *3.2. Advantages and disadvantages*

One of the main advantages of syngas fermentation is that it utilizes all the biomass components unlike saccharification fermentation where lignin cannot be fermented (Lewis et al., 2008; Phillips et al., 1994). Syngas fermentation can result in high yields (Bredwell et al., 1999; Vega et al., 1989; Worden et al., 1991). Syngas fermentation also occurs at ambient temperatures and pressures. . Further, microbial catalysts are not poisoned by trace amount of sulfur gases like metal catalysts during chemical conversion processes (Ahmed and Lewis, 2007). In addition, no xenobiotic products are expected to be formed during syngas fermentation (Worden et al., 1991).

The main disadvantages of syngas fermentation are (i) low solubility and mass transfer limitations of the CO and $H_2$ gaseous substrates, (ii) slow reactions resulting in long residence times, (iii) low metabolic energy is produced when the microorganisms grow on gaseous substrate instead of sugar substrates resulting in slow growth, low cell density and low solvent production (Barik et al., 1988; Vega et al., 1989).

### *3.3. Process development*

A schematic of the hybrid gasification-syngas fermentation process is shown in **Figure 4**. The hybrid conversion process involves gasification of biomass and other feedstocks followed by fermentation and purification of bioethanol. Industrial flue gases can be directly fed into the fermentor.

Non-food based biomass feedstocks can be partially combusted to produce syngas, which is cleaned and then fed into a fermentor to produce ethanol, acetate, and cell carbon in the presence of acetogenic biocatalysts (**Table 1**). Various process parameters such as temperature, pH, gas composition, gas partial pressures, medium components, reducing agents, and gas-liquid mass transfer affect the cell growth and product distribution during syngas fermentation (Abrini et al., 1994; Hurst and Lewis, 2010; Munasinghe and Khanal, 2010a). The ability to predict and control the onset of solventogenesis is important for improving ethanol yields and productivity. Even though the effect of some of the above operating parameters on the ethanol yield and productivity using different clostridia species was studied; there are

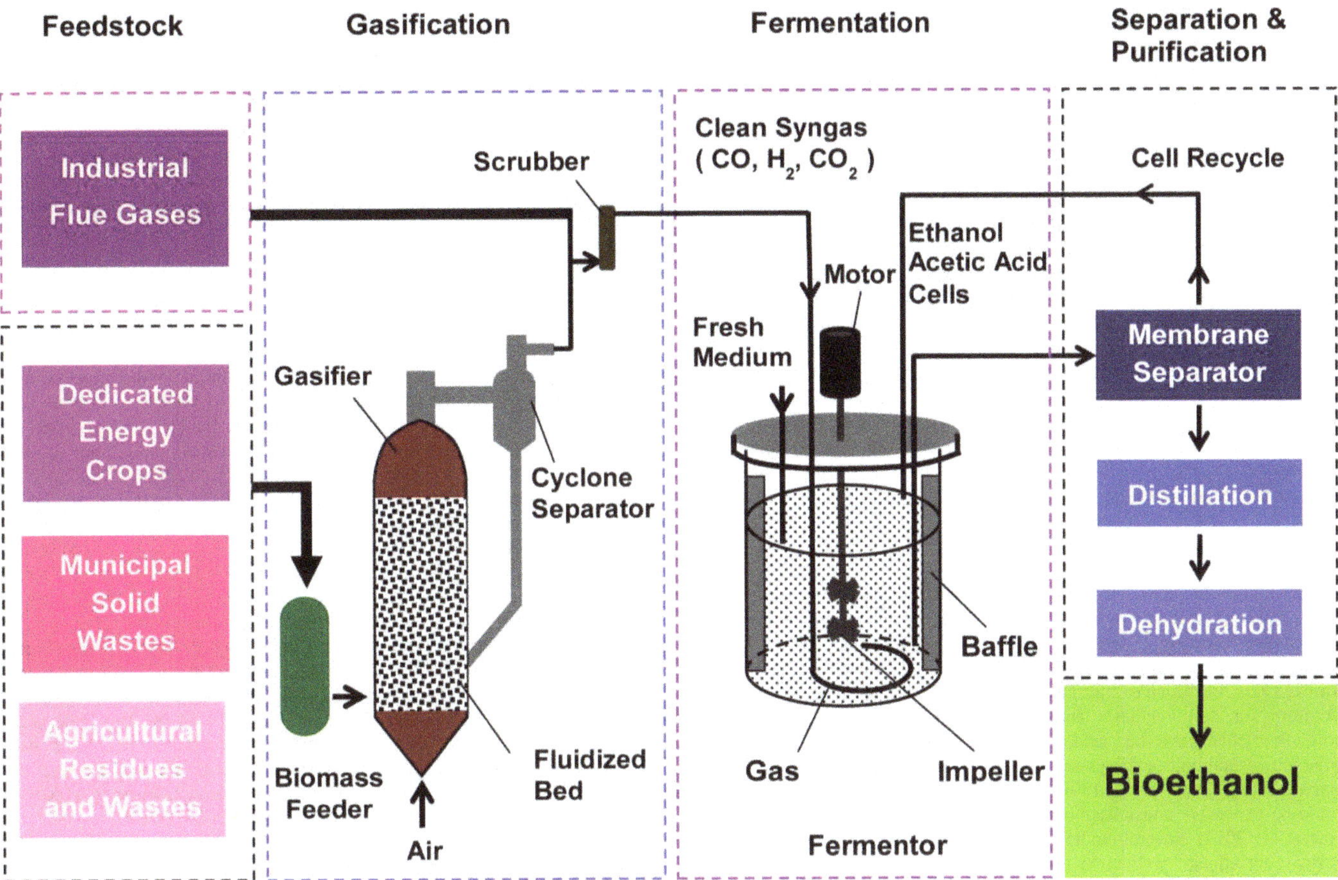


**Fig.4.** Bioethanol hybrid gasification-syngas fermentation conversion process for the production of ethanol and acetic acid from various feedstocks.

opportunities for further optimization of these parameters to make ethanol production from syngas more feasible at commercial scale.

### *3.3.1. Temperature*

Fermentation temperature impacts the cell growth, enzyme activities and gas solubility. Acetogenic species such as *C. ljungdahlii, C. ragsdalei, C. carboxidivorans,* and *A. bacchi* used in syngas fermentation are mesophiles with an optimum temperature between 37 and 40°C (Gaddy and Clausen, 1992; Huhnke et al., 2010; Liou et al., 2005). However, thermophiles such as *Carboxydocella sporoproducens, Moorella thermoacetica, M. thermoautotrophica* have an optimum temperature between 50 and 80°C (Daniel et al., 1990; Henstra et al., 2007; Savage et al., 1987; Slepova et al., 2006). Thermophilic conditions usually result in reduction of gas solubility, however the rate of gas transfer is considered to increase due to low viscosity of the medium (Munasinghe and Khanal, 2010a).

### *3.3.2. pH*

Fermentative bacteria maintain a pH gradient across the membrane and regulate the internal pH which is essential for stability and functioning of metabolic enzymes (Gutierrez, 1989) . Studies with *C. acetobutylicum* reported that when acetate and butyrate production decrease external pH, acids accumulate inside cells and lower their internal pH to maintain a constant pH gradient (Gottwald and Gottschalk, 1985). However, accumulation of high concentrations of undissociated acid inside the cells stresses them and decreases the pH gradient. Thus, the cells counteract by producing solvents (Ahmed, 2006; Gottschal and Morris, 1981; Gottwald and Gottschalk, 1985).

In syngas fermentation, the external pH in the fermentation medium is a widely studied physiological parameter to optimize cell growth and solvent production. The optimum external pH range for cell growth of most of the syngas fermenting microbes usually varies from 5.5 to 6.5 (Abrini et al., 1994; Liou et al., 2005; Tanner et al., 1993). The optimum external pH for solvent production was reported to be around 4.5 to 4.8 (Ahmed et al., 2006; Sakai et al., 2004; Worden et al., 1991). Recently a moderately alkaliphilic bacterium called *A. bacchi* has shown capabilities to grow on syngas at an optimum pH between 8 and 8.5 and produce ethanol at pH range between 6.5 and 7 (Allen et al., 2010; Liu et al., 2012). In syngas fermentation studies, the changes in external pH were correlated with the substrate metabolism and release of metabolic by-products (Devi et al., 2010; Hu, 2011; Kundiyana et al., 2011b; Liu et al., 2014a). However, future studies to understand how internal pH changes and the pH gradient across the cell membrane effect syngas fermentation are important to improve solvent production and reduce acid stress on cells.

### *3.3.3. Gas partial pressure*

The concentration of CO in syngas has a significant impact on the overall process efficiency and utilization of other syngas components (namely $CO_2$ and $H_2$). Hu (2011) reported that electron production from CO is thermodynamically favorable compared to $H_2$ independent of pH, ionic strength and gas partial pressure. In a syngas fermentation using *C. carboxidivorans*, the increase of CO partial pressure from 35.5 to 70.9 kPa and from 35.5 to 202.7 kPa was reported to decrease hydrogenase activity by 84% and 97 %, respectively (Hurst, 2005). In addition, CO partial pressure of 8.5 kPa was reported to inhibit hydrogenase activity in *C. ragsdalei* cells by 90% (Skidmore, 2010).

The decrease in hydrogenase activity and $H_2$ utilization results in a decrease in overall gas conversion efficiency. However, a study on effect of CO partial pressure using $CO:CO_2$ (molar ratios of 1.7 to 4) gas mixture without $H_2$ reported that *C. carboxidivorans* switched from non-growth related to growth related ethanol production and grew 440% more when the partial pressure of CO was increased from 35.5 to 202.79 kPa (Hurst and Lewis, 2010). When fructose in the medium was replaced with CO , *C. carboxidivorans* was reported to shift the molar ethanol to acetate ratio from 0.3 to 8 (Liou et al., 2005). The presence of CO, which may have acted as an effective electron source, enabled ethanol production rather than acetate production by acetogens (Tanner, 2008).

It should be noted that syngas produced during gasification contains $H_2$ along with CO and $CO_2$. Thus ideally, for high product yields and efficient gas utilization, CO and $CO_2$ should be used as carbon source and $H_2$ should be used as the sole electron source (Hu et al., 2011; Skidmore, 2010). In a batch culture with *C. ljungdahlii*, when the total pressure of syngas was varied from 81.1 to 182.4 kPa, the ethanol to acetate molar ratio of 5:1 was achieved at total syngas pressure of 162.1 and 182.4 kPa (Najafpour and Younesi, 2006; Younesi et al., 2005). Younesi et al. (2005) reported that $H_2$ and $CO_2$ consumption occurred after CO was exhausted indicating CO as a preferred substrate for cell growth.

### *3.3.4. Medium components*

Fermentation medium components such as vitamins, minerals, and metals act as cofactors or coenzymes that are necessary for enzymes to catalyze biochemical reactions (Phillips et al., 2014; Zabriskie and Mill, 1988). Additionally, syngas fermentation medium is often supplemented with yeast extract (YE) to provide the amino acids and nitrogenous compounds necessary for cell synthesis and with buffer solutions (such as 2-(N-morpholino)ethanesulfonic acid and [N-tris(hydroxymethyl)methyl]-3-aminopropanesulfonic acid) to maintain the medium pH (Liu et al., 2012; Saxena, 2008; Tanner et al., 1993). The addition of YE and buffer solution would be expensive and uneconomical for commercial syngas fermentation (Gao et al., 2013). Several studies were reported on the optimization of the nutrients for ethanol production using syngas fermentation. Studies with *C. ljungdahlii* showed that reducing or completely removing YE from fermentation medium increased ethanol concentration from 1 g/L to 48 g/L (Phillips et al., 1993; Vega et al., 1989). The increase in the concentrations of $Ni^{2+}$, $Zn^{2+}$, $SeO_4^-$ and $WO_4^-$ from 0.84 μM, 6.96 μM, 1.06 μM and 0.68 μM to 8.4 μM, 34.8 μM, 5.3 μM and 6.8 μM, respectively, improved ethanol production by *C. ragsdalei* by fourfold (Saxena and Tanner, 2011).

In another study with *C. ragsdalei*, limiting calcium pantothenate, vitamin $B_{12}$ and cobalt chloride in two-stage continuous bioreactor resulted in 15 g ethanol/g cell compared to 2.5 g ethanol/g cell in a single-stage bioreactor (Kundiyana et al., 2011a). Standard YE medium was replaced with defined minimal medium, cotton seed extract (CSE) and corn steep liquor (CSL) to reduce medium cost and improve ethanol production (Gao et al., 2013; Kundiyana et al., 2010; Maddipati et al., 2011; Phillips et al., 2014). CSL medium, which is rich in vitamins, minerals and amino acids was shown to produce 40% more ethanol using *C. ragsdalei* (Maddipati et al., 2011). Also, the use of a completely defined minimal medium was shown to result in 36% higher ethanol yield than in standard YE medium at 5% of the cost of the YE medium (Gao et al., 2013).

### *3.3.5. Reducing agents*

Reducing agents are artificial electron carriers that alter $NADH/NAD^+$ ratio. Reducing agents significantly decrease the redox potential of the fermentation medium (Frankman, 2009). Redox potential is a fermentation parameter that defines the ability of the solution to undergo oxidation reduction reaction (IFIS, 2009). In syngas fermentation using *C. ragsdalei*, a decreasing trend of redox potential during cell growth and increasing trend of redox potential during ethanol production was reported (Kundiyana et al., 2010; Maddipati et al., 2011). Solventogenesis is an electron intensive process that requires high levels of NADH (Rao et al., 1987). Addition of reducing agents was reported to increase the NADH levels in cells and direct electron flow towards ethanol production (Babu et al., 2010; Sim and Kamaruddin, 2008). Reducing agents such as neutral red were also reported to increase the activity of aldehyde dehydrogenase and alcohol dehydrogenase, which catalyze the aldehyde and ethanol production from acetyl-CoA intermediate (Girbal et al., 1995).

The addition of methyl viologen to the fermentation broth of *Thermoanaerobacter ethanolicus* and *C. acetobutylicum* resulted in the onset of ethanol production from glucose (Rao and Mutharasan, 1986; Rao et al., 1987). The addition of neutral red was reported to increase the activity of alcohol dehydrogenase and ethanol production from syngas by *C. carboxidivorans* (Ahmed et al., 2006). The addition of methyl viologen and dithiothreitol to fermentation medium with *C. ragsdalei* also showed enhancement in ethanol production (Babu et al., 2010; Panneerselvam et al., 2009).

### *3.4. Bioreactor design*

A bioreactor should provide a controlled environment to enhance cell growth, substrate conversion and productivity of the biological process, and minimize the overall cost of production of desired products (Wilkins and Atiyeh, 2012). Continuous stirred tank reactors, bubble columns, packed columns, air-lift, trickle beds and hollow fiber reactors are some of the bioreactor configurations studied for alcohol production using syngas fermentation (Datar et al., 2004; Hickey et al., 2011; Kimmel et al., 1991; Kundiyana et al., 2010; Mohammadi et al., 2012; Shen et al., 2014a). Further, these reactors can be operated in different fermentation modes such as batch, fed-batch, continuous with and without cell recycle (Cotter et al., 2009; Grethlein et al., 1991; Lewis et al., 2007; Maddipati et al., 2011; Phillips et al., 1993). Klasson et al. (1990a) used two STRs in series and reported a 30 fold increase in ethanol productivity using *C. ljungdahlii*. Bredwell and Worden (1998) showed that the use of a microsparger in a STR for production of acetate, ethanol and butyrate by *Butyribacterium methylotrophicum* increased the mass transfer by six times with 50% of the flow rate used without a microsparger. The highest ethanol concentration of 48 g/L was produced in a continuous syngas fermentation in a CSTR with cell recycle (Phillips et al., 1993). A list of reactors used for syngas fermentation, syngas composition, ethanol yield and productivity is shown in **Table 1**.

In addition, higher ethanol production (20-24 g/L) was achieved in a two stage CSTR and bubble column with cell recycle and in the hollow fiber membrane reactor (HFR) with biofilm formation (Richter et al., 2013; Shen et al., 2014a). The increase in cell mass density and mass transfer increased ethanol production. However, bubble columns and monolithic biofilm bioreactor only produced about 3 g/L ethanol (Shen et al., 2014b).

Efficient syngas fermentation bioreactor designs should (i) provide gas-liquid mass transfer that balances the cells' kinetic requirement without inhibiting the cells' metabolic activity, (ii) sustain biocatalyst viability and high concentration, (iii) reduce operating and maintenance cost, (iv) be easily scaled up.

The ability to maintain high cell concentrations and high gas transfer rates in the reactor enhances productivity and reduces required reactor size. Gas-liquid mass transfer can limit the rate of syngas fermentation due to the low solubility of CO and $H_2$ in fermentation medium (Bredwell et al., 1999). The rate of mass transfer (*dn/dt*) is given as follows (Sherwood et al., 1975):

$$\frac{1}{V} \cdot \frac{dn}{dt} = -k_L a \cdot (C_i - C_L) \tag{10}$$

where, *dn/dt* is the rate of mass transfer (*mmol/h*); $k_La$ is the overall mass transfer coefficient ($h^{-1}$); $C_i$ is the concentration of the gas in gas liquid interface (*mmol/L*); $C_L$ is the concentration of gas in the bulk liquid (*mmol/L*) and $V$ is the working volume of the reactor ($L$). The rate of gas transfer can be increased by either increasing the mass transfer coefficient ($k_La$) or by increasing the driving force ($C_i$-$C_L$). The driving force can be increased by operating the reactor at high CO partial pressures (Klasson et al., 1993b). However, high concentrations of CO could be inhibitory to the microorganisms (Munasinghe and Khanal, 2010a). The mass transfer limiting conditions occur when the concentration of CO in the liquid is zero, at which the reaction rate is a function of the gas transfer rate.

Mass transfer characteristics of various reactor configurations have compared by many researchers (Bredwell and Worden, 1998; Cowger et al., 1992; Jones, 2007; Klasson et al., 1990b; Klasson et al., 1991; Klasson et al., 1993a; Munasinghe and Khanal, 2010b; Munasinghe and Khanal, 2014; Orgill et al., 2013; Riggs and Heindel, 2006; Shen et al., 2014a; Yasin et al., 2014). In a STR, the mass transfer coefficient can be increased by increasing the agitation speed or the gas flow rate (Orgill et al., 2013). However, using high gas flow rates decreases the gas conversion efficiency. The increase in agitation speed has been widely used to increase the $k_La$ in STRs. The hydrodynamic shear generated by the impeller reduces the bubble size and increases the interfacial area for mass transfer (Bredwell et al., 1999). However, the use of high agitation speed increases the power requirement for large reactors.

Ungerman and Heindel (2007) reported a dual impeller scheme with axial flow impeller at the top and lower concave impeller that resulted in a similar $k_La$ and less power requirement compared to Rushton impellers. Bredwell and Worden (1998) used a microsparger that was shown to be energy efficient and increased the $k_La$ by six fold compared to conventional gas sparging. In the case of an air lift reactor, the use of a 20 μm bulb diffuser was reported to provide higher mass transfer coefficient (91 $h^{-1}$) than air lift reactor configurations with column diffusers, gas spargers with mechanical mixing (Munasinghe and Khanal, 2010b). Also, it was claimed that due to the simple reactor configuration and low energy requirements, the scale up of air lift reactors with a 20 μm bulb diffuser will be easy and cheap compared to a conventional STR (Munasinghe and Khanal, 2010b).

Performances of different syngas fermentation reactors were compared during the production of hydrogen and methane using a mixed culture of *R. rubrum*, *M. formicicum* and *M. barkeri* (Klasson et al., 1990b; Klasson et al., 1991; Klasson et al., 1992). The TBR was reported to have better $CH_4$ productivity, CO gas conversion and mass transfer capabilities than the packed bubble column reactor (PBR). The mass transfer coefficients of 3.5 $h^{-1}$ and 780 $h^{-1}$ were reported for PBR and TBR, respectively (Klasson et al., 1990b). The TBR showed better mass transfer capabilities than PBR and STR for the production of acetate from syngas by *P. productus*.

A comparison between STR, TBR and five different HFR modules showed that the polydimethylsiloxane (PDMS-HFR) provided better gas liquid mass transfer (1063 $h^{-1}$) followed by the TBR (421 $h^{-1}$) and STR (114 $h^{-1}$) (Orgill et al., 2013). In addition, the use of 0.3 wt% methyl-functionalized silica nanoparticles was reported to enhance the mass transfer of syngas components into the medium leading to a significant increase in the levels of biomass, ethanol and acetic acid production (Kim et al., 2014). Besides assessing the mass transfer capabilities of different reactor configurations, researchers have recently focused on developing new techniques to measure the dissolve concentrations of CO and $H_2$ gases in the liquid phase to determine the CO and $H_2$ mass transfer coefficients (Munasinghe and Khanal, 2014).

An accurate and reliable technique would be essential to adjust the fermentation parameters (such as agitation speeds, gas and liquid flow rates) in order to meet the cells kinetic requirement. The increase in gas flow rate beyond cells kinetic requirements would decrease gas conversion efficiency, while increasing agitation speed and liquid flow rate would have detrimental effects on the cell viability and costs associated with power consumption in large-scale reactors.

### *3.5. Commercialization and future prospective*

LanzaTech, Coskata, and INEOS Bio are among the companies that are currently pursuing commercialization of syngas fermentation for biofuels production (Liew et al., 2013). Coskata has a fully integrated demonstration facility in Madison, Pennsylvania (USA) and has recently isolated and patented a new strain *C. coskatii* (Zahn and Saxena, 2012). The company is focusing on fermentation of syngas produced from natural gas reforming or gasification of wood and coal (Coskata, 2011).

INEOS Bio has operated the first commercial cellulosic ethanol and power generation facility using syngas fermentation technology in Vero Beach, Florida (USA) since July 2013 (INEOS, 2013). However, soon after the commissioning of the plant it was stopped due to the very high sensitivity of the microorganisms to hydrogen cyanide in the syngas produced during gasification of vegetative matter (Lane, 2014). The company is currently installing scrubbers to reduce the hydrogen cyanide concentrations from 15 ppm to less than 5 ppm (Lane, 2014). The company utilizes patented bacteria to produce ethanol and generate power from vegetative and woody waste. The company was projected to produce 8 million gallons of ethanol per year and generate 6 MW of renewable electricity (INEOS, 2013).

LanzaTech is a New Zealand based company that utilizes CO-rich flue gases from steel making industries to produce ethanol using its proprietary *Clostridial* biocatalyst. It has a pilot plant facility in Glenbrook, New Zealand and a demonstration facility in Shanghai, China that has an operating capacity of 100,000 gallons ethanol per year. LanzaTech reported to expands to production of more products through syngas fermentation (LanzaTech, 2015).

The future of syngas fermentation technology depends on production of high value products beyond ethanol. Ethanol's low heating value, miscibility with water and inability to use the existing infrastructure for fuel transportation are just a few of the disadvantages that led to the focus towards advanced biofuels such as butanol and hexanol.

In addition, discovering new microorganisms, processes, and strain development, including synthetic biology are required to utilize the biological gas conversion technology to produce fuels and biobased products. Recent research indicates production of advanced biofuels such as butanol and hexanol from CO, $CO_2$ and $H_2$ through medium optimization (Phillips et al., 2015). Several studies also reported production of higher alcohols such as isopropanol, butanol, and hexanol using syngas fermentation (Liu et al., 2014b; Maddipati et al., 2011; Rajagopalan et al., 2002; Ramachandriya et al., 2011; Worden et al., 1991). In presence of CO as a reductant, *C. formicaceticum* and *M. thermoacetica* were reported to reduce acids to their corresponding alcohols (Fraisse and Simon, 1988; White et al., 1987). *C. acetobutylicum* was also shown to directly reduce acetate and butyrate to corresponding alcohols (Hartmanis et al., 1984).

It was recently reported that mono-cultures of *C. ljungdahlii* and *C. ragsdalei* as well as a mixed culture of *A. bacchi* and *C. propionicum* were able to convert added acids such as propionic, butyric, and hexanoic acids to their respective alcohols (Liu et al., 2014b; Perez et al., 2013). Additional development and optimization of biological gas conversion processes are expected to result in production of various products besides biofuels at commercial scale in the near future.

## 4. Conclusions

The production of ethanol using diverse conversion technologies and various renewable non-food feedstocks marks the beginning of sustainable energy future. Production of ethanol from sustainable non-food feedstocks in first generation biorefineries has been recently deployed at commercial scale. Biological conversion processes including hydrolysis-fermentation and syngas fermentation have been developed for the production of ethanol. Various process configurations are possible in the hydrolysis-fermentation route. Syngas fermentation is an indirect conversion process for production of alcohols and chemicals from CO, $CO_2$, and $H_2$. Advancement in metabolic engineering, strain and process development of syngas fermentation resulted in production of new products from syngas and enhanced product selectivity, productivity, and yields. Further research efforts should be focused on utilization of different types of non-food feedstocks, process integration, metabolic engineering, and discovering new highly productive microorganisms. Ultimately, the reduction in biofuels production cost improves their feasibility to become a viable alternative to fossil fuels.

## Acknowledgments

This research was supported by the Sun Grant Initiative through the U.S. Department of Transportation, USDA–NIFA Project No. OKL03005 and Oklahoma Agricultural Experiment Station.

## References

Abrini, J., Naveau, H., Nyns, E.J., 1994. *Clostridium autoethanogenum*, sp. Nov., an anaerobic bacterium that produces ethanol from carbon monoxide. Arch. Microbiol. 161(4), 345-351.

Abubackar, H.N., Veiga, M.C., Kennes, C., 2015. Carbon monoxide fermentation to ethanol by *Clostridium autoethanogenum* in a bioreactor with no accumulation of acetic acid. Bioresour. Technol. 186, 122-127.

Ahmed, A., 2006. Effects of biomass-generated syngas on cell-growth, product distribution and enzyme activities of *Clostridium carboxidivorans* P7$^T$, Ph.D. Dissertation, Oklahoma State University, pp. 229.

Ahmed, A., Cateni, B.G., Huhnke, R.L., Lewis, R.S., 2006. Effects of biomass-generated producer gas constituents on cell growth, product distribution and hydrogenase activity of *Clostridium carboxidivorans* P7$^t$. Biomass Bioenergy. 30(7), 665-672.

Ahmed, A., Lewis, R.S., 2007. Fermentation of biomass-generated synthesis gas: Effects of nitric oxide. Biotechnol. Bioeng. 97(5), 1080-1086.

Allen, T.D., Caldwell, M.E., Lawson, P.A., Huhnke, R.L., Tanner, R.S., 2010. *Alkalibaculum bacchi* gen. Nov., sp. Nov., a co-oxidizing, ethanol-producing acetogen isolated from livestock-impacted soil. Int. J. Syst. Evol. Microbiol. 60(10), 2483-2489.

Alvira, P., Tomás-Pejó, E., Ballesteros, M., Negro, M.J., 2010. Pretreatment technologies for an efficient bioethanol production process based on enzymatic hydrolysis: A review. Bioresour. Technol. 101(13), 4851-4861.

Babu, B., Atiyeh, H., Wilkins, M., Huhnke, R., 2010. Effect of the reducing agent dithiothreitol on ethanol and acetic acid production by *Clostridium* strain p11 using simulated biomass-based syngas. Biol. Eng. 3(1), 19-35.

Bai, F., Anderson, W., Moo-Young, M., 2008. Ethanol fermentation technologies from sugar and starch feedstocks. Biotechnol. Adv. 26(1), 89-105.

Barik, S., Prieto, S., Harrison, S., Clausen, E., Gaddy, J., 1988. Biological production of alcohols from coal through indirect liquefaction. Appl. Biochem. Biotechnol. 18(1), 363-378.

Bothast, R., Schlicher, M., 2005. Biotechnological processes for conversion of corn into ethanol. Appl. Microbiol. Biotechnol. 67(1), 19-25.

Bredwell, M.D., Worden, R.M., 1998. Mass-transfer properties of microbubbles. 1. Experimental studies. Biotechnol. Progr. 14(1), 31-38.

Bredwell, M.D., Srivastava, P., Worden, R.M., 1999. Reactor design issues for synthesis-gas fermentations. Biotechnol. Progr. 15(5), 834-844.

Brennan, L., Owende, P. 2010. Biofuels from microalgae—a review of technologies for production, processing, and extractions of biofuels and co-products. Renew. Sustainable Energy Rev. 14(2), 557-577.

Cardona Alzate, C.A., Sánchez Toro, O.J., 2006. Energy consumption analysis of integrated flowsheets for production of fuel ethanol from lignocellulosic biomass. Energy. 31(13), 2447-2459.

Carriquiry, M.A., Du, X., Timilsina, G.R., 2011. Second generation biofuels: Economics and policies. Energy Policy. 39(7), 4222-4234.

Chisti, Y., 2007. Biodiesel from microalgae. Biotechnol. Adv. 25(3), 294-306.

Coskata., 2011. Semi-commercial facility demonstrates two years of successful operation, Coskata, Inc.'s. Madison, PA. Available at http://www.coskata.com/company/media.asp?story=504B571C-0916-474E-BFFA-ACB326EFDB68 (accessed on 10 August 2015).

Cotter, J.L., Chinn, M.S., Grunden, A.M., 2009. Ethanol and acetate production by *Clostridium ljungdahlii* and *Clostridium autoethanogenum* using resting cells. Bioproc. Biosyst. Eng. 32(3), 369-380.

Cowger, J.P., Klasson, K.T., Ackerson, M.D., Clausen, E., Caddy, J.L., 1992. Mass-transfer and kinetic aspects in continuous bioreactors using *Rhodospirillum rubrum*. Appl. Biochem. Biotechnol. 34 (1), 613-624.

Cruz, V., Hernández, S., Martín, M., Grossmann, I.E., 2014. Integrated synthesis of biodiesel, bioethanol, isobutene, and glycerol ethers from algae. Ind. Eng. Chem. Res. 53(37), 14397-14407.

Daniel, S.L., Hsu, T., Dean, S., Drake, H., 1990. Characterization of the $h_2$-and co-dependent chemolithotrophic potentials of the acetogens *Clostridium thermoaceticum* and *acetogenium kivui*. J. Bacteriol. 172(8), 4464-4471.

Datar, R.P., Shenkman, R.M., Cateni, B.G., Huhnke, R.L., Lewis, R.S., 2004. Fermentation of biomass-generated producer gas to ethanol. Biotechnol. Bioeng. 86(5), 587-594.

Demirbas, A., Fatih Demirbas, M., 2011. Importance of algae oil as a source of biodiesel. Energy Convers. Manage. 52(1), 163-170.

Devi, M.P., Mohan, S.V., Mohanakrishna, G., Sarma, P., 2010. Regulatory influence of $co_2$ supplementation on fermentative hydrogen production process. *Int. J. Hydrogen* Energy. 35(19), 10701-10709.

Dien, B.S., Cotta, M.A., Jeffries, T.W. 2003. Bacteria engineered for fuel ethanol production: Current status. Appl. Microbiol. Biotechnol. 63(3), 258-266.

Drake, H.L., Daniel, S.L. 2004. Physiology of the thermophilic acetogen *Moorella thermoacetica*. Res. Microbiol. 155(10), 869-883.

Drake, H.L., Gößner, A.S., Daniel, S.L. 2008. Old acetogens, new light. Ann. N.Y. Acad. Sci. 1125(1), 100-128.

Dutta, A., Talmadge, M., Hensley, J., Worley, M., Dudgeon, D., Barton, D., Groenendijk, P., Ferrari, D., Stears, B., Searcy. E.M., Wright. C.T., Hess. J.R., 2011. Process design and economics for conversion of lignocellulosic biomass to ethanol: thermochemical pathway by indirect gasification and mixed alcohol synthesis. NREL/TP-5100-51400. Golden, CO: National Renewable Energy Laboratory, 2011. Available at http://www.nrel.gov/docs/fy11osti/51400.pdf. (accessed on 10 August 2015).

EIA., 2015a. Annual energy outlook, Energy Information Administration. Washington, D.C. Available at http://www.eia.gov/forecasts/aeo/MT_liquidfuels.cfm (accessed on 10 August 2015).

EIA., 2015b. Net petroleum imports data- 2014, U.S Energy Information Administration. Washington, D.C. Available at http://www.eia.gov/tools/faqs/faq.cfm?id=32&t=6 (accessed on 10 August 2015).

Fasahati, P., Woo, H.C., Liu, J.J., 2015. Industrial-scale bioethanol production from brown algae: Effects of pretreatment processes on plant economics. Appl. Energy. 139, 175-187.

Fischer, C.R., Klein-Marcuschamer, D., Stephanopoulos, G., 2008. Selection and optimization of microbial hosts for biofuels production. Metab. Eng. 10(6), 295-304.

Fontaine, F., Peterson, W., McCoy, E., Johnson, M.J., Ritter, G.J., 1942. A new type of glucose fermentation by *Clostridium thermoaceticum*. J. Bacteriol. 43(6), 701-715.

Foust, T., Aden, A., Dutta, A., Phillips, S., 2009. An economic and environmental comparison of a biochemical and a thermochemical lignocellulosic ethanol conversion processes. Cellulose. 16(4), 547-565.

Fraisse, L., Simon, H., 1988. Observations on the reduction of non-activated carboxylates by *Clostridium formicoaceticum* with carbon monoxide or formate and the influence of various viologens. Arch. Microbiol. 150(4), 381-386.

Frankman, A.W., 2009. Redox, pressure and mass transfer effects on syngas fermentation. in: *Department of Chemical Engineering*, MS Thesis, Brigham Young University, pp. 106.

Gaddy, J.L., Clausen, E.C., 1992. *Clostridiumm ljungdahlii*, an anaerobic ethanol and acetate producing microorganism, US, 5173429.

Gao, J., Atiyeh, H.K., Phillips, J.R., Wilkins, M.R., Huhnke, R.L., 2013. Development of low cost medium for ethanol production from syngas by *Clostridium ragsdalei*. Bioresour. Technol. 147, 508-515.

German, L., Schoneveld, G.C., Pacheco, P., 2011. The social and environmental impacts of biofuel feedstock cultivation: Evidence from multi-site research in the forest frontier. Ecol. Soc. 16(3), 24.

Girbal, L., Croux, C., Vasconcelos, I., Soucaille, P., 1995. Regulation of metabolic shifts in *Clostridium acetobutylicum* atcc 824. FEMS Microbiol. Rev. 17(3), 287-297.

Gottschal, J., Morris, J., 1981. The induction of acetone and butanol production in cultures of *Clostridium acetobutylicum* by elevated concentrations of acetate and butyrate. FEMS Microbiol. Lett. 12(4), 385-389.

Gottwald, M., Gottschalk, G., 1985. The internal ph of *Clostridium acetobutylicum* and its effect on the shift from acid to solvent formation. Arch. Microbiol. 143(1), 42-46.

Grethlein, A.J., Worden, R.M., Jain, M.K., Datta, R., 1991. Evidence for production of *n*-butanol from carbon monoxide by *Butyribacterium methylotrophicum*. J. Ferment. Bioeng. 72(1), 58-60.

Gutierrez, N.A., 1989. Role of motility and chemotaxis in solvent production by *Clostridium acetobutylicum* in: *Biotechnology*, Ph.D. Dissertation, Massey University, pp. 482.

Hahn-Hägerdal, B., Galbe, M., Gorwa-Grauslund, M.F., Lidén, G., Zacchi, G., 2006. Bio-ethanol – the fuel of tomorrow from the residues of today. Trends Biotechnol. 24(12), 549-556.

Hahn-Hägerdal, B., Karhumaa, K., Fonseca, C., Spencer-Martins, I., Gorwa-Grauslund, M., 2007. Towards industrial pentose-fermenting yeast strains. Appl. Microbiol. Biotechnol. 74(5), 937-953.

Hamelinck, C.N., Hooijdonk, G.v., Faaij, A.P., 2005. Ethanol from lignocellulosic biomass: Techno-economic performance in short-, middle- and long-term. Biomass Bioenergy. 28(4), 384-410.

Hartmanis, M.N., Klason, T., Gatenbeck, S., 1984. Uptake and activation of acetate and butyrate in *Clostridium acetobutylicum*. Appl. Microbiol. Biotechnol. 20(1), 66-71.

Harun, R., Danquah, M.K., Forde, G.M., 2010. Microalgal biomass as a fermentation feedstock for bioethanol production. J. Chem. Technol. Biotechnol. 85(2), 199-203.

Harun, R., Yip, J.W., Thiruvenkadam, S., Ghani, W.A., Cherrington, T.,

Danquah, M.K., 2014. Algal biomass conversion to bioethanol–a step-by-step assessment. Biotech. J. 9(1), 73-86.

Henstra, A.M., Sipma, J., Rinzema, A., Stams, A.J.M., 2007. Microbiology of synthesis gas fermentation for biofuel production. Curr. Opin. Biotechnol. 18(3), 200-206.

Hickey, R., Basu, R., Datta, R., Tsai, S.P., 2011. Method of conversion of syngas using microorganism on hydrophilic membrane, US,7923227.

Hu, P., 2011. Thermodynamic, sulfide, redox potential and ph effects on syngas fermentation. in: *Chemical Engineering*, Ph.D. Dissertation, Bringham Young University, pp. 206.

Hu, P., Bowen, S.H., Lewis, R.S., 2011. A thermodynamic analysis of electron production during syngas fermentation. Bioresour. Technol. 102(17), 8071-8076.

Huhnke, R., Lewis, R.S., Tanner, R.S., 2010. Isolation and characterization of novel clostridial species, US, 7704723.

Hurst, K.M., 2005. Effect of carbon monoxide and yeast extract on growth, hydrogenase activity and product formation of *Clostridium carboxidivorans* P7[t]. in: *Chemical Engineering*, Master of Science, Oklahoma State University, pp. 160.

Hurst, K.M., Lewis, R.S., 2010. Carbon monoxide partial pressure effects on the metabolic process of syngas fermentation. Biochem. Eng. J. 48(2), 159-165.

IFIS., 2009. Dictionary of food science and technology. **2 ed, (Ed.)** I.F.I. Service, Wiley-Blackwell. Singapore, pp. 488.

INEOS., 2013. Ineos bio produces cellulosic ethanol at commercial scale, INEOS Bio. Vero Beach, FL. Available at http://www.ineos.com/en/businesses/INEOS-Bio/News/INEOS-Bio-Produces-Cellulosic-Ethanol/?business=INEOS+Bio (accessed on 10 August 2015).

Ito, K., Hori, K., 1989. Seaweed: Chemical composition and potential food uses. Food Rev. Int. 5(1), 101-144.

John, R.P., Anisha, G.S., Nampoothiri, K.M., Pandey, A., 2011. Micro and macroalgal biomass: A renewable source for bioethanol. Bioresour. Technol. 102(1), 186-193.

Jones, S.T., 2007. Gas-liquid mass transfer in an external airlift loop reactor for syngas fermentation. in: *Chemical Engineering*, Ph.D. Dissertation, Iowa State University, pp. 378.

Jung, K.A., Lim, S.-R., Kim, Y., Park, J.M., 2013. Potentials of macroalgae as feedstocks for biorefinery. Bioresour. Technol. 135, 182-190.

Kim, Y.-K., Park, S.E., Lee, H., Yun, J.Y., 2014. Enhancement of bioethanol production in syngas fermentation with *Clostridium ljungdahlii* using nanoparticles. Bioresour. Technol. 159, 446-450.

Kimmel, D.E., Klasson, K.T., Clausen, E.C., Gaddy, J.L., 1991. Performance of trickle-bed bioreactors for converting synthesis gas to methane. Appl. Biochem. Biotechnol. 28 (1), 457-469.

Kitani, O., Hall, C.W., 1989. *Biomass handbook*. Gordon and Breach Science Publishers, New York, pp.963.

Klasson, K., Elmore, B., Vega, J., Ackerson, M., Clausen, E., Gaddy, J., 1990a. Biological production of liquid and gaseous fuels from synthesis gas. Appl. Biochem. Biotechnol. 24(1), 857-873.

Klasson, K., Cowger, J., Ko, C., Vega, J., Clausen, E., Gaddy, J., 1990b. Methane production from synthesis gas using a mixed culture of *R. Rubrum, M. Barkeri,* and *M. Formicicum*. Appl. Biochem. Biotechnol. 24(1), 317-328.

Klasson, K., Ackerson, M., Clausen, E., Gaddy, J., 1991. Bioreactor design for synthesis gas fermentations. Fuel. 70(5), 605-614.

Klasson, K., Gupta, A., Clausen, E., Gaddy, J., 1993a. Evaluation of mass-transfer and kinetic parameters for *Rhodospirillum rubrum* in a continuous stirred tank reactor. Appl. Biochem. Biotechnol. 39(1), 549-557.

Klasson, K.T., Ackerson, M.D., Clausen, E.C., Gaddy, J.L., 1992. Bioconversion of synthesis gas into liquid or gaseous fuels. Enzyme Microb. Technol. 14(8), 602-608.

Klasson, T.K., Ackerson, M.D., Clausen, E.C., Gaddy, J.L., 1993b. Biological conversion of coal and coal-derived synthesis gas. Fuel. 72(12), 1673-1678.

Kundiyana, D.K., Huhnke, R.L., Maddipati, P., Atiyeh, H.K., Wilkins, M.R., 2010. Feasibility of incorporating cotton seed extract in *Clostridium* strain p11 fermentation medium during synthesis gas fermentation. Bioresour. Technol. 101(24), 9673-9680.

Kundiyana, D.K., Huhnke, R.L., Wilkins, M.R., 2011a. Effect of nutrient limitation and two-stage continuous fermentor design on productivities during "*Clostridium ragsdalei*" syngas fermentation. Bioresour. Technol. 102(10), 6058-6064.

Kundiyana, D.K., Wilkins, M.R., Maddipati, P., Huhnke, R.L., 2011b. Effect of temperature, ph and buffer presence on ethanol production from synthesis gas by "*Clostridium ragsdalei*". Bioresour. Technol. 102(10), 5794-5799.

Lane, J., 2014. On the mend: Why ineos bio isn't producing ethanol in florida, Biofuels Digest. Available at http://www.biofuelsdigest.com/bdigest/2014/09/05/on-the-mend-why-ineos-bio-isnt-reporting-much-ethanol-production/ (accessed on 10 August 2015).

Lane, J., 2015. Abengoa bioenergy: Biofuels digest's 2015 5-minute guide, Biofuels Digest. Available at http://www.biofuelsdigest.com/bdigest/2015/01/19/abengoa-bioenergy-biofuels-digests-2015-5-minute-guide/ (accessed on 10 August 2015).

LanzaTech., 2015. Chemicals. Available at http://www.lanzatech.com/innovation/markets/chemicals/ (accessed on 10 August 2015).

Lewis, R.S., Tanner, R.S., Huhnke, R.L., 2007. Indirect or direct fermentation of biomass to fuel alcohol, US, 11441392.

Lewis, R.S., Frankman, A., Tanner, R.S., Ahmed, A., Huhnke, R.L., 2008. Ethanol via biomass-generated syngas. Int. Sugar J. 110(1311), 150-155.

Liew, F.M., Köpke, M., Simpson, S.D., 2013. Gas fermentation for commercial biofuels production, (Eds.) Fang, P.Z., IntechOpen. Rijeka, Croatia, pp. 125-173.

Lin, Y., Tanaka, S., 2006. Ethanol fermentation from biomass resources: Current state and prospects. Appl. Microbiol. Biotechnol. 69(6), 627-642.

Liou, J.S.C., Balkwill, D.L., Drake, G.R., Tanner, R.S., 2005. *Clostridium carboxidivorans* sp. Nov., a solvent-producing *clostridium* isolated from an agricultural settling lagoon, and reclassification of the acetogen *Clostridium scatologenes* strain sl1 as *Clostridium drakei* sp. Nov. Int. J. Syst. Evol. Microbiol. 55(5), 2085-2091.

Liu, K., Atiyeh, H.K., Tanner, R.S., Wilkins, M.R., Huhnke, R.L., 2012. Fermentative production of ethanol from syngas using novel moderately alkaliphilic strains of *Alkalibaculum bacchi*. Bioresour. Technol. 104, 336-341.

Liu, K., Atiyeh, H.K., Stevenson, B.S., Tanner, R.S., Wilkins, M.R., Huhnke, R.L., 2014a. Continuous syngas fermentation for the production of ethanol, n-propanol and n-butanol. Bioresour. Technol. 151, 69-77.

Liu, K., Atiyeh, H.K., Stevenson, B.S., Tanner, R.S., Wilkins, M.R., Huhnke, R.L., 2014b. Mixed culture syngas fermentation and conversion of carboxylic acids into alcohols. Bioresour. Technol. 152, 337-346.

Liu, K., Atiyeh, H.K., Pardo-Planas, O., Ezeji, T.C., Ujor, V., Overton, J.C., Berning, K., Wilkins, M.R., Tanner, R.S., 2015a. Butanol production from hydrothermolysis-pretreated switchgrass: Quantification of inhibitors and detoxification of hydrolyzate. Bioresour. Ttechnol. 189, 292-301.

Liu, K., Atiyeh, H.K., Pardo-Planas, O., Ramachandriya, K.D., Wilkins, M.R., Ezeji, T.C., Ujor, V., Tanner, R.S., 2015b. Process development for biological production of butanol from eastern redcedar. Bioresour. Technol. 176, 88-97.

Ljungdhal, L., 1986. The autotrophic pathway of acetate synthesis in acetogenic bacteria. Annu. Rev.Microbiol. 40(1), 415-450.

Lynd, L.R., Weimer, P.J., Van Zyl, W.H., Pretorius, I.S., 2002. Microbial cellulose utilization: Fundamentals and biotechnology. Microbiol. Mol Biol. R. 66(3), 506-577.

Lynd, L.R., Zyl, W.H.v., McBride, J.E., Laser, M., 2005. Consolidated bioprocessing of cellulosic biomass: An update. Curr. Opin. Biotechnol. 16(5), 577-583.

Lynd, L.R., Laser, M.S., Bransby, D., Dale, B.E., Davison, B., Hamilton, R., Himmel, M., Keller, M., McMillan, J.D., Sheehan, J., 2008. How biotech can transform biofuels. Nat. Biotechnol. 26(2), 169-172.

Maddipati, P., Atiyeh, H.K., Bellmer, D.D., Huhnke, R.L., 2011. Ethanol production from syngas by *Clostridium* strain p11 using corn steep liquor as a nutrient replacement to yeast extract. Bioresour. Technol. 102(11), 6494-6501.

Madigan, M., Martinko, J., Parker, J. 2003. Brock biology of microorganisms. Prentice Hall, New Jersey, pp. 1019.

Margeot, A., Hahn-Hagerdal, B., Edlund, M., Slade, R., Monot, F., 2009. New improvements for lignocellulosic ethanol. Curr. Opin. Biotechnol. 20(3), 372-380.

Mata, T.M., Martins, A.A., Caetano, N.S., 2010. Microalgae for biodiesel production and other applications: A review. Renew. Sustainable Energy Rev. 14(1), 217-232.

Menetrez, M.Y., 2014. Meeting the us renewable fuel standard: A comparison of biofuel pathways. Biofuel Res. J. 1(4), 110-122.

Mohammadi, M., Najafpour, G.D., Younesi, H., Lahijani, P., Uzir, M.H., Mohamed, A.R., 2011. Bioconversion of synthesis gas to second generation biofuels: A review. Renew. Sustainable Energy Rev. 15(9), 4255-4273.

Mohammadi, M., Younesi, H., Najafpour, G., Mohamed, A.R., 2012. Sustainable ethanol fermentation from synthesis gas by *Clostridium ljungdahlii* in a continuous stirred tank bioreactor. J. Chem. Technol. Biotechnol. 87(6), 837-843.

Mood, S.H., Golfeshan, A.H., Tabatabaei, M., Jouzani, G.S., Najafi, G.H., Gholami, M., Ardjmand, M., 2013. Lignocellulosic biomass to bioethanol, a comprehensive review with a focus on pretreatment. Renew. Sustainable Energy Rev. 27, 77-93.

Munasinghe, P.C., Khanal, S.K., 2010a. Biomass-derived syngas fermentation into biofuels: Opportunities and challenges. Bioresour. Technol. 101(13), 5013-5022.

Munasinghe, P.C., Khanal, S.K., 2010b. Syngas fermentation to biofuel: Evaluation of carbon monoxide mass transfer coefficient ($k_La$) in different reactor configurations. Biotechnol. Progr. 26(6), 1616-1621.

Munasinghe, P.C., Khanal, S.K. 2014. Evaluation of hydrogen and carbon monoxide mass transfer and a correlation between the myoglobin-protein bioassay and gas chromatography method for carbon monoxide determination. RSC Adv. 4(71), 37575-37581.

Mussatto, S.I., Dragone, G., Guimarães, P.M., Silva, J.P.A., Carneiro, L.M., Roberto, I.C., Vicente, A., Domingues, L., Teixeira, J.A., 2010. Technological trends, global market, and challenges of bio-ethanol production. Biotechnol. Adv. 28(6), 817-830.

Naik, S., Goud, V.V., Rout, P.K., Dalai, A.K., 2010. Production of first and second generation biofuels: A comprehensive review. Renew. Sustainable Energy Rev. 14(2), 578-597.

Najafpour, G., Younesi, H., 2006. Ethanol and acetate synthesis from waste gas using batch culture of *Clostridium ljungdahlii*. Enzyme Microb. Technol. 38(1–2), 223-228.

Nayak, B.K., Roy, S., Das, D., 2014. Biohydrogen production from algal biomass (anabaena sp. Pcc 7120) cultivated in airlift photobioreactor. Int. J. Hydrogen Energ. 39(14), 7553-7560.

Nigam, P., Singh, D., 1995. Enzyme and microbial systems involved in starch processing. Enzyme Microb. Technol. 17(9), 770-778.

Öhgren, K., Bengtsson, O., Gorwa-Grauslund, M.F., Galbe, M., Hahn-Hägerdal, B., Zacchi, G., 2006. Simultaneous saccharification and co-fermentation of glucose and xylose in steam-pretreated corn stover at high fiber content with *Saccharomyces cerevisiae* tmb3400. J. Biotechnol. 126(4), 488-498.

Orgill, J.J., Atiyeh, H.K., Devarapalli, M., Phillips, J.R., Lewis, R.S., Huhnke, R.L., 2013. A comparison of mass transfer coefficients between trickle-bed, hollow fiber membrane and stirred tank reactors. Bioresour. Technol. 133, 340-346.

Pandey, A., 2010. Handbook of plant-based biofuels. CRC Press, Boca Raton, FL, pp. 312.

Panneerselvam, A., Wilkins, M., DeLorme, M., Atiyeh, H., Huhnke, R., 2009. Effects of various reducing agents on syngas fermentation by *Clostridium ragsdalei*. Biol. Eng. 2(3), 135-144.

Park, H.R., Jung, K.A., Lim, S.-R., Park, J.M., 2014. Quantitative sustainability assessment of seaweed biomass as bioethanol feedstock. Bioenergy Res. 7(3), 974-985.

Perales, A.V., Valle, C.R., Ollero, P., Gómez-Barea, A., 2011. Technoeconomic assessment of ethanol production via thermochemical conversion of biomass by entrained flow gasification. Energy. 36(7), 4097-4108.

Percival Zhang, Y.H., Himmel, M.E., Mielenz, J.R., 2006. Outlook for cellulase improvement: Screening and selection strategies. Biotechnol. Adv. 24(5), 452-481.

Perez, J.M., Richter, H., Loftus, S.E., Angenent, L.T., 2013. Biocatalytic reduction of short-chain carboxylic acids into their corresponding alcohols with syngas fermentation. Biotechnol. Bioeng. 110(4), 1066-1077.

Pessani, N.K., Atiyeh, H.K., Wilkins, M.R., Bellmer, D.D., Banat, I.M., 2011. Simultaneous saccharification and fermentation of kanlow switchgrass by thermotolerant *Kluyveromyces marxianus* imb3: The effect of enzyme loading, temperature and higher solid loadings. Bioresour. Technol. 102(22), 10618-10624.

Phillips, J., Atiyeh, H., Huhnke, R., 2014. Method for design of production medium for fermentation of synthesis gas to ethanol by acetogenic bacteria. Biolog. Eng. Trans. 7(3), 113-128.

Phillips, J.R., Klasson, T.K., Clausen, E.C., Gaddy, J.L., 1993. Biological production of ethanol from coal synthesis gas. Appl. Biochem. Biotechnol. 39-40(1), 559-571.

Phillips, J.R., Clausen, E.C., Gaddy, J.L., 1994. Synthesis gas as substrate for the biological production of fuels and chemicals. Appl. Biochem. Biotechnol. 45(1), 145-157.

Phillips, J.R., Atiyeh, H.K., Tanner, R.S., Torres, J.R., Saxena, J., Wilkins, M.R., Huhnke, R.L., 2015. Butanol and hexanol production in *Clostridium carboxidivorans* syngas fermentation: Medium development and culture techniques. Bioresour. Technol. 190, 114-121.

POET-DSM., 2014. First commercial-scale cellulosic ethanol plant in the U.S. Opens for business, POET-DSM Advanced Biofuels, LLC. Emmetsburg, Iowa. Sept. 3rd, 2014. Available at http://www.dsm.com/corporate/media/informationcenter-news/2014/09/29-14-first-commercial-scale-cellulosic-ethanol-plant-in-the-united-states-open-for-business.html (accessed on 10 August 2015).

Posten, C., Schaub, G., 2009. Microalgae and terrestrial biomass as source for fuels—a process view. J. Biotechnol. 142(1), 64-69.

Ragsdale, S.W., 1997. The eastern and western branches of the wood/ljungdahl pathway: How the east and west were won. Biofactors. 6(1), 3-11.

Rajagopalan, S., P. Datar, R., Lewis, R.S., 2002. Formation of ethanol from carbon monoxide via a new microbial catalyst. Biomass Bioenergy. 23(6), 487-493.

Ramachandriya, K.D., Wilkins, M.R., Delorme, M.J., Zhu, X., Kundiyana, D.K., Atiyeh, H.K., Huhnke, R.L., 2011. Reduction of acetone to isopropanol using producer gas fermenting microbes. Biotechnol. Bioeng. 108(10), 2330-2338.

Rao, G., Mutharasan, R., 1986. Alcohol production by *Clostridium acetobutylicum* induced by methyl viologen. Biotechnol. Lett. 8(12), 893-896.

Rao, G., Ward, P., Mutharasan, R., 1987. Manipulation of end-product distribution in strict anaerobes. Ann. N.Y. Acad. Sci. 506(1), 76-83.

Rathmann, R., Szklo, A., Schaeffer, R., 2010. Land use competition for production of food and liquid biofuels: An analysis of the arguments in the current debate. Renew. Energy. 35(1), 14-22.

Rausch, K., Belyea, R., 2006. The future of coproducts from corn processing. Appl. Biochem. Biotechnol. 128(1), 47-86.

Richter, H., Martin, M.E., Angenent, L.T., 2013. A two-stage continuous fermentation system for conversion of syngas into ethanol. Energies. 6(8), 3987-4000.

Riggs, S.S., Heindel, T.J., 2006. Measuring carbon monoxide gas—liquid mass transfer in a stirred tank reactor for syngas fermentation. Biotechnol. Progr. 22(3), 903-906.

Rodríguez, L.F., Li, C., Khanna, M., Spaulding, A.D., Lin, T., Eckhoff, S.R., 2010. An engineering and economic evaluation of quick germ–quick fiber process for dry-grind ethanol facilities: Analysis. Bioresour. Technol. 101(14), 5282-5289.

Sakai, S., Nakashimada, Y., Yoshimoto, H., Watanabe, S., Okada, H., Nishio, N., 2004. Ethanol production from $H_2$ and $CO_2$ by a newly isolated thermophilic bacterium, *Moorella sp.* Huc22-1. Biotechnol. Lett. 26(20), 1607-1612.

Savage, M.D., Wu, Z., Daniel, S.L., Lundie, J., Leon, L., Drake, H.L., 1987. Carbon monoxide-dependent chemolithotrophic growth of *Clostridium thermoautotrophicum*. Appl. Environ. Microbiol. 53, 1902-1906.

Saxena, J., 2008. Development of an optimized and cost-effective medium for ethanol production by *Clostridium* strain P11, in: *Department of Botany and Microbiology*, Ph.D. Dissertation, University of Oklahoma. Ann Arbor, pp. 131.

Saxena, J., Tanner, R.S., 2011. Effect of trace metals on ethanol production from synthesis gas by the ethanologenic acetogen, *Clostridium ragsdalei*. *J. Ind.* Microbiol. Biotechnol. 38(4), 513-521.

Searchinger, T., Heimlich, R., Houghton, R.A., Dong, F., Elobeid, A.,

Fabiosa, J., Tokgoz, S., Hayes, D., Yu, T.H., 2008. Use of US croplands for biofuels increases greenhouse gases through emissions from land-use change. Science. 319(5867), 1238-1240.

Shen, Y., Brown, R., Wen, Z., 2014a. Syngas fermentation of *Clostridium carboxidivoran* P7 in a hollow fiber membrane biofilm reactor: Evaluating the mass transfer coefficient and ethanol production performance. Biochem. Eng. J. 85, 21-29.

Shen, Y., Brown, R., Wen, Z., 2014b. Enhancing mass transfer and ethanol production in syngas fermentation of *Clostridium carboxidivorans* P7 through a monolithic biofilm reactor. Appl. Energy. 136, 68-76.

Sherwood, T.K., Pigford, R.L., Wilke, C.R., 1975. *Mass transfer*. McGraw-Hill Inc., New York, pp. 512.

Sim, J.H., Kamaruddin, A.H., 2008. Optimization of acetic acid production from synthesis gas by chemolithotrophic bacterium– *Clostridium aceticum* using statistical approach. Bioresour. Technol. 99(8), 2724-2735.

Sims, R., Taylor, M., Saddler, J., Mabee, W., 2008. From 1$^{st}$-to 2$^{nd}$-generation biofuel technologies: An overview of current industry and RD&D activities, International Energy Agency. Paris, France, pp. 120. Available at http://environmentportal.in/files/2nd_Biofuel_Gen.pdf (accessed on 10 August 2015).

Skidmore, B.E., 2010. Syngas fermentation: Quantification of assay technoques, reaction kinetics and pressure dependencies of the *Clostridium* p11 hydrogenase, in: *Department of Chemical Engineering*, M.S. Thesis, Bringham Young University, pp. 136.

Slepova, T.V., Sokolova, T.G., Lysenko, A.M., Tourova, T.P., Kolganova, T.V., Kamzolkina, O.V., Karpov, G.A., Bonch-Osmolovskaya, E.A., 2006. *Carboxydocella sporoproducens* sp. Nov., a novel anaerobic CO-utilizing/$H_2$-producing thermophilic bacterium from a kamchatka hot spring. Int. J. Syst. Evol. Microbiol. 56(4), 797-800.

Stöcker, M., 2008. Biofuels and biomass-to-liquid fuels in the biorefinery: Catalytic conversion of lignocellulosic biomass using porous materials. Angew. Chem. Int. Ed. 47(48), 9200-9211.

Sun, Y., Cheng, J., 2002. Hydrolysis of lignocellulosic materials for ethanol production: A review. Bioresour. Technol. 83(1), 1-11.

Szczodrak, J., Fiedurek, J., 1996. Technology for conversion of lignocellulosic biomass to ethanol. Biomass Bioenergy. 10(5), 367-375.

Tanner, R.S., Miller, L.M., Yang, D., 1993. *Clostridium ljungdahlii* sp. Nov., an acetogenic species in clostridial rrna homology group i. Int. J Syst. Bacteriol. 43(2), 232-236.

Tanner, R.S., 2008. Production of ethanol from synthesis gas, in: Wall, J., Harwood, C., Demain A. (Eds.), Bioenergy. ASM Press, pp. 147-151.

Terrill, J., Wilkins, M., DeLorme, M., Atiyeh, H., Lewis, R., 2012. Effect of energetic gas composition on hydrogenase activity and ethanol production in syngas fermentation by C*lostridium ragsdalei*. Biol. Eng. Trans. 8, 87-96.

Tilman, D., Hill, J., Lehman, C., 2006. Carbon-negative biofuels from low-input high-diversity grassland biomass. Science. 314(5805), 1598-1600.

Twidell, J., Weir, T., 2003. Renewable energy resources. Taylor and Francis Group, New York, pp. 601.

Ukpong, M.N., Atiyeh, H.K., De Lorme, M.J., Liu, K., Zhu, X., Tanner, R.S., Wilkins, M.R., Stevenson, B.S., 2012. Physiological response of *Clostridium carboxidivorans* during conversion of synthesis gas to solvents in a gas-fed bioreactor. Biotechnol. Bioeng. 109(11), 2720-2728.

Ungerman, A.J., Heindel, T.J., 2007. Carbon monoxide mass transfer for syngas fermentation in a stirred tank reactor with dual impeller configurations. Biotechnol. Progr. 23(3), 613-620.

USGC, 2012. Ddgs user handbook, US Grains Council. Washington, DC, pp. 406. Available at http://www.ethanolrfa.org/page/-/rfa-association-site/studies/2012_DDGS_Handbook.pdf?nocdn=1 (accessed on 10 August 2015).

Vega, J.L., Prieto, S., Elmore, B.B., Clausen, E.C., Gaddy, J.L., 1989. The biological production of ethanol from synthesis gas. Appl. Biochem. Biotechnol. 20(1), 781-797.

Vega, J.L., Clausen, E.C., Gaddy, J.L., 1990. Design of bioreactors for coal synthesis gas fermentations. Resour. Conserv. Recycl. 3(2), 149-160.

White, H., Lebertz, H., Thanos, I., Simon, H., 1987. *Clostridium thermoaceticum* forms methanol from carbon monoxide in the presence of viologen dyes. *FEMS* Microbiol. Lett. 43(2), 173-176.

Wilkins, M.R., Atiyeh, H.K., 2011. Microbial production of ethanol from carbon monoxide. Curr. Opin. Biotechnol. 22(3), 326-330.

Wilkins, M.R., Atiyeh, H., 2012. Fermentation, in: Dunford, N.T. (Eds.), Food and industrial bioproducts and bioprocessing. John Wiley & Sons, Inc. Ames, Iowa, pp. 185.

Wood, H.G., Ragsdale, S.W., Pezacka, E., 1986. The acetyl-coa pathway of autotrophic growth. FEMS Microbiol. Lett. 39(4), 345-362.

Woolcock, P.J., Brown, R.C., 2013. A review of cleaning technologies for biomass-derived syngas. Biomass Bioenergy. 52, 54-84.

Worden, R.M., Grethlein, A.J., Jain, M.K., Datta, R., 1991. Production of butanol and ethanol from synthesis gas via fermentation. Fuel. 70(5), 615-619.

Wright, J., Wyman, C., Grohmann, K., 1988. Simultaneous saccharification and fermentation of lignocellulose. Appl. Biochem. Biotechnol. 18(1), 75-90.

Xu, D., Tree, D.R., Lewis, R.S., 2011. The effects of syngas impurities on syngas fermentation to liquid fuels. Biomass Bioenergy. 35(7), 2690-2696.

Xu, D., Lewis, R.S., 2012. Syngas fermentation to biofuels: Effects of ammonia impurity in raw syngas on hydrogenase activity. Biomass Bioenergy. 45, 303-310.

Yasin, M., Park, S., Jeong, Y., Lee, E.Y., Lee, J., Chang, I.S., 2014. Effect of internal pressure and gas/liquid interface area on the CO mass transfer coefficient using hollow fibre membranes as a high mass transfer gas diffusing system for microbial syngas fermentation. Bioresour. Technol. 169, 637-643.

Younesi, H., Najafpour, G., Mohamed, A.R. 2005. Ethanol and acetate production from synthesis gas via fermentation processes using anaerobic bacterium, *Clostridium ljungdahlii*. Biochem. Eng. J. 27(2), 110-119.

Zabriskie, D.W., Mill, T.O., 1988. Traders guide to fermentation media formulation. Traders Protein, pp. 60.

Zahn, J.A., Saxena, J., 2012. Ethanologenic clostridium species, *Clostridium coskatii*. US, 8143037.

Zeikus, J., 1980. Chemical and fuel production by anaerobic bacteria. Annu. Rev.Microbiol. 34(1), 423-464.

Zhang, M., Eddy, C., Deanda, K., Finkelstein, M., Picataggio, S., 1995. Metabolic engineering of a pentose metabolism pathway in ethanologenic *Zymomonas mobilis*. Science. 267(5195), 240-243.

# Effect of extrusion conditions and hydrolysis with fiber-degrading enzymes on the production of C5 and C6 sugars from brewers' spent grain for bioethanol production

Erick Heredia-Olea, Esther Pérez-Carrillo, Sergio O. Serna-Saldívar*

*Centro de Biotecnología FEMSA, Escuela de Ingeniería y Ciencias, Tecnológico de Monterrey, Avenida Eugenio Garza Sada 2501 Sur, CP 64849, Monterrey, Nuevo León, México.*

## HIGHLIGHTS

- Thermoplastic extrusion improved the enzymatic hydrolysis of brewers' spent grain.
- The extruder barrel temperature and screw speed affected the sugars yield.
- No enzymatic and yeast inhibitors were detected in all enzymatic hydrolyzates.
- Despite the high protein content after hydrolysis, low levels of FAN were achieved.
- *S. cerevisiae* fermented glucose into ethanol with a maximum yield after 48 h.

## GRAPHICAL ABSTRACT

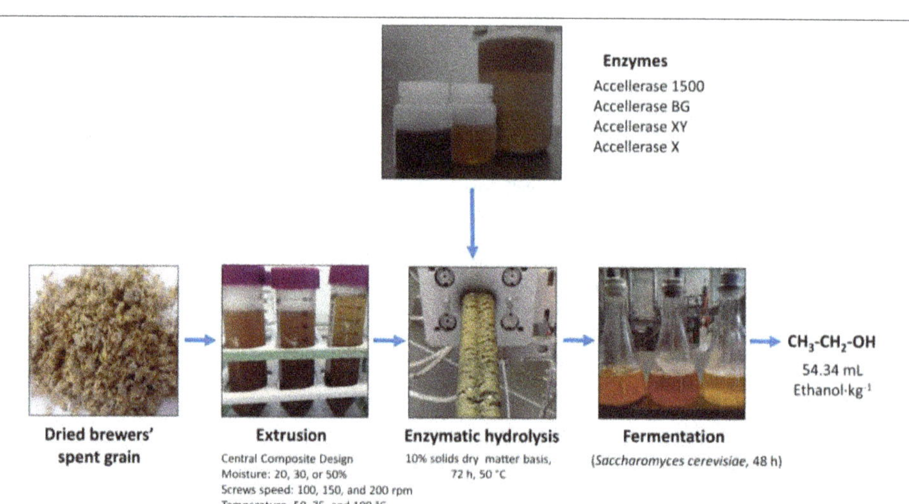


## ABSTRACT

The bioconversion of brewers' spent grain into bioethanol was investigated in the present study using thermoplastic extrusion and the use of fiber degrading enzymes. The extrusion conditions i.e. tempering moisture, screws speed, and temperature of last zone of the barrel were taken into account in order to optimize the yield of C5 and C6 sugars during the subsequent enzymatic hydrolysis step of the fibers. The most important variable that affected the sugar yield was the extrusion temperature, followed by the screws speed. The best extrusion conditions were 20% tempering moisture, 200 rpm and 50 °C. No enzymatic and yeast inhibitors were detected in any of the enzymatically-treated fiber hydrolyzates. The fermentation resulted in 5.43 mL bioethanol per 100g of extruded brewers' spent grain (dry weight basis). The only sugar consumed was glucose. The free amino nitrogen amount quantified in the hydrolyzates was as low as >20 mg $L^{-1}$, negatively affecting sugars consumption during the fermentation and consequently the ethanol yield.

**Keywords:**
Brewers' spent grain
Extrusion
Bioethanol
Fermentation
Free amino nitrogen

* Corresponding author
E-mail address: sserna@itesm.mx

## 1. Introduction

Brewers' spent grain (BSG) is the most abundant byproduct generated in the brewing industry comprising approximately 85% of the total waste materials (Aliyu and Bala, 2011). BSG is separated by lautering after mashing or starch extraction. This byproduct is composed primarily of protein and lignocellulose. The main use of BSG is for animal feed, particularly for ruminants. Moreover, since BSG is rich in protein and fibers, it has been used successfully as an ingredient for production of high-fiber snacks and breads and as a feedstock for protein hydrolysis and fractionation. It has also been used for the extraction of prebiotics rich in xylo-oligosaccharides (Forssell et al., 2008).

Shindo and Tachibana (2006) and White et al. (2008) investigated the conversion of BSG into bioethanol with steam explosion and acid hydrolysis, respectively. Both investigations were able to convert the pretreated feedstock into C5 and C6 sugars after hydrolysis with fiber-degrading enzymes. They proved the possibility of using the BSG as a potential feedstock for the production of second-generation bioethanol. However, most of the pretreatment technologies reported have also resulted in the production of enzyme and yeast inhibitors due to the harsh process conditions applied (Karunanithy and Muthukumarappan, 2010). Extrusion has been reported to provide a continuous thermophysical pretreatment for lignocellulose bioethanol production withough leading to hazardous materials generation (Lin et al., 2012). In addition, there is no published report regarding the use of extruded BSG for bioethanol production. The objective of this research was to study the extrusion conditions (i.e. temperature of the last zone of the barrel, screws speed, and feedstock moisture content) for preparation of BSG for its subsequent enzymatic hydrolysis with fiber-degrading enzymes for second-generation bioethanol production.

## 2. Material and method

### *2.1. Materials*

BSG was procured from Grupo Cuauhtémoc Moctezuma (Monterrey, México). The wet BSG was transported to the Tecnológico de Monterrey campus Monterrey and dried at 50-60 °C for 24 h.

### *2.2. Physical-Chemical characterization*

Moisture and protein contents were determined using the AACC standard assays 44-15 and 46-13, respectively. For the structural carbohydrates assay, the non-structural material was removed from the BSG to prevent any interferences. More specifically, a two-step extraction process was used to remove water and ethanol solubles according to the methods recommended by the National Renewable Energy Laboratory (NREL) (Sluiter et al., 2008a). Then, the insoluble fibers were hydrolyzed and filtered for structural sugars quantification with an high-performance liquid chromatography (HPLC) as described by Sluiter et al. (2008b).

### *2.3. Extrusion pretreatment*

A twin-screw co-rotating extruder (BTSM-30, Bühler AG, Uzwil, Switzerland) with a barrel composed of 5 zones, and two independent feeders for the solid raw material and water was used. The temperature of the fifth zone of the barrel was controlled by a heat exchanger device (Tool Temp, Bühler AG, Uzwil, Switzerland). The total length and outer diameter of the screws were 800 mm and 30 mm, respectively, and the L/D ratio was 20. A die with a single 4 mm hole was used. The screws configuration consisted of three different sections: inlet/conveying elements section (for the introduction and transport of the dry feedstock and water), mixing elements section, and the final work elements section consisting of kneading and reverse elements. The solid feed rate was set constantly at 7.3 kg $h^{-1}$ for all the conditions tested. The BSG extrusions were conducted as described in the section 2.3.1. The extruded BSG was dried at 50-60 °C for 6 h and stored in plastic bags.

#### *2.3.1. Experimental design and extrusion conditions*

A central composite design was used in the present study. The factors evaluated included the moisture inside the barrel, screws speed and the temperature in the last zone of the barrel. Each factor had a low or high level and a center point with intermediate conditions (**Table 1**). The center point was repeated four times and all treatments were performed in triplicate. As mentioned arlier, whole dried BSG was fed at a constant rate of 7.3 kg $h^{-1}$ for all the conditions investigated.

**Table 1.**
Central composite design used for the extrusion conditions using BSG.

| Level | Extrusion Variables | | |
|---|---|---|---|
| | Moisture [%] | Screws speed [rpm] | Temperature [°C] |
| Low | 20 | 100 | 50 |
| High | 40 | 200 | 100 |
| Center point | 30 | 150 | 75 |

### *2.4. Enzymatic hydrolysis*

After the extrusion, the pretreated dried BSG samples were enzymatically hydrolyzed. Untreated BSG was also enzymatically hydrolyzed as control. The hydrolysis assays were conducted in triplicate by using a total volume of 100 mL in 500 mL flasks in the presence of 10% solid BSG (dry matter basis). Citrate buffer (50 mM) was adjusted to pH 5 with 10 mM sodium azide and was used for all the hydrolyses. Enzymatic hydrolysis was carried out in an orbital shaking incubator (VWR Model 1575) set at 50 °C and 150 rpm for 72 h. Du Pont food grade enzymes i.e. Accellerase® 1500, Accellerase® BG, Accellerase® XC, and Accellerase® XY dosed at 0.25, 0.9, 0.05, and 0.125 mL $g^{-1}$ loaded solids were used, respectively. These enzymes consist in a fibrolytic mixture of exoglucanase, endoglucanase, beta-glucosidase, hemicellulase, and xylanase. The declared activity of each enzyme was as follows: for Accellerase® 1500: 2,200-2,800 carboxymethylcellulose activity units (CMC U)/g and 450-775 para-nitrophenyl-B-D-glucopyranoside units (pNPG U)/g, for Accellerase® BG: 3,000 para-nitrophenyl-B-D-glucopyranoside units (pNPG U)/g, for Accellerase® XC: 1,000-1,400 carboxymethylcellulose activity units (CMC U)/g and 2,500-3,800 acid birchwod xylanase units (ABXU)/g, and for Accellerase® XY: 20,000-30,000 acid birchwod xylanase units (ABXU)/g. The dosage used was the maximum enzyme concentration as recommended by the manufacturer.

### *2.5. Fermentation*

*Saccharomyces cerevisiae* ATCC® 20252™ capable of fermenting both C5 and C6 sugars was used. The extruded BSG at 20% tempering moisture, 200 rpm and 50°C was enzymatically hydrolyzed and fermented with the *S. cerevisiae* strain without nutrients supplementation. These hydrolyzates contained the highest sugars concentrations. The fermentations were conducted as recommended by Heredia-Olea et al. (2013). An aliquot of 15 mL at 0, 24 and 48 h was taken and stored for HPLC analysis (sugars and ethanol).

### *2.6. HPLC-based quantification of sugars and inhibitors*

The enzymatically-hydrolyzed samples were treated as described in our previous research (Heredia-Olea et al., 2012). Analytes were separated by a Shodex SH1011 column (300 × 7.8 mm) with a flow rate of 0.6 mL $min^{-1}$ of HPLC-grade water containing 5 mM $H_2SO_4$ for the quantification of inhibitors and ethanol. The sugar quantification was performed with a Shodex SP0810 column and a cation/anion deasher (Biorad). The column temperature was set at 60 and 85 °C for inhibitors and sugars, respectively, whereas the detector (refractive index detector Waters 2414) and the autosampler at 50 and 4°C, respectively. Standards of ethanol, cellobiose, D-glucose, D-xylose, L-arabinose, D-mannose, D-galactose, acetic acid, 5-hidroxymethylfurfural, and furfural (Sigma Chemical Co. St. Louis, MO) were used. The run times for sugars an inhibitors quantifications were 20 and 45 min, respectively.

#### *2.6.1. Free amino nitrogen determination*

A 100 mL aliquot was taken from each filtered hydrolyzate. Free amino nitrogen was quantified with the ninhydrin reaction assay (Lie, 1973).

*2.7. Statistical analysis*

The analysis of data generated by the central composite design and the response surfaces were performed with the software Statgraphics Centurion XVI with a statistical significance of a=0.05.

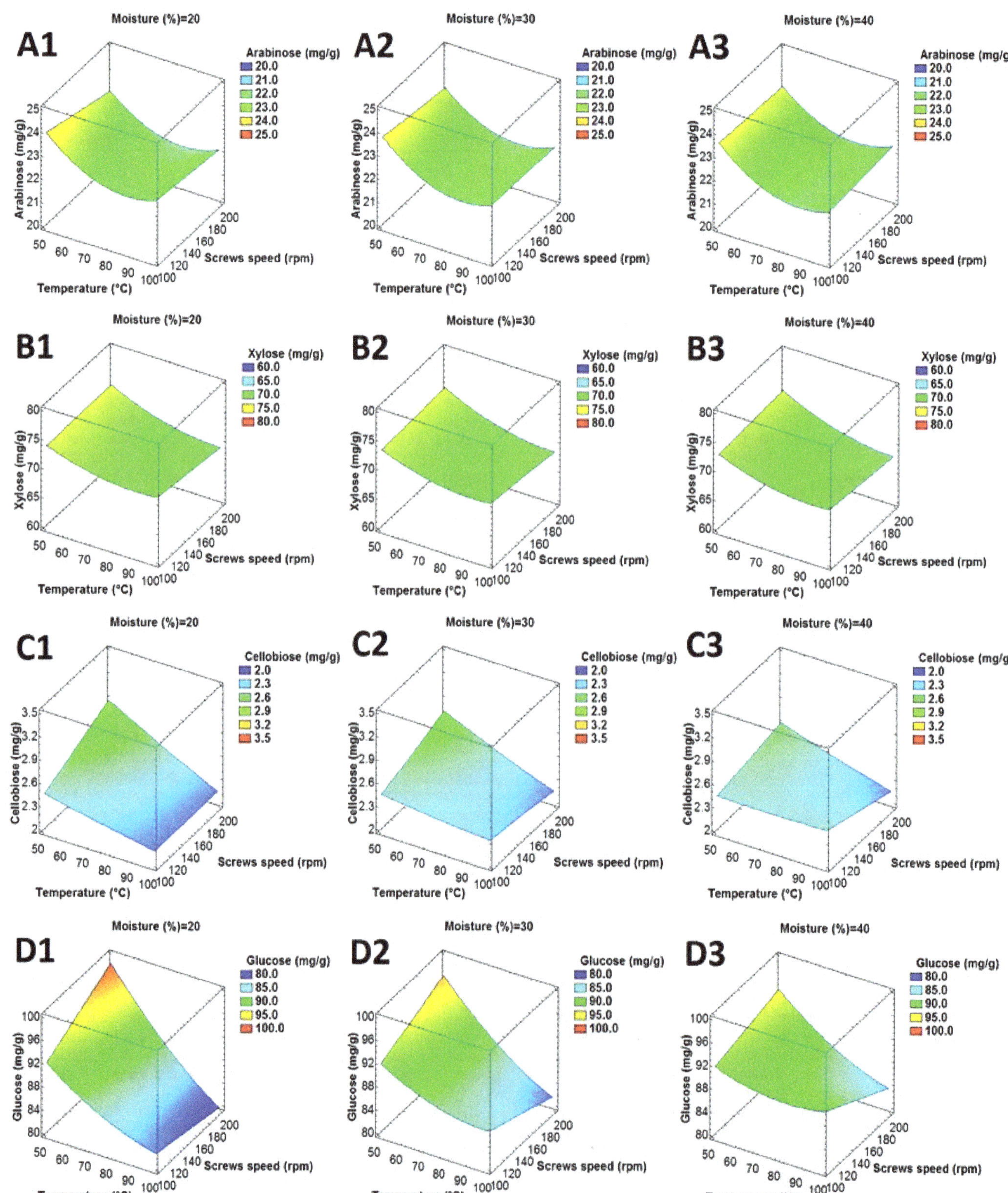


**Fig.1.** Response surfaces for arabinose (A1, A2, and A3), xylose (B1, B2, and B3), cellobiose (C1, C2, and C3), and glucose (D1, D2, and D3) generated from brewers' spent grain extruded at different temperatures (50, 75, and 100 °C), screws speeds (100, 150, and 200 rpm) and tempering moistures (20, 30, or 40 %) ad was further hydrolyzed with fiber-degrading enzymes. Sugars are expressed in mg $g^{-1}$ BSG (dry weight basis).

## 3. Results and discussion

### *3.1. Chemical characterization*

The physical-chemical characteristics of the BSG are tabulated in Table 2. BSG contained barley husk, pericarp and seed coat remains rich in polysaccharides and was not starchy. The BSG contained 36.59% structural sugars, out of which 18.03% was glucans and 18.56 C5 sugars (arabinans+xylans). Mannose and galactose were not detected. Such sugar composition indicated that both cellulose and hemicellulose were present in approximately equal amounts. Forssell et al. (2008) reported 45% carbohydrates and 21.5% protein in BSG. The BSG tested herein contained 41.15% carbohydrates (as sugars) and 22.77% protein. In a previously work, White et al. (2008) found a similar sugars content proportion in BSG. However, they reported higher xylan and lignin contents. The composition differences could be ascribed to differences in barley genotype, harvest time, and the malting and mashing conditions employed during brewing (Mussatto et al., 2006; Forssell et al., 2008).

### *3.2. Extrusion effect*

#### *3.2.1. C5 sugars hydrolysis*

The enzymatic release of both C5 sugars i.e. arabinose and xylose, was only temperature dependent (**Figure 1 A and B series**). The temperature was the only factor that affected the release of the C5 sugars embedded in the hemicellulose matrix. The yield of arabinose was nearly described by a quadratic modeling with a temperature dependent correlation coefficient of $R^2$=58.09% (**Fig. 1 A1, A2, and A3**). On the other hand, xylose had a linear modeling (**Fig. 1 B1, B2, and B3**) with a correlation coefficient of $R^2$=41.48%. In both cases, the maximum sugar amounts were released at 50 °C. At this temperature, 30.6 and 57.1% of the total arabinose and xylose were released, respectively. Normally, high pretreatment temperatures lead to higher hemicellulose hydrolysis rates making cellulose more accessible to enzymes (White et al., 2008). Karunanithy et al. (2013) found that extrusion temperatures of around 100°C had a positive effect over hemicellulose degradation into monomeric sugars by the following enzymatic hydrolysis, but when the extrusion temperature was increased less arabinose and xylose were enzymatically hydrolyzed. Although both sugars were generated after the enzymatic hydrolysis, apparently they were not significanly affected by the screws speed and tempering moisture. Nontheless, these extrusion parameters must be controlled because the temperature in the last zone of the barrel is affected by these factors (Moscicki, 2011).

#### *3.2.2. C6 sugars hydrolysis*

Based on the BSG composition (**Table 2**), glucose and the dimer cellobiose were the C6 sugars that were released after the enzymatic hydrolysis step. For the cellobiose hydrolysis, the temperature (P value=0.000), the interactions between temperature-screws speed (P value=0.001), temperature-moisture (P value=0.024), and screws speed-moisture (P value=0.044) were statistically significant. The cellobiose modeling was fitted as a linear regression ($R^2$=75.89%). The cellobiose released did not exceed 2.8 mg $g^{-1}$ of BSG (dwb). Beside the temperature, the interaction with the tempering moisture also had a negative effect over the release of cellobiose (**Fig. 1 C1, C2, and C3**). For the glucose, the modeling was fitted like a linear correlation ($R^2$=78.22%). The significant variables were the temperature (P value=0.000) and the interactions between temperature-screws speed (P value=0.003) and temperature-moisture (P value=0.011). Similar to cellobiose, the interaction between temperature-moisture significantly affected the generation of glucose. Interestingly, the application of higher tempering moistures decreased the glucose enzymatic recovery (**Fig. 1 D1, D2, and D3**).

Using the combination of 50 °C, 200 rpm, and 20% feedstock moisture generated the maximum glucose (98.51 mg·$g^{-1}$) which represented a total recovery of about 47.8%. In fact, the high hemicellulose and lignin contents (especially the latter for it absorbs cellulosic enzymes) negatively affected the enzymatic hydrolysis of cellulose and the removal of these components could result in higher glucose release (Mussatto et al., 2008). Forrssell et al. (2008) reached the same conclusion, suggesting an acid or alkaline treatment to enhance de hemicellulose hydrolysis or lignin removal. It is important to mention that the chemical hydrolysis could also lead to the production of inhibitory compounds which need to be removed before enzymatic hydrolysis and fermentation.

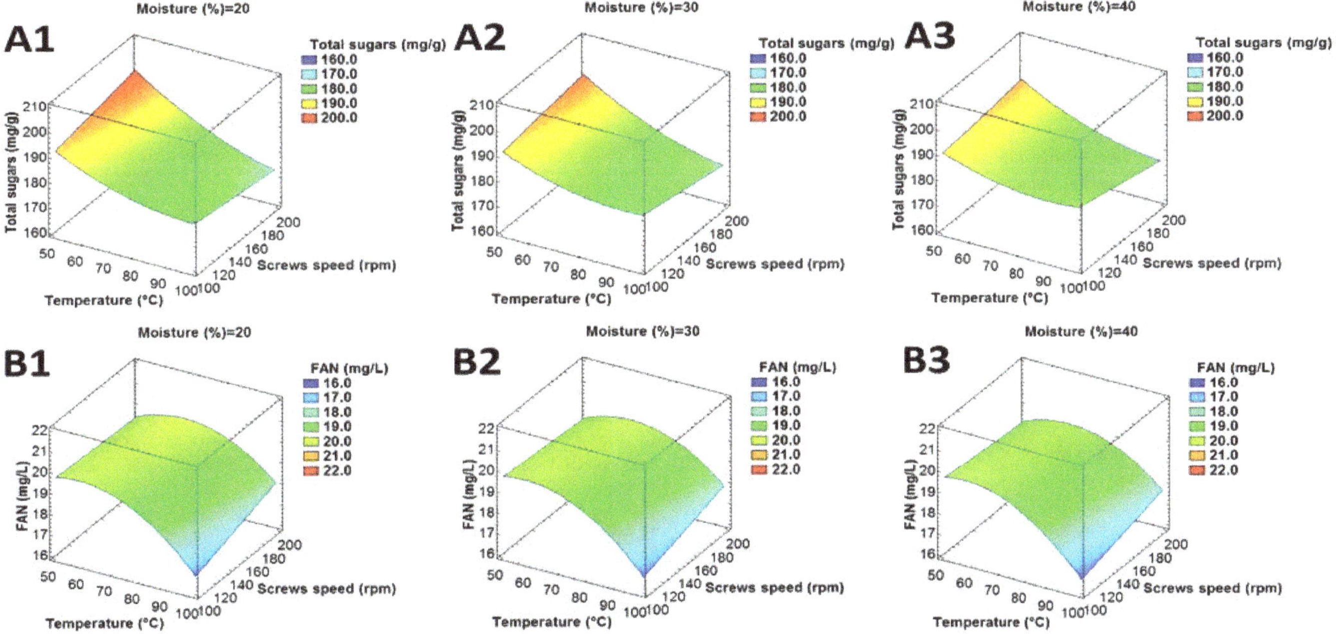


**Fig.2.** Response surfaces for total sugars (A1, A2, and A3) and free amino nitrogen (B1, B2, and B3) generated from brewers' spent grain extruded at different temperatures (50, 75, or 100 °C), screws speeds (100, 150, or 200 rpm) and operating moistures (20, 30, or 40%) that were further hydrolyzed with fiber degrading enzymes. Sugars are expressed in milligrams $g^{-1}$ of BSG and FAN is expressed in milligrams $L^{-1}$ (dry weight).

**Table 2.**
Physical-chemical characteristics of brewer's spent grain (proximate, fiber and carbohydrates composition analyses).

| Component | Amount (%) ± SD[1] |
|---|---|
| Protein | 22.77 ±0.02 |
| Water extractives | 7.61 ±0.51 |
| Ethanol extractives | 12.63 ±0.36 |
| Glucan | 18.56 ±0.11 |
| Xylan | 11.44 ±0.07 |
| Arabinan | 6.59 ±0.09 |
| Acetyl groups | 0.22 ± 0.01 |
| Acid insoluble lignin | 11.05 ±0.78 |
| Acid soluble lignin | 8.00 ±1.41 |

[1] Mean values are expressed on dry matter basis.

### 3.2.3. Total sugars hydrolysis

For total sugars hydrolysis, a linear regression with a regression coefficient of $R^2$=70.07% was the best model. The statistically significant extrusion variables were temperature (P value=0.000) and the interaction between temperature-screws speed (P value=0.036). The feedstock moisture was relevant for cellobiose and glucose yields. However, the effect of extrusion temperature was minimal in terms of total sugar production (**Fig. 2 A1, A2, and A3**).

The temperature had a negative effect over the total sugar generation. According to Kurananithy and Muthukumarappan (2010), the shear stress during extrusion disrupts the lignocellulose structure of BSG. The use of low temperatures and low tempering moisture caused a better shear stress inside the extruder barrel. The average maximum value was obtained using 50 °C, 200 screws rpm, and 20% moisture. These conditions generated an extruded feedstock that after enzyme hydrolysis resulted in 48.4% sugars based on the total fiber. The control treatment released 33.8% of the total sugars (139.03±2.26 mg total sugars $g^{-1}$ BSG). By using the extrusion pretreatment, the total sugars yield was improved by 14.6%. Forssell et al. (2008) tested various commercial fibrolytic cocktails on BSG and achieved 28% recovery of the total sugars. Likewise, White et al. (2008) treated the BSG with hot water at 121°C and after the hydrolysis with fiber-degrading enzymes achieved 184 mg total sugars $g^{-1}$. This yield was similar to the sugar amount reported in the present research, proving that physical pretreatments had a positive effect towards releasing sugars from BSG cellular walls during the hydrolysis step. The same authors obtained 150, 140 and 70 mg $g^{-1}$ of glucose, xylose and arabinose, respectively, by treating the BSG with diluted $HNO_3$ acid hydrolysis and fiber-degrading enzymes. Although they achieved 1.8 times more sugars, the proportion between C5 and C6 sugars was approximately similar to the ratio of sugars recorded herein.

The complex cell wall structure (cellulose, hemicellulose, lignin, and protein), is the most relevant barrier diminishing enzyme performance and sugar yields (Pandey et al., 2011). Amomng all, lignin is the most important barrier which reduces the releasing of sugars. The B contained about 18% lignin (**Table 2**). In fact, compared to other lignocellulosic byproducts such as corn stover (7%), wheat straw (17%), rice straw (12%), and sorghum bagasse (11%) (Reddy and Yang, 2005), the investigated BSG contained higher lignin content. In a previous work on extruded sweet sorghum bagasse hydrolyzed with fibrolytic enzymes, total sugar yield of 67.5% was accomplished (Heredia-Olea et al., 2013). This could be explained by the fact that the bagasse contained 14% lignin and thus the subsequent enzymatic hydrolysis step performed better. After enzymatic hydrolysis, the proteins remain intact or insoluble, acting as an important barrier against the subsequent steps of bioethanol production (Forsell et al., 2008; Faulds et al., 2009).

### 3.2.4. Free amino nitrogen

The free amino nitrogen (FAN) concentration was fitted as a quadratic model in the temperature axis (**Fig. 2 B1, B2, and B3**). FAN was affected by the temperature (P value=0.000), square temperature (P value=0.005), and the interaction between temperature-screws speed (P value=0.044). The FAN amounts in the hydrolyzates ranged from 17.6 to 19.9 mg $L^{-1}$. The low FAN yield occurred because of the high lignification degree of BSG. Forrssell et al. (2008) found that after enzymatic hydrolysis of BSG, the protein practically remained intact. During the malt mashing, some of the barley and the adjunct proteins are hydrolyze into low molecular weight peptides and amino acids (quantified as FAN), leaving the remaining proteins between the cellular walls layers (Celus et al. 2006). Even after the enzymatic degradation of cell walls, these proteins are mainly coated with lignin, preventing the protein release from the matrix. It has also been reported that the disulphide cross-linkings of proteins forms a coat on the surface layers of spent grain (Faulds et al., 2009). The concentration of FAN in the hydrolyzates was low for the key fermentation step and therefore, the hydrolyzates needed to be supplemented with yeast food rich in nitrogenous compounds (Thomas and Ingledew, 1990).

### 3.2.5. Inhibitory compounds

None of the hydrolysates contained significant amounts of inhibitory compounds such as acetic acid, furfural, or HMF. In fact, the harsh conditions involved in most commercial chemical pretreatments induce the rupture of hemicellulose and its subsequent dehydration (Pandey, 2011). While the extrusion conditions employed in this research did not generate these deleterious hydration compounds (Lin et al., 2012), especially at 50°C when the highest amount of sugars was produced. The null generation of inhibitors is advantageous for the ethanol production and from the environmental viewpoint because the chemical detoxification step of the process can be skipped. White et al. (2008) reported the existence of inhibitors in their BSG test (no amounts were reported), and they proposed adjusting the pH and inclusion of a detoxifying step for the hydrolyzates prior to the subsequent fermentation step. With the extrusion pretreatment, no detoxifying step was needed because no extreme conditions like acid or alkali pretreatments for example were applied.

### 3.3. Fermentation

During the first 24 h of fermentation, only 24.6% of the total sugars were consumed by the yeast generating 24.61 mg of ethanol $g^{-1}$ of the extruded BSG treated with the fiber-degrading enzymes (**Table 3**).

**Table 3.**
Different sugars and ethanol concentrations obtained from the extruded BSG at 50 °C after 48 h enzymatic saccharification and fermentation with *S. cerevisiae* 20252™ at 10% solids fraction[1,2,3].

| Fermentation Time (h) | Compound (mg $g^{-1}$) | | | | | | Cellulose Conversion (%) |
|---|---|---|---|---|---|---|---|
| | Arabinose | Cellobiose | Glucose | Xylose | Total Sugars | Ethanol | |
| 0 | 33.60 ±0.95 a | 3.43 ±0.20 a | 91.96 ±4.42 a | 72.00 ±3.20 a | 200.99 ±7.28 a | ND | - |
| 24 | 33.68 ±0.33 a | 3.78 ±0.14 b | 39.78 ±2.42 b | 74.31 ±1.71 a | 151.56 ±2.96 b | 24.61 ±0.18 a | 50.4% |
| 48 | 33.92 ±1.46 a | 3.94 ±0.06 b | 2.93 ±0.13 c | 73.77 ±1.92 a | 114.55 ±139 c | 42.88 ±0.77 b | 87.8% |

[1] Means with different letter within columns are significantly different.
[2] The compounds not detected by the HPLC were represented by the acronym ND.
[3] Modified from Dowe and McMillan (2008). % Cellulose Conversion =100×(Ethanol amount /(glucose amount+(0.51×(cellobiose amount×1.111))))

The cellulose conversion at this point of fermentation was 50.4%. However, after 48 h of fermentation, 43.0% of the total sugars were consumed and 42.88 mg of ethanol $g^{-1}$ was produced. This means that 87.8% of the glucose contained in the treated BSG was converted. At both fermentation times, only glucose was consumed, fermenting 56.7 and 96.8% of the total glucose at 24 and 48 h, respectively.

White et al. (2008) reported an ethanol yield of 8.3 g $L^{-1}$ after 48 h fermentation with *Pichia stipitis* strain using spent grain pretreated with diluted acid and hydrolyzed with fiber degrading enzymes. However, they also did not ferment the arabinose. In the present research, an ethanol yield of 4.28 g $L^{-1}$ was achieved by fermenting only glucose, reaching similar the fermentation yield reported by Whithe et al. (2008). Heredia-Olea et al. (2013) also used the same *S. cerevisiae* strain and xylose and arabinose were fermented. On the contrary, these sugars were not fermented in the present work,. According to Scheper (2007), in order to ferment C5 sugars into ethanol, right amount of oxygen is needed. This factor and the low FAN contained in the hydrolyzates (19.38±0.50 mg $L^{-1}$) could explain the low ethanol yield and the absence of C5 fermentation herein. Thus, nutrient supplementation and initial aeration might be required for a successful fermentation of C5 sugars and to further improve the ethanol yield.

## 4. Conclusion

The extrusion pretreatment disrupted the BSG structure and made the extruded feedstock more susceptible to the subsequent enzymatic hydrolysis step. The BSG extruded at 50 °C, 200 rpm screws speed, and 20% feedstock moisture, released 48.4% of the total sugars and more importantly did not generate yeast inhibitors. Relatively low and constant amount of FAN was recorded in the extruded and enzymatically-hydrolyzed BSG. Such amounts of FAN were not sufficient for successful fermentations. After 48 h fermentation and using 10% solid loading, the enzymatically-hydrolyzed BSG yielded 5.43 mL ethanol $L^{-1}$. The yeast only consumed glucose and therefore, low ethanol yields were achieved. The remaining spent material still contained high protein and lignin contents that could be considered for other bioprocesses.

## Acknowledgments

The authors would like to thank the Consejo Nacional de Ciencia y Tecnología and Escuela de Biotecnología y Alimentos for the support provided. We are grateful to Du Pont for providing the enzymes used in this work.

## References

Aliyu, S., Bala, M., 2011. Brewer`s spent grain: a review of its potentials and applications. Afr. J. Biotechnol. 10(3), 324-331.

Celus, I., Brijs, K., Delcour, J.A., 2006. The effects of malting and mashing on barley protein extractability. J. Cereal Sci. 44(2), 203-211.

Dowe, N., McMillan, J., 2008. SSF Experimental protocols- Lignocellulosic biomass hydrolysis and fermentation. Laboratory Analytical Procedure (LAP). National Renewable Energy Laboratory. Golden, Colorado.

Faulds, C.B., Collins, S., Robertson, J.A., Treimo, J., Eijsink, V.G.H., Hinz, S.W.A., Schols, H.A., Buchert, J., Waldron, K.W., 2009. Protease-induced solubilisation of carbohydrates from brewers' spent grain. J. Cereal Sci. 50(3), 332-336.

Forssell, P., Kontkanen, H., Schols, H. A., Hinz, S., Eijsink, V. G. H., Treimo, J., Robertson, J. A., Waldron, K. W., Faulds, C. B., Buchert, J., 2008. Hydrolysis of brewer's spent grain by carbohydrate degrading enzymes. J. Inst. Brew. 114(4), 306-314.

Heredia-Olea, E., Pérez-Carrillo, E., Serna-Saldívar, S. O., 2013. Production of ethanol from sweet sorghum bagasse pretreated with different chemical and physical processes and saccharified with fiber degrading enzymes. Bioresour. Technol. 134, 386-390.

Karunanithy, C., Muthukumarappan, K. 2010. Effect of extruder parameters and moisture content of switchgrass, prairie cord grass on sugar recovery from enzymatic hydrolysis. Appl. Biochem. Biotechnol. 162 (6), 17585-1803.

Karunanithy, C., Muthukumarappan, K., Gibbons, W.R., 2013. Effects of extruder screw speed, temperature, and enzyme levels on sugar recovery from different biomasses. ISRN Biotechnol. Article ID 942810.

Lie. S., 1973. The EBC-ninhydrin method for determination of free amino nitrogen. J. Inst. Brew. 79(1), 37-41.

Lin, Z., Liu, L., Li, R., Shi, J. 2012., Screw extrusion pretreatments to enhance the hydrolysis of lignocellulosic biomass. J. Microb. Biochem. Technol. 12, 5.

Moscicki, L., 2011. Extrusion-cooking techniques, applications, theory and sustainability. ed. WILEY-VCH Verlag and Co. KGaA, Weinheim, Germany.

Mussatto, S. I., Dragone, G., Roberto, I. C., 2006. Brewers' spent grain: generation, characteristics and potential applications. J. Cereal Sci. 43(1), 1-14.

Mussatto, S.I., Fernandes, M., Milagres, A.M.F., Roberto, I.C., 2008. Effect of hemicellulose and lignin on enzymatic hydrolysis of cellulose from brewers' spent grain. Enzyme Microb. Technol. 43(2), 124-129.

Pandey, A., 2011. Biofuels: alternative feedstocks and conversion processes. ed. Academic Press, Kidlington, Oxford; Burlington, MA.

Reddy, N., Yang, Y., 2005. Biofibers from agricultural byproducts for industrial applications. Trends Biotechnol. 23(1), 22-27.

Sluiter, A., Ruiz, R., Scarlata, C., Sluiter, J., Templeton, D. 2008a. Determination of extractives in biomass. Laboratory Analytical Procedure (LAP). National Renewable Energy Laboratory. Golden, Colorado.

Sluiter, A., Hames, B., Ruiz, R., Scarlata, C., Sluiter, J., Templeton, D., Crocker, D. 2008b. Determination of structural carbohydrates and lignin in biomass. Laboratory Analytical Procedure (LAP). National Renewable Energy Laboratory. Golden, Colorado.

Scheper, T., 2007. Biofuels. Advances in Biochemical Engineering/Biothecnology. ed. Springer, London.

Shindo, S., Tachibana, T., 2006. Production of bioethanol from spent grain –a by-product of beer production. Technical quarterly-Master Brewers Association of the Americas. 43(3), 189-193.

Thomas, K. C., Ingledew, W. M., 1990. Fuel alcohol production: effects of free amino nitrogen on fermentation of very-high-gravity wheat mashes. Appl. Environ. Microbiol. 56 (7), 2046-2050.

White, J.S., Yohannan, B.K., Walker, G.M., 2008. Bioconversion of brewer's spent grain to bioethanol. FEMS Yeast Res. 8(7), 1175-1184.

# 9

# Effect of various carbon-based cathode electrodes on the performance of microbial fuel cell

Mehrdad Mashkour, Mostafa Rahimnejad*

*Biofuel and Renewable Energy Research Center, Department of Chemical Engineering, Babol Noshirvani University of Technology, Babol, Iran.*

## HIGHLIGHTS

- The performance of different carbon-based cathodes was compared in a dual-chambered microbial fuel cell.
- A novel CNT/Pt-coated carbon paper was fabricated.
- The maximum current and power generated were 82.38 mA/m$^2$ and 16.26 mW/m$^2$, respectively.
- Aeration in the cathode compartment was found effective on power generation in MFCs.

## GRAPHICAL ABSTRACT

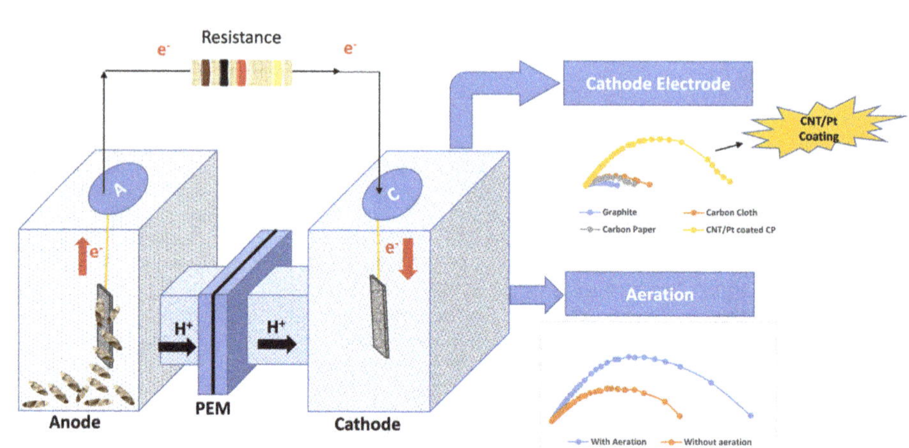


## ABSTRACT

Microbial fuel cell (MFC) is a prospective technology capable of purifying different types of wastewater while converting its chemical energy into electrical energy using bacteria as active biocatalysts. Electrode materials play an important role in the MFC system. In the present work, different carbon-based materials were studied as electrode and the effect of dissolved oxygen (aeration) in the cathode compartment using actual wastewater was also investigated. More specifically, the effect of different electrode materials such as graphite, carbon cloth, carbon paper (CP), and carbon nanotube platinum (CNT/Pt)-coated CP on the performance of a dual-chambered MFC was studied. Based on the results obtained, the CNT/Pt-coated CP was revealed as the best cathode electrode capable of producing the highest current density (82.38 mA/m$^2$) and maximum power density (16.26 mW/m$^2$) in the investigated MFC system. Moreover, aeration was found effective by increasing power density by two folds from 0.93 to 1.84 mW/m$^2$ using graphite as the model cathode electrode.

**Keywords:**
Microbial fuel cell
Cathode compartment
Graphite
Carbon cloth
CNT/Pt-coated Carbon paper

* Corresponding author
E-mail address: rahimnejad_mostafa@yahoo.com

## 1. Introduction

The world faces energy crises as the traditional energy sources like coal, gas, and petroleum are running out (Scott, 2005). This has turned looking for new energy supplies into an important requirement for life. Among the most promising renewable energy supplies with minimal environmental impacts are wind energy, solar energy, and microbial fuel cells (MFCs). The first and second categories have been developed as there are many giant wind turbines and solar power plants all over the world while MFCs are still to be further investigated before used commercially (Rahimnejad et al., 2014; Beurskens and Brand, 2015; Chaudhari and Deshmukh, 2015; Green et al., 2015).

The MFC technology is a promising method of transforming bacterial metabolic energy into electricity (Chae et al., 2009). Traditional MFCs consist of two chambers in which electrodes are placed while a membrane separates the chambers from each other (Rahimnejad et al., 2012). Microorganisms in the anode chamber consume the substrate and produce electrons and protons. The produced electron is transferred by an external circuit to the cathode compartment and the generated protons are transferred by the proton exchange membrane to the cathode chamber (Du et al., 2007).

Therefore, electrodes play a very important role in the efficiency of an MFC (Ghoreishi et al., 2014). In fact, the electrons produced by bacteria are transferred on the anode's surface while on the cathode's surface an electrochemical reduction reaction takes place (Du et al., 2007; Freguia et al., 2007). Hence, the surface of the electrodes is of significant importance in the overall performance of an MFC. More specifically, the higher electrode surface area leads to an enhanced MFC performance. Traditional electrode materials commonly used as anode and cathode include graphite, carbon cloth, and carbon paper (Ghasemi et al., 2013). Many studies are underway with a focus on the fabrication of novel electrode materials in order to improve MFCs' system performance (Guo et al., 2015; Kim et al., 2015).

It is worth quoting that among the reasons given for the very poor performance of the currently-available MFCs for large-scale application is the low efficiency of electricity generation caused by the limited surface of electrodes (Qiao et al., 2007). Progresses in nanotechnology and nanomaterial sciences have brought about evolutionary developments in the MFC technology. This is ascribed to the fact that materials at nano-scale exhibit different and unique properties in comparison with their macro-scale forms (Klabunde and Richards, 2009). Accordingly, many researchers have tried to apply nanomaterials in the fabrication of electrodes (Qiao et al., 2007; Xie et al., 2010; Xie et al., 2012; Wang et al., 2015). In fact, the problems associated with the low surface area of electrodes could be overcome by the application of nanomaterials owing to their extensive active surface area. The low performance of MFCs is also attributed to the low conductivity and insufficient velocity of the electron transfer process caused by the high electrical resistance of conventional electrode materials (Ghasemi et al., 2013; Rahimnejad et al., 2015). Various microorganisms are used in the anode chamber of MFCs including pure or mixed cultures (Logan, 2009). The latter is however preferred as resembles the practical conditions more (Izadi and Rahimnejad, 2014).

In the present study, a two-chamber MFC was inoculated by using a mixed culture, i.e., anaerobic sludge as biocatalyst in the anode chamber. Several conventional carbon-based electrodes including graphite, carbon cloth, carbon paper (CP), as well as a novel electrode, i.e., carbon nanotube platinum (CNT/Pt)-coated CP were investigated as cathodes. More specifically, CNT and Pt were used to simultaneously achieve both high active surface and high conductivity, respectively. Generated power and current in the presence of these cathodes were evaluated to determine the best carbon-based electrode. Moreover, the effect of aeration in the cathode chamber was also taken into account. The biofilm formed on the anode's surface was also electrochemically investigated.

## 2. Materials and methods

### *2.1. MFC construction and operation*

Two cubic and H-shaped chambers with a working volume of 760 ml each were constructed using Plexiglas material and were separated by a Nafion 117, acting as the proton exchange membrane (PEM). Both the cathode and the anode surface areas were 8 $cm^2$ and the MFC was operated at the ambient temperature and neutral pH (6.5-7) at the anode compartment. (Kim et al., 2002; Zhao et al., 2005). The pH was adjusted by the phosphate buffer solution. Graphite was used as the anode and cathodes were graphite, CP, carbon cloth, and CP coated by CNT/Pt (0.1 $mg/cm^2$). The MFC used is schematically shown in **Figure 1**.

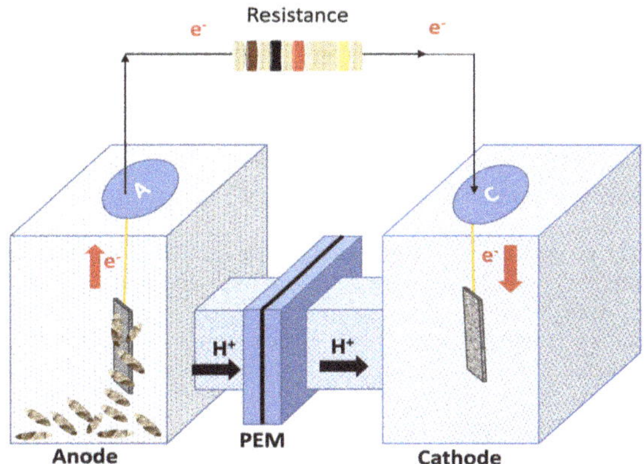


**Fig.1.** Schematic image of the fabricated MFC.

### *2.2. Microorganisms*

Anaerobic sludge collected from the anaerobic process tank of the Ghaemshahr wastewater treatment center (Mazandaran, Iran) was used as inoculums. The media used contained (g/L): glucose (20), yeast extract (3), and peptone (1 g).

### *2.3. CNT/Pt composite electrode*

A chemical reduction technique was used to fabricate the CNT/Pt. First, the CNT was ultrasonicated in nitric acid for approximately 3 h. Then, the sample was dried, washed with deionized water several times, and air-dried. The dried sample was then ultrasonicated again with acetone for 1 h and a 0.075 M $H_2PtCl_6$ solution was added slowly under stirring. After 24 h, the mixture was reduced using a 1 M NaOH and 0.1 M $NaBH_4$ solution. When the mixture was ready, it was washed by deionized water and dried at 80 °C for 6 h. The required amount of the CNT/Pt was added to a small amount of ethanol, dispersed properly, and then brushed onto the CP surface.

### *2.4. Analyses and calculations*

The current and power produced by the system were calculated by using the following equations (**Eqs. 1and 2**).

$$I=\frac{V}{R} \quad (1)$$

$$P=R\times I^2 \quad (2)$$

Where I is the current (ampere), V is the voltage (volt), R stands for the external resistance (ohm) and P denotes the power (watt) produced by the system.

The MFC's internal resistance was calculated by two methods:

A. Polarization slope method (V-I curve): the slope of the voltage-current curve represents the internal resistance.
B. Power density peak method: the external resistance at which the MFC power output reaches maximum amount is considered as the internal resistance of system (Logan, 2008).

An IVUM package (Ivium Technology, Netherland) was used to analyze cyclic voltammetry (CV). The CV test was performed to identify the oxidation and reduction potential of the substrate. Potentials ranging from -400 mV to 1000 mV at a scan rate of 50 mV/s were applied to conduct the CV experiments. In order to measure the oxidation and reduction peaks, the CP (NARA, Guro-GU, Seoul, Korea) was used as the working electrode and Platinum (Platinum, gauze, 100 mesh, 99.9% meta basis, Sigma Aldrich) was used as the counter electrode. Moreover, Ag/AgCl electrode (Ag/AgCl, sat KCl, Sensortechnik Meinsberg, Germany) was utilized as the reference electrode.

All the chemicals and reagents used for the experiments were of analytical grades and were supplied by Merck (Germany). A HANA 211 pH meter (Romania) was employed for measuring the pH values. The initial pH of the working solutions was adjusted by the addition of diluted $HNO_3$ or 0.1 M NaOH solutions.

## 3. Results and discussion

As mentioned earlier, microorganisms play a very effective role in the performance of MFCs. A mixed culture of microorganisms was used as the electron generator in this study. Prior to the MFC experiment and in order to evaluate the required time to create a steady state by the mixed culture, the growth kinetics were surveyed using glucose as the only carbon source. After the inoculation, sampling was performed every 2 or 4 h and the light absorption of the samples at 620 nm was determined. **Figure 2** presents the growth curve of the active biocatalyst used. As can be seen in the figure, the microorganisms used were capable of efficiently growing under anaerobic conditions. More specifically, the microbial growth initiated with a lag phase followed by a rapid and sharp tangent entry into the logarithmic phase. Then, after 26 h, the mixed culture reached the stationary phase (**Fig. 2**).

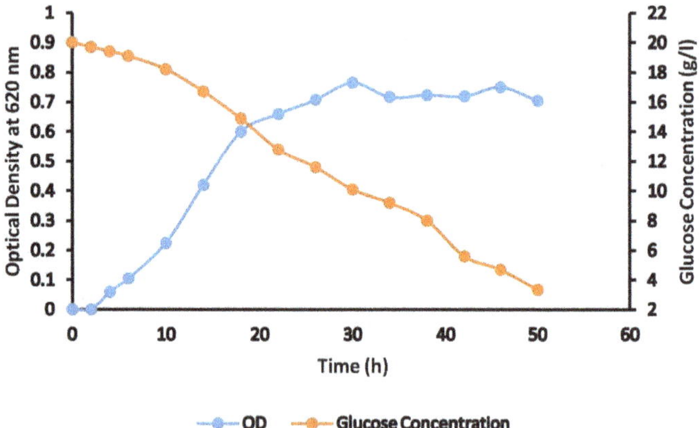


**Fig.2.** Growth curve of the used microorganisms under anaerobic conditions.

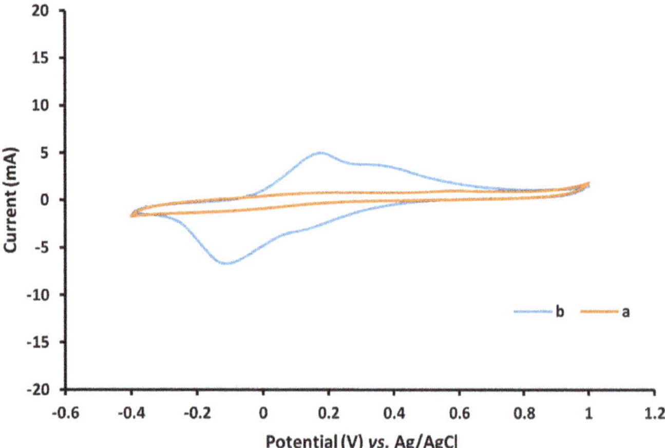


**Fig.3.** Investigation of the developed active biofilm on the anode's surface by electrochemical analysis at 0.05 V/s scanning rates, a) before, and b) after the development of the biofilm by the mixed culture used.

The capability of the mixed culture in producing biofilm on the surface of the anode electrode (graphite) was also examined by CV analysis and the results obtained are presented in **Figure 3**. As could be seen, no electrochemical activity was observed on the anode's surface before the development of an active biofilm. However, after the development of an active biofilm by the end of the process, electrochemical analysis revealed the existence of two oxuidation/reduction peaks confirming the suitability of the mixed culture used (**Fig. 3**).

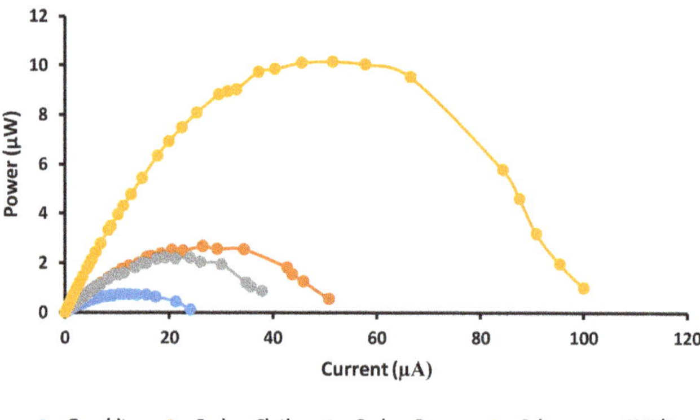


**Fig.4.** Polarization Curve of the graphite, carbon cloth, carbon paper (CP), and the CNT/Pt-coated CP electrodes without aeration.

In MFCs, electrodes are electron transferring sites and by selecting a suitable electrode, it is possible to maximize the output power of the system. In the present study, four carbon-based electrodes were used as cathode. Among the conventional electrodes used, the carbon cloth and the CP were respectively the most suitable cathode electrodes in terms of the generated power density (**Fig. 4**). More specifically, the maximum power density values obtained were 0.937, 2.76, and 3.35 $mW/m^2$ for the graphite, CP, and carbon cloth electrodes, respectively. It is worth quoting that the conductivity of the carbon cloth, the CP, and the graphite electrodes were approximately equal but the higher surface area of the carbon cloth and the CP electrodes compared with the graphite electrode led to higher efficiencies of the MFC system. The current density of the MFC using the graphite, CP, and carbon cloth electrodes was measured at 13.92, 23.9, and 33.07 $mA/m^2$, respectively.

The modification of the CP with the CNT/Pt coating resulted in significant changes. In more details, the power and current generated by the CNT/Pt-coated CP (16.26 $mW/m^2$ and 82.38 $mA/m^2$, respectively) increased by approximately six folds and three folds in comparison with the conventional CP (2.76 $mW/m^2$ and 23.9 $mA/m^2$, respectively). These improvements could be attributed to the high active surface area of the CNT and the high electrical conductivity of the Pt. In another words, the CNT and the Pt dispersed on the surface of the CP provided a desirable conductive porous surface for the electrons to rapidly react with oxygen. This could have consequently resulted in decreased proton accumulation and therefore, the pH of the anolyte and catholyte remained in an acceptable range for both metabolic activity of the microorganisms and the existence of a driving force for proton motion to the cathode chamber in which the reduction reaction occurred.

Besides the above-mentioned parameters, the internal resistance of the MFC was also studied. The internal resistance calculated by means of the V-I curve is shown in **Figure 5**. The slope of the linear parts of the V-I curve represents the internal resistance of the MFC (Ghasemi et al., 2013). It was found that by applying the carbon cloth, CP, and the CNT/Pt-coated CP as cathode electrodes, the internal resistance could be decreased by 20-25% (from 5.2 to 3.9 kΩ) compared with the graphite electrode. In fact, the CNT/Pt coating did not change the internal resistance of the CP while it increased the voltage and current generation significantly. This finding could be attributed to the voltage losses due to the higher current as a result of the coating applied (Logan, 2008). Calculations of the internal resistance were also carried out by the Power density peak method and similar results were obtained (data not shown).

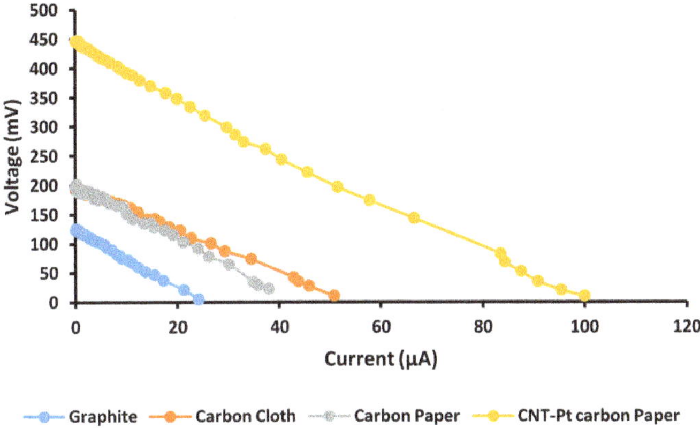


**Fig.5.** Voltage-current curves of the electrodes used as cathode.

**Figure 6** demonstrates the effect of aeration on the performance of the MFC using the graphite electrode as cathode. The graphite electrode was selected to solely take into account the impact of aeration and not the electrode composition. The aeration in the cathode chamber increased the maximum voltage by more than 50% (from 124 to 190 mV). The power and current generation also increased significantly from 0.9 to 1.84 mW/m$^2$ and from 15 to 21 mA/m$^2$, respectively. In fact, the aeration in the cathode chamber was shown to result in a more complete and faster reduction reaction and minimal potential losses owing to limited electron acceptor concentration level in the catholyte. This finding was in line with those of Rabaey and Verstraete (2005).

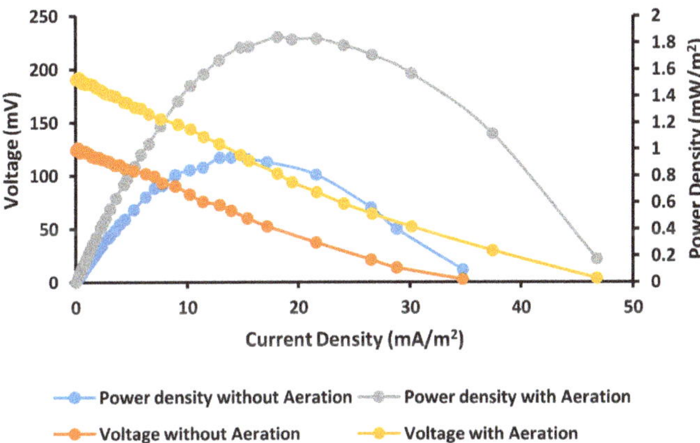


**Fig.6.** Polarization curve of the graphite electrode with and without aeration.

## 4. Conclusions

The performance of an MFC by using an active biocatalyst was investigated. The CV analysis demonstrated microbial biofilm production capabilities of the biocatalyst on the anode's surface. Different carbon-based materials as cathode electrode were also investigated. The best performance parameters were achieved by using the carbon paper electrode modified by CNT/Pt. This novel electrode led to significantly higher generated power (16.26 mW/m$^2$) and current (82.38 mA/m$^2$) compared to the other carbon-based cathode electrodes used, i.e., graphite, carbon cloth, and carbon paper. This was accomplished owing to the perfect electrical conductivity of the Pt and the high surface area of the CNT. It is worth mentioning that a separate experiment revealed that aeration in the cathode chamber led to a considerable increase in MFC performance (i.e., power density) by two folds.

## References


[1] Beurskens, H., Brand, A., 2015. Wind Energy, ECN Wind Energy, Netherlands.

[2] Chae, K.J., Choi, M.J., Lee, J.W., Kim, K.Y., Kim, I.S., 2009. Effect of different substrates on the performance, bacterial diversity, and bacterial viability in microbial fuel cells. Bioresour. Technol. 100 (14), 35183525.

[3] Chaudhari, S., Deshmukh, A., 2015. Studies on sewage treatment of industrial and municipal wastewater by electrogens isolated from microbial fuel cell. Int. J. Curr. Microbiol. Appl. Sci. 4(4), 118-122.

[4] Du, Z., Li H., Gu, T., 2007. A state of the art review on microbial fuel cells: a promising technology for wastewater treatment and bioenergy. Biotechnol. Adv. 25(5), 464-482.

[5] Freguia, S., Rabaey, K., Yuan, Z., Keller, J., 2007. Non-catalyzed cathodic oxygen reduction at graphite granules in microbial fuel cells. Electrochim. Acta. 53(2), 598-603.

[6] Ghasemi, M., Daud, W.R.W., Hassan, S.H., Oh, S.E., Ismail, M., Rahimnejad, M., Jahim, J.M., 2013. Nano-structured carbon as electrode material in microbial fuel cells: A comprehensive review. J. Alloys Compd. 580, 245-255.

[7] Ghasemi, M., Ismail, M., Kamarudin, S.K., Saeedfar, K., Daud, W.R.W., Hassan, S.H., Heng L.Y., Alam, J., Oh, S.E.,2013. Carbon nanotube as an alternative cathode support and catalyst for microbial fuel cells. Appl. Energy. 102, 1050-1056.

[8] Ghoreishi, K.B., Ghasemi, M., Rahimnejad, M., Yarmo, M.A., Daud, W.R.W., Asim, N., Ismail, M., 2014. Development and application of vanadium oxide/polyaniline composite as a novel cathode catalyst in microbial fuel cell. Int. J. Energy Res. 38(1), 70-77.

[9] Green, M.A., Emery, K., Hishikawa, Y., Warta, W., Dunlop, E.D., 2015. Solar cell efficiency tables (Version 45). Prog. Photovoltaics Res. Appl. 23(1), 1-9.

[10] Guo, K., Prévoteau, A., Patil, S.A., Rabaey, K. 2015. Engineering electrodes for microbial electrocatalysis. Curr. Opin. Biotechnol. 33, 149-156.

[11] Izadi, P., M. Rahimnejad, M., 2013. Simultaneous electricity generation and sulfide removal via a dual chamber microbial fuel cell. Biofuel Res. J. 1 (1), 34-38.

[12] Kim, H.J., Park, H.S., Hyun, M.S., Chang, I.S., Kim, M., Kim, B.H., 2002. A mediator-less microbial fuel cell using a metal reducing bacterium, Shewanella putrefaciens. Enzyme Microb. Technol. 30(2), 145-152.

[13] Kim, K.Y., Yang, W., Logan, B.E., 2015. Impact of electrode configurations on retention time and domestic wastewater treatment efficiency using microbial fuel cells. Water Res. 80, 41-46.

[14] Klabunde, K.J. and R.M. Richards, R.M., 2009. Nanoscale materials in chemistry, John Wiley & Sons.

[15] Logan, B.E., 2008. Microbial fuel cells, John Wiley & Sons.

[16] Logan, B.E., 2009. Exoelectrogenic bacteria that power microbial fuel cells. Nat. Rev. Microbiol. 7(5), 375-381.

[17] Qiao, Y., Bao, S.J., Li, C.M., Cui, X.Q., Lu, Z.S., Guo, J., 2007. Nanostructured polyaniline/titanium dioxide composite anode for microbial fuel cells. Acs Nano. 2(1), 113-119.

[18] Qiao, Y., Li, C.M., Bao, S.J., Bao, Q.L., 2007. Carbon nanotube/polyaniline composite as anode material for microbial fuel cells. J. Power Sources. 170(1), 79-84.

[19] Rabaey, K., Verstraete, W., 2005. Microbial fuel cells: novel biotechnology for energy generation. Trends Biotechnol. 23(6), 291-298.

[20] Rahimnejad, M., Adhami, A., Darvari, S., Zirepour, A., Oh, S.E., 2015. Microbial fuel cell as new technology for bioelectricity generation: A review. Alexandria Eng. J. 54, 745-756.

[21] Rahimnejad, M., Bakeri, G., Najafpour, G., Ghasemi, M., Oh, S.E., 2014. A review on the effect of proton exchange membranes in microbial fuel cells. Biofuel Res. J. 1(1), 7-15.

[22] Scott, D.S., 2005. Fossil sources:"running out" is not the problem. Int. J. Hydrogen Energy. 30(1), 1-7.

[23] Wang, L., Su, L., Chen, H., Yin, T., Lin, Z., Lin, X., Fu, D., 2015. Carbon paper electrode modified by goethite nanowhiskers promotes bacterial extracellular electron transfer. Mater. Lett. 141, 311-314.

[24] Xie, X., Hu, L., Pasta, M., Wells, G.F., Kong, D., Criddle, C.S., Cui, Y., 2010. Three-dimensional carbon nanotube– textile anode for high-performance microbial fuel cells. Nano Lett. 11(1), 291-296.

[25] Xie, X., Ye, M., Hu, L., Liu, N., McDonough, J. R., Chen, W., Alshareef, H., Criddle, C.S., Cui, Y., 2012. Carbon nanotube-coated macroporous sponge for microbial fuel cell electrodes. Energy Environ. Sci. 5(1), 5265-5270.

[26] Zhao, F., Harnisch, F., Schröder, U., Scholz, F., Bogdanoff, P., Herrmann, I., 2005. Application of pyrolysed iron (II) phthalocyanine and CoTMPP based oxygen reduction catalysts as cathode materials in microbial fuel cells. Electrochem. Commun. 7(12), 1405-1410.

# 10

# Glucoamylase production from food waste by solid state fermentation and its evaluation in the hydrolysis of domestic food waste

Esra Uçkun Kiran[1], Antoine P. Trzcinski[1], Yu Liu[1,2*]

[1]*Advanced Environmental Biotechnology Centre, Nanyang Environment & Water Research Institute, Nanyang Technological University, 1 Cleantech Loop, Singapore 637141, Singapore.*

[2]*Division of Environmental and Water Resources Engineering, School of Civil and Environmental Engineering, Nanyang Technological University, 50 Nanyang Avenue, Singapore 639798, Singapore.*

## HIGHLIGHTS

- *Various food wastes were evaluated to produce glucoamylase using SSF.*
- *Waste cake was the best substrate for glucoamylase production.*
- *The highest glucoamylase activity of 108.5 U/gd was achieved under optimal conditions.*
- *Increasing the initial pH to 7.9 enhanced the GA activity.*
- *The enzyme preparation can completely hydrolyzed domestic food waste to sugars.*

## GRAPHICAL ABSTRACT

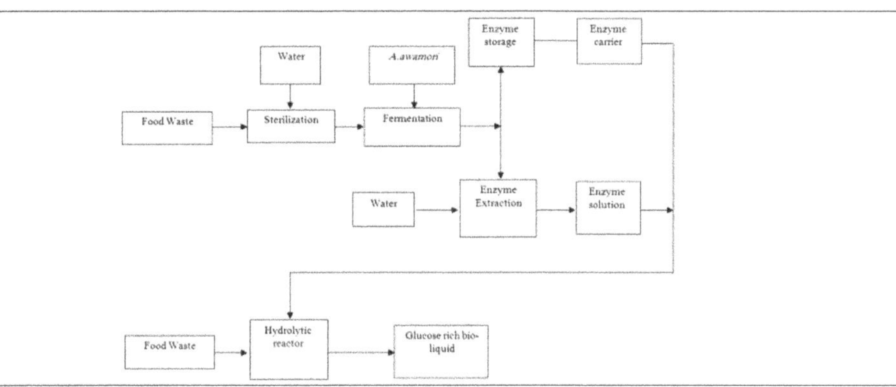


## ABSTRACT

In this study, food wastes such as waste bread, savory, waste cakes, cafeteria waste, fruits, vegetables and potatoes were used as sole substrate for glucoamylase production by solid state fermentation. Response surface methodology was employed to optimize the fermentation conditions for improving the production of high activity enzyme. It was found that waste cake was the best substrate for glucoamylase production. Among all the parameters studied, glucoamylase activity was significantly affected by the initial pH and incubation time. The highest glucoamylase activity of 108.47 U/gds was achieved at initial pH of 7.9, moisture content of 69.6% wt., inoculum loading of $5.2\times10^5$ cells/gram substrate (gs) and incubation time of 6 d. The enzyme preparation could effectively digest 50% suspension of domestic food waste in 24 h with an almost complete saccharification using an enzyme dose of only 2U/g food waste at 60°C.

**Keywords:**
Glucoamylase
*Aspergillus awamori*
Food waste
Saccharification
Solid state fermentation
Response surface methodology

* Corresponding author
E-mail address: CYLiu@ntu.edu.sg

## 1. Introduction

Food waste (FW) is a kind of organic waste discharged from households, cafeterias and restaurants. According toFAO (2012), one third of food produced for human consumption (nearly 1.3 billion tons) is lost or wasted globally throughout the food supply chain and it is increasing dramatically while almost 1 billion people worldwide are classified as starving. Besides, every tone of FW means 4.5 tons of $CO_2$ emissions (Smith et al., 2001). Currently, the majority of FW in Singapore is incinerated with other combustible municipal wastes for heat or energy production, while residual ash is then disposed of in landfills. However, incineration is an expensive waste conversion technique and can cause severe air pollution (El-Fadel et al., 1997). From an environmental viewpoint, there is an urgent need for appropriate management of FW. Due to its chemical complexity, high moisture content, easy degradation and nutrient rich composition, FW should be treated as a useful resource for higher value products, such as fuels and chemicals through fermentation. Recently, there is a growing interest in biochemicals production from FW (Han and Shin, 2004; Wang et al. 2005; Sakai and Ezaki 2006; Yang et al. 2006; Ohkouchi and Inoue, 2007; Koike et al., 2009; Zhang et al., 2010; Zhang et al. 2013a). Starch is an important biopolymer in foods, as such, it is a significant part of kitchen waste (Arooj et al., 2008). Hence, the saccharification of FW is a key step for its bioconversion into value-added products. To achieve this, commercial enzymes, particularly glucoamylases, were often used to promote the bioconversion of polymers to bioproducts. To produce lactic acid from FW, Sakai et al. (2004) used glucoamylase to saccharify the production medium. In other studies, commercial glucoamylase, alpha-amylase and cellulase solutions were used to saccharify the kitchen wastes for ethanol production (Kim et al., 2008; Uncu and Cekmecelioglu, 2011; Yan et al., 2012). If the enzymes could be produced *in situ* without downstream treatments and integrated with the biochemicals production, the cost of the process would be decreased (Merino and Cherry, 2007; Wang et al., 2010). Moreover, the transportation cost and enzyme inactivation during storage could be avoided. If the crude enzyme activity is high, it would be feasible and economical for it to be used directly without any recovery process. Such strategy has been explored by several researchers (Leung et al., 2012; Zhang et al., 2013b) who produced succinic acid from waste bread. *Aspergillus awamori* and *Aspergillus oryzae* produced an enzyme cocktail rich in amylolytic and proteolytic enzymes to hydrolyze waste bread in SSF. The resulting fermented solids were added directly to a bread suspension to generate a hydrolysate rich in glucose and free amino nitrogen. The bread hydrolyzate was then used as the sole feedstock for *A. succinogenes* fermentation.

The microorganisms reported to be active producers of amylolytic enzymes are *Aspergillus awamori, Aspergillus foetidus, Aspergillus niger, Aspergillus oryzae, Aspergillus terreus, Mucor rouxians, Mucor javanicus, Neurospora crassa, Rhizopus delmar, Rhizopus oryzae and Thermomucor indicae-seudaticae* (Norouzian et al., 2006). Although glucoamylases have been produced by submerged fermentation traditionally, solid state fermentation (SSF) processes have been increasingly applied for the production of this enzyme in recent years (Ellaiah et al., 2002). SSF has advantages over submerged fermentation in that it is simpler, requires less capital, has superior productivity, lower energy requirement, requires simpler fermentation media, does not require rigorous control of fermentation parameters, uses less water, produces less waste water, allows for the easy control of bacterial contamination, and has a lower downstream processing cost (Ellaiah et al., 2002; Anto et al., 2006; Melikoglu et al., 2013a). However, the scale up of the SSF is a great challenge due to hardship of mixing, difficulty of heat removal and restricted water content which cause rapid change of moisture.

In order to attain higher enzyme activities, a number of factors needs to be optimized. The statistical methods for optimization are gaining a growing interest and application as they have proved to be cost and time saving. Recently, several statistical experimental design methods have been employed for optimizing enzyme production (Soni et al., 2012). Among the optimization methods used, central composite design using response surface methodology (RSM) is a method suitable for identifying the effects of individual variables and for seeking the optimal conditions for a multivariable system efficiently. This approach reduces the number of experiments, improves statistical interpretation possibilities and reveals possible interactions among parameters. To develop a viable process it is important to determine the most appropriate substrate and to optimize the fermentation conditions.

Although there are some reports explaining the production of various enzymes from agro-industrial biomass, the effect of different FW constituents on glucoamylase production has not been investigated to date and the produced enzymes were not evaluated for their suitability to hydrolyze starch in FW and produce fermentable sugars. In this study, different FWs were evaluated to produce glucoamylase using solid state fermentation. The overall strategy was to find out the best substrate for glucoamylase production, and to optimize the yield in order to hydrolyze raw mixed food waste. This novel strategy could help to produce fermentable sugars for the production of biofuels or chemicals. The fermentation conditions such as particle size, initial moisture content, inoculum loading, pH and incubation time for high activity glucoamylase production were optimized statistically. Finally, as part of an integrated solution, the effect of the produced enzyme solution on the hydrolysis of domestic FW was evaluated.

## 2. Materials and methods

### 2.1. Materials

*Aspergillus awamori* obtained from ABM Chemicals Ltd (Cheshire, England) was used to produce glucoamylase (GA) in SSF through FW hydrolysis. The enzyme was stored and prepared according to the procedures explained by Wang et al. (2007). The waste cakes used in this study was collected from local caterings. The cake waste was ground, sieved and then stored at -20°C pending further experiments. The mixed FW (MFW) and domestic FW used in this study were collected from a cafeteria at Nanyang Technological University and a local food court, respectively. Potatoes, fruits and vegetables were obtained from a local supermarket. These were discarded from the packaging line due to low quality. The FWs were homogenized in a blender and directly stored in zipped plastic bags at -20°C pending use in experiments.

### 2.2. Methods

#### 2.2.1. Effect of particle size on SSF

To determine the effect of particle size, the substrate was sieved through the mesh numbers 5, 10, 16 and 230 corresponding to size cut-off of 0.6, 1.18, 2 and 4 mm, respectively (Endecotts Ltd., UK). After sieving, the moisture content was adjusted to 70% wt. and the SSF was carried out with an inoculum loading of $10^6$ spores/g substrate at neutral initial pH and 30°C for 4 d as these conditions were reported as optimum for GA production from *A. awamori* by SSF by Melikoglu et al. (2013a).

#### 2.2.2. Experimental design for enzyme production

A $2^4$ full factorial design was used in the optimization of GA production from cake waste. Initial pH ($X_1$), moisture content ($X_2$, %, w/w), inoculum loading ($X_3$, inoculum/g substrate) and time ($X_4$, day) were chosen as independent input variables as they are the most important parameters for enzyme production during SSF (Garg et al., 2011). The GA activity (Units/gram dry solid or U/gds) was used as dependent output variables. A total of 30 experiments including 16 cube points (runs 1-16), 8 star points (runs 17-24), and 6 replicas of the central point (runs 25-30) were performed to fit a second order polynomial model. The experimental range and the levels of the variables are defined and presented in Table 1. The ranges of variables used in this work were selected based on the literature (Ellaiah et al., 2002; Melikoglu et al., 2013a; Wang et al., 2009; Pandey, 1991).

#### 2.2.3. Solid state fermentation (SSF) and enzyme extraction

Substrates were moistened with the calculated amount of 0.1 M phosphate and citrate buffer solutions in 500 mL Erlenmeyer flasks depending on the targeted initial pHs. After sterilization by autoclaving (120°C for 20 min), the flasks were cooled down, inoculated with the inoculum to obtain a certain spore concentration and the contents were mixed thoroughly with a sterile spatula. Then, 10 g of the content was distributed into each Petri dish and incubated at 30°C under stationary conditions. Petri dishes, in duplicate, were withdrawn at regular time intervals and the content was extracted with 60 mL of distilled sterile water. This was then centrifuged at 6,000 rpm for 10 min and cell free supernatant was used for assaying the GA activity.

**Table1.**
The experimental range and the levels of the variables in the Central Composite Design.

| Variable | Low Axial (-α) | Low factorial (-1) | Center (0) | High factorial (+1) | High axial (+α) |
|---|---|---|---|---|---|
| pH | 5 | 6 | 7 | 8 | 9 |
| Moisture content (%) | 50 | 60 | 70 | 80 | 90 |
| Inoculum loading (per gs) | $10^3$ | $10^5$ | $5.5\times10^5$ | $10^6$ | $1.1\times10^6$ |
| Time (d) | 3 | 4 | 5 | 6 | 7 |

*2.2.4. GA assay*

The activity of GA was determined at 55°C using 2% (w/v) soluble starch (Sigma) in 100 mM sodium acetate buffer, pH5. The glucose concentration was determined with Optium Xceed blood glucose monitor (Abbott Diabetes Care, Oxon, UK) (Bahcegul, 2011). One unit (1 U) of GA activity was defined as the amount of enzyme releasing 1 micromole glucose equivalent per minute under the assay conditions.

*2.2.5. Statistical analysis*

The data obtained from the central composite design experiments were analyzed using Design Expert (Stat-Ease Inc., Minneapolis, USA) (Version 8.0.7.1) software, and response surface curves, corresponding contour plots, regression coefficients and *F* values were obtained. Analysis of variance (ANOVA) was applied for the response function. The effects of the variables were estimated by the following second-order quadratic equation:

$$Y = b_0 + \sum biXi + \sum bijXij + \sum bi^2Xi^2 + \text{error} \quad \text{(Eq. 1)}$$

Where $Y$ is the predicted response for GA activity (U/gds); $b_0$ is the intercept; $bi$ is the coefficient for linear direct effect; $bij$ is the coefficient for interaction effect; $bi^2$ is the coefficients for quadratic effect (a positive or negative significant value implies possible interaction between the medium constituents); $X_i$ and $X_{ij}$ are the independent variables. The quality of fit to the second order equation was expressed by the coefficient of determination ($R^2$) and its statistical significance was determined by the F-test. Variables with probability below 95% ($P > 0.05$) were regarded as not significant for the final model. Three dimensional surface plots were drawn to illustrate the main and interactive effects of the independent variables on the dependent variables. The influence of experimental error on the central composite design was assessed with six replications at the central point of the experimental domain. Experiments were carried out in triplicates. Results were presented as the average of three independent trials. To maximize the enzyme activity, numerical optimization was used for determination of the optimal levels of the four variables.

*2.2.6. Model validation*

One set of experiments was performed to validate the model. SSF were conducted using an initial pH of 7.9, moisture content of 69.6%, inoculum loading of $5.2\times10^5$/gs and incubation time of 6 d to obtain the highest GA activity. All experiments were performed in triplicate, and the mean and standard deviations of the triplicates were reported.

*2.2.7. Hydrolysis of domestic FW*

Twenty five mL of 10% suspension of domestic FWs from a local food court in 0.1 M phosphate buffer, pH 7.0 was mixed with GA produced *in situ* from Asp*ergillus awamori* with enzyme to substrate ratio of 2U/g FW. The mixture was incubated at 60°C in a water bath for 24 h. The extent of saccharification was calculated by estimating glucose concentrations, after centrifugation at 5000 rpm for 5 min. The degree of saccharification was determined in terms of the ratio of glucose formed and the theoretical obtainable glucose from starch actually degraded (in percentages). Theoretical glucose yield was calculated based on the equation: 1 g starch = 1.11 g glucose. The whole process is described in Figure 1.

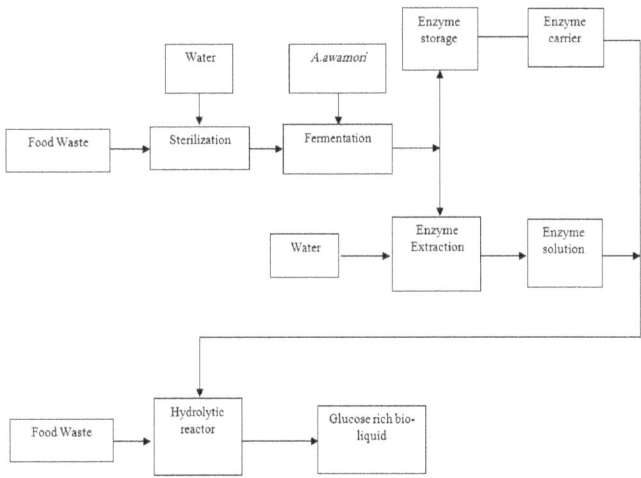


**Fig.1.** The process scheme of enzyme production and saccharification.

*2.2.8. Optimization of FW hydrolysis*

The hydrolysis of FW was optimized with respect to the main influencing parameters, i.e., the temperature, enzyme dose and FW concentration. All the experiments were performed at an enzyme to substrate ratio of 2U/g FW in the reaction mixtures made with 0.1 M phosphate buffer, pH 7.0 containing 10% of FW at 60°C for 24 h unless otherwise stated. The temperature levels of 50, 60, 70, 80, and 90°C, enzyme dosage levels of 2, 5, and 10 U/g FW, and FW concentration levels of 10, 20, 30, 40 and 50% w/v FW were used in the optimization of hydrolysis process.

*2.2.9. Analytical methods*

Moisture and ash contents were determined according to the analytical gravimetric methods (AOAC 2001). Crude protein content was determined using the HR Test'n tube TN kit (HACH, US) and calculated according to the the Kjeldahl method with a conversion factor of 6.25. Starch content was determined using the Megazyme's TN kit (Bray, Ireland). The lipid content was determined by the hexane/isopropanol (3:2) method (Hara and Radin, 1978). The glucose concentration was determined with the Optium Xceed blood glucose monitor (Abbott Diabetes Care, Oxon, UK) (Bahcegul, 2011). Reducing sugars were quantified to monitor the saccharification of FW according to the dinitrosalicylic acid (DNSA) method using glucose as standard (Miller, 1959).

## 3. Results and discussion

In order to understand the effects of different substrates, the wastes were characterized (Table 2). As seen in the table, the food wastes composed of different constituents. Bread waste had the highest starch content (71.6%) followed by potato (47.6%), cake waste (45.8%) and savory (45.7%). The reducing sugar content of cake waste (16.8%), fruit (11.7%) and potato (1.2%) were higher than that of the bread (1.5%).

**Table 2.**
Composition of different FWs.

| FW | Moisture (%) | TS (%) | VS/TS (%) | Starch (%, db) | RS (%, db) | Protein (%, db) | Lipid (% db) | Ash (%, db) |
|---|---|---|---|---|---|---|---|---|
| **Bread waste** | 34.4±0.2 | 65.6±0.2 | 96.7±0.0 | 71.6±0.5 | 0.5±0.1 | 8.6±2.1 | 3.9±2.6 | 3.2±0.0 |
| **Cake waste** | 29.9±1.9 | 70.1±1.9 | 96.0±0.3 | 45.8±3.0 | 16.8±0.5 | 14.1±0.8 | 16.1±7.5 | 3.9±0.2 |
| **Savory** | 37.8±0.4 | 62.2±0.4 | 96.6±0.3 | 45.7±2.8 | 0.3±0.0 | 2.3±1.1 | 22.1±0.3 | 3.3±0.4 |
| **Discarded Fruits** | 83.8±2.2 | 16.2±2.2 | 96.6±0.6 | 24.8±4.5 | 11.7±1.5 | 3.5±0.4 | 1.0±0.2 | 3.4±0.6 |
| **Discarded Potato** | 82.4±0.7 | 17.6±0.7 | 97.2±0.7 | 47.6±5.5 | 1.2±0.1 | 6.9±2.2 | 0.2±0.0 | 2.7±0.5 |
| **Discarded Vegetables** | 95.2±0.6 | 4.8±0.6 | 85.7±2.0 | 16.4±0.1 | 0.0±0.0 | 0.5±2.2 | 1.5±0.1 | 11.3±1.3 |
| **Mixed FW** | 80.3±1.1 | 19.7±1.1 | 95.2±0.4 | 19.0±1.3 | 0.7±0.0 | 15.4±2.4 | 19.4±0.1 | 4.7±0.4 |
| **Domestic FW** | 79.23±1.0 | 20.77±1.4 | 96.2±0.9 | 46.1±3.2 | 8.2±0.7 | 11.1±1.8 | 15.3±2.1 | 2.1±0.1 |

Total Solid, Starch, Reducing sugar (RS) Lipid, Protein and Ash Contents are given in wt% on the dry weight (db) basis. Volatile solid (VS) contents were given as the %VS ratio on total solid basis.

The influence of different FWs such as bread, cake, savory, vegetable, fruit, potato and MFW on GA production by *Aspergillus awamori* was investigated for 10 d (Fig. 2). Generally, the incubation time is governed by characteristics of the culture, its growth rate and enzyme production. Maximum GA production normally occurs after 2-5 d of incubation as reported by other researchers working with solid state cultures involving bacteria and fungi (Soni et al., 2003; Melikoglu et al., 2013a). The fungus used in the present study colonized well on the waste materials, and exhibited a good growth on the surface after 24 h. The high reducing sugars in cake, fruit and potato wastes may have triggered the GA production, so it was higher than savory and MFW on day 1. The growth and enzyme yields improved gradually, and the maximum activity of GA was obtained using waste cakes on the 4$^{th}$ day of fermentation (Fig. 2). The protein content of cake waste (14.1%) was also higher than that of bread (8.6%) which may have resulted in a better fungal growth and higher GA activity. To our knowledge this is the first study demonstrating that cake waste is a better substrate for GA production. The optimization of GA production from cake waste which resulted in the highest enzyme activity was afterward investigated.

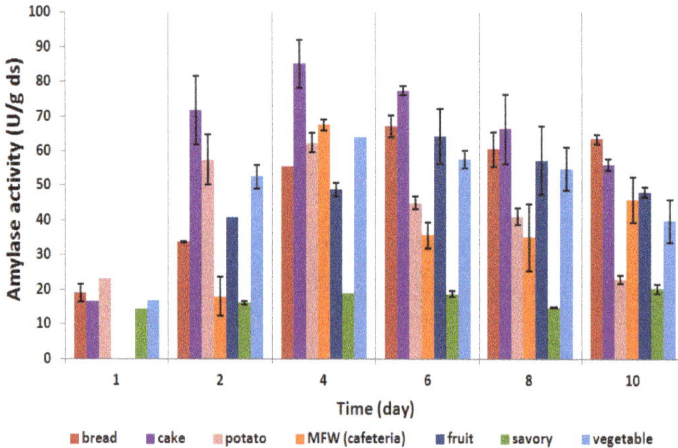


**Fig.2.** The effect of substrate on glucoamylase production using moisture content of 70% (wb), inoculum loading of $10^6$/g substrate at neutral initial pH and 30°C.

The utilization of the substrate during SSF by the fungi was not only influenced by its nutritional quality but also by the particle size of the solid substrate (Schmidt and Furlong, 2012). Experimental findings shown in Figure 3 validated that particle size had a direct effect on GA production during SSF. The highest GA activity of 63.06 U/gds was achieved with a particle size of 0.6≤X≤1.18 mm. In SSF, smaller particle size provides a larger contact area. However, reduction in particle size increases the packing density, which causes a reduction in the void space between the particles, which results in reduction in microbial growth and enzyme production (Ruiz et al., 2012). Therefore, there must be an optimum value for particle size. As the highest GA activity was obtained using 0.6≤X≤1.18, hence, particle size was adjusted to that particle range in the subsequent experiments.

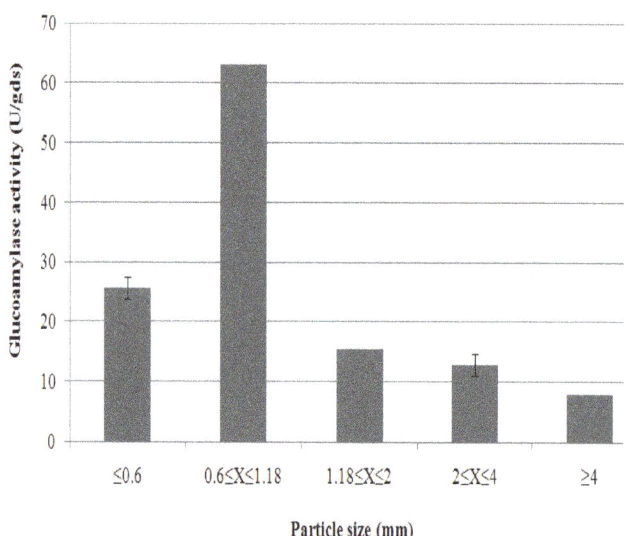


**Fig.3.** The effect of cake particle size on glucoamylase production using moisture content of 70% (wb), inoculum loading of $10^6$/g substrate at neutral initial pH and 30°C for 4 d.

To determine the optimum pH, moisture content, inoculum loading and time that would maximize GA activity, thirty experiments were designed using a central composite design. The experimental conditions and the responses are presented in Table 3. A quadratic model was chosen from several models and fitted to the results. The regression equation obtained after the analysis of variance (ANOVA) represented the level of enzyme activity as a function of initial pH, moisture content, inoculum loading and time.

On the basis of their P-value, $R^2$, SD and predicted sum of square values, the adequacy of the quadratic regression model was found to be significant for GA production. The statistical significance of the ratio of mean square variation due to regression and mean square residual error was tested using the ANOVA. The associated P-value was used to estimate whether F was large enough to indicate statistical significance. If P-value was lower than 0.05, it indicated that the model was statistically significant. The ANOVA result for the GA production system showed the model F-value of 21.96 indicating that the model was significant (Table 4).

**Table 3.**
Central composite design with observed and predicted responses of glucoamylase activities. Each row corresponds to a single experiment.

| Run | $X_1$ [a] | $X_2$ [b] | $X_3$ [c] | $X_4$ [d] | Experimental | Predicted |
|---|---|---|---|---|---|---|
| | Actual (coded) | Actual (coded) | Actual (coded) | Actual (coded) | | |
| 1 | 6 (-1) | 60 (-1) | 100000 (-1) | 4 (-) | 13.73 | 14.56 |
| 2 | 6 (-1) | 60 (-1) | 100000 (-1) | 6 (+) | 10.34 | 18.73 |
| 3 | 6 (-1) | 60 (-1) | 1000000 (+1) | 4 (-) | 4.2 | -4.59 |
| 4 | 6 (-1) | 60 (-1) | 1000000 (+1) | 6 (+) | 2.06 | 12.71 |
| 5 | 8 (+1) | 60 (-1) | 100000 (-1) | 4 (-) | 36.18 | 35.15 |
| 6 | 8 (+1) | 60 (-1) | 100000 (-1) | 6 (+) | 92.57 | 84.06 |
| 7 | 8 (+1) | 60 (-1) | 1000000 (+1) | 4 (-) | 4.14 | 3.73 |
| 8 | 8 (+1) | 60 (-1) | 1000000 (+1) | 6 (+) | 67.34 | 65.81 |
| 9 | 6 (-1) | 80 (+1) | 100000 (-1) | 4 (-) | 5.76 | 5.66 |
| 10 | 6 (-1) | 80 (+1) | 100000 (-1) | 6 (+) | 5.26 | 7.42 |
| 11 | 6 (-1) | 80 (+1) | 1000000 (+1) | 4 (-) | 7.34 | 17.60 |
| 12 | 6 (-1) | 80 (+1) | 1000000 (+1) | 6 (+) | 33.14 | 32.53 |
| 13 | 8 (+1) | 80 (+1) | 100000 (-1) | 4 (-) | 26.62 | 17.72 |
| 14 | 8 (+1) | 80 (+1) | 100000 (-1) | 6 (+) | 57.08 | 64.24 |
| 15 | 8 (+1) | 80 (+1) | 1000000 (+1) | 4 (-) | 27.46 | 17.43 |
| 16 | 8 (+1) | 80 (+1) | 1000000 (+1) | 6 (+) | 76.24 | 77.13 |
| 17 | 7 (0) | 50 (-α) | 550000 (0) | 5 (0) | 39.9 | 40.14 |
| 18 | 9 (+α) | 70 (0) | 550000 (0) | 5 (0) | 51.00 | 62.24 |
| 19 | 7 (0) | 70 (0) | 1000 (-α) | 5 (0) | 53.41 | 53.08 |
| 20 | 7 (0) | 70 (0) | 1100000 (+α) | 5 (0) | 6.58 | 15.70 |
| 21 | 7 (0) | 70 (0) | 550000 (0) | 3 (-α) | 88.78 | 79.54 |
| 22 | 7 (0) | 70 (0) | 550000 (0) | 7 (+α) | 93.20 | 88.44 |
| 23 | 5 (-α) | 70 (0) | 550000 (0) | 5 (0) | 8.44 | -2.92 |
| 24 | 7 (0) | 90 (+α) | 550000 (0) | 5 (0) | 42.88 | 42.52 |
| 25 | 7 (0) | 70 (0) | 550000 (0) | 5 (0) | 93.67 | 88.44 |
| 26 | 7 (0) | 70 (0) | 550000 (0) | 5 (0) | 74.86 | 88.44 |
| 27 | 7 (0) | 70 (0) | 550000 (0) | 5 (0) | 93.67 | 88.44 |
| 28 | 7 (0) | 70 (0) | 550000 (0) | 5 (0) | 81.38 | 88.44 |
| 29 | 7 (0) | 70 (0) | 550000 (0) | 5 (0) | 93.67 | 88.44 |
| 30 | 7 (0) | 70 (0) | 550000 (0) | 5 (0) | 117.86 | 88.44 |

[a] Coded values of initial pH
[b] Moisture content (%, w/w)
[c] Inoculum loading (inoculum/g substrate)
[d] Time (day).

There is only a 0.01% chance that a "Model F-Value'' this large could occur due to noise. Considering the P-values of parameters, the effect of terms of $X_1$, $X_4$, $X_{14}$, $X_{23}$, $X_{11}$, $X_{22}$, $X_{33}$ and $X_{44}$ were significant, whereas that of $X_2$, $X_3$, $X_{12}$, $X_{13}$, $X_{24}$ and $X_{34}$ were negligible. The coefficient of determination ($R^2$) for the enzyme activity was calculated as 0.9565, showing that the fitted model could explain 95.65% of variability in the response. Moreover, the high $R^2$ value indicates that the quadratic equation is able to represent the system under the given experimental domain. An adequate precision of 12.74 for the enzyme activity was recorded.

**Table 4.**
ANOVA for glucoamylase production as a function of initial pH ($X_1$), moisture content ($X_2$), inoculum loading ($X_3$) and time ($X_4$).

| Source | Sum of Squares | DF | Mean Square | F Value | p-value Prob > F |
|---|---|---|---|---|---|
| **Model** | 31703.19 | 14 | 2264.51 | 21.96 | < 0.0001 |
| **$X_1$- pH** | 6367.44 | 1 | 6367.44 | 61.75 | < 0.0001* |
| **$X_2$- moisture content** | 8.52 | 1 | 8.52 | 0.083 | 0.7780 |
| **$X_3$- inoculums loading** | 42.37 | 1 | 42.37 | 0.41 | 0.5318 |
| **$X_4$- time** | 6112.04 | 1 | 6112.04 | 59.28 | < 0.0001* |
| **$X_{12}$** | 72.25 | 1 | 72.25 | 0.70 | 0.4166 |
| **$X_{13}$** | 149.57 | 1 | 149.57 | 1.45 | 0.2484 |
| **$X_{14}$** | 2003.91 | 1 | 2003.91 | 19.43 | 0.0006* |
| **$X_{23}$** | 969.39 | 1 | 969.39 | 9.40 | 0.0084* |
| **$X_{24}$** | 5.66 | 1 | 5.66 | 0.055 | 0.8181 |
| **$X_{34}$** | 173.45 | 1 | 173.45 | 1.68 | 0.2156 |
| **$X_{11}$** | 5660.75 | 1 | 5660.75 | 54.90 | < 0.0001* |
| **$X_{22}$** | 3636.26 | 1 | 3636.26 | 35.27 | < 0.0001* |
| **$X_{33}$** | 3628.09 | 1 | 3628.09 | 35.19 | < 0.0001* |
| **$X_{44}$** | 2730.15 | 1 | 2730.15 | 26.48 | 0.0001* |
| **Residual** | 1443.52 | 14 | 103.11 | | |
| **Lack of Fit** | 1104.55 | 9 | 122.73 | 1.81 | 0.2660 |
| **Pure Error** | 338.97 | 5 | 67.79 | | |
| **Corrected total** | 33146.71 | 28 | | | |

*Significant variable; DF, degree of freedom; determination coefficient ($R^2$), 0.9565; adjusted determination coefficient ($R^2$adj), 0.9129; coefficient of variation (CV), 22.81; adequate precision ratio, 12.74.

A value greater than 4 is desirable in support of the fitness of the model (Muthukumar et al., 2003). The adjusted $R^2$ corrects the $R^2$ value for the sample size and the number of terms used in the selected model. If there are many terms in the model and the sample size is not large enough, the adjusted $R^2$ may be clearly smaller than $R^2$. The P-value was used to determine the significance of related coefficients. If the P-value is lower than 0.05, the model and the corresponding coefficient is statistically significant (Khuri and Cornell, 1987). The Coefficient of Variation (CV) indicates the degree of precision with which the treatments are compared. Usually, the higher the CV value, the lower the reliability of experiment is. In this study, a CV value of 22.81 indicates a great reliability of the experiments performed. The table also shows a term for residual error, which measures the amount of variation in the response data left unexplained by the model. The analysis revealed that the form of the model chosen to explain the relationship between the factors and the response was correct.

The equation (2) in terms of actual factors (confidence level above 95%) as determined by Design of expert software is given below:

$$\text{GA Activity (U/gds)} = -1366.16+184.69\,X_1+17.38\,X_2+8.02\times10^{-6}\,X_3+39.82\,X_4-0.21\,X_1X_2-6.79\times10^{-6}\,X_1X_3+11.19\,X_1X_4+1.73\times10^{-6}\,X_2X_3-0.06\,X_2X_4+7.32\times10^{-6}\,X_3X_4-14.7\,X_1^2-0.12\,X_2^2-1.11\times10^{-10}\,X_3^2-10.21\,X_4^2 \quad \text{(Eq. 2)}$$

Where $X_1$, $X_2$, $X_3$ and $X_4$ are independent variables representing the pH, moisture content, inoculum loading and time, respectively. The negative coefficients for $X_{12}$, $X_{13}$, $X_{24}$, $X_{11}$, $X_{22}$, $X_{33}$ and $X_{44}$ demonstrate the existence of quadratic and linear interaction effects that decrease the response quantity, while the positive coefficients for $X_{14}$, $X_{23}$ and $X_{34}$ expose the existence of quadratic interaction effects that enhance the activity of GA. Figure 4 shows the correlation between the experimental and predicted values of the response.

The points close to the line indicate a good fit between the experimental and predicted data.

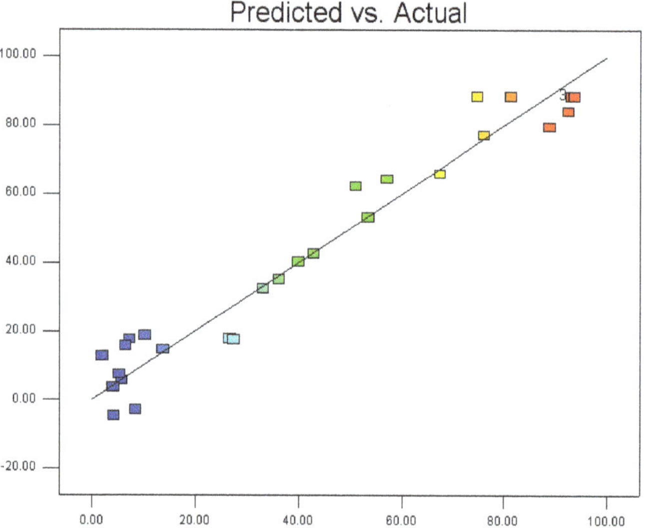


**Fig.4.** The observed (X axis) *vs.* the predicted (Y axis) glucoamylase activities under the experimental conditions.

The optima of the variables for which the responses were maximized are represented by the contour plots (Fig. 5). The contour plot of the moisture content and pH effect on the activity of GA illustrates that the neutral pHs led to higher enzyme activity using an initial moisture content of around 66-74% (wb) (Fig. 5A). The maximum activity of 92.92 U/gds was determined at pH 7.5 using initial moisture content of 69.6%. Lower initial moisture content provided lower solubility of the nutrients while higher moisture contents resulted in decreased porosity and gas exchange. The moisture content range was consistent with the levels reported in the literature for SSF of waste bread and wheat flour by *A. awamori* (Wang et al., 2009; Melikoglu et al., 2013a). Generally, the initial pH for GA production by *A. awamori* using SSF is adjusted to neutral pHs as the fungus grows well at such pHs. Since the maximum activity of 92.92 U/gds was determined at pH 7.5 using initial moisture content of 69.6%, these conditions were kept constant in the subsequent studies to find the optimum inoculum loading and incubation time.

The GA production increased by using an inoculum loading of $2\times10^5$ to $9\times10^5$/gs for 5 to 7 d and the maximum GA activity of 104.29 U/gds was obtained using $5.2\times10^5$/gs inoculum on the 6th day of the fermentation (Fig. 5B). Generally during fermentation, medium pH, nutrient concentration, temperature, moisture content, and physical structure of the raw material changes continuously. All these parameters affect microbial growth and enzyme production. According to Melikoglu et al. (2013a), the growth of *A. awamori* on bread pieces increased exponentially between the 3rd and 5th days and GA production reached its maximum level on the 6th day of the fermentation. However, as the medium pH was not controlled, the pH was decreasing during this period (Melikoglu et al., 2013a). They reported that pH decreased to 3.8 on the 5th day of the fermentation. This may be one of the major causes of deceleration of the growth and enzyme production after 6th day of the fermentation. Therefore, the effect of initial pH was evaluated using the optimized parameters and it was predicted that the GA activity increased from 90.69 to 107.1 U/gds using the initial pH of 7.9 instead of pH 7.0 (Fig. 5C). The pH value reached 4.5 after the 5th day of the fermentation when the initial pH was at 8 and 9. On the other hand, the pH value decreased to 3.5 and 4 when the initial pH was adjusted to 6 and 7, respectively. This explains why the microbial growth and GA production was enhanced using an initial pH of 7.9.

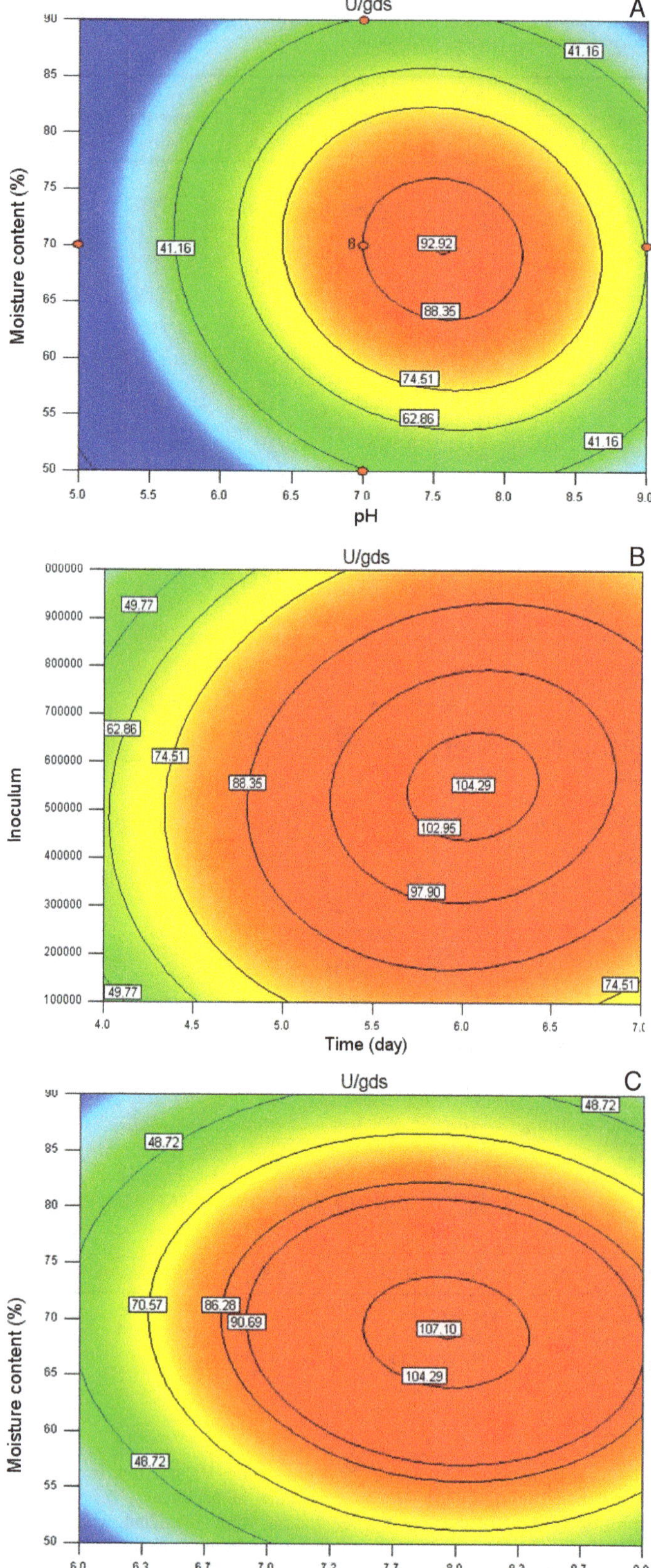

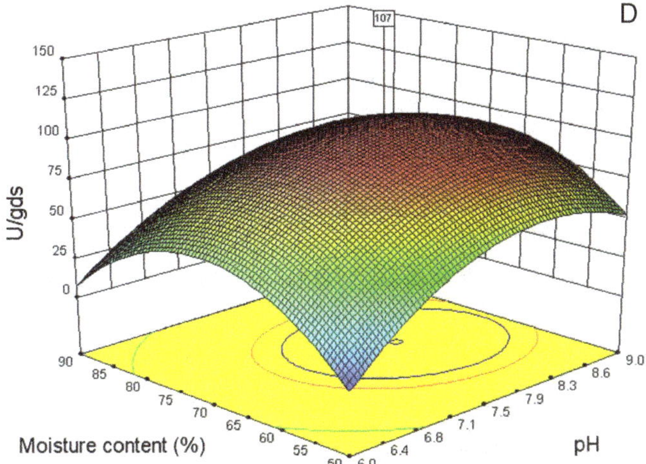


**Fig.5**. Contour plots, described by Eq. (2), representing the effect of initial pH and moisture content using inoculum loading of $5\times10^5$/g substrate for 5 d (5A); inoculum loading and incubation time using the initial moisture content of 69.6% and pH of 7.5 (5B); inoculum loading and pH using initial moisture content of 69.6% for 6 d (5C); initial pH and moisture content using inoculum loading of $5.2\times10^5$/g substrate for 6 d (5D) on glucoamylase activity from cake waste.

To evaluate the accuracy of the quadratic polynomial model, a verification experiment was conducted under the predicted optimal conditions and the result was 108.47 U/gds which was 1.37% higher than the predicted value. This is higher than values reported by Wang et al. (2009) for the same fungus using wheat flour and similar to those reported by Melikoglu et al. (2013a) on bread pieces. However, higher activities were reported with *A. niger* (695 U/g), but the enzymatic assay was carried out at pH 4.5 and the substrate was wheat bran (Silveira et al., 2006). This high degree of accuracy obtained confirmed the validity of the model with minor discrepancy due to the slight variation in the experimental conditions. The activity obtained was 1.4 fold higher than the yield obtained by cake wastes on the 6th day of the fermentation without optimization suggesting the important role of RSM for rapid screening of important process variables in optimization studies.

Many factors affect enzymatic hydrolysis including temperature, enzyme dose, substrate concentration and duration. The effect of reaction temperature on domestic FW (10%, w/v) hydrolysis using *in situ* produced GA was evaluated between 50 and 90°C (Fig. 6). During the first 6 h, the glucose production was the highest at 70°C (6.59 g $L^{-1}$) and then it slowed down (Fig. 6A). After 6 h, the glucose production at 50°C and 60°C was higher than that of 70°C. This might be because of enzyme denaturation at temperatures higher than 60°C. These findings are similar to the results reported in the literature. Melikoglu et al. (2013b) evaluated the kinetics of the GA using the same microorganism and reported that the maximum enzyme activity (12 U/ml) was obtained at 60°C and started to decrease at higher temperatures which was due to thermal deactivation of the enzyme. The highest glucose concentration of 10.4 g $L^{-1}$ corresponding to a saccharification degree of 97.9% was obtained at 60°C after 24 h. Hence, the subsequent studies were conducted at 60°C for 24 h.

The enzyme concentration also affected the enzymatic hydrolysis. FW hydrolysis speeded up with an increase in enzyme concentration especially in the first 6 h of hydrolysis. The glucose concentration obtained using 2 and 5 U/g FW was similar to the concentration obtained using 10U/g FW after 24 h (Fig. 6B). The effect of substrate loading was also evaluated using FW suspensions within the range of 10 and 50% (w/v) (Fig. 6C). Glucose production increased with an increase in substrate concentration. Among the various concentrations investigated, 50% (w/v) FW yielded the highest glucose concentration (52.3 g $L^{-1}$ with a saccharification degree of 98.4%) compared to the lower FW concentrations showing that there was no substrate inhibition.

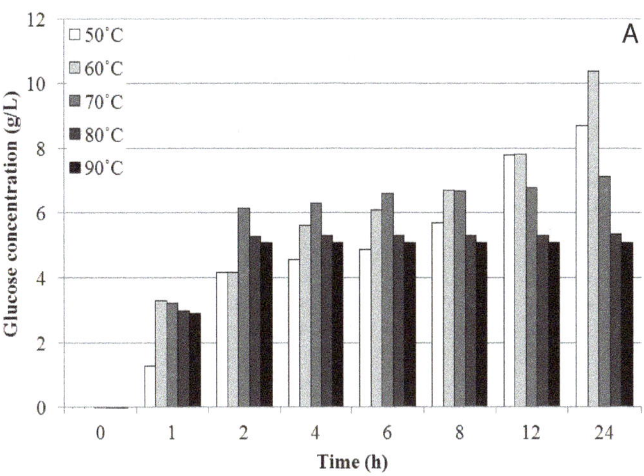


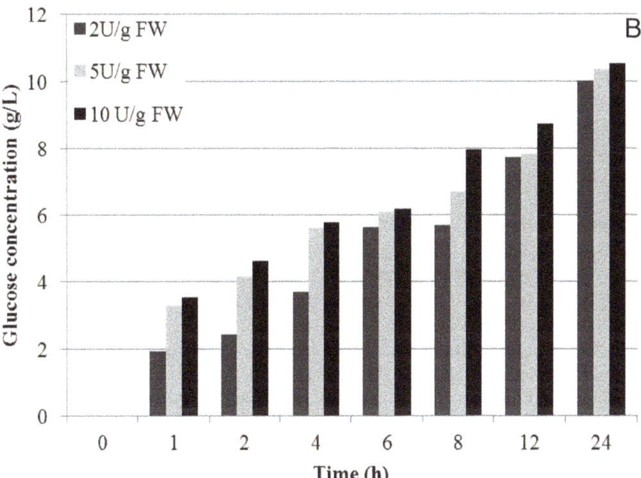


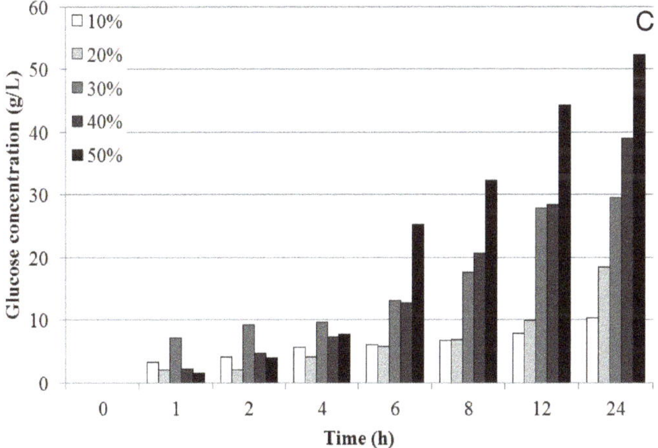


**Fig.6**. Effect of (A) temperature, (B) enzyme dose and (C) substrate concentration on reducing sugar formation during the hydrolysis of domestic food waste with the produced GA preparation. Data points show the averages from duplicate analyses for which the standard deviation was <1%.

## 4. Conclusion

This study demonstrated the feasibility of effective production of GA with SSF using FWs as sole nutrient source. GA with the highest activity was produced from cake waste using SSF by *A. awamori*. The optimum conditions for GA production from cake waste were determined as initial pH of 7.9, initial moisture content of 69.6%, inoculum loading of $5.2\times10^5$/gs and incubation time of 6 d. Under these conditions, GA activity of 108.47 U/gds was obtained. This study showed that waste cakes could be ideal raw materials for production of high-activity enzymes through SSF. The produced enzyme solution can be a potential candidate for the saccharification of FW, so it can significantly reduce the process cost for commercial enzymes are not purchased. The saccharification degree obtained during the hydrolysis may be one of the best reported to date and the glucose concentration obtained was sufficient enough for the production of various kinds of biofuels.

## 5. Acknowledgements

The authors gratefully acknowledge the financial support by National Environmental Agency, Singapore, grant number ETRP 1201 105.

## References

Anto, H., Trivedi, U.B., Patel, K.C., 2006. Glucoamylase production by solid-state fermentation using rice flake manufacturing waste products as substrate. Bioresour. Technol. 97(10), 1161-1166.

AOAC, 2001. Official Methods of Analysis, Association of Official Analytical Chemists, Washington, DC.

Arooj, M.F., Han, S.K., Kim, S.H., Kim, D.H., Shin, H.K., 2008. Continuous biohydrogen production in a CSTR using starch as a substrate. Int. J. Hydrogen Energy. 33(13), 3289-3294.

Bahcegul, E., Tatli, E., Haykir, N.I., Apaydin, S., Bakir, U., 2011. Selecting the right blood glucose monitor for the determination of glucose during the enzymatic hydrolysis of corncob pretreated with different methods. Bioresour. technol. 102, 9646-9652.

El-Fadel, M., Findikakis, A.N., Leckie, J.O., 1997. Environmental impacts of solid waste landfilling. J. Environ. Manage. 50(1), 1-25.

Ellaiah, P., Adinarayana, K., Bhavani, Y., Padmaja, P., Srinivasulu, B., 2002. Optimization of process parameters for glucoamylase production under solid state fermentation by a newly isolated *Aspergillus* species. Process Biochem. 38(4), 615-620.

FAO, 2012. Towards the future we want: End hunger and make the transition to sustainable agricultural and food systems, Food and agriculture organization of the United Nations Rome.

Garg, G., Mahajan, R., Kaur, A., Sharma, J., 2011. Xylanase production using agro-residue in solid-state fermentation from *Bacillus pumilus* ASH for biodelignification of wheat straw pulp. Biodegradation 22(6), 1143-1154.

Han, S.K., Shin, H.S.,2004. Biohydrogen production by anaerobic fermentation of food waste. Int. J. Hydrogen Energy. 29(6), 569-577.

Hara, A., Radin, N.S., 1978. Lipid extraction of tissues with a low toxicity solvent. Anal. Biochem. 90, 420–426.

Khuri, A.I., Cornell, J.A., 1987. Response Surfaces: Design and Analysis, Marcel Dekker, New York.

Kim, J.K., Oh, B.R., Shin, H., Eom, C., Kim, S.W., 2008. Statistical optimization of enzymatic saccharification and ethanol fermentation using food waste. Process Biochem. 43(11), 1308-1312.

Koike, Y., An, M.Z., Tang, Y.Q., Syo, T., Osaka, N., Morimura, S., Kida, K., 2009. Production of fuel ethanol and methane from garbage by high-efficiency two-stage fermentation process. J. Biosci. Bioeng. 108(6), 508-512.

Leung, C.C.J., Cheung, A.S.Y., Zhang, A.Y.Z., Lam, K.F. and Lin, C.S.K., 2012. Utilisation of waste bread for fermentative succinic acid production. Biochem. Eng. J. 65, 10-15.

Melikoglu, M., Lin, C.S.K., Webb, C., 2013a. Stepwise optimisation of enzyme production in solid state fermentation of waste bread pieces. Food Bioprod. Process. 91(4), 638-646.

Melikoglu, M., Lin, C.S.K., Webb, C., 2013b. Kinetic studies on the multi-enzyme solution produced via solid state fermentation of waste bread by *Aspergillus awamori*. Biochem. Eng. J. 80, 76-82.

Merino, S.T., Cherry, J., 2007. Progress and challenges in enzyme development for biomass utilization. Adv. Biochem. Eng./Biotechnol. 108, 95-120.

Miller, G.L., 1959. Use of dinitrosalicylic acid reagent for determination of reducing sugars. Anal. Chem. 426-428.

Muthukumar, M., Mohan, D., Rajendran, M., 2003. Optimization of mix proportions of mineral aggregates using Box Behnken design of experiments. Cem. Concr. Compos. 25, 751-758.

Norouzian, D., Akbarzadeh, A., Scharer, J.M., Moo Young, M., 2006. Fungal glucoamylases. Biotechnol. Adv. 24(1), 80-85.

Ohkouchi, Y., Inoue, Y., 2007. Impact of chemical components of organic wastes on l(+)-lactic acid production. Bioresour.Technol. 98(3), 546-553.

Pandey, A.,1991. Effect of particle size of substrate on enzyme production in solid-state fermentation. Bioresour. Technol. 37, 169-172.

Ruiz, H.A., Rodriguez-Jasso, R.M., Rodriguez, R., Contreras-Esquivel, J.C., Aguilar, C.N., 2012. Pectinase production from lemon peel pomace as support and carbon source in solid state fermentation column-tray bioreactor. Biochem. Eng . 65, 90-95

Sakai, K., Taniguchi, M., Miura, S., Ohara, H., Matsumoto, T., Shirai, Y., 2004. Making plastics from garbage: A novel process for poly-L-lactate production from municipal food waste. J. Ind. Ecol. 7(3-4), 63-74.

Sakai, K., Ezaki, Y., 2006. Open L-lactic acid fermentation of food refuse using thermophilic *Bacillus coagulans* and fluorescence in situ hybridization analysis of microflora. J.Bioscie. Bioeng. 101(6), 457-463.

Silveira, S.T., Oliveira, M.S., Costa, J.A.V., Kalil, S.J., 2006. Optimization of glucoamylase production by *Aspergillus niger* in solid-state fermentation. Appl. Biochem. Biotechnol. 128, 131-139.

Smith, A., Brown, K., Ogilvie, S., Rushton, K., Bates, J., 2001. Waste Management Options and Climate Change: Final Report.

Soni, S.K., Kaur, A. and Gupta, J.K.,2003. A solid-state fermentation based bacterial alpha-amylase, fungal glucoamylase system, its suitability for hydrolysis of wheat starch. Process Biochem. 39, 158-192.

Soni, S.K., Goyal, N., Gupta, J.K., Soni, R., 2012. Enhanced production of α-amylase from bacillus subtilis subsp. spizizenii in solid state fermentation by response surface methodology and its evaluation in the hydrolysis of raw potato starch. Starch/Staerke. 64(1), 64-77.

Uncu, O.N., Cekmecelioglu, D., 2011. Cost-effective approach to ethanol production and optimization by response surface methodology. Waste Manage. 31(4), 636-643.

Wang, Q., Wang, X., Wang, X., Ma, H., Ren, N., 2005. Bioconversion of kitchen garbage to lactic acid by two wild strains of *Lactobacillus* species. J. Environ. Sci. Health., Part A. Toxic/Hazardous Substances and Environmental Engineering. 40(10), 1951-1962.

Wang, R., Ji, Y., Melikoglu, M., Koutinas, A., Webb, C., 2007. Optimization of innovative ethanol production from wheat by response surface methodology. Process Saf. Environ. Prot. 85(5B), 404-412.

Wang, R., Godoy, L.C., Shaarani, S.M., Melikoglu, M., Koutinas, A., Webb, C., 2009. Improving wheat flour hydrolysis by an enzyme mixture from solid state fungal fermentation. Enzyme Microb. Technol. 44(4), 223-228.

Wang, Q.H., Liu, Y.Y., Ma, H.Z., 2010. On-site production of crude glucoamylase for kitchen waste hydrolysis. Waste Manag. Res. 28, 539-544.

Yan, S., Chen, X., Wu, J., Wang, P., 2012. Ethanol production from concentrated food waste hydrolysates with yeast cells immobilized on corn stalk. Appl. Microbiol. Biotechnol.. 94(3), 829-838.

Yang, S.Y., Ji, K.S., Baik, Y.H., Kwak, W.S., McCaskey, T.A., 2006. Lactic acid fermentation of food waste for swine feed. Bioresour. Technol. 97(15), 1858-1864.

Zhang, C., Xiao, G., Peng, L., Su, H., Tan, T., 2013a. The anaerobic co-digestion of food waste and cattle manure. Bioresour. Technol. 129, 170-176.

Zhang, A.Y., Sun, Z., Leung, C.C.J., Han, W., Lau, K.Y., Li, M. and Lin, C.S.K., 2013b. Valorisation of bakery waste for succinic acid production. Green Chem. 15(3), 690-695.

Zhang, M., Shukla, P., Ayyachamy, M., Permaul, K., Singh, S., 2010. Improved bioethanol production through simultaneous saccharification and fermentation of lignocellulosic agricultural wastes by *Kluyveromyces marxianus* 6556. World J. Microbiol. Biotechnol. 26(6), 1041-1046.

11

# Simultaneous electricity generation and sulfide removal via a dual chamber microbial fuel cell

Paniz Izadi, Mostafa Rahimnejad*

*Biotechnology Research Lab., Faculty of Chemical Engineering, Babol University of Technology, Babol, Iran*

## HIGHLIGHTS

- Transfer of produced electrons to anode is one of the main parts in MFCs.
- Some MFCs needs artificial electron acceptors in their aerobic cathode compartment.
- Cyclic voltammetry was used to study of anodic electrochemistry analysis.
- Hexacyanoferrate was used as cathodic solution in different concentrations.
- Maximum generated voltage and power density were 988.9145 mV, 346.746 mW.m$^{-2}$, respectively.

## GRAPHICAL ABSTRACT

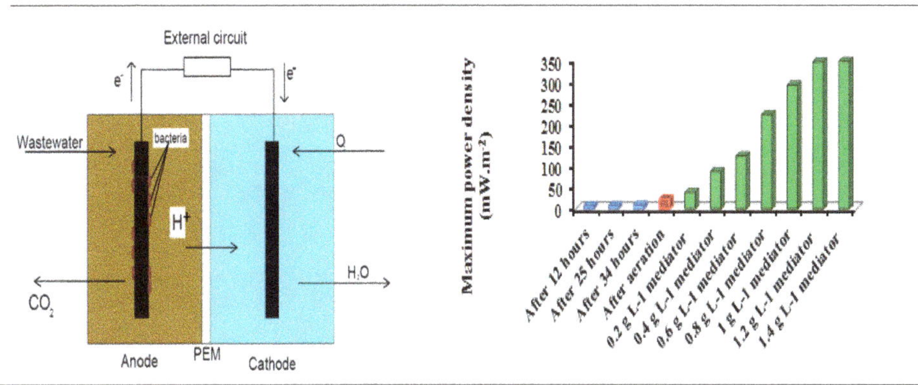


## ABSTRACT

Microbial fuel cells (MFCs) have recently been used to alter different sources of substrates to produce bioelectricity. MFCs can also be used for wastewater treatment and electricity generation simultaneously. Sulfur compounds such as sulfides commonly exist in wastewater and organic waste. In this study a dual chamber MFC was constructed for power production. Sulfide was used as the electron donor in the anaerobic anode compartment. A mixed culture of microorganisms was used as an active biocatalyst to convert the substrate into electricity. The obtained experimental results illustrated that the MFC can successfully alter sulfide to elementary sulfur while generating power. The initial concentration of sulfide in the anode compartment was 0.4 g l$^{-1}$ and it was completely removed after 3 days of MFC operation. The influence of oxygen was examined in the cathode chamber and the cell voltage gradually increased during aeration, reaching 480 mV after 1200 s. Hexacyanoferrate was added to the cathodic solution in different concentrations and its effects were investigated. The maximum generated voltage, power and current density were 988.9145 mV, 346.746 mW.m$^{-2}$, 1285.64 mA.m$^{-2}$, respectively and they were obtained in the presence of 1.4 g l$^{-1}$ of mediator.

**Keywords:**
Fuel cell
Bioelectricity
Power density
Sulfide
Dual chamber

## 1. Introduction

The reduction in the earth's fossil fuels and the environmental problems associated with them are two important issues which have attracted researcher's attention to fuel cells (Bagotzky et al., 2003). Fuel cells are a suitable alternative to fossil fuels; however their use has some disadvantages such as harsh reaction conditions and expensive catalysts (Rodrigo et al., 2007). Accordingly, renewable bioenergy is viewed as one of the ways to decrease the current crisis. A microbial fuel cell (MFC) is a biological system which converts biodegradable organic substances to electricity using bacteria as its biocatalyst (Katuri and Scott, 2011). An MFC generates electricity from bio-convertible substrates. Bacteria change natural electron acceptors, such as oxygen into an insoluble acceptor, such as the MFC anode (Rabaey and Verstraete, 2005). The advantages of this technology include non-pollution, high energy efficiency, mild operating conditions, strong biocompatibility and great application potential in various areas, which have received a great deal of attention from scientists.

In many studies, oxygen was used as the final electron acceptor in the

* Corresponding author
E-mail address: rahimnejad_mostafa@yahoo.com (M. Rahimnejad).

cathode compartment (Reddy et al., 2010). The consumption of electrons and protons that are combined with oxygen, eventually create water, and end this transfer cycle. Oxidized mediators, can also accelerate the reaction of water formation in the cathode chamber. Substrate as nutrient source of the cell played an important biological role and other factors such as mediator, microorganisms and nutrients affected the bioelectricity production pattern (Sun et al., 2010).

MFCs are capable of producing clean energy, apart from their effective treatment of waste material such as wastewater (Liu et al., 2004). The active microorganisms in the anaerobic anode compartment have the ability to use the organic matter which exists in the wastewater as a source of energy and generate protons and electrons, through which electricity can be recovered (Zhang and Liu, 2010). Wastewater treatment by MFC was first done by Habermann and Pommer (1991). Wastewaters include various organic compounds, so they can be consumed in MFCs as fuel. After Habermann and Pommer, many other researchers also used this application of MFC but with different substrates. Food processing wastewater, hygienic wastes, and swine wastewater were used as biomass sources for MFCs because of their rich organic matters (Liu et al., 2004; Min et al., 2011; Oh and Logan, 2005). MFC can be used to treat organic and inorganic matters as well as oxidized metal pollutants in the cathode and anode chambers and many studies have been related to this topic (Du et al., 2007). Table 1 shows a list of MFCs which have been examined for the elimination of different pollutants and electricity generation.

**Table 1**
Different elimination of substrate via MFC.

| Elimination of | Cathode Material | Anode Material | Maximum power | Removal at | Removal efficiency (%) | Reference |
|---|---|---|---|---|---|---|
| Carbon | Granular graphite | Granular graphite | 34.6 W.m-3 | Anode chamber | 100 | (Virdis et al., 2008) |
| COD | Graphite felt | Graphite felt | 7.6 mW | Anode chamber | 90 | (Moon et al., 2005) |
| Dye | Graphite plate | Graphite plate | 15.73 mW.m-2 | Anode chamber | 93.15 | (Yadav et al., 2012) |
| Nitrogen | Granular graphite | Granular graphite | 34.6 W.m-3 | Cathode chamber | 67 | (Virdis et al., 2008) |
| Copper | Graphite plate | Graphite plate | 339 mW.m-3 | Cathode chamber | 96 | (Tao et al., 2011) |

Organic molecules in wastewaters can be biodegraded in the anode compartment of MFCs (Du et al., 2007). Sulfide is a common organic molecule which is found in wastewaters by the action of anaerobic bacteria on organic substances. High levels of sulfide ions in waters are perilous for living creatures and so should be treated from wastewater before they are released into the environment (Rahimnejad et al., 2012). Exploring new energy sources is an important issue in achieving sustainable development and circulation economy (Najafpour et al., 2011). Techniques such as physical, chemical and electrochemical methods have been used for sulfide treatment but they are expensive and high operated (Lee et al., 2012). Microbial fuel cell is a new technology which can be used for power production and sulfide removal simultaneously. Many literatures have focused on sulfide treatment in MFCs (Cai and Zheng, 2012). Sulfide removal in MFC using certain microbes such as Alcaligenes sp., Paracoccus sp. and Pseudomonas sp. C27 was investigated before (Lee et al., 2012); but few studies have considered using this process in mixed cultures. Microbial communities used in MFCs for power generation and sulfide alteration have been probed (Rahimnejad et al., 2012). Sulfide can be oxidized to different sulfur species depending on its surrounding condition and redox potential. Polysulfides, sulfates, thiosulfates and elemental sulfur are the results of sulfide oxidation in different redox potential (Sun et al., 2009). Also it has been indicated that sulfide oxidation plays a key role in power production in sedimentary MFCs (Tender et al., 2002).

In this study, a two chamber MFC was designed and fabricated for power production and sulfide removal. A mixed culture of microorganisms was used for this aim. The effects of different concentrations of oxidizer agent and aeration on MFC performance, power generation, and voltage and sulfide removal efficiency of MFC were investigated.

## 2. Materials and methods

### *2.1. Material*

The MFC used in this paper was constructed by using two 750 ml cubic Plexiglas chambers as its anode and cathode compartments (figure 1). The two chambers were separated by a PEM (3.0cm× 3.0 cm, Nafion-117, Sigmae Aldrich, USA) and assembled using stainless steel studding, nuts and washers. Graphite plates were used as anode and cathode electrodes. The anode was placed in the center of the anodic chamber (area, 21cm$^2$) and it was connected to the cathode (area, 18 cm$^2$) to provide the connections for the external circuit with wire copper. Both electrodes were parallel to the PEM. The voltage was measured by using data logger (fabricated analog digital data acquisition to record data point) and recorded through a personal computer. The MFC was operated in the fed-batch mode and at room temperature between 16 °C and 22 °C. The cathode solution contained 500 ml of deionized water.

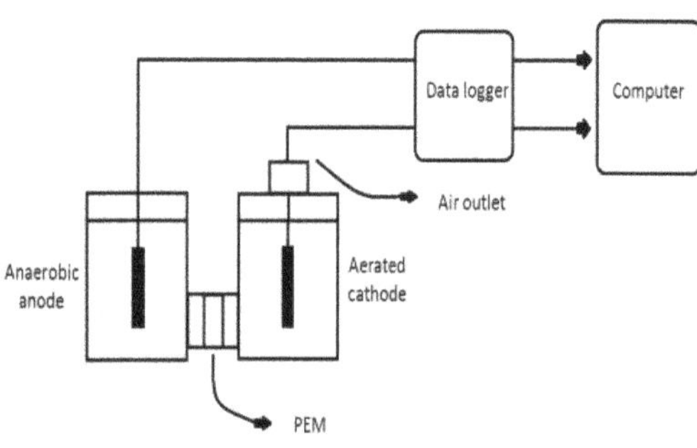


**Fig.1.** Schematic diagram of the dual chamber MFC for bioelectricity production

### *2.2. Experimental procedure*

Activated sludge was used as inoculums and it was collected from the anaerobic process tank operated at the Ghaemshahr wastewater treatment center located in northern Iran. Sulfide was used as the sole electron donor. The conductivity and pH of the wastewater were 1093 μS cm$^{-1}$ and 6.5, respectively. Wastewater containing 0.4 g l$^{-1}$ sulfide was used as the anode solution and the total liquid volume was 600 ml. The anode chamber was purged with nitrogen gas for 10 min to remove dissolved oxygen so as to maintain anaerobic conditions.

### *2.3. Analysis*

All chemicals and reagents used for the experiments were analytical grades and supplied by Merck (Darmstadt, Germany). The sulfide concentration was determined by using electrochemical methods. A cyclic voltammetry was performed at room temperature using the potentiostat/galvanostat electrochemical analysis system Ivium (Netherlands, V11100) with a voltammetry cell in a three electrodes configuration. An anode electrode, platinum wire and Ag|AgCl|KCl (3 M) were used as working, auxiliary and reference electrodes, respectively. Scanning electronic microscopy (SEM, VEGA2-TESCAN) techniques were applied to provide surface and morphological information for the used electrodes in the anode compartment.

The voltage was measured by using data logger and polarization curves were obtained through an adjustable external resistance, measured when the voltage was kept constant. The power and current were calculated based on the following equations:

$$P = I^2.R \quad (1) \qquad P = E^2/R \quad (2)$$

In these equations P represents produced power, E is measured cell voltage; R is external resistance and I indicate the produced current. The current and power generation density was normalized to the geometric area of membrane (9 $cm^2$).

## 3. Results and discussion

### *3.1 MFC performance*

Microbial fuel cell (MFC) is a new technology that can produce electricity and treat wastewater concurrently. After inoculation of 0.4 $g.l^{-1}$ sulfide in the anode chamber with an active mixed culture as electrically active bacteria, data logger was used to obtain data in the form of an open circuit voltage until reaching a steady state condition.

The fabricated MFC was operated in the batch mode at room temperature. The performance of the microbial fuel cell was then evaluated by using a polarization curve. A maximum power density of 7.867 $mW.m^{-2}$ was obtained for MFC after 12 hours. For the latter, the power density remained around 8.134 $mW.m^{-2}$ until the 25th hour. The steady-state conditions were achieved after 34 hours of operation time. At the steady-state conditions, the maximum produced power and current density were 9.758 $mW.m^{-2}$ and 134.487 $mA.m^{-2}$, respectively. The maximum power and current density of the MFC in these 34 hours have been shown in table 2.

**Table 2**
Maximum power and current density generated in 34 hours after incubation of MFC by active microorganisms.

| Time (h) | Maximum power ($mW.m^{-2}$) | Maximum current ($mA.m^{-2}$) |
|---|---|---|
| 12 | 7.867 | 108.890 |
| 25 | 8.134 | 121.191 |
| 34 | 9.758 | 134.487 |

Power density and polarization curves of the MFC used during the steady state condition have been presented in Figure 2. The maximum power density obtained without aeration and optimization were 9.758 $mW.m^{-2}$ (see Figure.2).

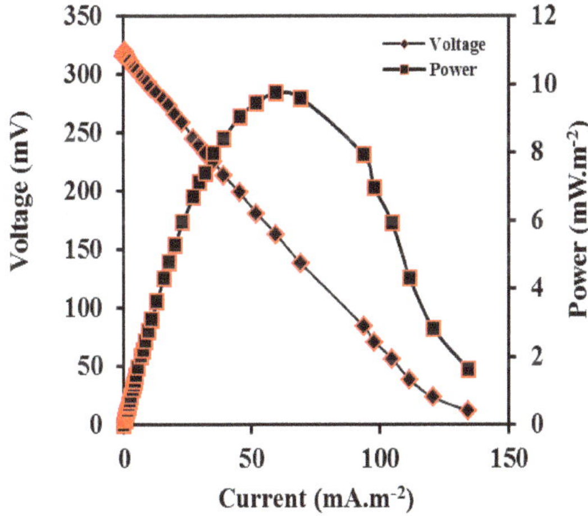


**Fig. 2.** Polarization curve of MFC in batch mode (34 hours after MFC operation).

According to prior studies, oxygen is an appropriate electron acceptor; therefore an increment of oxygen concentration in MFC can improve its performance (Gil et al., 2003). In this study, the influence of oxygen in the cathode chamber on OCV and the generated bioelectricity in the steady-state condition has been examined. While OCV results showed a stable value, the air supply with a constant flow rate was used to aerate the cathode chamber.

Figure 3 shows the performance of the MFC in terms of OCV improvement and with respect to time in the presence and absence of aeration in the cathode compartment. It can be observed that during aeration, the cell voltage rapidly increased, reaching 480 mV, leading to an increase in the MFC's performance and power generation. The maximum produced power and current density in the presence of aeration in the cathode chamber were 23.78 $mW.m^{-2}$ and 212.77 $mA.m^{-2}$, respectively.

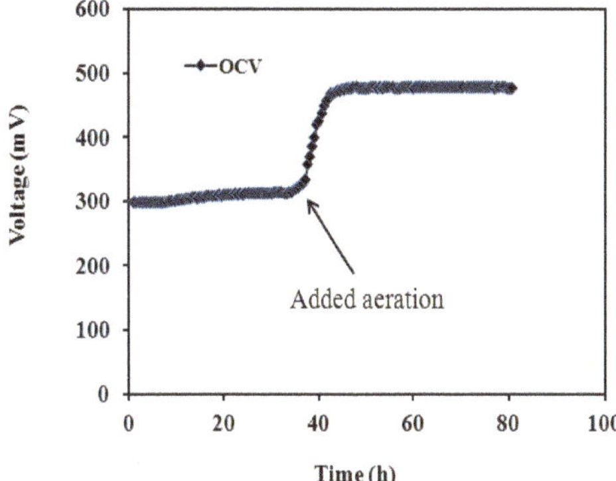


**Fig.3.** Open circuit voltage produced in MFC at steady-state condition and after aeration.

In each power supply, the main parameter is to enhance power and then to acquire the highest current density under maximum power density. The effect of mediators in the cathode compartment on MFC performances was investigated. Hexacyanoferrate was used as a mediator in the cathode chamber. The effect of aeration and lowest concentration of oxidizer agent on MFC performance have been shown in figure 4. When 0.2 g $l^{-1}$ hexacyanoferrate was added to the cathode solution, the produced power density increased up to 38.773 $mW.m^{-2}$.

Hexacyanoferrate increases the conductivity of the system and is a suitable oxidizer agent in the MFC. This is the main reason behind the improvement seen in the performance of the MFC after the addition of a mediator (Jang et al., 2004). Also mediators such as hexacyanoferrate play a significant role in decrement of internal resistance (Zhang and Liu, 2010). Internal resistance is an important parameter in MFC performance and affects the output power.

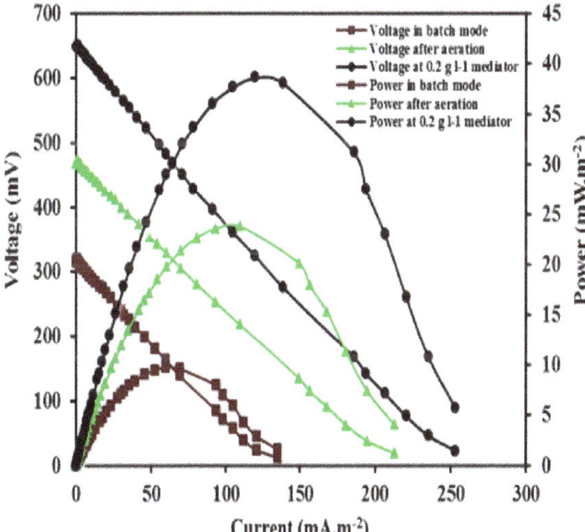


**Fig.4.** Variation of polarization curves in batch mode, after aeration and while using 0.2 g $L^{-1}$ hexacyanoferrate as mediator in cathode chamber.

Different concentrations of hexacyanoferrate were examined in the cathode solution. At each concentration the polarization data was obtained when the voltage output became stable (after 2 h). Maximum power densities and open circuit voltages have been shown in table 3. Concentrations of oxidizer agent those were greater than 1.2 g $l^{-1}$ showed no positive impact for the additional current and power (See Table 3).

**Table 3**
Maximum power and current density and OCV of MFC in different concentration of hexacyanoferrate in cathode chamber as mediator.

| Hexacyanoferrate concentration in cathode chamber (g l-1) | OCV (mV) | Maximum power (mW.m-2) | Maximum current (mA.m-2) |
|---|---|---|---|
| 0 | 480.3073 | 9.782 | 99.392 |
| 0.2 | 648.1525 | 38.773 | 253.268 |
| 0.4 | 814.0657 | 88.727 | 442.238 |
| 0.6 | 836.7671 | 124.732 | 588.015 |
| 0.8 | 929.2631 | 221.841 | 855.273 |
| 1 | 968.9145 | 291.127 | 1063.141 |
| 1.2 | 984.9145 | 346.220 | 1257.51 |
| 1.4 | 988.9145 | 346.746 | 1285.64 |

Figure 5 shows the polarization, power density and open circuit voltage curves obtained after addition of the highest concentration of the mediator (1.4 g $l^{-1}$) in the MFC.

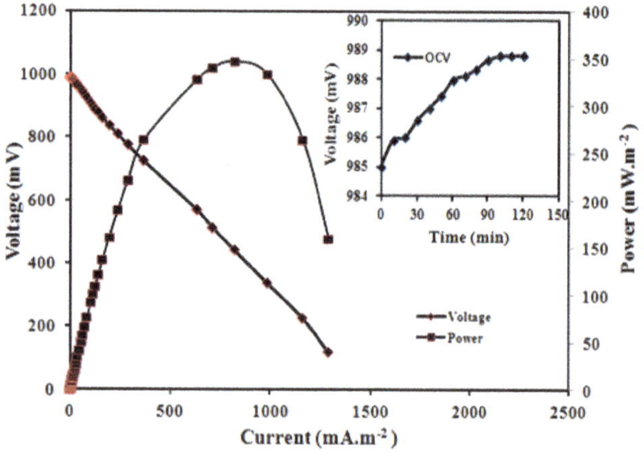


**Fig.5.** Polarization and power density curves after addition of 1.4 g $L^{-1}$ hexacyanoferrate in the cathode compartment. Inset: open circuit potential curve in two hours after addition of 1.4 g $L^{-1}$ hexacyanoferrate.

The initial concentration of sulfide in wastewater was determined. 0.4 g $l^{-1}$sulfide was consumed by the microorganisms to produce bioelectricity. Sulfide was detected after 3 days and almost 98% of the sulfide disappeared from the MFC. Our obtained results were very similar to the findings of previous researches; Zhao and coworkers used sulfur compounds as substrates in the MFC and sulfur compounds removal was attained (Zhao et al., 2009). Ryckelynck et al. produced electricity in the MFC by sulfide oxidation and they too achived 98% sulfide removal (Rabaey et al., 2006). Nearly complete sulfide removal was also obtained in the past e (Cai and Zheng, 2012).

### *3.2 Electrochemical behavior of MFC*

Cyclic voltammetry (CV) is an electrochemical analysis method which can investigate the electrochemical behavior and activity of microbes in MFC (Zhang et al., 2009). In this study, cyclic voltammograms have been plotted using platinum wire and Ag/AgCl as auxiliary and reference electrodes, respectively. As can be seen, Figure 6 is depicted for anolyte in three different compositions. The electrochemical activity in anolyte has been examined with no inoculums, with just microorganism and with a mixture of inoculums and sulfide.

When there were no inoculums in the anode chamber, voltammograms had no sensible peak. They confirmed that there were no electrochemical activities in the anode chamber. Voltammogram (b) had one oxidation-reduction peak at approximately 0.265 V (vs Ag/AgCl) which indicated the electrochemical activity of the attached microorganisms on the anode electrode. Voltammograms (c) had two oxidation-reduction peaks which were approximately at -0.19 V (vs Ag/AgCl) and 0.29 V (vs Ag/AgCl). The first peak was attributed to sulfide oxidation and the second, implied the microorganisms' activity.

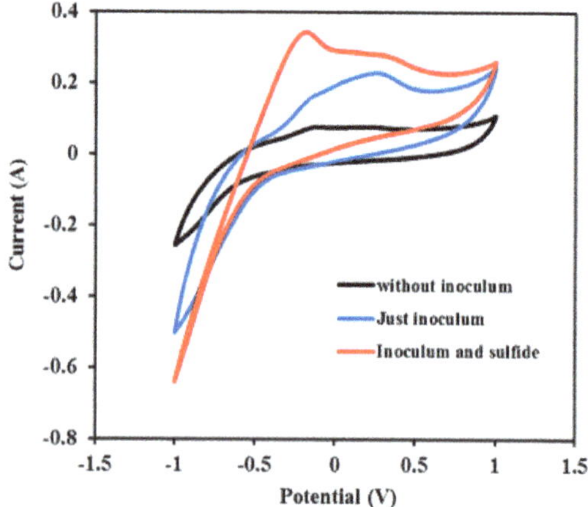


**Fig.6.** CV scans in the anode solution.

As can be seen in figure 7 scanning electronic microscopy demonstrates surface and morphological information about the used electrode in the anode compartment. A piece of the anode electrode (1×1cm) was analyzed by scanning electronic microscope before (7.a) and after (7.b) usage in the two chambers of the MFC with magnification of 2000. These obtained images demonstrated microorganisms grown on the graphite surface

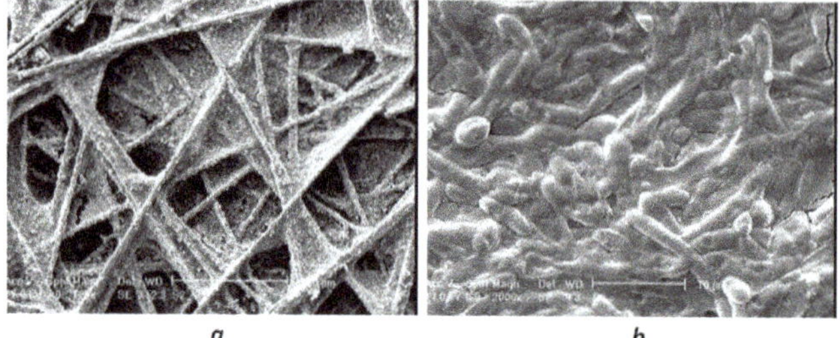

*a* *b*

**Fig.7.** SEM image of carbon paper before (a) and after (b) using in MFC.

## 4. Conclusions

In this study a fabricated MFC was successfully operated to treat sulfide in the wastewater and simultaneously generate bioelectricity. The system used sulfide as a substrate at a concentration of 0.4 g $l^{-1}$. This was almost completely removed from the wastewater during the MFC operation and oxidized to elemental sulfur. Aeration in the cathode chamber increased the power density approximately 2 times. Hexacyanoferrate was used with several concentrations as an oxidizing agent in the cathode chamber to enhance the performance of the MFC. The maximum obtained power and current density were 346.746 mW.$m^{-2}$ and 1285.64 mA.$m^{-2}$, respectively in the presence of 1.4 g $l^{-1}$ of the mediator. The electrochemical activity of microorganisms and sulfide oxidation were confirmed using cyclic voltammetry.

**Acknowledgements**

The authors wish to acknowledge the Biotechnology Research Center, Babol Noshirvani University of Technology, Babol, Iran, for the facilities provided to accomplish this research.

**References**

Bagotzky, V., Osetrova, N., Skundin, A., 2003. Fuel cells: state-of-the-art and major scientific and engineering problems. Russ. J. Electrochem. 39, 919-934.

Cai, J. Zheng, P., 2013. Simultaneous anaerobic sulfide and nitrate removal in microbial fuel cell. Bioresource Technol. 128, 760-764.

Du, Z., Li, H., Gu, T., 2007. A state of the art review on microbial fuel cells: a promising technology for wastewater treatment and bioenergy. Biotechnol. Adv. 25, 464-482.

Gil, G.C., Chang, I.S., Kim, B.H., Kim, M., Jang, J.K., Park, H.S., Kim, H.J., 2003. Operational parameters affecting the performance of a mediator-less microbial fuel cell. Biosen. Bioelectron. 18, 327-334.

Habermann, W., Pommer, E., 1991. Biological fuel cells with sulphide storage capacity. Appl. Microbiol. Biot. 35, 128-133.

Jang, J.K., Chang, I.S., Kim, B.H., 2004. Improvement of cathode reaction of a mediatorless microbial fuel cell. J. Microbiol. Biotechn. 14, 324-329.

Katuri, K.P., Scott, K., 2011. On the dynamic response of the anode in microbial fuel cells. Enzyme Microb. Tech. 48, 351-358.

Lee, C.Y., Ho, K.L., Lee, D.J., Su, A., Chang, J.S., 2012. Electricity harvest from nitrate/sulfide-containing wastewaters using microbial fuel cell with autotrophic denitrifier *Pseudomonas* sp. C27", Int. J. Hydrogen Energy 37, 15827-15832.

Liu, H., Ramnarayanan, R., Logan, B.E., 2004. Production of electricity during wastewater treatment using a single chamber microbial fuel cell. Environ. Sci. Technol. 38, 2281-2285.

Min, B., Kim, J., Oh, S., Regan, J.M., Logan, B.E., 2005. Electricity generation from swine wastewater using microbial fuel cells. Water Res. 39, 4961-4968.

Moon, H., Chang, I.S., Jang, J.K., Kim, B.H., 2005. Residence time distribution in microbial fuel cell and its influence on COD removal with electricity generation. Biochem. Eng. J. 27, 59-65.

Najafpour, G., Rahimnejad, M., Ghoreishi, A., 2011. The Enhancement of a Microbial Fuel Cell for Electrical Output Using Mediators and Oxidizing Agents. Energ. Source. 33, 2239-2248.

Oh, S. Logan, B.E., 2005. Hydrogen and electricity production from a food processing wastewater using fermentation and microbial fuel cell technologies. Water Res. 39, 4673-4682.

Rabaey, K., Van de Sompel, K., Maignien, L., Boon, N., Aelterman, P., Clauwaert, P., De Schamphelaire, L., Pham, H.T., Vermeulen, J., Verhaege, M.,2006. Microbial fuel cells for sulfide removal. Environ. Sci. Technol. 40, 5218-5224.

Rabaey, K., Verstraete, W., 2005. Microbial fuel cells: novel biotechnology for energy generation. Trends Biotechnol. 23, 291-298.

Rahimnejad, M., Ghasemi, M., Najafpour, G.D., Ismail, M., Mohammad, A.W., Ghoreyshi, A.A., Hassan, S.H.A., 2012. Synthesis, characterization and application studies of self-made Fe3O4/PES nanocomposite membranes in microbial fuel cell. Electrochim. Acta. 85, 700-706.

Reddy, L.V., Kumar, S.P., Wee, Y.J., 2010. Microbial fuel cells (MFCs)–a novel source of energy for new millennium. Appl. Microbiol. Biot. 956-964.

Rodrigo, M., Canizares, P., Lobato, J., Paz, R., Sáez, C., Linares, J., 2007. Production of electricity from the treatment of urban waste water using a microbial fuel cell. J. Power Sources. 169, 198-204.

Sun, M., Mu, Z.X., Chen, Y.P., Sheng, G.P., Liu, X.W., Chen, Y.Z., Zhao, Y., Wang, H.L., Yu, H.Q., Wei, L., 2009. Microbe-assisted sulfide oxidation in the anode of a microbial fuel cell. Environ. Sci. Technol. 43, 3372-3377.

Sun, M., Tong, Z.H., Sheng, G.P., Chen, Y.Z., Zhang, F., Mu, Z.X., Wang, H.L., Zeng, R.J., Liu, X.W., Yu, H.Q., 2010. Microbial communities involved in electricity generation from sulfide oxidation in a microbial fuel cell. Biosens. Bioelectron. 26, 470-476.

Tao, H.C., Liang, M., Li, W., Zhang, L.J., Ni, J.R., Wu, W.M., 2011. Removal of copper from aqueous solution by electrodeposition in cathode chamber of microbial fuel cell. J. Hazard. Mater. 189, 186-192.

Tender, L.M., Reimers, C.E., Stecher, H.A., Holmes, D.E., Bond, D.R., Lowy, D.A., Pilobello, K., Fertig, S.J., Lovley, D.R., 2002. Harnessing microbially generated power on the seafloor. Nat. Biotechnol. 20, 821-825.

Virdis, B., Rabaey, K., Yuan, Z., Keller, J., 2008. Microbial fuel cells for simultaneous carbon and nitrogen removal. Water Res. 42, 3013-3024.

Yadav, A.K., Dash, P., Mohanty, A., Abbassi, R., Mishra, B.K., 2012. Performance assessment of innovative constructed wetland-microbial fuel cell for electricity production and dye removal. Ecol. Eng. 47, 126-131.

Zhang, B., Zhao, H., Shi, C., Zhou, S., Ni, J., 2009. Simultaneous removal of sulfide and organics with vanadium (V) reduction in microbial fuel cells. J. Chem. Technol. Biot. 84, 1780-1786.

Zhang, P.Y., Liu, Z.L., 2010. Experimental study of the microbial fuel cell internal resistance. J. Power Sources 195, 8013-8018.

Zhao, F., Rahunen, N., Varcoe, J.R., Roberts, A.J., Avignone-Rossa, C., Thumser, A.E., Slade, R.C., 2009. Factors affecting the performance of microbial fuel cells for sulfur pollutants removal. Biosen. Bioelectron. 24, 1931-1936.

# 12

# Pyrolysis characteristic of kenaf studied with separated tissues, alkali pulp, and alkali lignin

Yasuo Kojima[1,*], Yoshiaki Kato[1], Minami Akazawa[2], Seung-Lak Yoon[3], Myong-Ku Lee[4]

[1] *Department of Applied Biological Chemistry, Faculty of Agriculture, Niigata University, 2-8050 Ikarashi, Nishi-ku, Niigata, 950-2181, Japan.*

[2] *Graduate School of Science and Technology, Niigata University, Niigata, 950-2181, Japan.*

[3] *Department of Interior Materials Engineering, Gyeongnam National University of Science and Technology, 150 Chiram-Dong, Jinju, Gyeongnam, 660-758, Korea.*

[4] *Department of Paper Science & Engineering, Kangwon National University, 192-1 hyoja 2-dong, Chuncheon, 200-701, Korea.*

## HIGHLIGHTS

- Lignin and pulp prepared from kenaf were pyrolyzed to produce chemicals.
- Pyrolysates from the alkali lignin of kenaf contained valuable phenols.
- Pyrolysis of the alkali pulp of kenaf produced levoglucosan in high yields.

## GRAPHICAL ABSTRACT

## ABSTRACT

To estimate the potential of kenaf as a new biomass source, analytical pyrolysis was performed using various kenaf tissues, i.e., alkali lignin and alkali pulp. The distribution of the pyrolysis products from the whole kenaf was similar to that obtained from hardwood, with syringol, 4-vinylsyringol, guaiacol, and 4-vinylguaiacol as the major products. The phenols content in the pyrolysate from the kenaf core was higher than that from the kenaf cuticle, reflecting the higher lignin content of the kenaf core. The ratios of the syringyl and guaiacyl compounds in the pyrolysates from the core and cuticle samples were 2.79 and 6.83, respectively. Levoglucosan was the major pyrolysis product obtained from the kenaf alkali pulp, although glycol aldehyde and acetol were also produced in high yields, as previously observed for other cellulosic materials. Moreover, the pathways for the formation of the major pyrolysis products from alkali lignin and alkali pulp were also described, and new pyrolysis pathways for carbohydrates have been proposed herein. The end groups of carbohydrates bearing hemiacetal groups were subjected to ring opening and then they underwent further reactions, including further thermal degradation or ring reclosing. Variation of the ring-closing position resulted in the production of different compounds, such as furans, furanones, and cyclopentenones.

**Keywords:**
Kenaf
Analytical pyrolysis
Valuable phenols
Levoglucosan

* Corresponding author
E-mail address: koji@agr.niigata-u.ac.jp

## 1. Introduction

Bio-oil produced from the fast pyrolysis of biomass is considered as a new resource and a substitute for fuel oil or diesel in many static applications such as boilers, furnaces, engines, and turbines used in electricity generation and chemical production (Bridgwater, 2003). Many types of biomass have been examined for bio-oil production, and pilot plants have been established worldwide (Bridgwater and Peacocke, 2000; Zhang et al., 2007).

For instance, switch grass is considered to be a good candidate for biofuel production because of its robustness in poor soils and climatic conditions, low fertilization and herbicide requirements, and high biomass yields (McLaughlin et al., 1999). Bernhard and Joanna (2012) showed that direct combustion of switch grass led to the emission of undesirable secondary products, such as alkali, chlorine, and sulfur components, and may increase the corrosion rate of boilers. Therefore, bio-oil production as an intermediate process is important for switch grass utilization. In addition, many researchers have studied the pyrolysis conversion of switch grass and have evaluated the obtained bio-oil with respect to its chemical composition (Charles et al., 2008), energy-conversion efficiency (Boateng et al., 2007), and effect of milling (Bridgman et al., 2007). Several studies on the pyrolysis conversion of crops other than switch grass, such as wheat straw (Fidalgo et al., 1993), rice husks (Gai et al., 2013), tobacco residue (Cardoso and Ataide, 2013), orange waste (Lopez-Velazquez et al., 2013), giant cane (Temiz et al., 2013), microalgae (Wang et al., 2013), and bamboo (Kato et al., 2014), have also been reported. These plants consist of cellulose, hemicellulose, and lignin. Therefore, the bio-oil obtained from these components is a mixture of pyrolysates.

It is worth quoting that various individual components obtained from biomass have also been examined to determine their pyrolysis products. Jiang et al. (2010) investigated the effect of temperature on the composition of pyrolysis products from alcell lignin (produced from hardwood) and Asian lignin (produced from wheat straw and Sarkanda grass). They concluded that for both lignins, the maximum yield of phenolic compounds obtained at 600 °C, was higher than that obtained directly from the whole biomass (Jiang et al., 2010). They also examined the pyrolysis kinetics of various lignin preparations and reported that the pyrolysis of alkali, hydrolytic, and organosolv lignins displayed first-order kinetics, while that of kraft lignin had 1.5-order kinetics. Analytical studies of pyrolysates obtained from isolated lignins are usually performed using pyrolysis gas chromatography mass spectrometry (Py-GC/MS), high-performance liquid chromatography (HPLC), nuclear magnetic resonance (NMR), and Fourier transform infrared spectroscopy (FT/IR). Lignin from bamboo was prepared by enzymatic/mild acidolysis followed by pyrolysis using an analytical flash pyrolyzer and gas chromatography mass spectrometry (GC/MS) (Lou et al., 2010; Lou and Wu, 2011). In a different study, Huang et al. (2012) prepared corncob acid hydrolysis residue (acid insoluble lignin) and analyzed it using thermogravimetric-FT/IR, Py-GC/MS, and scanning electron microscopy (SEM). They detected phenolic compounds which were mainly generated from the cracking of the aryl glycerol-β-ether (β-O-4) linkages in the samples. These phenolic compounds were suggested to be formed *via* multiple pathways involving further cleavage and degradation of macromolecular compounds at high temperature.

Ye et al. (2012) prepared cornstalks as a lignin model by enzymatic hydrolysis and pyrolyzed the lignin in a stainless steel autoclave reactor with various reaction (residence) times (30–180 min). The product distributions were analyzed by GC/MS, revealing that the increased residence times resulted in increased yields of 4-ethylphenol, 4-ethylguaiacol, and syringol and decreased yields of 2,3-dihydrobenzofuran and vinylguaiacol. Kraft lignin has also been pyrolyzed in order to estimate the optimized conditions for the production of useful bio-oil and to model the pyrolysis steps. For instance, Choi et al. (2013) reported that the pyrolysis temperature and catalyst loading level affected the yields of char, non-condensable gas, and crude bio-oil obtained from kraft. In a different investigation, Gooty et al. (2014) pyrolyzed kraft lignin in a modified bubbling bed reactor and concluded that the maximum yield of bio-oil was obtained at 550 °C. Recently, Guo et al. (2015) described the pyrolysis of kraft lignin as a two-step process, proposing that the first step is governed by the Boltzmann distribution of each bond (bond cleavage), while the second step follows a Monte Carlo algorithm (radical coupling).

The pyrolysis of various carbohydrates has also been examined in order to determine the pyrolysis products obtained from biomass. Yang et al. (2006) investigated the role of three different components (cellulose, hemicellulose, and lignin) in pyrolysis using thermogravimetric analysis (TGA) and observed negligible interactions between them. By studying the product distribution obtained following the fast pyrolysis of glucose-based carbohydrates, e.g., cellulose, cellobiose, maltose, glucose, and dextran, Patwardhan et al. (2009) found that levoglucosan was the most abundant product, whose yield decreased with decreasing chain length of the carbohydrates. As a carbohydrate resource, waste paper was subjected to pyrolysis for the production of bio-oil by Li et al. (2005). They characterized the products by HPLC, NMR, FTIR, and UV analyses and the results achieved revealed that there were four main compounds in the produced bio-oil: anhydrosugars, carboxyl compounds, carbonyl compounds, and aromatic compounds.

Kenaf (*Hibiscus cannabinus*) is a plant in the Malvaceae family and is grown in many parts of the tropics, sub-tropics, and other warm areas for its bark, used as a substitute for jute in cordage and sacking. It has been cultivated for centuries in countries such as India, Bangladesh, Pakistan, China, Sudan, Cuba, Brazil, Thailand, Argentina, Italy, Russia, and Hungary (Bahtoee et al., 2012). Kenaf has a high potential biomass (22 $t \cdot ha^{-1}$), stem yield (18 $t \cdot ha^{-1}$), and a high growth rate (180–220 $kg \cdot ha^{-1} \cdot day^{-1}$) throughout the growing season. Therefore, the cuticle of kenaf, which is located on the outer parts of the shoot and is composed of long fibers, has become an important resource for fiber and paper production (Danalatos and Archontoulis, 2010). However, the inner part of the kenaf shoot, referred to as the core, cannot be used because of the characteristics of the plant tissues present in it, and thus, is considered a waste material. This material can therefore be a potential biomass source. Soda pulping of whole kenaf has been investigated, and the potential for use of the core when blended with cuticle parts (bast fiber) has been reported (Ohtani et al., 2001; Khristova et al., 2002; Hemmasi, 2012).

The aim of the present study was to estimate the potential of kenaf and its separate parts as resources for the production of bio-oil intended for use as an energy source. The mechanisms of the formation of pyrolysis products and bio-oil from the kenaf core and cuticle samples as well as from kenaf alkali lignin and alkali pulp were also investigated.

## 2. Materials and methods

### 2.1. Raw materials and alkali pretreatment

The kenaf used in this study was grown at a plantation located in Jollabuk-do, Korea. The matured kenaf was harvested 152 d after planting and dried at ambient temperature for more than three months. A portion of each dried kenaf plant was then debarked to separate the core and the cuticle. All samples were cut to a length of 10–15 mm.

Alkali pulping of the kenaf samples was performed using a 10% NaOH solution with a liquor ratio of 10, a pulping temperature of 93 °C, and a reaction time of 2 h. After cooking, the alkali pulp was separated from the black liquor and washed with water. The black liquor was acidified with sulfuric acid to pH 4.0, and then, the resultant precipitate was filtered and washed with water to obtain the alkali lignin.

### 2.2. Analytical pyrolysis

The kenaf samples were pyrolyzed using a Py-GC/flame ion detector (Py-GC/FID) consisting of an EGA/PY-3030D multi-shot pyrolyzer (Frontier Lab) and a GC-2010 Plus (Shimadzu) as well as by using a Py-GC/MS consisting of a JCI-22 (JAi) Curie point pyrolyzer and a JMS T-100GCV GC-TOFMS instrument (JEOL). The samples (0.5 mg) were pyrolyzed at 590 °C using the Py-GC/MS. The GC was equipped with an Rtx-Wax capillary column (RESTEK, 60 m × 0.25 mm i.d.; 0.25 μm film thickness). The injector temperature was maintained at 250 °C, and split injection was performed at a 1:100 split rate. The column oven temperature was held at 40 °C for 5 min and was then increased to 250 °C at a rate of 4 °C/min. An NIST mass spectral library was employed for the identification of each peak. Each peak of the chromatogram resulting from the Py-GC/FID analysis was then further identified by comparison of its retention time with that of an authentic sample (methane, methanol, acetic acid, furfural, 2(5H) furanone, o-guaiacol, 4-methylguaiacol, phenol, eugenol, 5-hydroxymethylfurfural, vanillin, coniferylalcohol, and levoglucosan) and by using the results of the Py-GC/MS analysis.

## 3. Results and discussion

### *3.1. Physical and chemical properties of the kenaf*

Photographs of the kenaf used in the present study are shown in **Figure 1**. The average diameter and length of the kenaf shoots were 22 mm (16–25 mm) and 2.7 m (2.4–3.1 m), respectively. The center of the inner portion was in a cavity, and the surface area (1.8–2 mm) of the kenaf was determined by the cuticle (bast fiber). The core accounted for a large percentage of the kenaf tissues. SEM analysis revealed that the cell walls in the kenaf core are thinner than wood cell walls and, therefore, the density of kenaf is very low. In addition, the shapes of the cuticle and core are different: fibrous bundles and chip-like, respectively.

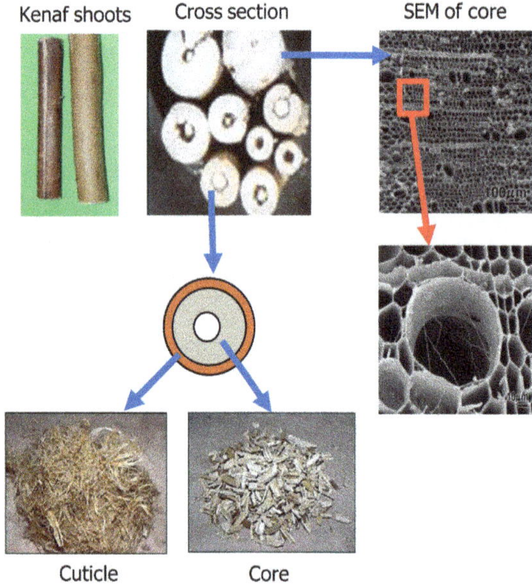


**Fig.1.** Shape and structure of the kenaf samples used in this study.

The proximate and chemical properties of the kenaf samples are listed in **Table 1**. These values are typical of lignocellulose materials without component values.

**Table 1.**
Proximate and chemical properties of kenaf.

| (Wt.% dry) | Content (%) |
|---|---|
| **Proximate analysis** | |
| Volatile mater | 80.4 |
| Fixed carbon | 17.2 |
| **Chemical composition** | |
| Ash | 2.4 |
| $SiO_2$ | 0.7 |
| Extractives | 1.2 |
| Holocellulose | 81.3 |
| Lignin | 10.3 |
| Cellulose | 42.6 |
| **Elemental composition** | |
| Carbon | 44.6 |
| Hydrogen | 5.7 |
| Nitrogen | 0.1 |
| Oxygen | 47.7 |
| Calorific value (HHV), kJ/g-dry | 17.8 |

The lignin and higher holocellulose contents in kenaf are lower than those in woody materials. These results are similar to those that have been previously reported (Bassam, 2010; Hemmasi, 2012). Khristova et al. (2002) found that the lignin contents in kenaf cores and cuticles were 19.6% and 8.1%, respectively. Ohtani et al. (2001) also obtained similar results. These combined results suggest that the physical and chemical properties of the core and cuticle of kenaf are significantly different.

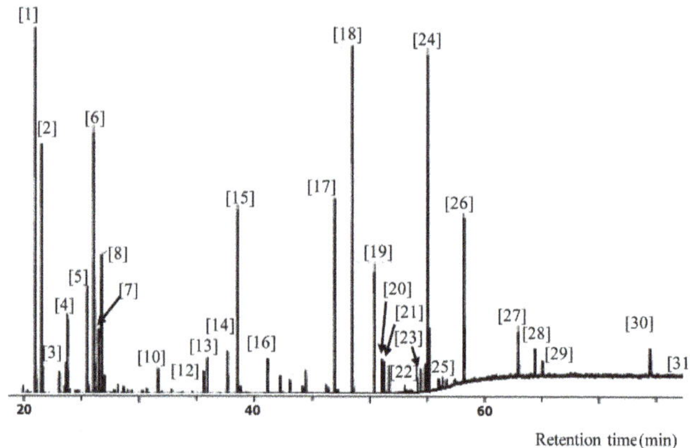


**Fig.2.** TIC of the py-GC/MS from whole kenaf.

### *3.2. Pyrolysis of whole kenaf*

The total ion chromatography (TIC) images of the Py-GC/MS of the whole kenaf is shown in **Figure 2**. Thirty-one compounds were identified as pyrolysis products from the whole kenaf as listed **Table 2**. Guaiacol [**Table 2**; peak No. 15] and 4-vinyl guaiacol [peak No. 17] were detected as major products from the guaiacyl unit of kenaf lignin, while syringol [peak No.18], 4-methylsyringol [peak No. 19], 4-vinylsyringol [peak No. 24], and 4-(1-propenyl)-syringol [peak No. 26] were major products obtained from the syringyl unit of kenaf lignin.

**Table 2.**
Yield (%) of pyrolysates from whole, core and cuticle of kenaf.

| Peak No. | R.T. | Product | Whole | Core | Cuticle |
|---|---|---|---|---|---|
| 1 | 21.0 | Acetol | 9.71 | 8.24 | 14.99 |
| 2 | 21.6 | Glycolaldehyde | 7.87 | 8.60 | 9.30 |
| 3 | 23.8 | 1-Hydroxy-2-butanone | 1.22 | 0.55 | 1.20 |
| 4 | 24.0 | 1-Hydroxy-3-propanal | 1.56 | 1.83 | 1.91 |
| 5 | 25.7 | Butanedial | 1.97 | 2.92 | 2.28 |
| 6 | 26.1 | Acetic acid | 9.26 | 8.16 | 10.50 |
| 7 | 26.7 | Methyl pyruvate | 1.84 | 1.77 | 2.90 |
| 8 | 27.0 | Furfural | 2.73 | 2.31 | 3.29 |
| 9 | 29.2 | Propionic acid | 0.61 | 0.31 | 0.53 |
| 10 | 31.9 | Ethanediol | 1.72 | 1.10 | 2.59 |
| 11 | 33.0 | Furfuryl alcohol | 0.18 | - | 0.49 |
| 12 | 35.9 | 2(5H)-Furanone | 0.80 | 1.12 | 1.13 |
| 13 | 36.2 | 2-Hydroxy-2-cyclopenten-1-on | 1.66 | 0.76 | 1.36 |
| 14 | 37.9 | 2-Hydroxy-4-methyl-2-cyclopenten-1-on | 1.88 | 0.74 | 2.00 |
| 15 | 38.8 | Guaiacol | 3.44 | 2.58 | 0.53 |
| 16 | 41.3 | 4-Methylguaiacol | 0.73 | 0.38 | - |
| 17 | 47.2 | 4-Vinylguaiacol | 4.56 | 4.51 | 0.60 |
| 18 | 48.7 | Syringol | 8.22 | 9.84 | 4.27 |
| 19 | 50.6 | 4-Methylsyringol | 3.95 | 4.18 | 1.04 |
| 20 | 51.4 | 4-Vinylphenol | 1.07 | 3.29 | - |
| 21 | 51.8 | 4-Ethylsyringol | 0.76 | 0.69 | 0.71 |
| 22 | 54.6 | (*E*)-4-(2-Propenyl) syringol | 0.99 | 1.03 | 0.49 |
| 23 | 54.9 | Vanillin | 0.62 | 0.80 | - |
| 24 | 55.0 | 4-Vinylsyringol | 7.76 | 7.99 | 3.37 |
| 25 | 56.9 | Guaiacyl-2-propanone | - | 0.44 | - |
| 26 | 58.2 | (*E*)-4-(1-Propenyl) syringol | 4.82 | 4.34 | 2.97 |
| 27 | 62.9 | Syringaldehyde | 1.81 | 1.69 | 0.80 |
| 28 | 64.4 | Acetosyringon | 1.06 | 0.79 | - |
| 29 | 65.0 | Syringyl-2-propanone | 3.53 | 1.03 | - |
| 30 | 74.5 | Coniferylalcohol | 2.17 | 4.17 | - |
| 31 | 78.0 | Levoglucosan | 2.21 | 1.78 | 5.34 |

In general, syringyl products were the dominant compounds observed in the pyrolysates obtained following whole kenaf pyrolysis, indicating that kenaf lignin is composed mainly of syringyl units. Along with low-molecular-weight aliphatic compounds, furfural [peak No. 8], furfuryl alcohol [peak No. 11], furanone [peak No. 12], cyclopentanone derivatives [peak No. 13 and 14], and levoglucosan [peak No. 31], which were derived from carbohydrates, were also detected in the pyrolysate of the whole kenaf.

### 3.3. *Pyrolysis of separated core and cuticle kenaf tissues*

The TICs of the Py-GC/MS of the core and cuticle kenaf samples are shown in **Figure 3**. Twenty-eight products from the kenaf core and 22 products from the kenaf cuticle were identified. It can be seen that the TIC for the Py-GC/MS of the core was similar to that of the whole kenaf. This result is not surprising since the core has been reported to account for 78–82% of the dry weight of the kenaf stem (Gutiérrez et al., 2014).

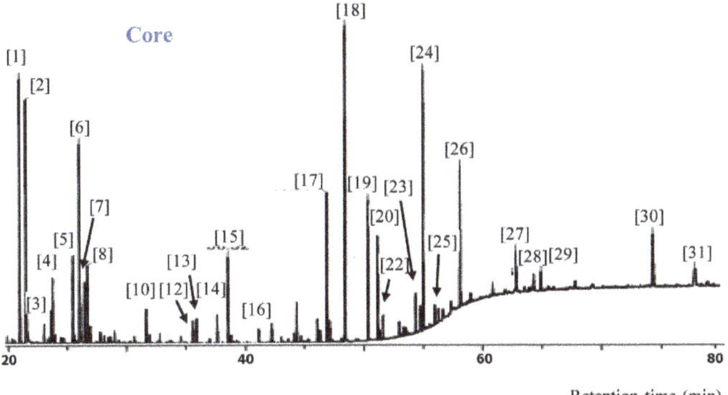


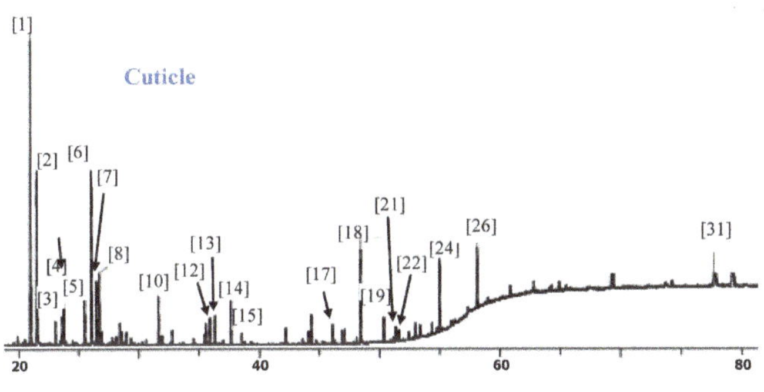


**Fig.3.** TIC of the py-GC/MS from kenaf's core and cuticle.

On the other hand, the TIC profile of the cuticle tissue was different from the TIC profiles of the whole kenaf and core. The peak area percentages for the major products obtained following the pyrolysis of whole kenaf, the core, and the cuticle are listed in **Table 2**. Following kenaf core pyrolysis, syringyl compounds such as syringol [**Table 2**; peak No. 18], vinyl syringol [peak No. 24], (E)-4-(1-propenyl) syringol [peak No. 26], 4-methylsyringol [peak No. 19], and syringaldehyde [peak No. 27] were identified as the major products. Some guaiacyl compounds including vinyl guaiacol [peak No. 17], guaiacol [peak No. 15], coniferyl alcohol [peak No. 30], and guaiacyl-2-propanone [peak No. 25] were also identified. Vinyl phenol [peak No. 20], a product derived from p-hydroxyl compounds, was also identified as a minor product. All these products were generated during the pyrolysis of the lignin fraction in the kenaf core tissue. Low-molecular-weight aliphatic compounds, including acetol, glycol aldehyde, and acetic acid, were identified as major pyrolysis products derived from carbohydrates.

Pyrolysis of the kenaf cuticle provided syringol [peak No. 18], vinyl syringol [peak No. 24], and syringyl propenes [peak No. 22 and 26], as well as guaiacol. The yields of these phenols were lower than those obtained from the kenaf core. Hemmasi (2012) reported that the lignin content in kenaf cuticles was lower than that in kenaf cores at 11.5% and 18.9%, respectively. Therefore, the low yields of phenols in the pyrolysates from kenaf cuticle samples investigated in the present study are not surprising. On the other hand, aliphatic low-molecular-weight products were obtained in high yields. Seca et al. (1998) analyzed kenaf cores and cuticles using various methods and reported that there were strong structural differences in the bark (cuticle) and core lignins in kenaf. Nishimura et al. (2002) also analyzed separated kenaf bast, inner bast, and core samples prepared from the top, upper middle, lower middle, and bottom of kenaf stems. Their results indicated that both the lignin content and molar ratio of total syringyl to total vanillyl (S/V) ratio varied according to the position in the kenaf stem.

**Table 3.**
Chemical components (%) of kenaf's core and cuticle (Jin et. al. 2006).

| | | Extractives | Klason | Glucose | Xylose | Arabinose | S/V ratio |
|---|---|---|---|---|---|---|---|
| Cuticle | Upper | 8.9 | 9.1 | 42.6 | 6.8 | 2.7 | 5.7 |
| | Lower | 6.2 | 11.7 | 42.7 | 6.5 | 3.3 | 5.0 |
| Core | Upper | 7.9 | 11.4 | 50.7 | 7.6 | 1.1 | 1.5 |
| | Lower | 7.4 | 22.1 | 40.3 | 14.3 | 0.1 | 3.2 |

Jin et al. (2006) analyzed the bast (cuticle) and core tissues from kenaf (*H. cannabinus*) and reported that the cuticle tissue had a high lignin content and a high S/V ratio, as shown in **Table 3**. Gutierrez et al. (2004) also reported that the S/V ratio of kenaf fiber (cuticle) was as high as 5.4. In the present study, the molar ratios of syringyl to guaiacyl (S/G) were calculated using the ratios of the peak percentages for syringol/guaiacol and vinylsyringol/vinylguaiacol. The results are tabulated in **Table 4**. The S/G ratios of the cuticle and core were 8.54 and 3.81, respectively, as determined by using the TIC peak percentages for syringol and guaiacol, while they stood at 5.62 and 1.77, respectively, using the TIC peak percentages for vinylsyringol and vinylguaiacol. These data combined resulted in average values for the S/G ratios of 6.83 for the cuticle and 2.79 for the core.

**Table 4.**
Syringyl and guaiacyl ratio (S/G) calculated based on the corresponding products from cuticle of kenaf.

| | Syringol/Guaiacol | Vinylsyringol/Vinylguaiacol | Average of S/G |
|---|---|---|---|
| Cuticle | 8.54 | 5.62 | 6.83 |
| Core | 3.81 | 1.77 | 2.79 |

### 3.4. *Pyrolysis of kenaf alkali lignin and alkali pulp obtained from whole kenaf*

The TIC image of the Py-GC/MS of the kenaf alkali lignin is shown in **Figure 4**. Sixteen compounds were identified as pyrolysis products, including four additional products not obtained directly from the whole kenaf or its parts: phenol, 2-methycatechol, acetoguaiacone, and catechol. The structures of these chemicals are illustrated next to their peaks in **Figure 4**.

The TIC of the Py-GC/MS of the kenaf alkali pulp is shown in **Figure 5**. As presented, eighteen compounds were identified as pyrolysis products derived from carbohydrates, and once again, four additional products not detected for the other samples were observed: 5-acetyldihydro-2-furanone, 3-hydroxy-butyrolactone, 3-hydroxy-5-methyl-2-furanone, and 5-hydroxymethyl furfural. Notably, levoglucosan [**Table 2**; peak No. 31] was detected as one of the major products obtained from the pyrolysis of alkali pulp. It has been previously demonstrated that the fast pyrolysis of cellulosic materials resulted in the production of levoglucosan as a major product (Ronsee et al., 2012; Shra'ah et al., 2014).

The peak areas of the pyrolysis products obtained from the alkali lignin and alkali pulp are listed in **Table 5** and compared with the pyrolysates

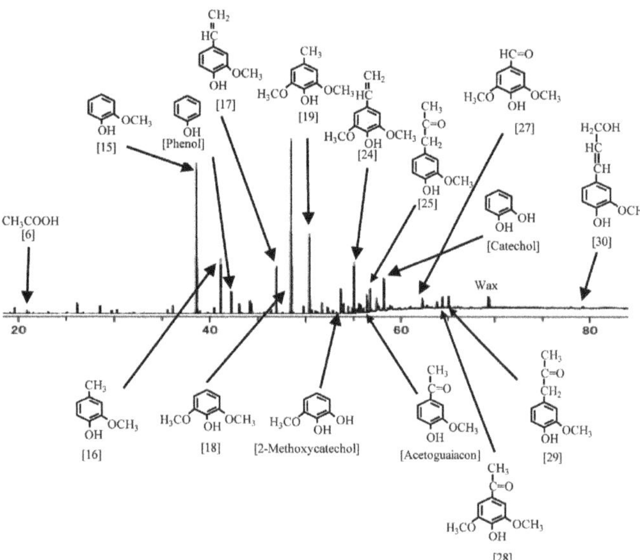

Fig.4. TIC of the py-GC/MS alkali lignin from kenaf.

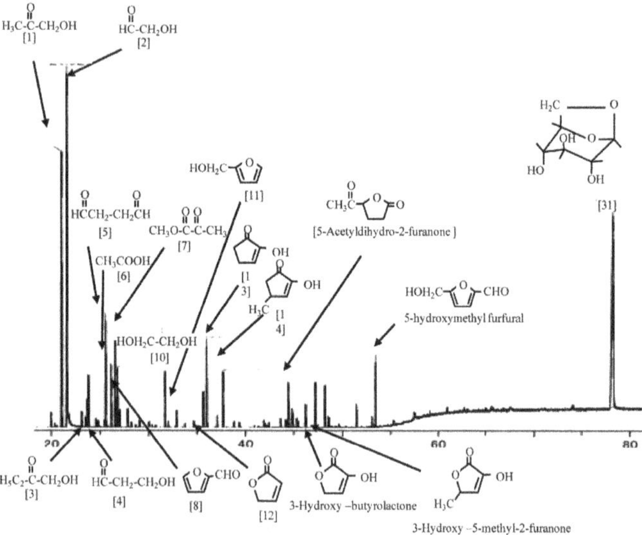

Fig.5. TIC of the py-GC/MS alkali pulp from kenaf.

obtained from the whole kenaf. Alkali lignin pyrolysis resulted in the production of typical phenol compounds in high yields, and the ratio of syringyl and guaiacyl compounds in the pyrolysate was 1.10. This result indicated that some syringyl lignin in the kenaf tissues degraded to low-molecular-weight compounds, which were not recovered from the black liquor as alkali lignin. In addition, the yields of guaiacol, syringol, 4-methylguaiacol, and 4-methylsyringol obtained from the alkali lignin were remarkably increased compared to those obtained from the pyrolysis of the whole kenaf. This result suggested that aryl ether linkages, which are the major linkages in lignin, were cleaved, leading to an increase in the content of phenolic hydroxyl groups in the alkali lignin. As a result, electron-transfer reactions from the α-hydroxyl groups of these compounds and *via* quinone methide intermediates were accelerated, as shown in **Figure 6**.

The pathways for the formation of cyclic monomers, such as furans, anhydrous sugars, and cyclopentanones, from carbohydrate polymers are as yet unknown, but the pathways for the formation of low-molecular-weight products from glucose have been reported in detail by Vinu et al. (2012). The pyrolysis products obtained from the alkali pulp were formed *via* more complex pathways, and the proposed mechanisms are illustrated in **Figures 7–9**. The formation of levoglucosan [peak No. 31], which is a heterocyclic compound or anhydrosugar, is initiated by radical cleavage of a glycoside bond in an intermolecular sugar residue. The C1 radical then couples with a C6-O radical to form a new ring structure. This

**Table 5.**
Yield (%) of pyrolysates from alkali lignin and alkali pulp from kenaf.

| Peak No. | R.T. | Product | Whole | Alkali lignin | Alkali pulp |
|---|---|---|---|---|---|
| 1 | 21.0 | Acetol | 9.71 | - | 11.80 |
| 2 | 21.6 | Glycolaldehyde | 7.87 | - | 19.02 |
| 3 | 23.8 | 1-Hydroxy-2-butanone | 1.22 | | 0.94 |
| 4 | 24.0 | 1-Hydroxy-3-propanal | 1.56 | - | 1.73 |
| 5 | 25.7 | Butanedial | 1.97 | - | 4.17 |
| 6 | 26.1 | Acetic acid | 9.26 | - | 2.69 |
| 7 | 26.7 | Methyl pyruvate | 1.84 | 1.57 | 3.16 |
| 8 | 27.0 | Furfural | 2.73 | - | 2.10 |
| 9 | 29.2 | Propionic acid | 0.61 | - | - |
| 10 | 31.9 | Ethanediol | 1.72 | - | 2.36 |
| 11 | 33.0 | Furfuryl alcohol | 0.18 | - | - |
| 12 | 35.9 | 2(5H)-Furanone | 0.80 | - | 1.42 |
| 13 | 36.2 | 2-Hydroxy-2-cyclopenten-1-on | 1.66 | - | 3.20 |
| 14 | 37.9 | 2-Hydroxy-4-methyl-2-cyclopenten-1-on | 1.88 | - | 1.81 |
| 15 | 38.8 | Guaiacol | 3.44 | 17.65 | - |
| 16 | 41.3 | 4-Methylguaiacol | 0.73 | 5.66 | - |
| 17 | 47.2 | 4-Vinylguaiacol | 4.56 | 5.62 | - |
| 18 | 48.7 | Syringol | 8.22 | 21.67 | - |
| 19 | 50.6 | 4-Methylsyringol | 3.95 | 9.10 | - |
| 20 | 51.4 | 4-Vinylphenol | 1.07 | - | - |
| 21 | 51.8 | 4-Ethylsyringol | 0.76 | - | - |
| 22 | 54.6 | (*E*)-4-(2-Propenyl) syringol | 0.99 | - | - |
| 23 | 54.9 | Vanillin | 0.80 | - | - |
| 24 | 55.0 | 4-Vinylsyringol | 7.76 | 5.47 | - |
| 25 | 56.9 | Guaiacyl-2-propanone | - | 2.74 | - |
| 26 | 58.2 | (*E*)-4-(1-Propenyl) syringol | 4.82 | 3.78 | - |
| 27 | 62.9 | Syringaldehyde | 1.81 | - | - |
| 28 | 64.4 | Acetosyringon | 1.06 | 1.99 | - |
| 29 | 65.0 | Syringyl-2-propanone | 3.53 | 2.08 | - |
| 30 | 74.5 | Coniferylalcohol | 2.17 | - | - |
| 31 | 78.0 | Levoglucosan | 2.21 | - | 29.53 |

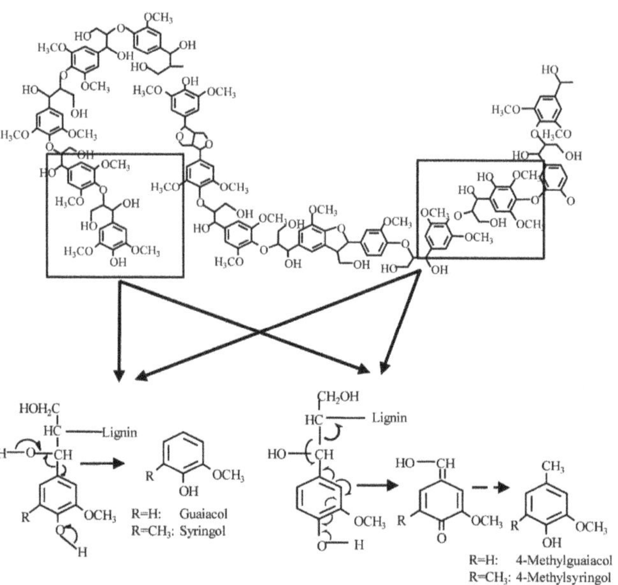

Fig.6. Proposed formation pathway of abundant pyrolysis products from lignin polymer.

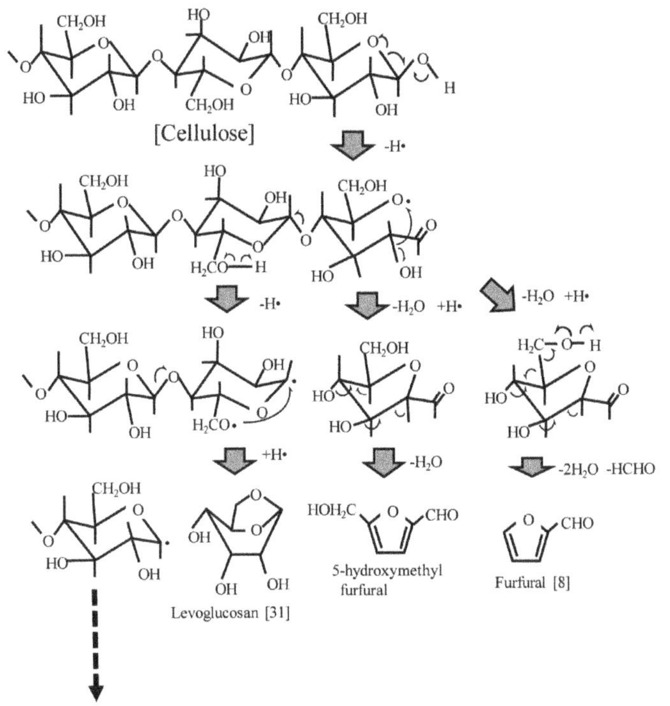

**Fig.7.** Proposed formation pathway of levoglucosan, 5-hydroxymethy-furfural, and furfural from alkali pulp.

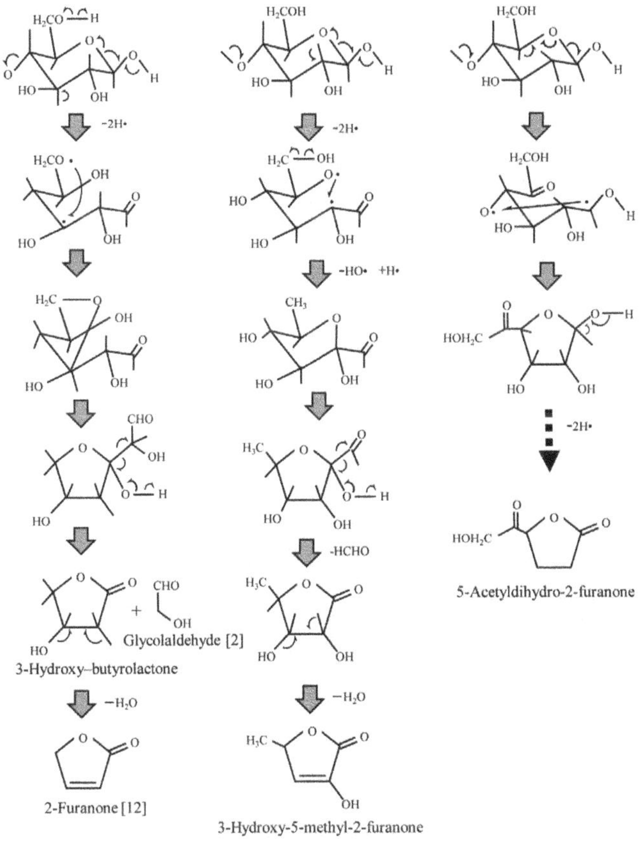

**Fig.8.** Proposed formation pathway of lactone and furanones from alkali pulp.

radical is then propagated as a chain reaction in carbohydrate molecules. End groups of carbohydrates with hemiacetal groups are subjected to ring opening and further react, undergoing thermal degradation or ring reclosing. Variation of the ring-closing position yields a range of compounds, including furans, furanones, and cyclopentenones. The formation of furfurals is initiated by ring opening at the hemiacetal site, followed by coupling of the C2 radical with a C5-O radical. Dehydration then results in the formation of 5-hydroxymethyl-furfural and subsequent elimination of the C6 hydroxymethyl group as formaldehyde generates furfural, as shown in **Figure** 7.

The formation of 2-furanone [peak No. 12] occurs *via* ring opening followed by coupling of the C3 and C6-O radicals, while 2-hydroxy-5-methyl-2-furanone is formed by the coupling of the C2 and C5-O radicals after ring opening, and 5-acetyldihydro-2-furanone is produced *via* the coupling of the C1 and C4-O radicals, followed by the donation of an H radical from the C1-OH group, as shown in **Figure 8**.

Hypothetical pathways for the formation of cyclopentenones are illustrated in **Figure** 9. The end sugar residue of carbohydrates is subjected to ring opening, followed by coupling of the C1 and C5 radicals to form a cyclopentane structure. Hydroxyl radical elimination from the C6 hydroxymethyl group of this cyclopentane intermediate is accompanied by the elimination of formaldehyde, yields 2-hydroxy-4-methyl-2 cyclopenten-1-one [peak No. 14]. The elimination of two molecules of formaldehyde at the C1 and C6 positions results in the production 2-hydroxy-2-cyclopenten-1-one [peak No. 13].

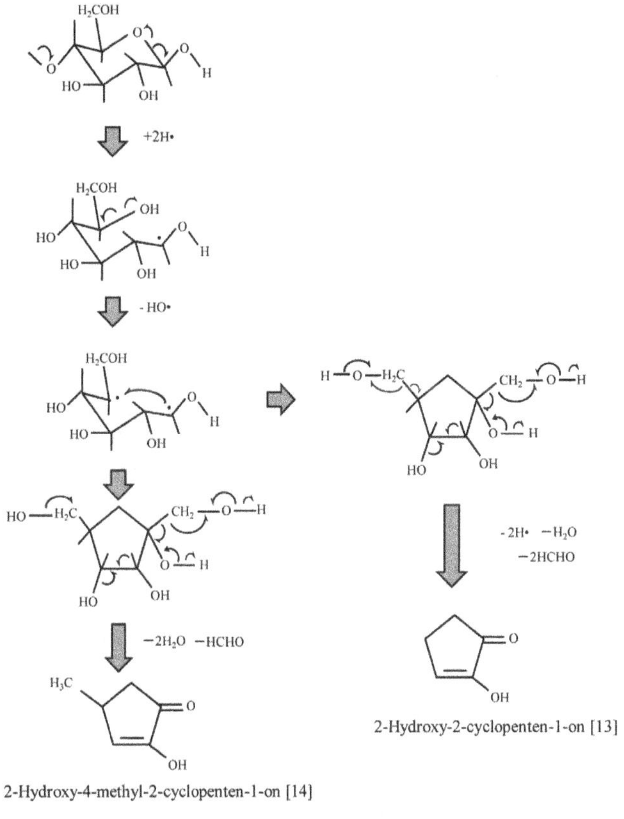

**Fig.9.** Proposed formation pathway of cyclopentenones from alkali pulp.

## 4. Conclusions

The distribution of pyrolysis products obtained from the whole kenaf was found similar to that from hardwood, with syringol, 4-vinylsyringol, guaiacol, and 4-vinylguaiacol as major products. The phenols content of the pyrolysate from the kenaf core was higher than that of the kenaf cuticle, reflecting its higher lignin content. The ratios of syringyl and guaiacyl compounds in the pyrolysates from the core and cuticle were measured at 2.79 and 6.83, respectively. Levoglucosan was the major pyrolysis product obtained from the kenaf alkali pulp, but glycol aldehyde and acetol were also formed in high yields. Notably, many pyrolysis products were detected that have the potential to be converted into more useful chemicals. Pathways for the formation of the major pyrolysis products from the alkali lignin and alkali pulp were presented along with newly-proposed pyrolysis pathways for carbohydrates.

**References**


[1] Bahtoee, A., Zargari, K., Baniani, E., 2012. An investigation on fiber production of different kenaf. World Appl. Sci. J. 16, 63-66.

[2] Bassam, E.L.N., 2010. Handbook of Bioenergy Crops: a complete reference to species, development and applications. Earthscan Ltd., Dunstan House, ISBN 978-1-84407-854-7 London.

[3] Bernhard, P., Joanna, S.O., 2012. Simultaneous prediction of potassium chloride and sulphur dioxide emissions during combustion of switchgrass. Fuel. 96, 29-42.

[4] Boateng, A.A., Daugaard, D.E., Goldberg, N.M., Hicks, K.B., 2007. Bench-scale fluidized-bed pyrolysis of switchgrass for bio-oil production. Ind. Eng. Chem. Res. 46, 1891-1897.

[5] Bridgeman, T.G., Darvell, L.I., Jones, J.M., Williams, P.T., Fahmi, R., Bridgwater, A.V., Barraclough, T., Shield, I., Yates, N., Thain, S.C., Donnison, I.S., 2007. Influence of particle size on the analytical and chemical properties of two energy crops. Fuel. 6, 60-72.

[6] Bridgwater, A.V., 2003. Renewable fuels and chemicals by thermal processing of biomass. Chem. Eng. J. 91, 87-102.

[7] Bridgwater, A.V., Peacocke, G.V.C., 2000. Fast pyrolysis processes for biomass. Renew. Sust. Energy Rev. 4, 1-73.

[8] Cardoso, C.R., Ataide, C.H., 2013. Analytical pyrolysis of tobacco residue: Effect of temperature and inorganic additives. J. Anal. Appl. Pyrol. 99, 49-57.

[9] Charles, A, Mullen, CA, Boateng, AA., 2008. Chemical composition of bio-oils produced by pyrolysis of two energy crops. Energy Fuels. 22, 2104-2109.

[10] Choi, H.S., Meier, D., 2013. Fast pyrolysis of kraft lignin-vapor cracking over various fixed-bed catalysts. J. Anal. Appl. Pyrol. 100, 207-212.

[11] Danalatos, N.G., Archontoulis, S.V., 2010. Growth and biomass productivity of kenaf [*Hibiscus cannabinus* L.] under different agricultural inputs and management practices in central Greece. Ind. Crops Prod. 32, 231-240.

[12] Fidalgo, M.L., Terron, M.C., Martinez, A.T., Gonzalez, A.E., Gonzalez-Vila, F.J., Galletti, G.C., 1993. Comparative study of fraction from alkaline extraction of wheat straw through chemical degradation, analytical pyrolysis, and spectroscopic technology. J. Agric. Food Chem. 41, 1621-1626.

[13] Gai, C., Dong, Y., Zhang, T., 2013. The kinetic analysis of the pyrolysis of agricultural residue under non-isothermal conditions. Bioresour. Technol. 127, 298-305.

[14] Gooty, A.T., Li D., Berruti, F., Briens C., 2014. Kraft-lignin pyrolysis and fractional condensation of its bio-oil vapors. J. Anal. Appl. Pyrol. 106, 33-40.

[15] Guo, X., Liu, Z., Liu, Q., Shi, L., 2015. Modeling of kraft lignin pyrolysis based on bond dissociation and fragment coupling. Fuel Process. Technol. 135, 133-149.

[16] Gutiérrez, A., Rodríguez, I.M., Río, J.C., 2004. Chemical characterization of lignin and lipid fractions in Kenaf bast fibers used for manufacturing high-quality papers. J. Agric. Food Chem. 52, 4764-4773.

[17] Hemmasi, A. H., 2012. Producing Paper from Iranian Kenaf by Soda and Soda-Anthraquinone Processes. American-Eurasian J. Agric. Environ. Sci. 12, 886-889.

[18] Huang, Y., Wei, Z., Qju, Z., Yin, X., Wu, C., 2012. Study on structure and pyrolysis behavior of lignin derived from corncob acid hydrolysis residue. J. Anal. Appl. Pyrol. 93, 153-159.

[19] Jiang, G., Nowakowski. D.J., Bridgwater, A., 2010. Effect of the temperature on the composition of lignin pyrolysis products. Energy Fuels. 24, 4470-4475.

[20] Jiang, G., Nowakowski, D.J., Bridgwater, A.A, 2010. Systematic study of the kinetics of lignin pyrolysis. Thermochimi. Acta. 498, 61-66.

[21] Jin, G., Nakagawa-izumi, A., Shimizu, K., Ohi, H., 2006. Chemical characterization and kraft pulping response of *Hibiscus cannabinus* bast. J. Wood Sci. 52, 107-112.

[22] Kato, Y., Kohnosu, T., Enomoto, R., Akazawa, M., Yoon, S.L., Kojima, Y., 2014. Chemical properties of bio-oils produced by fast pyrolysis of bamboo. Trans. Mat. Res. Soc. Japan. 39, 491-498.

[23] Khristova, P., Kordsachia, O., Patt, R., Khider, T., Karrar, I., 2002. Alkaline pulping with additives of kenaf from Sudan. Ind. Crops Prod. 15, 229-235.

[24] Li, L., Zhang, H., Zhuang, X., 2005. Pyrolysis of waste paper: characterization and composition of pyrolysis oil. Energy Sources. 27, 867-873.

[25] Lopez-Velazquez, M.A., Santes, V., Balmaseda, J., Torres-Garcia, E., 2013. Pyrolysis of orange waste: A thermo-kinetic study. J. Anal. Appl. Pyrol. 99, 170-177.

[26] Lou, R. Wu, S.B., Gao, J.L., 2010. Effect of condition on fast pyrolysis of bamboo lignin. J. Anal. Appl. Pyrol. 89, 191-196.

[27] Lou, R., Wu, S.B., 2011. Product properties from fast pyrolysis of enzymatic/mild acidolysis lignin. Appl. Energ. 88, 316-322.

[28] McLaughlin, S.B., Bouton, J., Bransby, D., Conger, B.V., Ocumpaugh, W.R., Parrish, D.J., Taliaferro, C., Voge, K.P.l., Wullschleger, S.D., 1999. Developing switch grass as a bioenergy crop, in: Janick, J. (Ed.), Perspectives on new crops and new uses. ASHS Press, Alexandria, VA, pp. 282–299.

[29] Nishimura, N., Izumi, A., Kuroda, K., 2002. Structural characterization of kenaf lignin: differences among kenaf varieties. Ind. Crops Prod. 15, 115-122.

[30] Ohtani, Y., Mazumder, B.B., Samejima, K., 2001. Influence of the chemical composition of kenaf bast and core on the alkaline pulping response. J. Wood Sci. 47, 30-35.

[31] Patwardhan, P.R., Satrio, J.A., Brown, R.C., Shanks, B.H., 2009. Product distribution from fast pyrolysis of glucose-based carbohydrates. J. Anal. Appl. Pyrol. 86, 323-330.

[32] Ronsee, F., Bai, X., Prins, W., Brown, R.C., 2012. Secondary reactions of levoglucosan and char in the fast pyrolysis of cellulose. Environ. Prog. Sustainable Energy. 31, 256-260.

[33] Seca, A.M.L., Cavaleiro, J.A.S., Domingues, F.M.J., Silvestre, A.J.D., Evtuguin, D., Neto, C.P., 1998. Structural characterization of the bark and core lignins from kenaf (*Hibiscus cannabinus*). J. Agric. Food Chem. 46, 3100–3108.

[34] Shra'ah, A.A, Helleur, R., 2014. Microwave pyrolysis of cellulose at low temperature. J. Anal. Appl. Pyrol. 105, 91-99.

[35] Temiz, A., Akbas, S., Panov, D., Terziev, N., Alma, S.P., Kose, G., 2013. Chemical composition and Efficiency of bio-oil obtained from giant cane (*Arundo donax* L.) as a wood preservative. Bioresouces. 8, 2084-2098.

[36] Vinu, R., Broadbelt, L.J., 2012. A mechanistic model of fast pyrolysis of glucose-based carbohydrates to predict bio-oil composition. Energy Environ. Sci. 5, 9808-9826.

[37] Wang, K., Brown, R.C., Homsy, S., Martinez, L., Sidhu, S.S., 2013. Fast pyrolysis of microalgae remnants in a fluidized bed reactor for bio-oil and biochar production. Bioresour. Technol. 127, 494-499.

[38] Ye, Y., Fan, J., Chang, J., 2012. Effect of reaction conditions on hydrothermal degradation of cornstalk lignin. J. Anal. Appl. Pyrol. 94, 190-195.

[39] Yang, H., Tan, R., Chen, H., Zheng, C., Lee, D.H., Liang, D.T., 2006. In-depth investigation of biomass pyrolysis based on three major components: hemicellulose, cellulose and lignin. Energy Fuel. 20, 388-393.

[40] Zhang, Q., Chang, J., Wang, T.J., Xu, Y., 2007. Review of biomass pyrolysis oil properties and upgrading research. Energy Convers. Manage. 48, 87-92.

# 13

# Iron effect on the fermentative metabolism of *Clostridium acetobutylicum* ATCC 824 using cheese whey as substrate

Victoria Rosalía Durán-Padilla[1], Gustavo Davila-Vazquez[2], Norma Angélica Chávez-Vela[1], José Raunel Tinoco-Valencia[3], Juan Jáuregui-Rincón[1*]

[1] *Departamento de Ingeniería Bioquímica, Universidad Autónoma de Aguascalientes (UAA), Av. Universidad 940 Ciudad Universitaria, Aguascalientes, Aguascalientes, México.*

[2] *Tecnología Ambiental, Centro de Investigación y Asistencia en Tecnología y Diseño del Estado de Jalisco, A.C.(CIATEJ), Normalistas 800, Colinas de La Normal, Guadalajara, Jalisco, México.*

[3] *Escalamiento y Planta piloto, Instituto de Biotecnología, Universidad Nacional Autónoma de México (UNAM), Av. Universidad 2001, Chamilpa, Cuernavaca, Morelos, México.*

## HIGHLIGHTS

- *Addition of iron was found essential to make cheese whey suitable for ABE fermentation.*
- *Lack of ferredoxin led to lactic acid production instead of the desired solvents.*
- *Addition of $FeSO_4$ improved butanol production by 65% compared to $FeCl_3$.*

## GRAPHICAL ABSTRACT

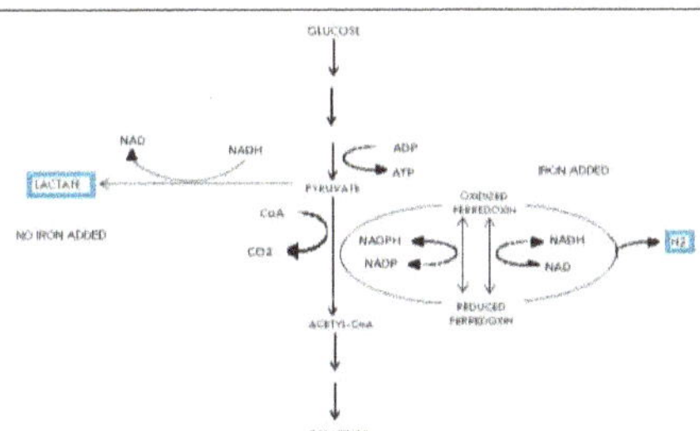


## ABSTRACT

Butanol is considered a superior liquid fuel that can replace gasoline in internal combustion engines. It is produced by acetone-butanol-ethanol (ABE) fermentation using various species of solventogenic clostridia. Performance of ABE fermentation process is severely limited mostly by high cost of substrate, substrate inhibition and low solvent tolerance; leading to low product concentrations, low productivity, low yield, and difficulty in controlling culture metabolism. In order to decrease the cost per substrate and exploit a waste generated by dairy industry, this study proposes using cheese whey as substrate for ABE fermentation. It was observed that the addition of an iron source was strictly necessary for the cheese whey to be a viable substrate because this metal is needed to produce ferredoxin, a key protein in the fermentative metabolism of *Clostridium acetobutylicum* serving as a temporary electron acceptor. Lack of iron in the cheese whey impedes ferredoxin synthesis and therefore, restricts pyruvate-ferredoxin oxidoreductase activity leading to the production of lactic acid instead of acetone, butanol and ethanol. Moreover, the addition of $FeSO_4$ notably improved ABE production performance by increasing butanol content (7.13 ± 1.53 g/L) by 65% compared to that of $FeCl_3$ (4.32 ± 0.94 g/L) under the same fermentation conditions.

**Keywords:**
Biobutanol
Cheese whey
ABE fermentation
Ferredoxin
Ferrous sulfate

* Corresponding author
E-mail address: jjaureg@correo.uaa.mx

## 1. Introduction

Butanol is an important industrial chemical considered as a superior liquid fuel with a potential to replace gasoline (Jang et al., 2012). It can be produced from petroleum, mineral fuel and biomass; the latter is conveniently denoted as biobutanol although it has the same characteristics as butanol from petroleum (Shapovalov and Ashkinazi, 2008). Compared to ethanol, the traditional biofuel, butanol has the following advantages: a) higher energy content (29.2 MJ/L, very similar to that of gasoline: 32 MJ/L), b) vapor pressure 11 times lower than that of ethanol, and is therefore safer to use (Rajchenberg-Ceceña et al., 2009), c) non hygroscopic, d) less corrosive, e) can be applied in pure form or blended in any proportions with gasoline or diesel, f) can be used in any automobile engine without modifications, g) easy to preserve and distribute, h) can be used with the existing infrastructure (Rajchenberg-Ceceña et al., 2009; Tashiro and Sonomoto, 2010), i) heat of vaporization (0.43 MJ/Kg) slightly higher than that of gasoline (0.36 MJ/Kg) avoiding cold start problems (Ranjan and Moholkar, 2009), and j) at combustion, does not produce sulfur and nitrogen oxides, advantageous from the environmental viewpoint (Shapovalov and Ashkinazi, 2008). Furthermore, it can be converted to valuable chemical compounds: acrylate, methacrylate esters, glycol ethers, butyl acetate, and *etc.* (Tashiro and Sonomoto, 2010).

Butanol has traditionally been produced by acetone-butanol-ethanol (ABE) fermentation of sugar substrates using various species of solventogenic clostridia. The fermentation occurs in two stages; the first is a growth stage in which acetic and butyric acids are produced and the second stage is characterized by acid re-assimilation into ABE solvents. During this stage, growth slows down, the cells accumulate granulose and form endospores (Green, 2011). The regular ratio of ABE solvents produced by *C. acetobutylicum* is 3:6:1 with 20 g/L being the maximum concentration achieved so far (Ranjan and Moholkar, 2009). Performance of butanol fermentation process using solventogenic clostridia is severely limited by: a) high substrate costs, b) substrate inhibition, c) low solvent tolerance (max 20 g/L of solvent), d) sluggish growth, and e) low cell density achievable during solventogenic fermentation (Jang et al., 2012). These limitations result in low final butanol concentration caused by butanol inhibition, low butanol productivity due to low cell density, low yield due to hetero-fermentation and high downstream processing cost for butanol recovery (Tashiro and Sonomoto, 2010). Unless these limitations are addressed, biological production of butanol cannot compete economically with petrochemical synthesis of these solvents (Gheshlaghi et al., 2009; Tashiro and Sonomoto, 2010; Cooksley et al., 2012; Sabra et al., 2014). Researchers have tried to solve these problems thorough various studies including: a) microbial strain development for improved butanol titer and tolerance, b) development of efficient *in situ* product recovery technologies to overcome the butanol toxicity to fermenting microorganisms, c) application of fermentation strategies to increase cell density, butanol titer, yield and productivity, and d) exploration of more economic alternative substrates (Jang et al., 2012).

The conventional ABE fermentation substrates include starch and molasses from maize, wheat, and rye. In order to reduce the cost of the product, a variety of more economic substrates have been tested including: sucrose (Parekh and Blaschek, 1999; Tashiro and Sonomoto, 2010), lignocellulosic biomass (Ranjan and Moholkar, 2009; Jang et al., 2012), domestic, agricultural and industrial waste (Jang et al., 2012; Niemistö et al., 2013), glycerol (exclusively in fermentations with *C. pasteurianum*) (Dabrock et al., 1992; Tashiro and Sonomoto, 2010; Jang et al., 2012; Sabra et al., 2014), algal biomass (Ranjan and Moholkar, 2009; Jang et al., 2012), wheat straw, corn fiber, liquefied corn starch, apple pomace, Jerusalem artichokes (Ranjan and Moholkar, 2009), cheese whey (Maddox, 1980; Bahl et al., 1986; Ennis and Maddox, 1989; Ranjan and Moholkar, 2009), among others.

Solventogenic *Clostridium* species are capable of fermenting a wide range of carbohydrates; lignocellulosic biomass has been identified as a potential substrate for inexpensive production of ABE and other fine chemicals, however, bioconversion of lignocellulosic biomass is currently plagued by a number of limitations, notably generation of microbial inhibitory compounds during pretreatment and hydrolysis of lignocellulose to mixed sugars, and inefficient utilization of the generated mixed sugars by fermenting microorganisms due to carbon catabolite repression (Ezeji and Blaschek, 2008), therefore, other economic and readily utilizable substrates, whose applications in fermentation do not require pretreatment, may prove to be a more cost-effective and efficient substrates than lignocelluloses.

Among the non-lignocellulosic substrates, cheese whey is the most widely researched for ABE production, mostly due to its abundance and high biological oxygen demand (BOD), which constitutes a major disposal problem to the dairy industry. Because of its relatively low sugar content (4-5% lactose) this waste is unsuitable for many fermentation processes without prior concentration, while considered satisfactory for the butanol fermentation where product toxicity limits the amount of sugar utilization (Maddox, 1980). All over the world cheese production and consumption has increased rapidly generating more wastes and consequently pollution problems, and this has increased the interest to use it for the production of fuels (Foda et al., 2010).

Previously published studies (Maddox, 1980; Bahl et al., 1986; Foda et al., 2010) characterize whey permeate as a substrate which can be used for ABE fermentation. In fact, Maddox (1980) and Bahl et al. (1986) described that lactose metabolism favors butanol production over acetone, adding economic incentives for butanol production. Other authors studied the specific effect of nutrient supplementation or limitation on the fermentative activity of Clostridia, including: iron (Bahl et al., 1986; Dabrock et al., 1992; Peguin and Soucaille, 1995; Vasileva et al., 2012), phosphate (Bahl et al., 1986; Dabrock et al., 1992), CO (Dabrock et al., 1992), ferredoxin substitutes (Peguin and Soucaille, 1995), flavonoids (Wang et al., 2014), reducing cofactors (Li et al., 2014), inter alia. However, fermentative conditions and nutritional requirements for optimal use of cheese whey have not been fully elucidated and further work is required.

Trying to recycle wastes of the local dairy industry, this study investigated the suitability of using cheese whey as fermentation broth for ABE production emphasizing on covering the nutritional requirements of *C. acetobutylicum*, especially iron supplementation (no iron, $FeSO_4$ and $FeCl_3$), for routing metabolism toward the production of the desired solvents.

## 2. Material and method

### 2.1. Strains and culture maintenance

*C. acetobutylicum* ATCC 824 was grown anaerobically at 37 °C for 48 h in the *C. acetobutylicum* medium as described by Atlas (2004) with modifications. The medium contained 40 g potato flakes, 6 g glucose, 2 g $CaCO_3$, 0.5 g L-cysteine, 1 mg resazurin and 15 g bacteriological agar, diluted to 1 L of distilled water, pH 7.0. Cultures were kept at 4 °C and reseeded into fresh media every 14 d.

### 2.2. Fermentation

Cheese whey was used as fermentation medium (lactose being the main carbon source with an initial concentration of approximately 55 g/L of reducing sugars). More specifically, it was first deproteinized followed by adjusting pH value to 5.2 with 1M HCl and autoclaving (14 psi-115 °C/15 min). The cold whey was then filtered through cheesecloth and filter paper (10 µm mesh) under sterile conditions. Yeast extract (5 g/L) and $CaCO_3$ (18 g/L) were added to the filtrate, and the pH value was readjusted to 7.0 using 1 M NaOH. Three sets of fermentation experiments were performed: Group 0: without added iron; Group $FeSO_4$: added with 20 mg/L $Fe^{+2}$, supplemented as $FeSO_4$-$7H_2O$; and Group $FeCl_3$: added with 20 mg/L $Fe^{+3}$, supplemented as $FeCl_3$-$6H_2O$. Prepared media was stored at 4°C. Anaerobiosis was achieved by sparging nitrogen to remove dissolved oxygen. Inocula were generated by seeding the bacteria in 80 mL of fermentation medium contained in 125 mL flasks covered with rubber septa and incubated under anaerobic conditions in an orbital shaker (125 rpm) at 37 °C for 72 h. After 72 h, the content of each flask was emptied into a fermentation bottle. One inoculum was required for each batch fermentation, which was carried out under anaerobic conditions using capped glass bottles of 500 mL, containing 320 mL of fresh media for a total fermentation volume of 400 mL. Fermentations were incubated at 37 °C in orbital shaking at 125 rpm for 7 d. A total of 3 fermentations were carried out per each tested condition. Samples were taken periodically every 12 or 24 h; a 2 mL sample of the broth was taken from each experiment using sterile syringes. The broth samples where centrifuged at 11,920 ×g for 10 min in a micro17TR microcentrifuge (Hanil Science Industries, South Korea), the supernatant was then filtered through a 45 µm acrodisc (Merck Millipore, Darmstadt, Germany). Filtrated samples were cooled and stored at 4 °C in sealed vials until analysis.

*2.3. Analytical methods*

*2.3.1. Lactose concentration*

Total sugar concentration was determined using dinitro salicylic acid (DNS) reagent (Sumner, 1921) with lactose as standard.

*2.3.2. pH*

Direct measurement on the sample of culture broth at each sampling time was performed using a potentiometer pH/mV meter UB1-10 ultrabasic (Denver Instruments, Colorado, US).

*2.3.3. Solvents and acids quantification*

An HP 6890 series gas chromatograph (Hewlett Packard, California, US) equipped with a flame ionization detector (FID) and a HP-Innowax capillary column (30 m long; 0.53 mm Ø; 1.00 µm film) (Agilent J&W GC Columns, California, US) was used for solvents and acids quantification. For the assays, 1 µL of each sample was injected into the gas chromatograph, then heated from 60 to 150°C (10 °C/min) and maintained at 150 °C/5min. Injector and detector temperatures were maintained at 250°C. Gas injection was constant at 4 mL/min, and gas pressures were: He 60 psi; $H_2$ 40 psi; dry air 60 psi. Concentration of each product was calculated by integrating the area under the curve of the peak generated using the software Peak Simple 3.21® (SRI Instruments, California, US).

*2.4. Statistical analyses*

One-way ANOVA was performed to compare butanol and total ABE production using different iron sources. $P< 0.05$ was considered as significant. LSD Multiple Range Tests (at 95.0 % confidence level) were performed to detect significant differences between the productive capacities. All analyses were performed using STATGRAPHICS Centurion XVI® (Statpoint Technologies, Virginia, US).

**3. Results and discussion**

Cheese whey fermentations supplemented with iron (Group $FeSO_4$ and Group $FeCl_3$) showed the typical behavior of a two-stage ABE fermentation (Fig. 1 and 2, respectively). On the first stage, lasting approximately 48 h, lactose uptake occured and acetic and butyric acids were produced lowering the pH value until it was substantially constant around a value of 5.5. Between 48 and 60 h of fermentation the second stage started, diminishing lactose uptake and promoting re-assimilation of acids into the desired solvents: acetone, butanol and ethanol, reaching maximum solvents concentration after 168 h of fermentation.

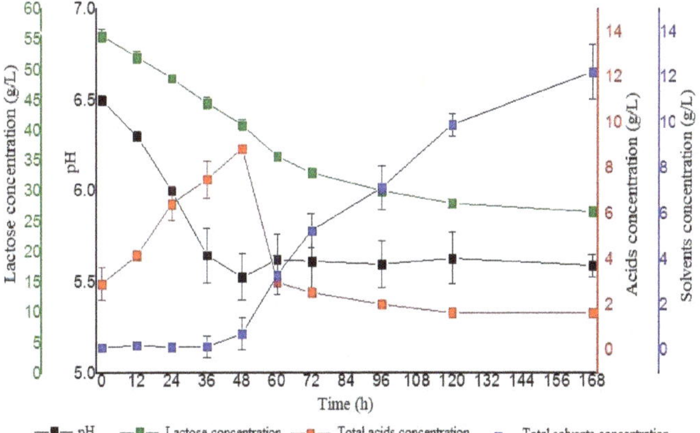


**Fig.1.** Time course profiles of average pH value, lactose concentration, total acids (acetic and butyric) and total solvents (acetone, butanol and ethanol) production from Group $FeSO_4$ fermentations. Bars show standard deviation (n=3).

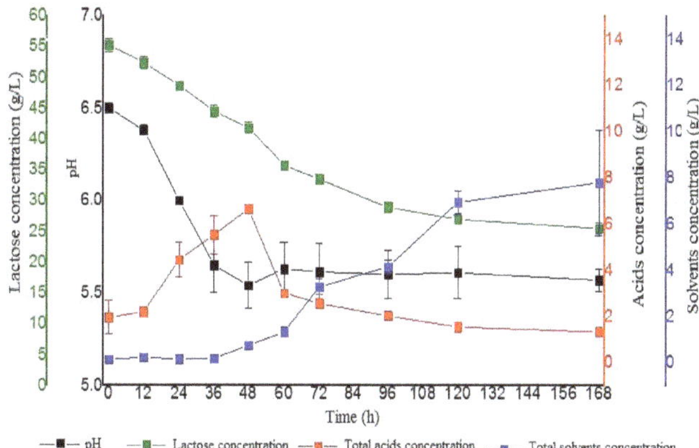


**Fig.2.** Time course profiles of average pH value, lactose concentration, total acids (acetic and butyric) and total solvents (ABE) production from Group $FeCl_3$ fermentations. Bars show standard deviation (n=3).

The main product of the fermentation was butanol, followed by ethanol, and almost negligible amounts of acetone were detected in the fermentation broths. This could be ascribed to two facts: 1) it has been proved that the use of lactose as the carbon source favors butanol production over acetone (Maddox, 1980; Bahl et al., 1986), and 2) culture conditions, specifically agitation and temperature (37° C), promoted acetone volatilization making it difficult to detect in the culture broth.

Time required to complete the fermentation (7 d) and for inoculum preparation (72 h) was longer than that reported by Foda et al. (2010) (75 h) and Napoli et al. (2008) (96 h) due to the presence of citrate in the culture medium (≈10 g/L). Citrate at high concentrations (> 2.5 g/L) inhibits *C. acetobutylicum* growth; the presence of citrate contributes to unusual prolonged time required to complete whey fermentation (Bahl et al., 1986).

Group 0 fermentations did not show the typical two-stage behavior (Fig. 3). Lactose uptake occurred but acetic and butyric acid productions were remarkably low (less than 4.5 g/L combined), instead, substantial amounts of lactic acid were produced (5.49 ± 0.69 g/L) (Table 1). No re-assimilation of acids was observed but ethanol production occurred.

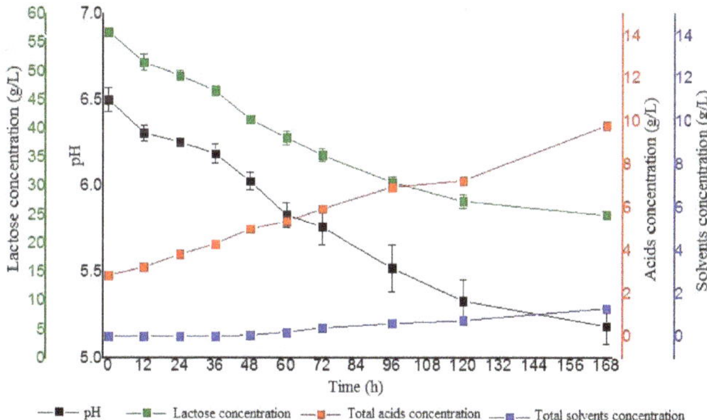


**Fig.3.** Time course profiles of average pH value, lactose concentration, total acids (lactic, acetic and butyric) and total solvents (ABE) production from Group 0 fermentations. Bars show standard deviation (n=3).

These results differ from those presented by Maddox (1980), who claimed that using cheese whey supplemented only with yeast extract, *C. acetobutylicum* N.C.I.B. 2951 could produce up to 15 g/L of butanol while requiring 5 d to reach this maximum concentration. Also cheese whey without any supplementation could achieve a maximum butanol concentration of 13 g/L within 7 d. However, in later studies, Ennis and Maddox (1986), using a different strain of *C. acetobutylicum* (P262), prepared semi-synthetic media simulating cheese whey, supplemented with ferrous sulfate and other components, despite having yeast extract added, but no explanation for these additions were given.

**Table 1.**
Performance of ABE fermentations (168 h).

| Product | Group 0 | Group $Fe^{+2}$ | Group $Fe^{+3}$ |
|---|---|---|---|
| Acetate (g/L) | 2.61 ± 0.34 | 0.972 ± 0.08 | 1.44 ± 0.28 |
| Butyrate (g/L) | 1.70 ± 0.53 | 0.650 ± 0.12 | 1.27 ± 0.35 |
| Lactate (g/L) | 5.49 ± 0.69 | Not detected | Not detected |
| Acetone (g/L) | 0.04 ± 0.01 | 0.00 ± 0.00 | 0.01 ± 0.00 |
| Butanol (g/L) | 0.06 ± 0.10 | 7.13 ± 1.53 | 4.32 ± 0.94 |
| Ethanol (g/L) | 1.20 ± 0.10 | 5.11 ± 1.65 | 3.46 ± 1.37 |
| Total solvents (g/L) | 1.30 ± 0.21 | 12.24 ± 3.18 | 7.78 ± 2.30 |
| Lactose utilization (%) | 56.24 ± 0.65 | 52.28 ± 0.62 | 53.80 ± 0.60 |
| ABE yield (gABE/g lactose consumed) | 0.04 ± 0.01 | 0.42 ± 0.10 | 0.25 ± 0.07 |
| A:B:E Ratio | 0.31:0.46:9.23 | 0.0:5.87:4.13 | 0.0:5.61:4.39 |

Moreover, Bahl et al. (1986), Peguin and Soucaille (1995), and Vasileva et al. (2012) agree on the importance of iron in the fermentative metabolism of *C. acetobutylicum*, asserting that in an iron-deficient environment lactate was the main product obtained from the fermentation, observing no re-consumption of acetic and butyric acids.

As shown in Figure 4, *C. acetobutylicum* has the ability to break down lactose by hydrolyzing it into glucose and galactose which are then metabolized by the Embden Meyerhof Pathway (EMP) to generate pyruvate. Then pyruvate has to be oxidized to produce acetyl CoA which plays a central role in the metabolism of *C. acetobutylicum* as it serves as a precursor for the generation of all desired products. Oxidation of pyruvate occurs in a reaction that is coupled to the reduction of ferredoxin, using hydrogen as the ultimate electron acceptor. Sufficient iron is required to produce enough ferredoxin to complete this oxidation (Lee et al., 2008). In an iron deficient environment formation of molecular hydrogen does not take place and metabolism changes from pyruvate to lactate instead of producing acetyl CoA (Gheshlaghi et al., 2009).

Bahl et al. (1986) indicated that limiting the amount of iron in the culture broth helped increase the proportion of butanol/acetone (from 2:1 to 8:1). They argued that ferrous ions had the most drastic effect on that ratio; however, at iron-limited conditions lactate became the main product, instead of producing acetate and butyrate, adversely affecting the amount, yield and productivity of butanol. Increasing the butanol/acetone proportion could facilitate the recovery and purification processes of the product, but iron-limitation backfires by decreasing the amount of butanol produced.

Moreover, Peguin and Soucaille (1995) suggest that limiting the available iron in glucose culture broths, using *C. acetobutylicum* ATCC824, helps modulate carbon and electron flows. For desirable fermentation results, a compound (methyl viologen) that replaces the functions of ferredoxin must be provided to the culture media, because ferredoxin could not be properly synthesized due to iron deficiency. By limiting the amount of iron present in the medium, dehydrogenase activity involved in the conversion of acetyl-CoA into β-hydroxybutyrate is limited. Methyl viologen function as a better substrate for ferredoxin-$NAD(P)^+$ reductase than ferredoxin itself, creating an artificial electron transport chain. Both effects: reduction of the dehydrogenase activity and increase of ferredoxin-$NAD(P)^+$ reductase activity made possible to obtain higher yields of butanol that those obtained with regular ABE fermentation, reaching a maximum 13 g/L butanol concentration. Nevertheless adding methyl viologen causes long lag growth phase affecting the fermentation productivity. Increasing the production of alcohols by limiting iron may sound favorable, but to be obligated to supplement compounds who can serve as electron acceptors to prevent the formation of lactate may be more costly for the process and more problematic for the recovery of products. In this paper, it has been shown that by adding adequate amounts of iron (20 mg /L) butanol production improvement could be achieved, favoring the desired normal behavior of the metabolic pathway.

Metabolic development of Group 0 fermentations demonstrate the effect of iron deficiency when cheese whey is used as substrate, effect that was not observed by Napoli et al. (2008) while they also worked with synthetic lactose medium simulating cheese whey enriched with yeast extract and $CaCO_3$ and without any additional iron source, reporting the typical two-stage ABE fermentation behavior with a 3 g/L butanol production and no lactate detected, results that were reproduced in our laboratory too (data not shown).

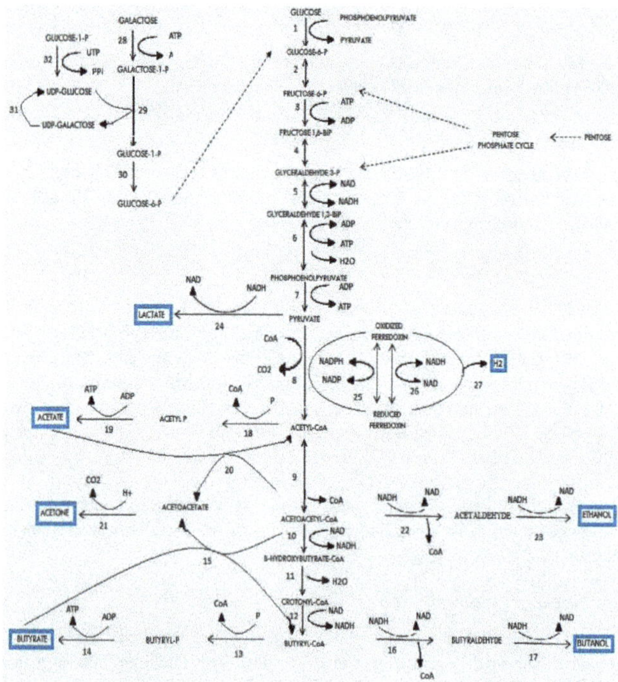


**Fig.4**. Metabolic pathways of glucose and galactose in *C. acetobutylicum*. Extracellular products are highlighted in blue boxes. Numbers indicate the enzymes involved: 1: phosphotransferases system; 2: glucose-6-phosphate isomerase; 3: phosphofructo kinase; 4: aldolase; 5: glyceraldehyde-3-phosphate dehydrogenase; 6: phosphoglycerate kinase 7: pyruvate kinase; 8: pyruvate-ferredoxin oxidoreductase; 9: thiolase; 10: β-hydroxybutyryl CoA dehydrogenase; 11: crotonase; 12: butyryl-CoA dehydrogenase; 13: phosphotransbutyrylase; 14: butyrate kinase; 15: acetoacetyl-CoA Butyrate-CoA transferase; 16: butyraldehyde dehydrogenase; 17: butanol dehydrogenase; 18: phosphotransacetylase; 19: acetate kinase; 20: acetoacetyl-CoA acetate-CoA transferase; 21: acetoacetate decarboxylase; 22: acetaldehyde dehydrogenase; 23: ethanol dehydrogenase ; 24: lactate dehydrogenase; 25: ferredoxin-NADP reductase NADPH-ferredoxinoxidorreductase; 26: ferredoxin-NAD reductase NADH-ferredoxinoxidorreductase; 27: ferredoxin hydrogenase; 28: galactokinase; 29: UDP-glucose glucose-1-P uridyl transferase; 30: phosphoglucomutase; 31: UDP-galactose epimerase; 32: UDP-glucose phosphotransferase.

In fact, yeast extract contains iron in amounts up to 20 mg/100 g that may be sufficient to produce enough ferredoxin for routing of metabolism to the production of acetyl CoA and later generation of acetone, ethanol and butanol when it is added to a synthetic fermentation media; but, when cheese whey is used an extra amount of iron is necessary because of the presence of lactoferrin (20- 200 μg/ml) (Law and Reiter, 1977), an iron-binding protein that binds to iron reversibly with a high affinity. This blocks iron and impedes ferredoxin formation and, therefore, restricts solvents production.

The fermentation performance of each group is summarized in Table 1. Analyses of variance were conducted to assess whether the factor studied (iron source) had an effect on ABE production. LSD Multiple Range Test was conducted to assess which fermentation condition tested was better than the others in order to obtain a higher butanol concentration and higher total solvents production, being butanol as the main product. One way ANOVA indicated statistically significant differences in the production of total solvents between fermentations with added iron and unsupplemented ($p< 0.005$). Multiple range tests showed that the total production of solvents presents statistically significant difference between cheese whey without iron supplementation and the same substrate with an extra iron source, however, no statistically significant difference existed when comparing the two different iron sources i.e. $FeSO_4$ and $FeCl_3$.

Evaluating only butanol production, one way ANOVA indicated that there is a statistically significant difference between the total amount of butanol produced by each fermentation condition tested, ($p<0.001$). LSD Multiple Range Test indicated that, for butanol production, there is statistically significant difference between the three groups. Both iron sources tested showed practically nil production of acetone; while butanol proportion was higher using $FeSO_4$ (0.0:5.87: 4.13) than $FeCl_3$ (0.0: 5.61: 4.39). Of particular interest is the fact that the use of $FeSO_4$ as iron source increased butanol production by 65% compared to what achieved by employing $FeCl_3$. This could be explained by the presence of iron and sulfur, both essential components of ferredoxin.

Additionally Group $FeSO_4$ showed higher ABE yield ($0.42 \pm 0.10$) than Group $FeCl_3$ ($0.25\pm0.07$), despite almost equal sugar consumptions. These results showed that sugar conversion to acids and acids reassimilation developed best when the iron source also included sulfur, under conditions tested.

These results showed that whey supplemented appropriately, can serve as a good substrate for ABE fermentation. Cheese whey is a readily available and widely-produced inexpensive substrate which requires low pretreatment to serve as fermentation substrate. These make it more economic than other substrates used such as lignocellulosic materials, algal biomass or different industrial and domestic wastes. Cheese whey tends to favor the production of butanol over acetone and therefore by improving fermentative conditions to fully utilize lactose in whey could lead ABE fermentation to economic competitiveness.

## 4. Conclusion

Cheese whey is a suitable substrate for ABE fermentation as long as it is properly prepared and supplemented. Addition of an iron source is strictly necessary for cheese whey to be a viable substrate and the lack of iron in cheese whey impedes ferredoxin synthesis and restricts pyruvate-ferredoxin oxidoreductase activity leading to the production of lactic acid instead of ABE. Furthermore, the addition of $FeSO_4$ improved butanol production by 65% reaching a concentration of $7.13 \pm 1.53$ g/L, compared to what obtained with $FeCl_3$ under the same fermentation conditions. This indicates that $FeSO_4$ is an ideal iron source for improving butanol production under conditions tested.

## References

Atlas, R.M., 2004. Handbook of Microbiological Media, third ed. CRC Press, USA.

Bahl, H., Gottwald, M., Kuhn, A., Rale, V., Andersch, W., Gottschalk, G., 1986. Nutritional Factors Affecting the Ratio of solvents Produced by *Clostridium acetobutylicum*. Appl. Environ. Microbiol. 52 (1), 169-172.

Cooksley, C.M., Zhang, Y., Wang, H., Redl, S., Winzer, K., Minton, N.P., 2012. Targeted mutagenesis of the *Clostridium acetobutylicum* acetone–butanol–ethanol fermentation pathway. Metab. Eng. 14(6), 630-641.

Dabrock B., Bahl, H., Gottschalk, G., 1992. Parameters Affecting Solvent Production by *Clostridium pasteurianum*. Appl. Environ. Microbiol. 58 (4), 1233-1239.

Ennis, B.M., Maddox, I.S., 1989. Production of solvents (ABE fermentation) from whey permeate by continuous fermentation in a membrane bioreactor. Bioprocess. Eng. 4, 27-34.

Ezeji, T.C., Blaschek, H., 2008. Fermentation of dried distillers´soluble (DDGS) hydrolysates to tolvents and value-added products by solventogenic Clostridia. Bioresour. Technol. 99 (12), 5232-5242.

Foda, M.I., Joun, H., Li, Y., 2010. Study the Suitability of Cheese Whey for Bio-Butanol Production by Clostridia. J. Am. Sci. 6(8), 8.

Gheshlaghi, R., Scharer, J.M., Moo-Young, M.,Chou, C.P., 2009. Metabolic pathways of clostridia for producing butanol. Biotechnol. Adv. 27(6), 764-781.

Green, E.M., 2011. Fermentative production of butanol the industrial perspective. Curr. Opin. Biotechnol. 22(2), 337-343.

Jang, Y.S., Malaviya, A., Cho, C., Lee, J., Lee, S.Y., 2012. Butanol production from renewable biomass by clostridia. Bioresour. Technol. 123, 653-663.

Law B.A., Reiter B., 1977. The isolation and bacteriostatic properties of lactoferrin from bovine milk whey. J. Dairy Res. 44 (3), 595-599.

Lee, S.Y., Park, J.H., Jang, S.H., Nielsen, L K., Kim, J., Jung, K.S., 2008. Fermentative butanol production by clostridia. Biotechnol. Bioeng. 101 (2), 209-228.

Li, T. Yan, Y., He, J., 2014. Reducing cofactors contribute to the increase of butanol production by wild-type *Clostridium* so. Strain BOH3. Bioresour. Technol. 155, 220-228.

Maddox, I.S., 1980. Production of n-butanol from whey filtrate using *Clostridium acetobutylicum* N.C.I.B. 2951. Biotechnol. Lett. 2 (11), 493-498.

Napoli, F., Olivieri, G., Marzocchella, A., Salatino, P., 2008. Assessment of Kinetics for Butanol Production by *Clostridium acetobutylicum*, in: Briens, C., Berruti, F., Al-Dahhan M. (Eds.), Bionergy-II: Fuels and Chemicals from Renewable Sources. ECI Symposium Series, New York, pp. 10-17.

Niemistö, J., Saavalainen, P., Isomäki, R., Kolli, T., Huuhtanen, M., Keiski, R.L., 2013. Biobutanol Production from Biomass, Biofuel Technologies. Springer Berlin Heidelberg, pp. 443-470.

Parekh, M., Blaschek, H., 1999. Butanol production by hypersolvent-producing mutant *Clostridium beijerinckii* BA101 in corn steep water medium containing maltodextrin. Biotechnol. Lett. 21(1), 45-48.

Peguin, S., Soucaille, P., 1995. Modulation of Carbon and Electron Flow in *Clostridium acetobutylicum* by Iron Limitation and Methyl Viologen Addition. Appl. Environ. Microbiol. 61(1), 403-405.

Rajchenberg-Ceceña, E., Rodríguez-Ruíz, J.A., Juárez, K., Martínez, A., Morales, S., 2009. Producción Microbiológica de Butanol. BioTecnología. 13, 26-37.

Ranjan, A., Moholkar, V.S., 2009. Biobutanol: a Viable Gasoline Substitute through ABE Fermentation. Proc. World Acad. Sci: Eng. Technol. 51, 497-503.

Sabra, W., Groeger, C., Sharma, P.N., Zeng, A.P., 2014. Improved n-butanol production by a non-acetone producing *Clostridium pasteurianum* DSMZ 525 in mixed substrate fermentation. App. Microbiol. Biotechnol. 98 (9), 4267-4276.

Shapovalov, O.I., Ashkinazi, L.A., 2008. Biobutanol: Biofuel of Second Generation. Russ. J. Appl. Chem. 81 (12), 2232-2236.

Sumner, J.B., 1921. Dinitrosalicylic acid: a reagent for the estimation of sugar in normal and diabetic urine. J. Biol. Chem. 47, 5-9.

Tashiro, Y., Sonomoto, K., 2010. Advances in butanol production by clostridia, in: Mendez-Villas, A. (Ed.), Current Research, Technology Education Topics in Applied Microbiology and Microbial Technology. Formatex Research Center, Spain, pp.1383-1394.

Vasileva, D., Janssen, H., Hönicke, D., Ehrenreich, A., Bahl, H., 2012. Effect of iron limitation and *fur* gene inactivation on the transcriptional profile of the strict anaerobe *Clostridium acetobutylicum*. Microbiology. 158, 1918-1929.

Wang, L., Xia, m., Zhang, L., Chen, H., 2014. Promotion of the *Clostridium acetobutylicum* ATCC 824 growth and acetone-butanol-ethanol fermentation by flavonoids. World J. Microbiol. Biotechnol. 30(7), 1969-1976.

# 14

# Optimization of alkali catalyst for transesterification of jatropha curcus using adaptive neuro-fuzzy modeling

Vipan K Sohpal[1*], Amarpal Singh[2]

[1]*Department of Chemical Engineering, Beant College of Engineering & Technology, Post Box No 13, Gurdaspur Punjab, India*

[2]*Department of Electronics & Communication Engineering Beant College of Engineering & Technology, Post Box No 13, Gurdaspur Punjab, India*

## HIGHLIGHTS

➢ Biodiesel production from non edible oil through transesterification in batch reactor is highly effective technique for kinetic analysis.

➢ Temperature, molar ratio, mixing intensity and catalyst influenced the biodiesel production and kinetic. Alkaline catalysts are more efficient in nature as compare to acid and base catalyst.

➢ This paper particularly focuses on the impact of NaOH catalyst on transesterification process and yield of butyl ester production. ANFIS used to modeling and assess the output for large domain. In addition, K-S statistical tests are used for analysis of the ANFIS modeling.

## GRAPHICAL ABSTRACT

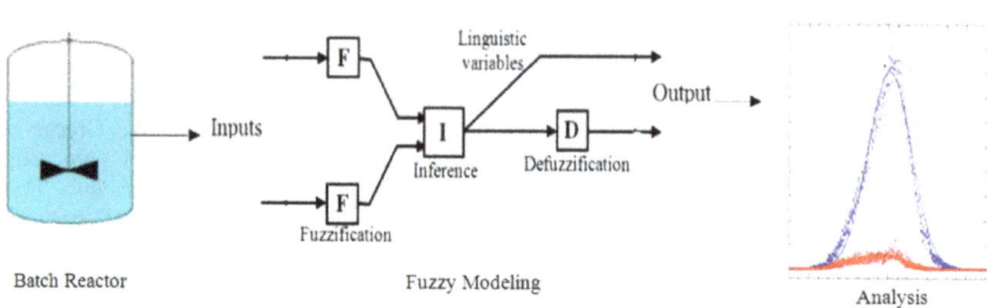


## ABSTRACT

Transesterification of Jatropha curcus for biodiesel production is a kinetic control process, which is complex in nature and controlled by temperature, the molar ratio, mixing intensity and catalyst process parameters. A precise choice of catalyst is required to improve the rate of transesterification and to simulate the kinetic study in a batch reactor. The present paper uses an Adaptive Neuro-Fuzzy Inference System (ANFIS) approach to model and simulate the butyl ester production using alkaline catalyst (NaOH). The amounts of catalyst and time for reaction have been used as the model's input parameters. The model is a combination of fuzzy inference and artificial neural network, including a set of fuzzy rules which have been developed directly from experimental data. The proposed modeling approach has been verified by comparing the expected results with the practical results which were observed and obtained through a batch reactor operation. The application of the ANFIS test shows which amount of catalyst predicted by the proposed model is suitable and in compliance with the experimental values at 0.5% level of significance.

**Keywords:**
Biodiesel
Jatropha curcus
Transesterification
Batch reactor
NaOH catalyst
Adaptive neuro fuzzy inference

## 1. Introduction

Biodiesel production from non-edible oils involves different chemical reactions, but transesterification with alcohol has long been a preferred method for producing biodiesel. The process of transesterification is controlled by various process parameters such as catalyst type & concentration, molar ratio of alcohol to oil, type of alcohol, reaction time & temperature and mixing intensity. But the most challenging part of the biodiesel industry is the high cost of raw materials due to the low availability of non-edible oils (Jatropha Curcas). The price per unit for building up the catalyst alone makes up about 50% of the total production cost. So the catalysts used for the transesterification process are a rate determining parameter. The transesterification rate can be controlled using three different categories of catalysts. Sodium and potassium hydroxide are more effective

* Corresponding author
E-mail address: vipan752002@gmail.com (V.K. Sohpal).

among alkali catalysts, while acid catalyzed transesterification functions in the presence of HC l and $H_2SO_4$. Enzymatic catalysts especially lipases are able to effectively catalyze the transesterification of triglycerides in both aqueous or non-aqueous systems.

Alkali catalysts like sodium hydroxide, sodium methoxide, potassium hydroxide and potassium methoxide are more effective (Zhang et al, 2003) when lower levels of free fatty acids and water are present. Conversely acid catalyzed transesterification is suitable, if the base oil has high free fatty acid content and more water. Sulfuric acid, phosphoric acid, hydrochloric acid or organic sulfonic acids are the various types of acidic catalysts. A methanolysis of beef tallow was studied with NaOH and NaOMe used as catalysts. The study found that NaOH was significantly better than NaOMe (Ma et al, 1998). In addition, it was found that high quality oil is required when NaOMe catalyst is used, which itself is a limiting factor (Ahn et al, 1995). 1% NaOH and 0.5% NaOMe with a higher molar ratio, exceeding the 6:1 ratio, perform almost the same conversion after 60 minutes run (Freedman et al, 1986). Although methanolysis of soybean oil with 1% potassium hydroxide as catalyst has given the best yields but higher viscosities of the esters causes a problem in the separation process (Tomasevic and Marinkovic, 2003). A study of the catalytic activity of magnesium oxide, calcium hydroxide, calcium oxide, calcium methoxide, barium hydroxide, compared to sodium hydroxide during the transesterification of rapeseed oil was carried out (Gryglewicz, 1999) and (Noureddini and D. Zhu, 1997). The study concluded that sodium hydroxide exhibited the highest catalytic activity in this process. The acid catalyzed transesterification wa also studied with waste vegetable oil (Canakci and Van Gerpen, 1999) and showed that the same concentration of HCl and $H_2SO_4$ in the presence of 100% excess alcohol decreases the viscosity. $H_2SO_4$ has superior catalytic activity in the range of 1.5-2.25 M concentration. In additional to conventional catalysts, enzymatic catalysts like lipases are suitable substitutes for catalyzing the transesterification of triglycerides (Dorado et al, 2002). However the cost of a lipase catalyst is significantly greater than that of an alkaline one. The transesterification reaction of Jatropha curcas oil using a sodium hydroxide catalyst at 105°C temperature, 250 rpm and molar ratio of Butanol to Jatropha curcas oil (11:1) in batch reactor was also studied (Sohpal et al 2011). Fuzzy models were developed using adaptive neurofuzzy inference systems to evaluate and compare the results (Sohpal et al 2011). The response surface methodology (RSM) can also be used to determine the optimum condition for the transesterification reaction (Pinzi et al 2010). FWM and WDM was analyzed using ANFIS and comparative analysis, showing that the ANFIS approach was close to the real situation (Amarpal et al, 2009).This demonstrates the possible use of a relatively new soft computing technique called adaptive neuro-fuzzy inference system (ANFIS) for predicting uniaxial compressive Strength (UCS) of granites (Yesiloglu-Gultekin et al, 2012).

The present paper aims to optimize the amount of catalyst (NaOH) used in commercial scale, for techno-economy viability of biodiesel production by introducing the fuzzy model. In this study other process parameters (temperature, rpm and molar ratio of Butanol to Jatropha curcas oil) were fixed while catalyst wt and reaction time were two independent variables.

## 2. Fuzzy modeling of batch reactor based on ANFIS

The modeling of batch reactor for transesterification has been ANFIS, considering input parameters such as weight of the catalyst, time for reaction, and output as esters of butyl (%w/w). In ANFIS, the parameters associated with the membership functions change through the learning process. The computation of these parameters is facilitated by the gradient vector, which provides a measure of how well the fuzzy inference system (FIS) is modeling the input and output data for a given set of parameters. Once the gradient vector is obtained, any of the various optimization routines can be applied in order to accommodate the parameters so as to minimize some measure of error. Some data points are located and have been used as an input for the training of the FIS.

### *2.1. Architecture of the ANFIS*

The fuzzy logic approach has the potential to produce a simplified control for various chemical engineering applications. The rule-based features of fuzzy models allow for a model interpretation in a way that is similar to the one humans use for describing reality. Conventional methods for statistical validation based on numerical data can be complemented by human knowledge which usually involves heuristic knowledge and intuition. A multi-input single output (MISO) fuzzy model of batch reactor for transesterification has been developed using ANFIS by considering two input parameters and one output variable in order to predict the product concentration. This technique provides a method for the fuzzy modeling method to understand information about a data set, in order to evaluate the membership function parameters that best provide the corresponding FIS to track the given input/output data. This learning process works similarly to that of neural networks. The parameters associated with the membership functions will change through the learning process. This method is based on the Sugeno-type fuzzy interface system and can simulate and analyze the mapping relation between the input and output data through hybrid learning to determine the optimal allocation of membership functions. The ANFIS architecture of the type from Takagi and Sugeno has been shown in Figure 1.

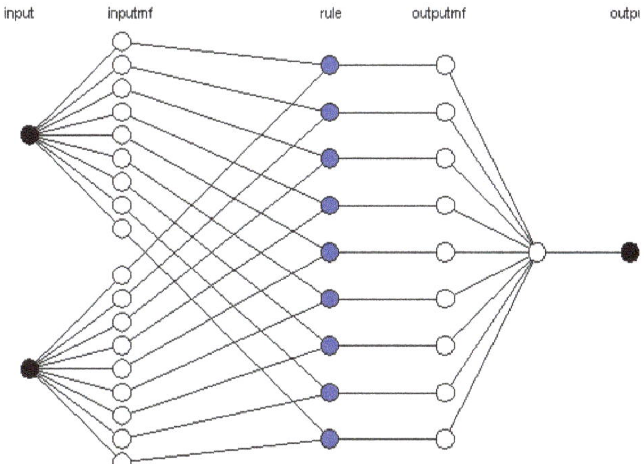


**Fig.1.** ANFIS Model with sub clustering

This interference system is composed of five layers. Each layer involves several nodes, which are described by the node function. The output signals from the nodes in the previous layers will be accepted as the input signals in the present layer. After manipulation by the node function in the present layer, the output will serve as the input signal for the next layer. To simply explain the mechanism of the ANFIS, we consider two inputs, x and y, and one output f in the FIS. Hence, the base rules will be fuzzy "if-then" rules as follows:

Rule 1: If x is A1 and y is B1, then f = p1x + q1y + r1

Rule 2: If x is A2 and y is B2, then f = p2x + q2y + r2

### *2.2. Fuzzy Interference System (FIS)*

The core of a fuzzy logic controller/modeling is the inference engine, which contains information of the control strategy in the form of "if-then" rules. Since the fuzzy logic, rules require linguistic variables. Inputs and outputs of a process are generally continuous crisp values, therefore the conversion of crisp values into fuzzy values and vice versa are required. The initial step of the fuzzy modeling approach is to determine the input and output variables of the fuzzy logic controller. Sugeno type FIS is used for this purpose. A typical direct fuzzy logic control system is shown in Fig. 2. The ANFIS editor is used to create, train, and assess the Sugeno fuzzy logic. This FIS system is designed for the MISO system. The MISO system includes two inputs and one output.

### *2.3. Identification of input and output variables*

The fuzzy logic is based on the identification of the fuzzy set that represents the possible values of the variables. In Figure 2, a block diagram of the fuzzy control process along with the physical system of the batch reactor for transesterification has been shown. In particular, Figure 3 shows real inputs and real outputs. The fuzzy model described in this article is a MISO system with two input parameters of catalyst weight, and chemical reaction time as well as the output parameter of butyl ester production.

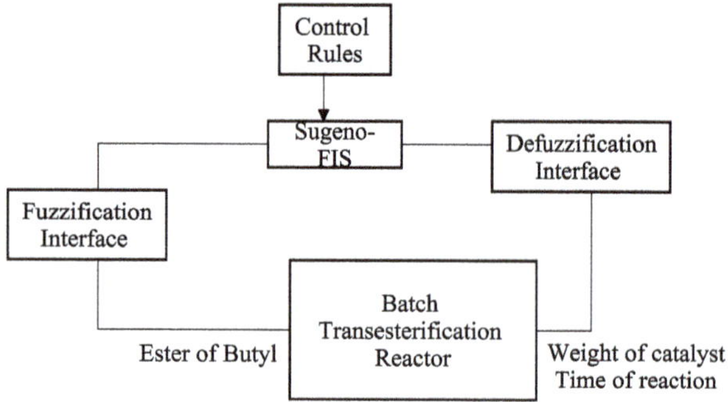


**Fig.2.** Block diagram for fuzzy control system of batch transesterification reactor.

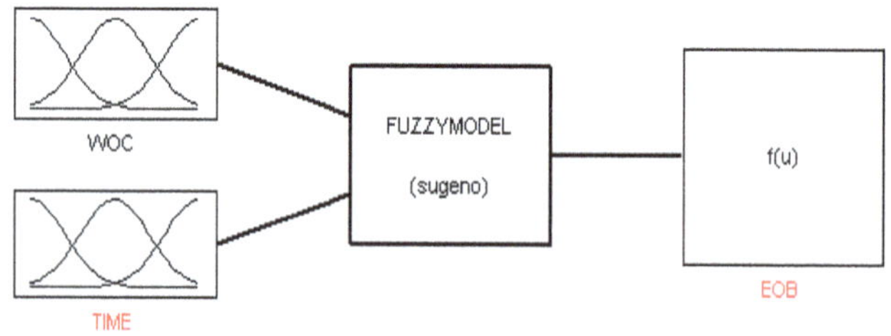


**Fig.3.** Fuzzy model of batch transesterification reactor showing inputs and output.

The possible universe of discourse for the input parameters has been given below:

**Input parameters:**

Weight of Catalyst (NaOH) = 20 gm to 40 gm
Time for chemical reaction (min) = 20 min to 60 min
Output parameter:
Ester of butyl (Product) = Predicted as %w/w
(Depending upon input parameters)

Membership Functions for the Input and Output Variables for ANFIS Modeling with sub-clustering and without sub-clustering.

In this process linguistic values were assigned to the variables using fuzzy subsets and their associated membership functions. A membership function assigns numbers between 0 and 1. Zero membership value means a non-member of the fuzzy set, while one represents full membership. A membership function can have any shape but the standard shapes for the membership function include trapezoids, triangles and bell shapes. Modeling without sub-clustering involves three membership functions that are produced for each input variable of catalyst weight (WOC), and reaction time based on the ANFIS. The in1mf1, in1mf2, in1mf3 are three linguistic levels for the amount of catalyst and in2mf1, in2mf2, in2mf3 are for reaction time which have been illustrated in Figures 4 (a), 4(b) and 4(c).

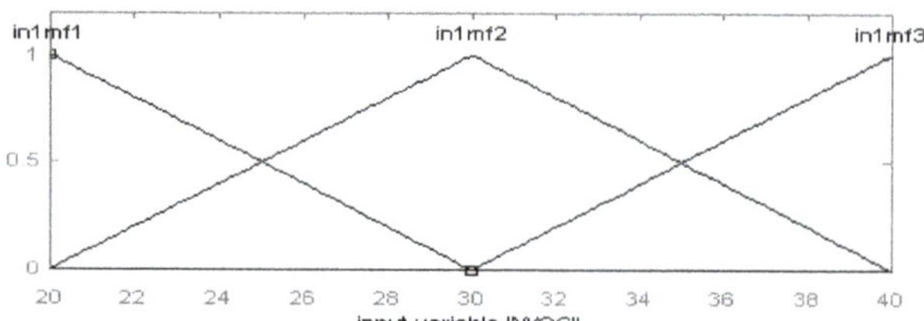


**Fig. 4.** (a) Membership function plots of input variable weight of catalyst.

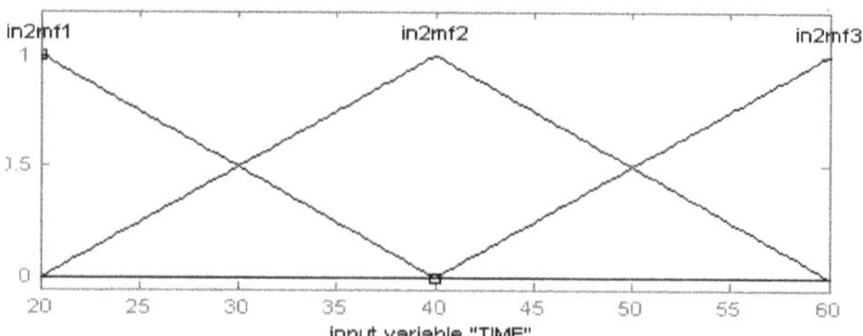


**Fig. 4.**(b) Membership function plots of input variable time of reaction.

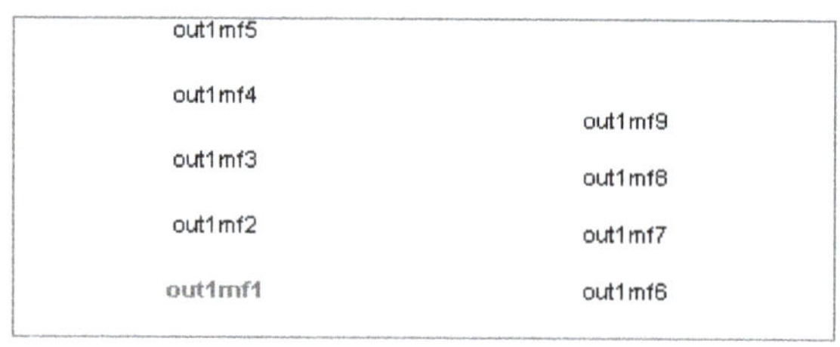


**Fig.4.** (c) Membership function plots of output variable ester of butyl.

On the other hand, in ANFIS modeling with sub-clustering, nine membership functions were generated for input variables of catalyst weight (WOC), and nine membership functions for reaction time based on the ANFIS. The in1cluster1, in1cluster2, in1cluster3, in1cluster4, in1cluster5, in1cluster6, in1cluster7, in1cluster8, in1cluster9 are nine linguistic levels for the catalyst weight variable and in2cluster1, in2cluster2, in2cluster3, in2cluster4, in2cluster5, in2cluster6, in2cluster7, in2cluster8, in2cluster9 in2mf1, in2mf2, in2mf3, in2mf4 are reaction time variables over a given universe of discourse as shown in Figures 5 (a) and 5(b). The output variable of butyl ester production (EOB) also has four membership functions as shown in Figure 5(c). The span of each function has been tuned within the specified range. Tests were conducted to evaluate the response parameters, and the span was varied accordingly for improvement. After a few iterations, the final membership functions for the system were determined as shown in the respective figures.

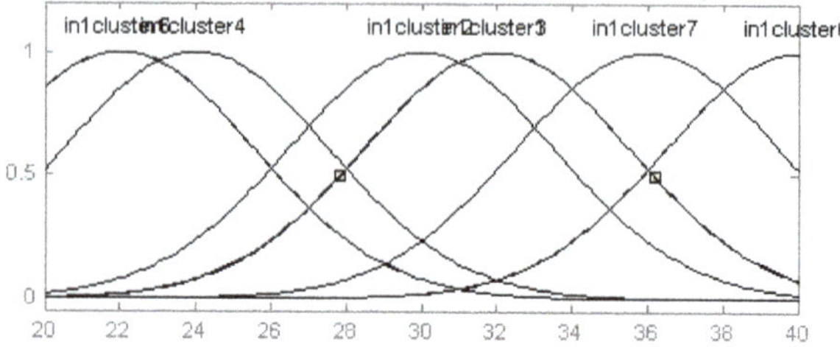


**Fig.5.** (a) Membership function plots of input variable weight of catalyst.

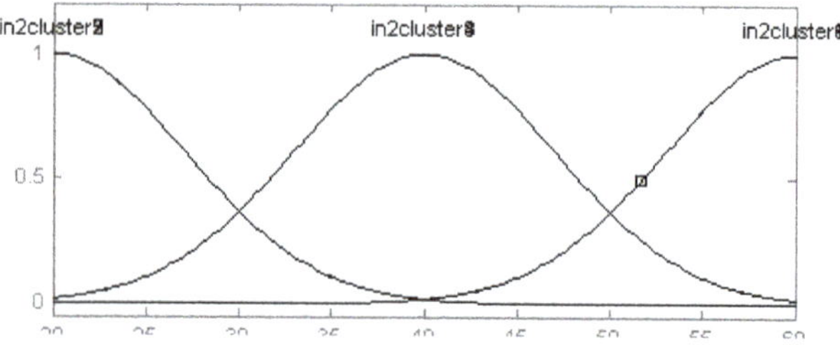


**Fig.5.** (b) Membership function plots of input variable time of reaction.

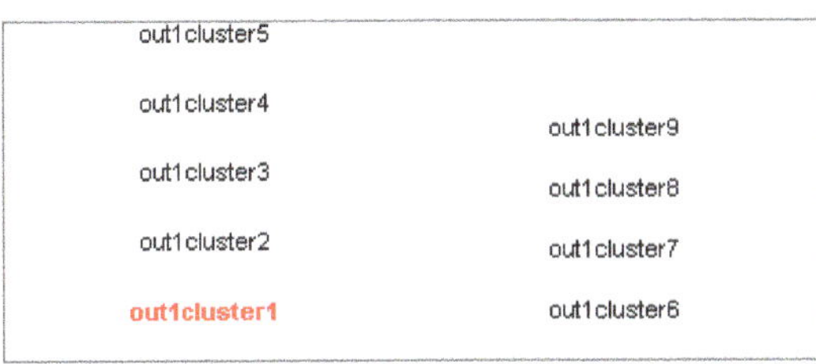


**Fig.5.** (c) Membership function plots of output variable ester of butyl.

*2.4. FIS Rules Employed in the Model*

The fuzzy model of the batch transesterification has been used to design the MISO fuzzy model with two inputs, i.e., weight of catalyst and time for reaction , each of which were determined for 3 linguistic variables using the ANFIS modeling method without sub-clustering. These variables generated 9 numbers of conditional statements as "if-and-then" rules of the model. The formulated set of rules of the model has been outlined below:

1. If (WOC is in1mf1) and (Time is in2mf1) then (EOB is out1mf1) (1)
2. If (WOC is in1mf1) and (Time is in2mf2) then (EOB is out1mf2) (1)
3. If (WOC is in1mf1) and (Time is in2mf3) then (EOB is out1mf3) (1)
4. If (WOC is in1mf2) and (Time is in2mf1) then (EOB is out1mf4) (1)
5. If (WOC is in1mf2) and (Time is in2mf2) then (EOB is out1mf5) (1)
6. If (WOC is in1mf2) and (Time is in2mf3) then (EOB is out1mf6) (1)
7. If (WOC is in1mf3) and (Time is in2mf1) then (EOB is out1mf7) (1)
8. If (WOC is in1mf3) and (Time is in2mf2) then (EOB is out1mf8) (1)
9. If (WOC is in1mf3) and (Time is in2mf3) then (EOB is out1mf9) (1)

Similarly the weight of the catalyst and the time for the reactions, were both variables that were also used for linguistic variables using ANFIS modeling with sub-clustering. These variables generated 9 numbers of conditional statements as "if-and-then" rules of the model. The formulated set of rules of the model has been outlined below:

1. If (WOC is in1cluster1) and (Time is in2cluster1) then (EOB is out1cluster1) (1)
2. If (WOC is in1cluster2) and (Time is in2cluster2) then (EOB is out1cluster2) (1)
3. If (WOC is in1cluster3) and (Time is in2cluster3) then (EOB is out1cluster3) (1)
4. If (WOC is in1cluster4) and (Time is in2cluster4) then (EOB is out1cluster4) (1)
5. If (WOC is in1cluster5) and (Time is in2cluster5) then (EOB is out1cluster5) (1)
6. If (WOC is in1cluster6) and (Time is in2cluster6) then (EOB is out1cluster6) (1)
7. If (WOC is in1cluster7) and (Time is in2cluster7) then (EOB is out1cluster7) (1)
8. If (WOC is in1cluster8) and (Time is in2cluster8) then (EOB is out1cluster8) (1)
9. If (WOC is in1cluster9) and (Time is in2cluster9) then (EOB is out1cluster9) (1)

Figure 6 (a) and 6 (b) indicates rule viewer that shows the values of the various input to the model and computed outputs.

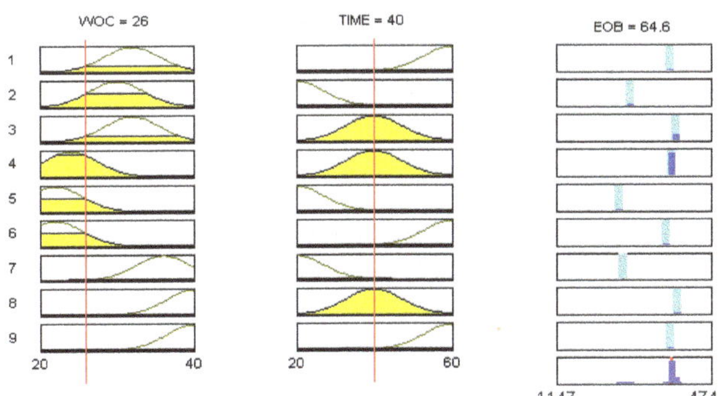


**Fig.6.** (a) Ruler view of input variable (weight of catalyst and time of reaction) with output variable ester of butyl using without sub clustering.

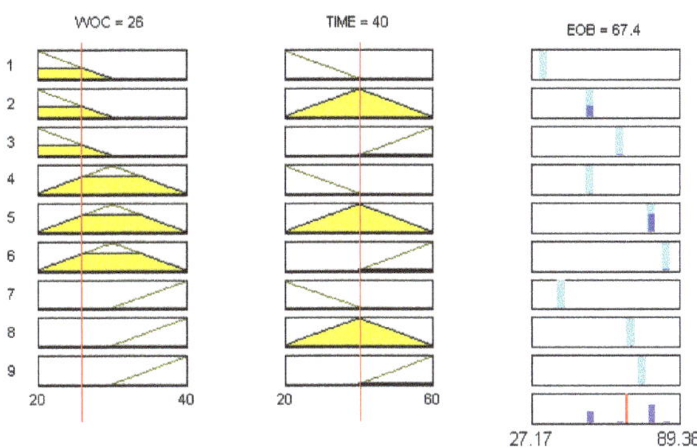


**Fig.6.** (b) Ruler view of input variable (weight of catalyst and time of reaction) with output variable ester of butyl using with sub clustering.

Here the ester of butyl (output) can be predicted by varying the input parameters weight of catalyst and time for chemical reaction. Figure 6 shows a particular instance with sub clustering having input values given to the system 26 gm for weight of catalyst, and 40 min for time for chemical reaction. The output generated by the system for ester of butyl production is shown as 64.6 %w/w.

A similar output was generated with sub clustering having the ester of butyl concentration 67.4 %w/w. Likewise; this fuzzy model has generated other values of output variable for different sets of data points in the specified range of input variables. Figure 7(a) and (b) show two different views of control surfaces, which indicate the results predicted by the fuzzy model for different sets of data points.

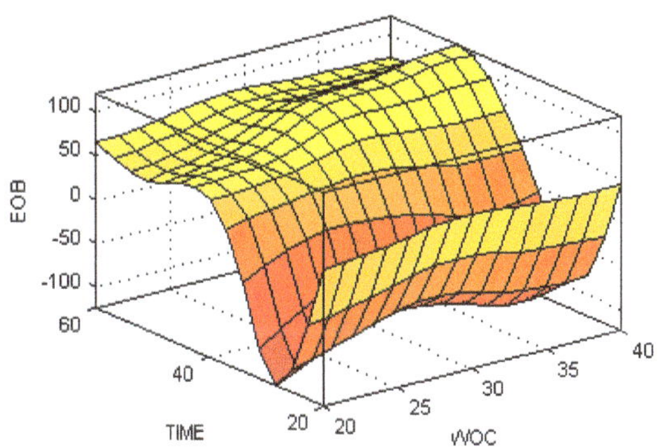


**Fig.7.** (a) and 7(b) having two different views of control surface of the fuzzy model. (a) Control surface view of the fuzzy model with sub clustering.

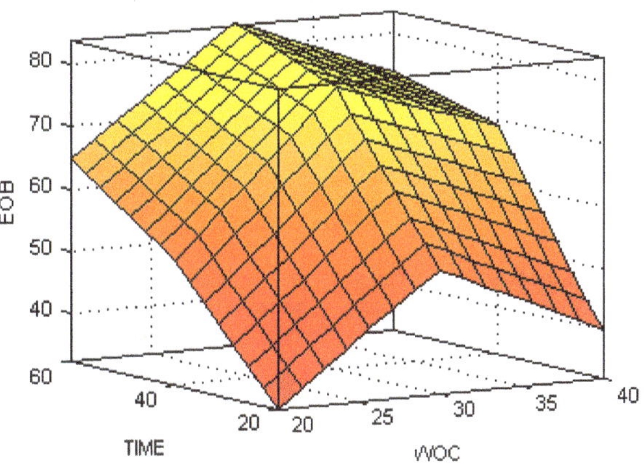


**Fig.7.** (b) Control surface view of the fuzzy model without sub clustering.

These control surfaces have been shown given the interdependency of input and output parameters guided by the various rules in the given universe of discourse. It has already been finalized that there are nine rules predicting the ester of butyl production depending upon the input parameters, weight of catalyst, and time of reaction, for MISO fuzzy model. These rules have been implemented in the MATLAB environment using the Sugeno type of FIS in fuzzy logic toolbox. The results predicted from this fuzzy model of batch transesterification of Jatropha curcas have been compared with the experimental results for its validation in the latter part of the article.

## 3. Experimental Setup & Methodology

Figure 8 shows the batch transesterification reaction system used in the experiments to validate the results of the designed fuzzy model.

A 1500 ml glass reactor equipped with control driven mechanical stirrer, thermocouple, condensing coil and sample port have been used in all kinetic experiments. The thermocouple was immersed in constant volume batch reactor, which was capable of controlling the temperature to within deviation of ±0.50C .A mechanical stirrer fitted with stainless steel propeller provided the mixing requirement. Thirty reactions were carried out over the entire duration of experimental work. The conditions of temperature, molar ratio and mixing intensity were fixed, where catalyst weight and time of reaction were considered as independent variable.

**Fig.8.** Batch transesterification reactor.

(i) The reactor was initially charged with Jatropha oil depending upon the required molar ratio of oil to Butanol. The reactor was then placed under constant heat to reach the desired temperature. (ii) A measured amount of Butanol and sodium hydroxide stock solution, were separately heated to the reaction temperature, and then added to the reactor. A mechanical stirrer was used as per required temperature. The reaction was timed as soon as the mechanical stirrer was turned on. (iii) During the experiment the samples were taken at 20 minute intervals. Approximately 20 to 25 samples were collected during the course of each reaction (60minutes). (iv) Samples were collected in 10ml test tubes filled with 4ml of distilled water. The test tubes were kept in an ice bath at about 50C prior to use. (v) Samples of (2ml) were quenched in the test tubes by being immediately placed in ice baths following their removal from the glass reactor. The test tubes were then shaken to stop the reactions. (vi) After measuring their residual weight, the upper and the lower portions were analyzed for the composition by using gas liquid chromatography (GLC).

## 4. Results and Discussions

A study of literature reveals that alkali metal alkoxides are found to be more effective transesterification catalysts compared to acidic catalysts. So the base catalyst (NaOH) was used for transesterification of Jatropha oil in the present work. The reaction was carried out at different catalyst weights, ranging from 20-40 gm in the presence of excess alcohol, results of which have been shown in Figure 9. It can be seen that the ester of butyl production through transesterification (EOB) of Jatropha oil with Butanol is approximately a logarithmic function of catalyst up to 30gm.

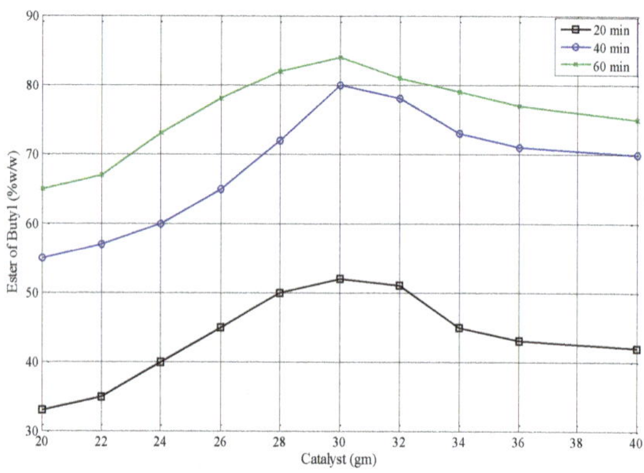


**Fig.9.** Output of product (EOB) versus catalyst under different reaction time in batch transesterification reactor.

The formation of 65 (%w/w) butyl ester was found to take 60 minutes. On increasing the weight of the catalyst a maximum of up to 30 gm of catalyst and 85% of butyl ester formation was observed. However on further increasing the amount of catalyst, soap was formed, thus reducing the transesterification rate. Moreover in all reaction times studied, 30 gm of catalyst was found to give the best and optimum result.

Figure 10 (a), 10(b) and 10 (c) compares the predicted production of ester of butyl using the fuzzy model and the data reported for batch transesterification reactor experiments. Figure 10 (a) shows the plot between productions of ester of butyl and time of reaction (20 minutes) and Fig. 10(b) shows the plot between the production of ester of butyl and time of reaction (40 minutes), and Fig. 10 (c) shows the plot between production of ester of butyl and time of reaction (60 minutes).

An increase in the production rate of the butyl ester, in the initial stages of the curve, along with the increase in time is due to more residence time. After that, the decrease in the ester production rate caused by an increase in time and weight of the catalyst, is due to soap formation and hinders the reaction. This trend seen in rate of butyl ester production is closely followed by the outcome of the designed fuzzy model. Out of the various outputs generated by the fuzzy model, only 2% of the data cross the experimental results, when the FIS system is used without sub clustering. FIS with sub clustering has only 0.1% point deviate from the experimental analysis. With the average error being 0.055%, the mean accuracy of the model comes out to be 99.95%. In the present study, the total number of data points involved was 90. Thus, it can be concluded that there is a close relationship between the simulated results and the practical results obtained at similar reactor conditions for predicting butyl ester production as shown in Figures 10 (a), 10(b) and 10 (c).

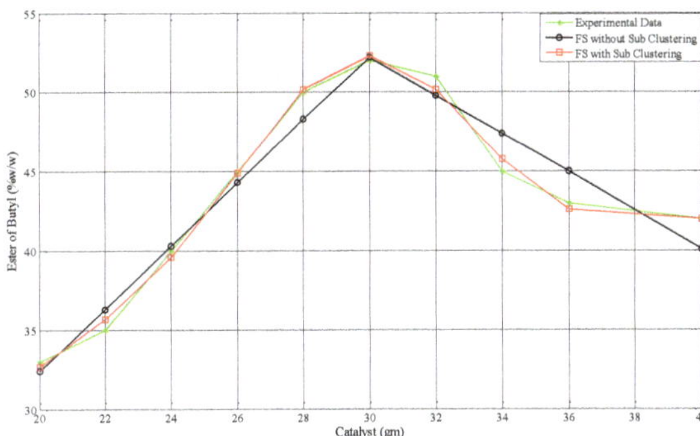


**Fig.10.** (a) Output of product (EOB) versus catalyst at the reaction time of 20 min in batch transesterification reactor.

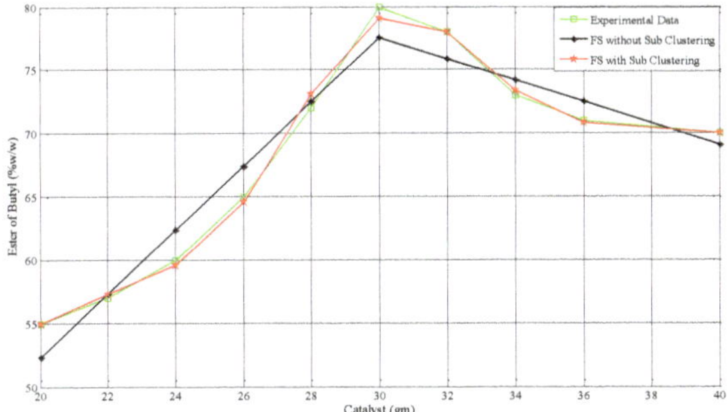


**Fig.10.** (b) Output of product (EOB) versus catalyst at the reaction time of 40 min in batch transesterification reactor.

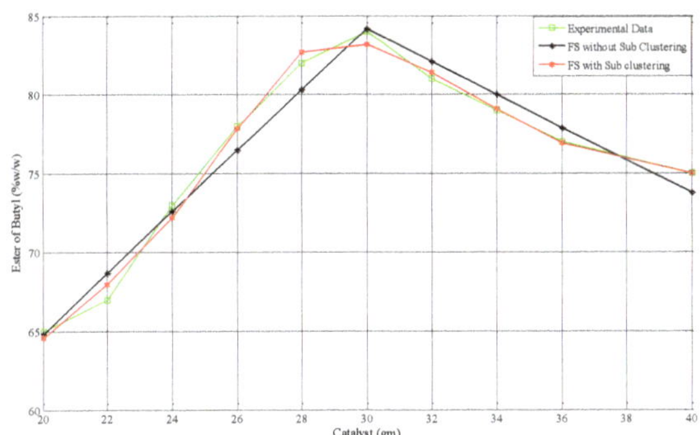


**Fig.10.** (c) Output of product (EOB) versus catalyst at the reaction time of 60 min in batch transesterification reactor.

### *4.1 Kolmorgov-Smirnov (K-S) test for significance of the ANFIS model*

The difference between the experimental and theoretical values of the fuzzy model can be evaluated using the K-S test of statistical methods. This test is one of the most valuable non parametric statical methods, for comparing two samples, as it is sensitive to differences in both location and shape of the empirical cumulative distribution functions of the two samples. The calculated value of maximum difference between the cumulative distributions (D), probabilities (P) and significance confidence level limits above 95% have been listed in Table 1. Since the calculated values of D and its corresponding P values are around 0.985 to 0.996, there is no significant difference between the ester of butyl production values generated by the fuzzy model and experimental values under null hypothesis.

**Table 1**
Kolmorgov-Smirnov (K-S) test output for experimental data/ FIS models.

| K-S test | Parameters | 20 minutes | 40 minutes | 60 minutes |
|---|---|---|---|---|
| Experiment Data/ FIS without sub clustering | D | 0.1818 | 0.1818 | 0.1591 |
| | P | 0.985 | 0.985 | 0.996 |
| | >95 Confidence (EOB % w/w) | 42.34 through 52.45 | 71.54 through 73.65 | 78.10 through 84.40 |
| Experiment Data/ FIS with sub clustering | D | 0.1818 | 0.1818 | 0.1591 |
| | P | 0.985 | 0.985 | 0.996 |
| | >95 Confidence (EOB % w/w) | 50.24 through 52.45 | 75.54 through 79.75 | 79.32 through 83.69 |

Moreover the difference between the ANFIS models (with and without sub clustering) have also evaluated using K-S test. It has been found that most of the data spread into a large fraction of the plot. This is a sign of a normal distribution of the data. The KS test has also found the data to be consistent with a log normal distribution P= 0.90 and normal distribution: P= 0.95. Thus making the maximum difference in the cumulative fraction is D=.0909 as shown in Figure 11. Hence the probability of generating accurate values for the rate ester of butyl production by using the fuzzy model is 99.90%.

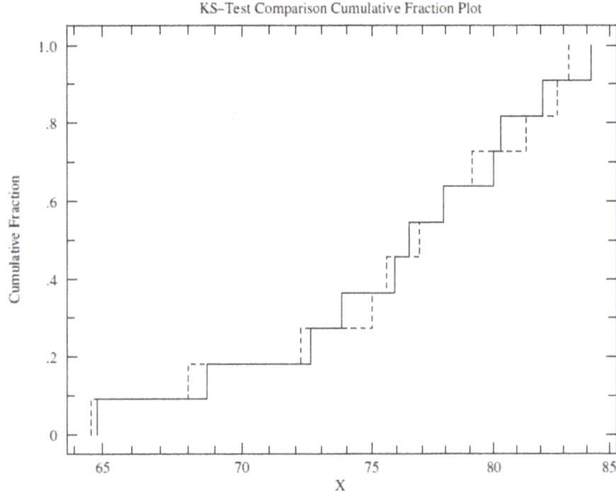


**Fig.11.** Cumulative fraction plot of FIS model with sub-clustering and without sub-clustering for 60 minutes operation.

## 5. Conclusions

This paper deals with the application of the fuzzy logic for predicting the production rate of the ester of butyl in the batch transesterification reactor. In this article, the MISO fuzzy model is developed using ANFIS and validated through experimental results for given conditions. With more than 99% average accuracy, it has been found that the results generated by the designed fuzzy models are close to the experimental results. From statistical analysis, it has been concluded that the maximum differences between the cumulative distributions (D) is in the range of 0.1591 to 0.1818 in comparison to the experimental data obtained using FIS. With this low deviation, the accuracy of the model can be used by the process engineer who would like to get quick answers for online intelligent control and/or optimization. The optimum amount of catalysts required in this particular reaction is 28.5-30 gm, and in its current state, the model is limited to the amount of catalyst and reaction time. This study supports the idea that the fuzzy logic technique can be used as a viable alternative for carrying out analysis. Moreover, the Fuzzy logic allows for the modeling and control problem to be treated simultaneously.

**References**

Y. Zhang, M.A. Dube, D.D. McLean and M. Kates, Biodiesel production from waste cooking oil: process design and technological assessment, Bioresource Technology. 89 (2003), pp. 1-16.

F. Ma, L.D. Clements and M.A. Hanna, The effect of catalyst, free fatty acids, and water on transesterification of beef tallow, Trans ASAE 41 (5) (1998), pp. 1261–1264.

E. Ahn, M. Mittelbach and R. Marr, A low waste process for the production of biodiesel, Sep Sci Technology 30 (1995), pp. 2021-2033

B. Freedman, R.O. Butterfield and E.H. Pryde, Transesterification kinetics of soybean oil, J Am Oil Chem Soc 63 (10) (1986), pp. 1375-1380.

A.V. Tomasevic and S.S. Marinkovic, Methanolysis of used frying oils, Fuel Process Technol 81 (2003), pp. 1-6.

S. Gryglewicz, Rapeseed oil methyl esters preparation using heterogeneous catalysts, Bioresource Technology 70 (1999), pp. 249-253

H. Noureddini and D. Zhu, Kinetics of transesterification of soybean oil, J Am Oil Chem Soc 74 (11) (1997), pp. 1457-1463.

M.Canakci and J.H. VanGerpen, Biodiesel production via acid catalysis, Trans. ASAE 42 (1999) (5), pp. 1203-1210.

M.P. Dorado, E. Ballesteros, J.A. Almeida, C. Schellet, H.P. Lohrlein and R. Krause, An alkali-catalyzed transesterification process for high free fatty acid oils, Trans ASAE 45 (3) (2002), pp. 525-529

Sohpal, V.K., Singh, A. and Dey, A. A comparative study of molar ratio effect on transesterification of Jatropha Curcas using kinetic and fuzzy techniques, Int. J. Oil, Gas and Coal Technology, 4(3) (2011) pp.296-306.

Sohpal, V.K., Singh, A. and Dey, A. Fuzzy Modeling to Evaluate the Effect of Temperature on Batch Transesterification of Jatropha Curcas for Biodiesel Production, Bulletin of Chemical Reaction Engineering & Catalysis, 6 (1) (2011) pp 31-38.

S. Pinzi, F.J. L-Gimenez, J.J. Ruiz, M.P. Dorado. Response surface modeling to predict biodiesel yield in a multi-feedstock biodiesel production plant. Bioresources Technology 101 (2010), pp 9587-9593.

Amarpal Singh, Ajay K Sharma, T S Kamal, The Effect of Phase Matching Factor on Four Wave Mixing in WDM Optical Communication Systems: Fuzzy and Analytical Analysis" International Journal of Computer Applications in Technology (IJCAT), 34 (3) (2009) pp. 165-171.

Amarpal Singh, Ajay K Sharma T S Kamal and Manju Sharma, Comparative study of FWM in WDM Optical Systems Using OptSim and ANFIS, International Journal for Information & Systems Sciences 5(1) (2009) pp 72-82.

Yesiloglu-Gultekin, N., Ebru Akcapinar Sezer, Candan Gokceoglu, and H. Bayhan., An application of adaptive neuro fuzzy inference system for estimating the uniaxial compressive strength of certain granitic rocks from their mineral contents., Expert Systems with Applications 40, 3 (2013), pp 921-928.

# 15

# Performance of an enzymatic packed bed reactor running on babassu oil to yield fatty ethyl esters (FAEE) in a solvent-free system

Aline S. Simões[1], Lucas Ramos[1], Larissa Freitas[1], Julio C. Santos[1], Gisella M. Zanin[2], Heizir F. de Castro[1,*]

[1]*Engineering School of Lorena, University of São Paulo Estrada Municipal do Campinho s/n, 12602-810 Lorena, São Paulo, Brazil.*

[2]*State University of Maringa, Department of Chemical Engineering, Av. Colombo 5790, E-46, 87020-900, Maringa – PR, Brazil.*

**HIGHLIGHTS**

- The performance of packed bed reactor running on babassu oil to yield ethyl esters was assessed.
- *Burkholderia cepacia* lipase immobilized on SiO2-PVA composite was used as a catalyst.
- The molar ratio of babassu oil and ethanol was found as a critical parameter for attaining high yields.
- High yield (96.0 ± 0.9%) and productivity (41.1 ± 1.6 $mg_{ester}{\cdot}g_{catalyst}^{-1}{\cdot}h^{-1}$) were achieved.

**GRAPHICAL ABSTRACT**

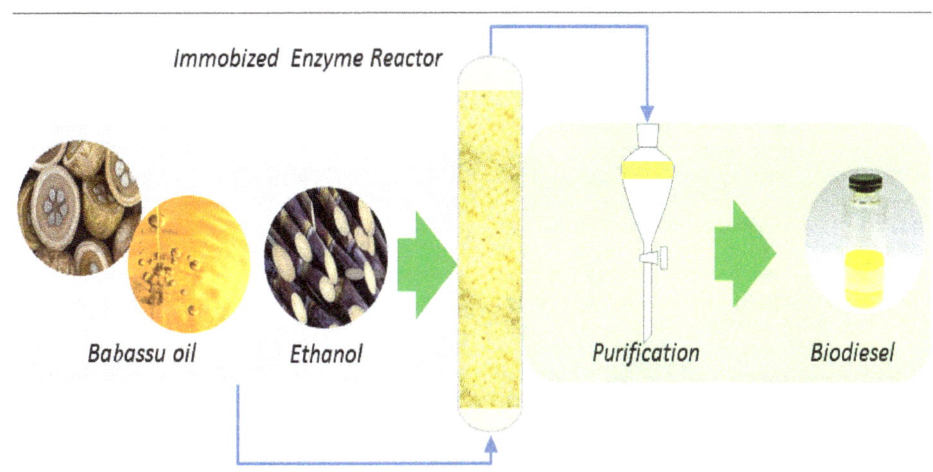


**ABSTRACT**

The transesterification reaction of babassu oil with ethanol mediated by *Burkholderia cepacia* lipase immobilized on $SiO_2$-PVA composite was assessed in a packed bed reactor running in the continuous mode. Experiments were performed in a solvent-free system at 50 °C. The performance of the reactor (14 mm ×210 mm) was evaluated using babassu oil and ethanol at two molar ratios of 1:7 and 1:12, respectively, and operational limits in terms of substrate flow rate were determined. The system's performance was quantified for different flow rates corresponding to space times between 7 and 13 h. Under each condition, the impact of the space time on the ethyl esters formation, the transesterification yield and productivity were determined. The oil to ethanol molar ratio was found as a critical parameter in the conversion of babassu oil into the correspondent ethyl esters. The highest transesterification yield of 96.0 ± 0.9% and productivity of 41.1 ± 1.6 $mg_{ester}{\cdot}g_{catalyst}^{-1}{\cdot}h^{-1}$ were achieved at the oil to ethanol molar ratio of 1:12 and for space times equal or higher than 11 h. Moreover, the immobilized lipase was found stable with respect to its catalytic characteristics, exhibiting a half-life of 32 d.

**Keywords:**
Biodiesel
Transesterification
Babassu oil
Packed bed reactor
Immobilized lipase
Continuous mode

* Corresponding author
E-mail address: heizir@dequi.eel.usp.br

## 1. Introduction

The transesterification of triglycerides (TGs) with alcohol in the presence of chemical or biochemical catalysts leads to the formation of alkyl ester, also known as biodiesel. Enzymatic transesterification using lipase (E.C. 3.1.1.3) is attractive and encouraging due to easy product separation, minimal wastewater treatment needs, easy glycerol recovery and the absence of side reactions, unlike with chemical catalysts (Jegannathan et al., 2008; Christopher et al., 2014).

Conventionally, methanol is used to produce fatty acid methyl esters (FAME), but production of fatty acid ethyl esters (FAEE) with bioethanol is considered more advantageous due to the expected improved sustainability of this type of biodiesel. More specifically, ethanol has a superior dissolving power in vegetable oils and low toxicity compared to methanol (Stamenković et al., 2011; Brunschwig et al., 2012). Additionally, unlike methanol (which is generally derived from fossil sources), ethanol is produced mainly from renewable sources *via* fermentation processes, and because its large scale production as a substitute fuel for gasoline already exists, the supply of bioethanol for the industrial production of biodiesel can be easily achieved (Brunschwig et al., 2012).

A biodiesel production process based on immobilized lipase-catalysis involves a multi-phase system throughout the reaction, including the insoluble phase of the biocatalyst, lipid phase and a polar phase when alcohol exceeds the solubility limit (or when the by-product glycerol is present). Stirred tank reactors (STRs) and packed bed reactors (PBRs) are often used as reactors for studying this type of reaction at different scales (Christopher et al., 2014).

Studies of the process conditions in reactors are necessary for the technical and economic viability of enzymatic biodiesel production. In most of the works published, transesterification is conducted in batch reactors, in which the enzyme is dispersed in the reaction mixture by agitation in STRs (Balcão et al., 2006). PBRs can be considered as a better choice than STRs for immobilized enzymes, mainly due to the lower shear stress imposed on the catalyst particles and the possibility of continuous operation (Yusuf, 2006). In addition, PBR may be operated with no need for biocatalyst separation from the reaction products and can be relatively easily scaled up (Yusuf, 2006).

The present study was set to investigate the efficiency of a PBR system through the transesterification reaction of babassu oil and ethanol mediated by lipase from *Burkholderia cepacia.* The enzyme was covalently immobilized on a hybrid matrix silica-polyvinyl alcohol ($SiO_2$-PVA) and the reactor was operated in a continuous mode. Variables, such as flow rate and molar ratio of the reactants, were assessed to define the operating parameters, aiming at establishing the potentials and challenges in scaling up the process based on the experimental results obtained. Babassu oil was selected as the TGs source because the tree from which it is extracted is abundant in certain countries, such as Brazil, and its primary use is for nonfood purposes (Teixeira, 2008; Carvalho et al., 2013).

## 2. Materials and methods

### *2.1. Materials*

Refined, bleached and deodorized babassu oil was provided by BASF (Jacarei, SP-Brazil) with its properties summarized in **Table 1**. Commercial virgin olive oil (0.3% acidity), purchased from a local market, was used to determine the hydrolytic activity of the biocatalysts. A commercial lipase from *Burkholderia cepacia* (Batch number: 01022TD) was purchased from Amano Pharmaceuticals (Nagoya, Japan) and was used as received without further purification. Tetraethoxysilane (TEOS) and epichlorohydrin (99%) were acquired from Aldrich Chemical Co. (Milwaukee, WI, USA). Polyvinyl alcohol (MW 88000, 88%) was obtained from Acros Organic (USA). Hydrochloric acid (36%), anhydrous ethanol (99.8%), *tert*-butanol, polyethylene glycol (MM 1500) and hexane were supplied by Reagen (Rio de Janeiro, RJ, Brazil). A deep blue liposoluble dye (organic synthetic pigment) was obtained from Glitter Ind. Com. Imp. Exp. Ltd (Carapicuiba, SP, Brazil) and was used as a tracer. All solvents and reagents for analyses were of chromatographic or analytical grade.

**Table 1.**
The properties of the babassu oil used in this study.

| Property | Value |
|---|---|
| Acid number (mg KOH·$g^{-1}$) | 0.65 |
| Peroxide value (mEq·$kg^{-1}$) | 1.82 |
| Iodine number (g $I_2$·$g^{-1}$) | 25 |
| Saponification number (mg KOH·$g^{-1}$) | 238 |
| Average molecular weight of TGs (g·$mol^{-1}$) | 709.90 |
| **composition of fatty acids (% wt)** | |
| Caprylic acid (C8:0) | 4.5 |
| Capric acid (C10:0) | 3.5 |
| Lauric acid (C12:0) | 44.7 |
| Myristic acid (C14:0) | 17.5 |
| Palmitic acid (C16:0) | 9.7 |
| Stearic acid (C18:0) | 3.1 |
| Oleic acid (C18:1) | 15.2 |
| Linoleic acid (C18:2) | 1.8 |

### *2.2. Support synthesis and lipase immobilization*

A polysiloxane-polyvinyl alcohol composite ($SiO_2$-PVA) was prepared, activated with epichlorohydrin and used to immobilize the lipase according to the methodology reported by Da Rós et al. (2010). To perform this work, ten batches of immobilized derivatives were prepared, and the average measured hydrolytic activity was 1950 ± 120 IU/g biocatalyst. One international unit (IU) of enzyme activity was defined as the amount of enzyme that liberates 1 μmol of free fatty acid per min under the assay conditions (37 °C, pH 7.0, 150 rpm). The properties of the immobilized derivative were as follows: diameter (0.175 mm); average pore diameter (29.42 Å); surface area (337 $m^2$·$g^{-1}$); porous volume (0.25 $cm^3$·$g^{-1}$); and density (1.865 g·$cm^{-3}$). The biochemical, kinetic properties, thermal stability and operational stability of this immobilized lipase preparation were described elsewhere (Da Ros et al., 2010).

### *2.3. Continuous runs of the PBR*

The ethanolysis of the babassu oil was conducted in a PBR including a glass column (internal diameter: 14 mm; height: 210 mm; and total volume: 32 $cm^3$) with a water jacket connected to a circulating water bath to maintain the temperature at 50 °C. The continuous operation was started by loading the reactor with the biocatalyst, and the substrates were continuously pumped using a peristaltic pump (Perista Pump SJ-1211, Atto Bioscience & Biotechnology, Tokyo, Japan) from a reservoir to the bottom end of the bioreactor at the required flow rate through Marprene tubing (Watson Marlow 913.AJ05.016). A reflux condenser system was connected to the feeding vessel to avoid ethanol loss. Heating tapes containing a thermostatic electrical resistance (25 W) were used to avoid heat loss in the inlet and outlet tubing.

For each run, 24.3 g (d= 1.865 g·$cm^{-3}$) of the biocatalyst corresponding to a working volume of 19.0 $cm^3$ was used. The substrate was prepared at a molar ratio of oil to ethanol of 1:7 and 1:12, respectively, and flow rates ranging from 1.5 to 2.8 mL·$h^{-1}$ were imposed to determine the performance limit of the reactor for both ratios. **Figure 1** shows the experimental setup of the reaction system. During the continuous runs, samples were periodically taken from the reactor vessel for analysis of the relevant variables, such as the ethyl esters concentration, viscosity and density. For each tested space time, the parameters were determined when the concentrations within the reactor remained relatively constant over a period corresponding to at least three space times. The space time was calculated according to Levenspiel (1997) as described by the following equation (**Eq. 1**).

$$\tau = \frac{V}{v_0} \qquad \text{(Eq. 1)}$$

Where τ is the space time (h), $V$ is the working volume of the reactor (mL) and $v_0$ is the flow rate ($mL \cdot h^{-1}$).

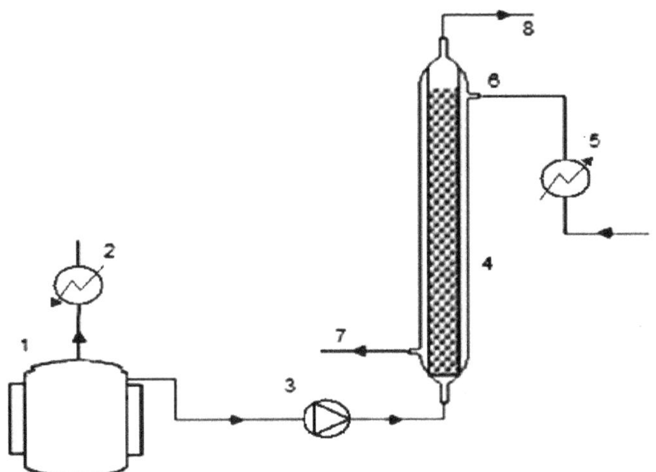


**Fig.1.** Schematic diagram of the packed bed experimental apparatus: 1) substrate reservoir; 2) reflux condenser; 3) peristaltic pump; 4) column; 5) thermostatic bath; 6) water in; 7) water out; 8) product output.

*2.4. Operational stability of the immobilized derivative*

Transesterification The biocatalyst stability was assessed by measuring the hydrolytic activity of the immobilized derivatives at the end of each continuous run, taking into account the original activity as 100%. The recovered immobilized lipase was then washed with tert-butanol to remove any substrate or product retained in the matrix. Hydrolytic activity was determined by the olive oil emulsion method according to the modification proposed by Soares et al. (1999). The inactivation constant (kd) and half-life (t1/2) for the immobilized lipase were calculated as described by Costa-Silva et al., (2014).

*2.5. Purification of biodiesel*

The volume of sample collected from the bioreactor was transferred into a decanting funnel, in which the same quantity of distilled water was added. Then, vigorous agitation was performed, and the mixture was allowed to stand for 6 h for phase separation. This procedure was performed three times in sequence. The upper phase, consisting of FAEE (biodiesel) was evaporated by a rotatory-evaporator. Then, the solution was dried with sodium sulfate, and the lower phase, consisting of glycerol and wastewater, was disposed of.

*2.6. Hydrodynamic characterization of the PBR system*

The Tracer response analysis was used to characterize and model the flow through the reactor. The PBR was filled with 24.5 g of previously denatured immobilized *B. cepacia* lipase. Then, the system was operated under a continuous flow rate of substrate (5.3 $mL \cdot h^{-1}$), which corresponded to 3.9 h of spatial time.

A deep blue liposoluble dye (15 wt %) was solubilized in a substrate and was used as a tracer. A pulse input experiment was performed, and the residence time distribution (RTD) was experimentally determined by injecting the tracer into the reactor at time zero and then by spectrophotometrically measuring the tracer concentration, C, at 650 nm in the output stream as a function of time using a Varian Cary 50 UV/Vis Spectrophotometer (Varian Australia Pty Ltd, Mulgrave, VIC, Australia). The experiments were performed in duplicate.

The RTD function, $E(t)$, is defined using the following equation (**Eq. 2**) (Fogler, 1992).

$$E(t) = \frac{C(t)}{\int_0^\infty C(t)dt} \qquad \text{(Eq. 2)}$$

Where $C(t)$ is the tracer concentration at time $t$.
The mean residence time, $t_m$, for a constant volumetric flow, was calculated using the **Equation 3** (Fogler, 1992).

$$t_m = \int_0^\infty t.E(t)dt \qquad \text{(Eq. 3)}$$

The integration in the Equation 3 was calculated using the ORIGIN 8.0 software (OriginLab Corporation, MA, USA).

The variance, σ, or the square of the standard deviation, is defined by the **Equation 4** (Fogler, 1992).

$$\sigma^2 = \int_0^\infty (t - t_m)^2 \cdot E(t)dt \qquad \text{(Eq. 4)}$$

The skewness, $s^3$, is defined by the **Equation 5** (Fogler, 1992).

$$s^3 = \frac{1}{\sigma^{3/2}} \int_0^\infty (t - t_m)^3 \cdot E(t)dt \qquad \text{(Eq. 5)}$$

To characterize the pipe flow through the enzyme particles in the column, the Reynolds number was calculated according to Lide (2007) (**Eq. 6**).

$$\mathrm{Re} = \frac{dp \cdot \upsilon \cdot \rho}{\mu} \qquad \text{(Eq. 6)}$$

Where $d_p$ is the enzyme particle diameter (0.175 mm), υ represents the fluid velocity calculated as the flow/cross area of the column, and, ρ denotes the fluid specific mass (848 and 863 $kg \cdot m^{-3}$ using substrates at oil to ethanol molar ratios of 1:7 and 1:12, respectively), $\mu$ is the fluid viscosity of the reactant mixture (8.12 and $6.95 \times 10^{-3}$ $kg \cdot m^{-1} \cdot s^{-1}$ using substrates at oil to ethanol molar ratios of 1:7 and 1:12, respectively) measured as 1 atm at 40°C with a Brookfield Viscometer–LVDVII (Brookfield Viscometers, England). The flow was hereafter characterized as turbulent if $R_e \geq 4000$ or laminar if $R_e \leq 2000$ (Lide, 2007).

*2.7. Analytical Procedure*

The FAEE formed in the transesterification reaction were analyzed by FID gas chromatography (Varian CG 3800, Inc. Corporate Headquarters, Palo Alto, CA, USA) using a 5% DEGS CHR-WHP 80/100 mesh 6 ft 2.0 mm ID and 1/8" OD column (Restek Frankel Commerce of Analytic Instruments Ltd, SP, Brazil) following previous established conditions (Urioste et al., 2008). The theoretical ester concentrations were calculated by taking into account the fatty acid composition and its initial weight mass in the reaction medium, and the transesterification yield (%) was defined as the ratio between the produced and theoretical esters concentrations (Carvalho et al., 2013).

The absolute viscosity of the purified biodiesel was determined by a Brookfield Viscometers model LVDVII (Brookfield Viscometers Ltd, England) using the cone CP 42. All assays were performed at 40 °C using a 0.5 mL aliquot of the sample. The biodiesel density was determined by a digital densimeter model DMA 35N EX (Anton Paar, Graz, Austria). In this test, all assays were performed at 20 °C using a 2.0 mL aliquot of each sample (Carvalho et al., 2013).

## 3. Results and discussion

### *3.1. Reactor performance at oil to ethanol molar ratio of 1:7*

The reactor was filled with immobilized lipase to form a well packed column, and the substrate at a oil to ethanol molar ratio of 1:7 was fed at increasing flow rates ranging from 1.5 to 2.8 mL·h$^{-1}$, corresponding to 13 to 7 h space times, respectively, for a total period of 30 d. **Figure 2** shows the concentration of ethyl esters for different space times.

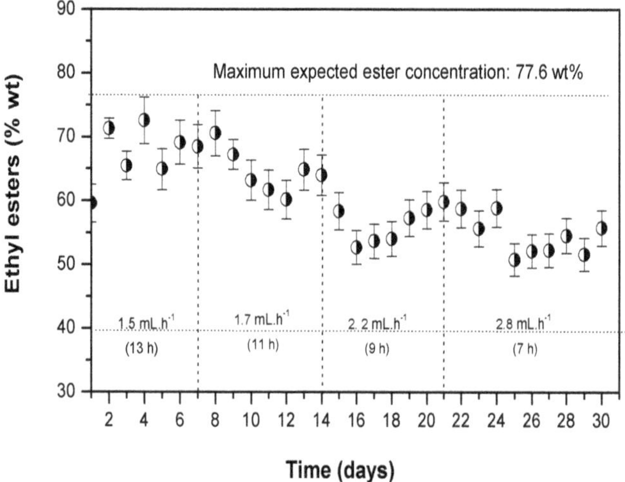


**Fig.2.** Ethyl esters concentrations from the transesterification of babassu oil performed in a packed bed reactor using immobilized *B. cepacia*. The reactions were carried out at a babassu oil to ethanol molar ratio of 1:7, at 50 °C, at feed flow rates ranging from 1.5 to 2.8 mL·h$^{-1}$ (corresponding to space times of 7 and 13 h).

Changes in the space time along the continuous run interfered with the formation of ethyl esters, resulting in a variation in their concentration from 52.8 to 69.5 wt%, corresponding to transesterification yields between 68.1 and 89.5%. The average values of the ethyl esters concentration and productivity achieved for each space time are shown in **Table 2**. Among the range of the flow rates studied, the best reactor performance was found at a space time of 13 h (flow rate = 1.5 mL.h$^{-1}$). Under such conditions, 89.5% of the fatty acids present in the babassu oil were converted into ethyl esters, attaining an average productivity of 37.0 ± 1.3 $mg_{ester}.g_{catalyst}^{-1}.h^{-1}$.

**Table 2.**
The average values of the ethyl esters concentration, transesterification yield and productivity obtained in the continuous ethanolysis of babassu oil in a PBR using lipase *B. cepacia* immobilized on $SiO_2$-PVA under different space times (oil to ethanol molar ratio of 1:7).

| Parameter | Space time (h) | | | |
|---|---|---|---|---|
| | 7 | 9 | 11 | 13 |
| Ethyl ester concentration (% wt.) | 52.8 ± 1.9 | 56.6 ± 2.6 | 62.7 ± 3.0 | 69.5 ± 3.6 |
| FAEE yield (%) | 68.1 ± 2.4 | 72.9 ± 3.8 | 80.8 ± 3.5 | 89.5 ± 3.5 |
| Productivity ($mg_{ester}.g_{catalyst}^{-1}.h^{-1}$) | 53.5 ± 1.9 | 44.9 ± 2.1 | 40.1 ± 1.9 | 38.3 ± 2.0 |

These results were favorable when compared to those previously reported by several researchers (Royon et al., 2007; Wang et al., 2011; Dors et al., 2012) using PBRs running on different feedstocks in the presence or absence of solvents. Nevertheless, the studied PBR exhibited lowered biodiesel yields compared to the batchwise reactors reported previously (Freitas et al., 2009; Carvalho et al., 2013). For instance, the performance of a STR used by Carvalho et al. (2013) to conduct lipase-catalyzed transesterification of babassu oil was superior in terms of the conversion achieved (at the same molar ratio) compared to that of the PBR studied herein. The decrease in yield in the present reactor system may be ascribed to the high viscosity of the substrate medium, which prevented a uniform forced flow through the inter-particle spaces within the support matrix. Another reason may be the steady state adsorption of glycerol on the surface, whose amount might have been larger compared to what absorbed in a non-steady state batch run. The FAEE yield at the steady state under the conditions of the lowest volumetric flow rate (1.5 mL.h$^{-1}$) reached 89.5%, and the maximum expected ethyl ester concentration of 77.6 wt% was not achieved.

To reduce the feedstock lost in the process, substrate containing a higher excess of ethanol was used, i.e. the molar ratio of babassu oil to ethanol was increased from 1:7 to 1:12. In addition to the investigation of the possible limitations of the process pertaining to the concentration of ethanol, the tolerance of the *B. cepacia* lipase to substrates containing high ethanol concentration was also checked. In fact, the substrate containing higher ethanol concentrations had a lower viscosity value obviously and facilitated the flow through the bed, avoiding reactor operational limitations caused by occasional obstructions. The average viscosity values of the substrate at oil to ethanol molar ratios of 1:7 and 1:12 were 7.0 and 8.5 x $10^{-5}$ kg m$^{-1}$. s$^{-1}$, respectively.

### *3.2. Reactor performance at oil to ethanol molar ratio of 1:12*

Experiments were conducted using the babassu oil to ethanol molar ratio of 1:12 while keeping the other operating conditions constant. The substrate was fed at increasing flow rates and the reactor was operated for 28 d. Oscillations were observed in the ethyl esters formed as seen in **Figure 3**.

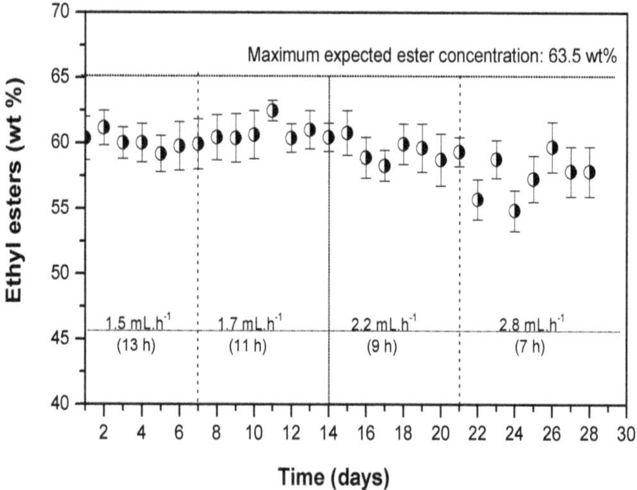


**Fig.3.** Ethyl esters concentrations through the transesterification of babassu oil performed in a packed bed reactor using immobilized *B. cepacia*. The reactions were carried out at a babassu oil to ethanol molar ratio of 1:12, at 50 °C, at feed flow rates ranging from 1.5 to 2.8 mL·h$^{-1}$ (corresponding to space times of 7 and 13 h).

In spite of this, the average concentration of ethyl esters attained throughout the process revealed lower loss of feedstock (minimum yield of 90.4%), independent of the flow rate of the input stream. In this way, the substrate composed of babassu oil and ethanol at a molar ratio of 1:12 promoted greater conversion of the fatty acids present in the babassu oil into ethyl esters.

**Figure 4** presents the average productivity and transesterification yield values for the tested spatial times. Based on these results, for the substrate containing a higher amount of excess ethanol (babassu oil to ethanol molar ratio of 1:12), space times greater than or equal to 9 h provided transesterification yields higher than 92.2 ± 3.3%. The highest yield of 96.0 ±0.9% was obtained for the reactor operating at a flow rate of 1.8 mL·h$^{-1}$ (space time = 11 h), corresponding to a productivity of 41.1 ± 1.6 $mg_{ester}.g_{catalyst}^{-1}.h^{-1}$.

### *3.3. Evaluation of the PBR parameters*

The RTD is an essential tool to assess fluid velocity patterns, which can be used to describe the flow regime in a reactor. Because the flow regime affects the reactor performance, its description may enable better process control

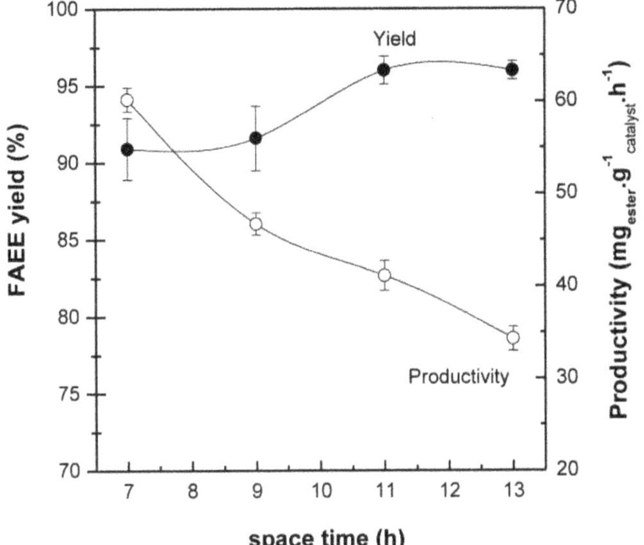


**Fig.4.** The average values of the transesterification yield and productivity attained in the continuous ethanolysis of babassu oil in a PBR using immobilized *B. cepacia* under different space times (oil to ethanol molar ratio of 1:12).

(Fogler, 1992; Levenspiel, 1997). In this work, for the experimental determination of the mean residence time of the reactants in the reactor, as well as the characterization of the reaction mixture in the bed, a pulse input experiment was performed.

The results obtained allowed the RTD curve E (t) to be traced as a function of the time, according to the **Equation 2** (data not shown). Using the **Equation 3**, the mean residence time was calculated at 4.1 ± 0.8 h. This value was approximately 5.9% higher than the value calculated by taking into account the experimental conditions based on the **Equation 1** and can be considered acceptable in this type of test. In better words, this finding indicated that no preferential paths or backmixing took place in the bed demonstrating the good quality of the packing.

The variance, or square of the standard deviation, indicating the spread of the distribution, was calculated using the **Equation 4**. The greater that value, the greater the distribution spread. **Equation 5** is related to the skewness and measures the extent that a distribution is skewed in one direction in reference to the mean. The main parameters of the RTD are shown in **Table 3**.

**Table 3.**
Mean residence time, variance, and asymmetry coefficient obtained through the pulse input experiment for the PBR system.

| Parameter | Value |
|---|---|
| Mean residence time, $t_m$ (h) | 4.1 |
| Variance, $\sigma^2$ ($h^2$) | 0.6 |
| Asymmetry coefficient, $s^3$ ($h^{3/2}$) | -3.2 $\times 10^2$ |

Parameters such as the column length-to-diameter ratio, substrate flow rate and the characterization of the pipe flow through the enzyme particle in the column were also evaluated. According to Damstrup et al. (2007), a column length-to-diameter ratio of less than 25 does not interfere with the transfer of the fluid mixture through the column. In this work, the column length-to-diameter ratio obtained was 15. Damstrup et al. (2007) reported that the highest yield was achieved with a column length-to-diameter ratio in the range of 13-14.

To characterize the pipe flow through the enzyme particles in the column, the Reynolds number was calculated at $4.3\times10^{-5}$ and $5.0\times10^{-5}$ for the experiments performed using the oil to ethanol molar ratio of 1:7 and 1:12, respectively, indicating a laminar flow behavior (Re $\leq$ 2000) for both experimental conditions (Fox et al., 2013). Therefore, the flow was dominated by viscous force with uniform non-turbulent flow in parallel layers and little mixing between layers (Lide, 2007).

### *3.4. Overall performance of the reactor running on babassu oil and ethanol*

Many processes can only reach high productivity values at the expense of high residual substrate levels (low substrate conversion rate) and/or low concentration of product formed (Rica et al., 2009). At the industrial scale, the ideal goal to be achieved is the adoption of a system that has a good productivity combined with low loss of the feedstock. For biodiesel production, an additional parameter should be attained according to the technical standards which is a minimum level of alkyl esters at 96.5% wt. In this context, the productivity data analysis took into account both economic and technical criteria. A summary of the obtained results is tabulated in **Table 4**, in which the average values achieved for both sets of experiments are compared.

**Table 4.**
Comparative performance of the ethanolysis of babassu oil in a PBR using lipase *B. cepacia* immobilized on $SiO_2$-PVA under different oil to ethanol molar ratios (biocatalyst initial activity of 1950 IU·g$^{-1}$).

| Parameter | Substrate molar ratio (oil to ethanol) | |
|---|---|---|
| | 1:7 | 1:12 |
| Biocatalyst (residual activity, IU·g$^{-1}$) | 1229 ± 42 | 1092 ± 51 |
| Total operating time (d) | 30 | 28 |
| Space time (h) | 13 | 11 |
| Deactivation coefficient ($k_d$, d$^{-1}$) $\times 10^{-2}$ | 1.6 ± 0.4 | 2.2 ± 0.5 |
| Biocatalyst half-life (days) | 44.7 ± 1.1 | 32.1 ± 3.4 |
| Ethyl ester concentration (% wt.) | 69.5 ± 3.6 | 60.9 ± 0.6 |
| Productivity (mg$_{ester}$·g $_{catalyst}$$^{-1}$·h$^{-1}$) | 38.3 ± 2.0 | 41.1 ± 1.6 |
| Transesterification yield (%) | 89.5± 3.5 | 96.0 ± 0.9 |
| Kinematic viscosity (mm$^2$·s$^{-1}$) | 8.2±0.8 | 4.3 ± 0.7 |

Regarding the influence of the molar ratio on the continuous synthesis of biodiesel, the use of equimolar amounts of ethanol to the number of fatty acids (FA) residues is enough to obtain complete conversion of the FA residues to their corresponding FAEE. However, ethanol in excess could result in an increase in yield. Moreover, an excess of alcohol can also be advantageous because it can contribute to medium homogeneity, minimizing the diffusion limitation that could result in low yields, mainly when immobilized systems are used (Antczak et al., 2009; Christopher et al., 2014). Based on the results obtained in the present study, the best reactor performance was attained for runs in which the oil to alcohol molar ratio of 1:12 was used. Under such condition, and at space time greater than or equal to 11 h, an average transesterification yield of 96.0 ± 0.9% and a productivity of 41.1 ± 1.6 mg$_{ester}$.g$_{catalyst}$$^{-1}$·h$^{-1}$ were achieved. This also resulted in biodiesel samples with viscosity values (average 4.3± 0.7 mm$^2$·s$^{-1}$) complying with the international standard for biodiesel viscosity i.e. ASTM 6751-02 (1.0<kinematic viscosity of B100< 6.0 mm$^2$·s$^{-1}$.).

The feasibility of enzymatic processes can also be determined by investigating the biocatalyst half-life, which depends on a series of factors, such as linkage of the enzyme with the support, obstruction of the pores by both sludge and by-products, as well as support loss by friction and obstruction of the fixed-bed causing bypass (Zanin and Moraes, 2004). In both sets of experiments, the biocatalyst half-life was measured by determining the biocatalyst activity at the end of each run. The results obtained revealed a loss of 37.2% of the initial enzyme activity for the reactor running on babassu oil and ethanol at a molar ratio of 1 to 7. The deactivation coefficient (*kd*) and half-life ($t_{½}$) were measured at 0.016 d$^{-1}$ and 44.7 d, respectively. Using the higher level excess ethanol (i.e. oil to ethanol molar ratio of 1:12), a lower operational stability of the biocatalyst was observed (32.1 ±6.4 d).

This is an expected behavior because the support used for immobilizing the lipase from *B. cepacia* (i.e. $SiO_2$-PVA particles) has a strong affinity to adsorb glycerol molecules (formed as a byproduct) due to the presence of hydroxyl groups on both the organic (PVA) and inorganic (silanol groups) parts of the hybrid matrix. Using the dyeing method (adsorption of Amaranth, a food grade pigment), Xu et al. (2011) verified the strong affinity of glycerol molecules produced by the ethanolysis of rapeseed oil on a silica particle support used to immobilize *Thermomyces lanuginosus* lipase. In that study, the authors reported a good correlation between the transesterification yield and the amount of glycerol adsorbed on the silica particles. According to the literature, glycerol tends to adsorb on the microenvironment of the support forming a hydrophilic layer, which in turn makes the lipases inaccessible to hydrophobic substrates, such as oil droplets (Dossat et al., 1999; Watanabe et al., 2000). Based on the results of the present study, the transesterification reaction performed using a higher oil to ethanol ratio led to FAEE yields higher than 95%, and this must have resulted in the formation of a more hydrophilic layer on the microenvironment of the biocatalyst and consequently lower operational stability. It is worth quoting that such a limitation could be overcome implementing a strategy to remove the soluble glycerol. For instance, Hama et al., (2011) proposed continuous biodiesel synthesis integrated with a glycerol-separating system through the incorporation of a column packed with resin such as Lewatit GF 202 to absorb glycerol.

## 4. Conclusions

The findings of the present study confirmed the feasibility of continuous enzymatic production of biodiesel from babassu oil and ethanol in the absence of any solvents using a PBR. The babassu oil and ethanol molar ratio was found as a critical parameter for attaining high FAEE yields. More specifically, the highest performance was achieved when the babassu oil to ethanol molar ratio of 1:12 was used. At this molar ratio and at space times greater than or equal to 11 h, high substrate conversion (transesterification yield of 96.0 ± 0.9%) and productivity value (average productivity 41.1 ± 1.6 $mg_{ester}{\cdot}g_{catalyst}^{-1}{\cdot}h^{-1}$) were obtained. Higher yields were also found to have negatively affected the operational stability of the biocatalyst caused by higher rate of glycerol production and its consequent destabilizing effects on the biocatalyst particles by forming a hydrophilic layer. Therefore, further investigation would be required to overcome the limitations imposed by glycerol to improve the overall performance of the proposed system.

## 5. Acknowledgments

The authors gratefully acknowledge the financial support provided by the Conselho Nacional de Desenvolvimento Científico e Tecnológico (CNPq) and the Coordenação de Aperfeiçoamento de Pessoal de Nível Superior (CAPES).

## References


Antczak, M.S., Kubiak, A., Antczak, T., Bielecki, S., 2009. Enzymatic biodiesel synthesis – Key factors affecting efficiency of the process. Renew. Energy. 34, 1185-1194.

Balcão, V.M., Paiva, A.L., Malcata, F.X., 1996. Bioreactors with immobilized lipases: State of the art. Enzyme Microb. Technol. 18, 392-416.

Brunschwig, C., Moussavou, W., Blin, J., 2012. Use of bioethanol for biodiesel production. Prog. Energ. Combust. 38, 283-301.

Carvalho, A.K.F., Da Rós, P.C.M., Teixeira, L.F., Andrade, G.S.S., Zanin, G.M., De Castro, H.F., 2013. Assessing the potential of non-edible oils and residual fat to be used as a feedstock source in the enzymatic ethanolysis reaction. Ind. Crop. Prod. 50, 485-493.

Christopher, L.P., Kumar, H., Zambare, V.P., 2014. Enzymatic biodiesel: Challenges and opportunities. Appl. Energ. 119, 497-520.

Costa-Silva, W., Teixeira, L.F., Carvalho, A.K.F., Mendes, A.A., De Castro, H.F., 2014. Influence of feedstock source on the biocatalyst stability and reactor performance in continuous biodiesel production. J. Ind. Eng. Chem. 20, 881-886.

Da Rós, P.C.M., Silva, G.A.M., Mendes, A.A., Santos, J.C., De Castro, H.F., 2010. Evaluation of the catalytic properties of *Burkholderia cepacia* lipase immobilized on non-commercial matrices to be used in biodiesel synthesis from different feedstocks. Bioresour. Technol. 101, 5508-5516.

Damstrup, M.L., Kill, S., Jensen, A.D., Sparso, F.V., Xu, X., 2007. Process development of continuous glycerolysis in an immobilized enzyme-packed reactor for industrial monoacylglycerol production. J. Agric. Food Chem. 55, 7786-7792.

Dors, G., Freitas, L., Mendes, A.A., Furigo Jr, A., De Castro, H.F., 2012. Transesterification of palm oil catalyzed by *Pseudomonas fluorescens* lipase in a packed- bed reactor. Energ Fuel. 26, 5977-5982.

Dossat, V., Combes, D., Marty, A., 1999. Continuous enzymatic transesterification of high oleic sunflower oil in a packed bed reactor: influence of the glycerol production. Enzyme Microb. Technol. 25, 194-200.

Fogler, H.S., 1992. Elements of chemical reaction engineering. Prentice Hall: Englewood Cliffs, NJ.

Fox, R.W., Pritchard, P.J., McDonald, A.T., 2013. Introdução à mecânica dos fluídos, (7th ed.). Tradução e Revisão Técnica: Koury, R.N.M., Machado, L. Rio de Janeiro: LTC.

Freitas, L., Da Rós, P.C.M., Santos, J.C., De Castro, H.F., 2009. An integrated approach to produce biodiesel and monoglycerides by enzymatic interesterification of babassu oil (*Orbinya* sp.). Process Biochem. 44, 1068-1074.

Freitas, L., Silva, G.S., Santos, J.C., Oliveira, P.C., De Castro, H.F., 2011. Strategies to remove water formed as by-product on the monoolein synthesis by enzymatic esterification performed on packed bed reactor. Eur. Food Res. Technol. 233, 743-750.

Halim, S.F.A., Kamaruddin, A.H., Fernando, W.J.N., 2009. Continuous biosynthesis of biodiesel from waste cooking palm oil in a packed bed reactor: Optimization using response surface methodology (RSM) and mass transfer studies. Bioresour. Technol. 100, 710-716.

Hama, S., Tamalampudi, S., Yoshida, A., Tamadani, N., Kuratani, N., Noda, H., Fukuda, H., Kondo, A., 2011. Process engineering and optimization of glycerol separation in a packed-bed reactor for enzymatic biodiesel production. Bioresour. Technol.102, 10419-10424.

Levenspiel, O., 1972. Chemical Reaction Engineering, 2nd ed., John Wiley, New York.

Lide, D.R., 2006. Definition of scientific terms. In Handbook of Chemistry and Physics, (87th ed.). CRC Press: Boca Raton, FL.

Jegannathan, K.R., Abang, S.A., Poncelet, D., Chan, E.S., Ravindra, P., 2008. Production of biodiesel using immobilized lipase—a critical review. Crit. Rev. Biotechnol. 28, 253-264.

Kawakami, K., Oda,Y., Takahashi, R., 2011. Application of a *Burkholderia cepacia* lipase immobilized on silica monolith to batch and continuous biodiesel production with a stoichiometric mixture of methanol and crude *Jatropha* oil. Biotechnol. Biofuels. 42, 1-11.

Ricca, E., Paola, M.G., Calabro, V., Curcio, S., Iori, G., 2009. Olive husk oil transesterification in a fluidized bed reactor with immobilized lipases. Asian Pac. J. Chem. Eng. 4, 365-368.

Royon, D., Daz, M., Ellenrieder, G., Locatelli, S., 2007. Enzymatic production of biodiesel from cotton seed oil using *t*-butanol as a solvent. Bioresour. Technol. 98, 648-653.

Soares, C.M.F., De Castro, H.F., De Moraes, F.F., Zanin, G.M., 1999. Characterization and utilization of *Candida rugosa* lipase immobilized on controlled pore silica. Appl. Biochem. Biotechnol. 79, 745-757.

Stamenkovic, O.S., Velickovic, A.V., Veljkovic, B.V., 2011. The production of biodiesel from vegetable oils by ethanolysis: Current state and perspectives. Fuel. 90, 3141-3155.

Teixeira, M.A., 2008. Babassu—A new approach for an ancient Brazilian biomass. Biomass Bionerg. 32, 857-864.

Urioste, D., Castro, M.B.A., Biaggio, F.C., De Castro, H.F., 2008. Synthesis of chromatographic standards and establishment of a method for the quantification of the fatty ester composition of biodiesel from babassu oil. Quim. Nova. 31, 407-412.

Yusuf, C. Bioreactor design. In: Ratledge, C., Kristiansen, B., 2006. Basic Biotechnology,( 3 ed.) Cambridge: Cambridge University Press, pp. 181-200.

Watanabe, Y., Shimada, Y., Sugihara, A., Noda, H., Fukudac, H., Tominaga, Y., 2000. Continuous production of biodiesel fuel from vegetable oil using immobilized *Candida antarctica* lipase. J. Am. Oil Chem. Soc. 77, 355-360.

Zanin, G.M., Moraes, F.F., 2004. Enzimas Imobilizadas. In: Said, S., Pietro, R.C.L.R. Enzimas como agentes biotecnológicos. Ribeirão Preto: Legis Summa,. pp. 35-85.

# Efficient Conversion of Carbohydrates to 5-Hydroxymethylfurfural (HMF) Using $ZrCl_4$ Catalyst in Nitromethane

Raju S. Thombal, Vrushali H. Jadhav *

*Department of Organic Chemistry, National Chemical Laboratory (CSIR-NCL), Pune-411008, India.*

## HIGHLIGHTS

➢A promising process for 5-Hydroxymethyl furfural (HMF) formation from biomass.

➢Efficient synthesis HMF from carbohydrates by $ZrCl_4$ in nitromethane

➢Fructose and inulin led to highest HMF yield of > 70% with 100% selectivity

## GRAPHICAL ABSTRACT

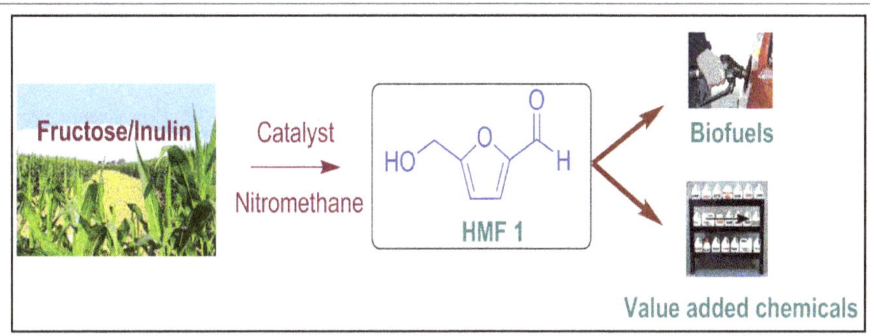


## ABSTRACT

Solvent nitromethane along with a variety of metal chloride and mineral acids as catalyst were studied for the synthesis of 5-Hydroxymethylfurfural (HMF), a key precursor in the formation of alternative fuel 2,5-dimethylfuran (DMF) and other value added chemicals. Reaction time, temperature and catalyst concentration were also systematically studied to achieve highest HMF formation. Among the carbohydrates studied for HMF synthesis, D-fructose and inulin were found particularly most productive yielding >70% and with 100% selectivity using ZrCl4 in nitromethane at 100 °C during 3h. Readily available reagents, solvents, and simple reaction conditions could mark this process promising for HMF formation from biomass.

**Keywords:**
Biomass
Carbohydrates
Alternative fuel
5-Hydroxymethylfurfural (HMF)
2,5-dimethylfuran (DMF)
Nitromethane

## 1. Introduction

Non renewable fossil sources such as natural gas, oil and coal meet around 85% of the world's energy demands today. With increase in world's population, there are growing concerns about diminishing fossil fuel resources, global warming, and environmental pollution; hence, there is a need to search for renewable resources to bridge the gap between the supply and demand of energy and chemicals. In this respect, biomass is the only widespread, abundant, inexpensive, and sustainable resource which can be an ideal substitute for fossil-based resources. Carbohydrates comprise about 75% of the annual renewable biomass (Roper., 2007). Hence, there is need to

* Corresponding author
E-mail address: vrushalijadhav.jadhav@gmail.com (V.H. Jadhav).

develop efficient and environmentally and ecofriendly methods to convert carbohydrates into useful chemicals e.g. alternative fuels.

5-Hydroxymethylfurfural (HMF) is a valuable biomass-derived platform chemical. HMF is a versatile intermediate between biomass-based carbohydrate chemistry and petroleum-based organic chemistry. HMF and its derivatives could replace petroleum-based building blocks and could also used to make polymers and valuable chemicals (Sutton et al., 2013). As presented in Figure 1, HMF can be a common intermediate for 2,5-dimethyl furan (DMF), which is proposed to be an alternative for petroleum-based liquid fuels (Román-Leshkov et al., 2007). It also could be used for the production of 5-chloromethylfurfural, a precursor to 5-ethoxymethylfurfural (potential bio-diesel candidate) (Mascal and Nikitin, 2008), 2,5-furan dicarboxylic acid (FDCA) (Pan et al., 2013), caprolactum (Buntara et al., 2011), maleic anhydride (Du et al., 2011), levulinic acid (Alonso et al., 2013) , adipic acid (Arias et al., 2013) and many other value added chemicals (Huber et al., 2005).

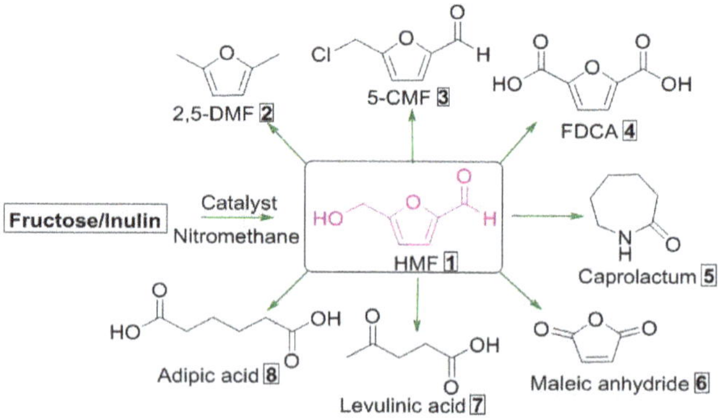

**Scheme 1.** HMF serves a key precursor for a number of fuels and value-added chemicals. 1: HMF; 2: 2,5-dimethyl furan (DMF); 3: 5-chloromethylfurfural (CMF); 4: 2,5-furan dicarboxylic acid (FDCA); 5: Caprolactum; 6: Maleic anhydride; 7: Levulinic acid and 8: Adipic acid.

As for HMF production, achieving a high yielding and highly selective process is still a challenge. In fact, high production cost currently limits the availability and use of HMF industrially. Hence the challenge faced by chemists/engineers is to develop an economically feasible process for synthesis of HMF from biomass. To date, numbers of methods have been developed for synthesis of HMF from various sustainable reactants (Yang et al., 2011). As reported in the literature, many acidic catalysts have been used for the HMF synthesis from D-fructose. Homogeneous acids such as HCl (Román-Leshkov et al., 2006) and $H_2SO_4$ (Binder et al., 2009) were used to catalyze the dehydration of D-fructose which resulted in 40–85% yield of HMF in solvents like dimethyl sulfoxide (DMSO), dimethylacetamide (DMA), and 2,5-dimethylfuran (DMF). Solvent systems studied for dehydration of D-fructose include organic solvents (DMSO, DMF, DMA, Sulfolane, γ-Valerolactone (GVL) (Van Putten et al., 2013), ionic liquids (Zhao et al., 2007), aqueous systems (Rosatella et a., 2011), nitromethane-water system (Karimi and Mirzaei, 2013) and biphasic mixtures (Pagan-Torres et al., 2012). The high boiling point and the instability of DMF, DMSO, DMA solvents at high temperatures limits their use in synthesis of HMF. Raines and co-workers (Caes and Raines, 2011) reported sulfolane as a solvent for synthesis of HMF, while recently Dumesic and co-workers reported GVL as a solvent for synthesis of HMF (Gallo et al., 2013). Ionic liquids are also used extensively in synthesis of HMF, but they suffer drawbacks of being expensive and are deactivated in presence of small amounts of water. Use of aqueous systems favor formation of more by-products i.e., decomposition of HMF to levulinic acid and formic acid (Girisuta et al., 2006). Biphasic mixtures are hard to handle for separation. The fact still remains that for the production of HMF to be made commercial for the purpose of biofuel production, economic and readily available solvents, as well as economic acid catalysts (capable of quick conversions of D-fructose to HMF with limited side reactions leading to HMF degradation) are to be addressed.

From this point of view, we strived to use nitromethane as solvent and metal chlorides/mineral acids as catalyst as a superior combination for HMF formation from biomass. Nitromethane is polar, protic (bp. 101 °C) solvent and is used in variety of industrial applications (Deshpande et al., 1998). The lower boiling point serves as an additional advantage leading to its easy removal from the reaction medium. Nitromethane is found stable at higher temperatures and acidic conditions. Herein, we report the formation of HMF from D-fructose in high yields using the combination of an efficient solvent; nitromethane in presence of different acid catalysts.

## 2. Experimental

### *2.1. Materials*

D-Fructose, D-glucose, starch, sucrose and inulin were purchased from Sigma Aldrich Chemicals. Metal chlorides, acids, and nitromethane were purchased from a local manufacturer. 5-Hydroxymethylfurfural (99%) was purchased from Aldrich Chemicals and was used establishing the calibration curve. 5-Chloromethylfurfural was synthesized based on the literature suggested by Kumari et al. (2011).

### *2.2. Experimental procedure*

In a typical reaction protocol for the dehydration of carbohydrates, a 10 wt% carbohydrate/nitromethane solution containing carbohydrates (1 g) was charged into a 50 mL flask, followed by the addition of 3 different concentrations (10, 50 and 100 mol%) of different catalysts. The carbohydtaes investigated included D-fructose, starch, sucrose, inulin, glucose, and cellulose. The catalyst investigated included chlorides (i.e. $ZrCl_4$, $SnCl_2$, $CrCl_3$, $FeCl_3$, $BiCl_3$, $CuCl_2$, $CaCl_3$, $CeCl_3$, $MgCl_2$, $LaCl_3.7H_2O$, $HgCl_2$, $NH_4Cl$, $ZnCl_2$, $AlCl_3$, $CrCl_2$, $SnCl_2.2H_2O$, $CoCl_2.6H_2O$, LiCl.Anh, CuCl, $BaCl_2.2H_2O$, $NiCl_2$, $TiCl_4$ , $ZrO_2Cl_2$, and $RuCl_3$) and mineral (HCl, HBr, $H_2SO_4$, TFA, and $H_3PO_4$) acid catalysts. The reaction was then heated at different temperatures (60, 80 and 100 °C) for a specific time (1-5 h, with measurements performed in 30 min intervals), while stirred with a magnetic stirrer. The reaction mixtures were diluted with a known mass of deionized water, stirred for 5 min to remove any insoluble products and were analyzed to determine the amounts of HMF and CMF by a gas chromotography (Agilent 6890N) equipped with a HP-5 5% phenyl methyl column (I.D: 320 mm, length 30 m, film thickness 1 mm) and a flame ionisation detector (280 °C max.). Furfuraldehyde was used as the internal standard. The temperature was initially set at 100 °C (1 min) and was then increased with a rate of 20 °C/min (10 min) upto 280 °C. Moreover, different carbohydrate sources including starch, sucrose, inulin, glucose and cellulose were investigated using the most efficient catalyst.

## 3. Results and Discussion

A variety of acid catalysts for HMF formation from D-fructose in solvent nitromethane was investigated and the reaction conditions i.e. reaction time, temperature and catalyst concentration were optimized. In fact, reaction time and temperature are very crucial factors for HMF formation for under un-optimized conditions, HMF decomposition occurs. It was observed that for all the catalysts investigated the reaction proceeded at lower rates at 60 and 80 °C in comparison with that of 100 °C. Figures 2 presents the effect of temperature and time on the rate of reaction or in another word, HMF yield in presence of D- Fructose using $ZrCl_4$ as catalyst.

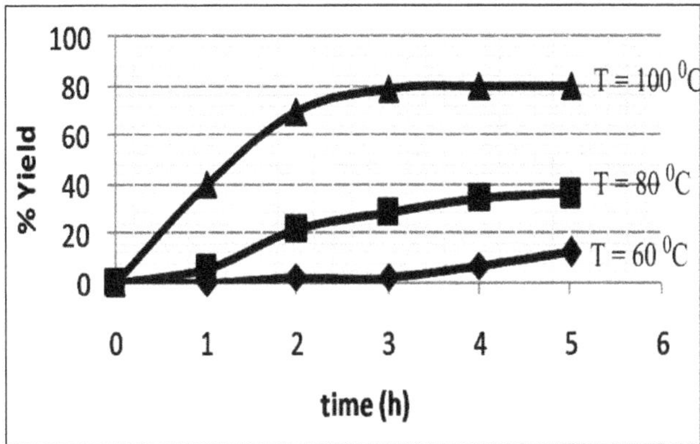


**Figure 2.** HMF formation from D- Fructose using $ZrCl_4$ as catalyst at different temperatures.

The catalyst concentration resulting in maximal HMF formation was also investigated. Catalyst concentration of 10 mol% was shown to lead to the highest HMF yield for all the catalysts tested i.e. both the chlorides and mineral acid catalysts. As an example, Figure 3 compares the impact of catalyst concentration on HMF yield in case of $CrCl_3$ as catalyst. As shown, the HMF yield is negatively correlated with catalyst concentration. More specifically, it was found that by increasing the concentration of catalysts, HMF yield decreased while CMF concentration increased and also the formation of some by-products was detected. Hence, to achieve highest HMF yields, catalyst concentration of 10 mol% was found. The same trend was also observed for all the catalysts investigated (data not shown).

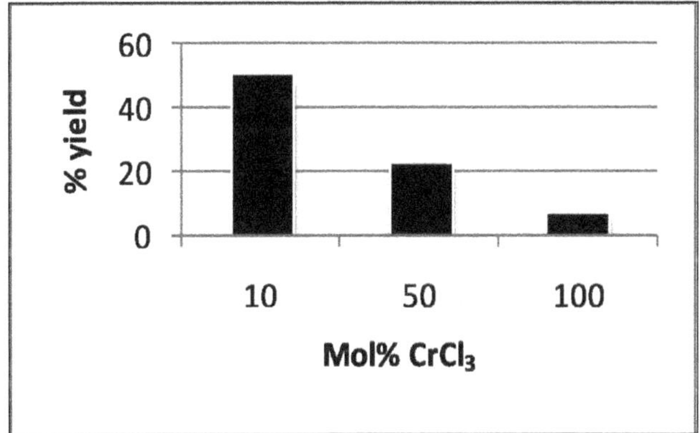


**Fig.3.** HMF formation from D- Fructose using CrCl3 as catalyst at different catalyst concentration (Mol%).

As for time, HMF formation reached its maximum value after 3 and 1.5 h for metal chloride and mineral acid catalysts, respectively. Tables 1 and 2, tabulates HMF and CMF yields from D-Fructose (10 wt %) in presence of chloride and mineral acid catalysts under optimal conditions, respectively

Among the chloride catalysts investigated, $ZrCl_4$ resulted in highest HMF yield of 82% with 100% selectivity of HMF. This catalyst was followed by $CeCl_3$ which achieved an HMF yield of 73% $ZrCl_4$. The same order was also obtained when HMF was isolated and yields were compared with $ZrCl_4$ and $CeCl_3$ ranking first and second (73 and 65% HMF yield). $RuCl_3$ also yielded moderately at about 53% (GC-based yield) and 48% (after isolation) with 100% selectivity but it was advantageous for only traces of levulinic acid and formic acid were formed during the reaction. As shown in Table 1. HMF isolation was only performed when GC-based HMF yield was above 15%.

**Table 1.**
HMF synthesis from D-Fructose (10 wt %) using metal chlorides as catalyst in nitromethane.

| No. | Catalyst Type [a] | HMF yield (%)[b] | CMF yield (%)[b] | HMF yield (%)[c] |
|---|---|---|---|---|
| 1 | $ZrCl_4$ | 82 | - | 73 |
| 2 | $SnCl_2$ | 15 | - | - |
| 3 | $CrCl_3$ | 55 | 5 | 43 |
| 4 | $FeCl_3$ | 7 | - | - |
| 5 | $BiCl_3$ | 3 | - | - |
| 6 | $CuCl_2$ | 56 | 10 | 44 |
| 7 | $CaCl_2$ | 0 | - | - |
| 8 | $CeCl_3$ | 73 | 6 | 65 |
| 9 | $MgCl_2$ | 3 | - | - |
| 10 | $LaCl_3.7H_2O$ | 10 | - | - |
| 11 | $HgCl_2$ | 6 | - | - |
| 12 | $NH_4Cl$ | 15 | - | - |
| 13 | $ZnCl_2$ | 0 | - | - |
| 14 | $AlCl_3$ | 26 | - | 18 |
| 15 | $CrCl_2$ | 25 | 3 | 20 |
| 16 | $SnCl_2.2H_2O$ | 6 | - | - |
| 17 | $CoCl_2.6H_2O$ | 4 | - | - |
| 18 | LiCl.Anh | 0 | - | - |
| 19 | CuCl | 8 | - | - |
| 20 | $BaCl_2.2H_2O$ | 0 | - | - |
| 21 | $NiCl_2$ | 19 | - | 11 |
| 22 | $TiCl_4$ | 39 | 21 | 13 |
| 23 | $ZrO_2Cl_2$ | 56 | 6 | 44 |
| 24 | $RuCl_3$ | 53 | - | 48 |

[a] Catalyst concentration was set at the optimum value of 10 mol.%.
(Optimum reaction time and temperature were at 3 h and 100 °C, respectively)
[b] Yields are based on GC analysis.
[c] Isolated yields.

**Table 2.**
HMF synthesis from D-Fructose (10 wt %) using mineral acids catalysts in nitromethane.

| No. | Catalyst Type [a] | HMF yield (%)[b] | CMF yield (%)[b] | HMF yield (%)[c] |
|---|---|---|---|---|
| 1 | HCl | 83 | 8 | 72 |
| 2 | HBr | 49 | - | 40 |
| 3 | $H_2SO_4$ | 80 | 7 | 70 |
| 4 | TFA | 10 | - | - |
| 5 | $H_3PO_4$ | 64 | - | 56 |

[a] Catalyst concentration was set at the optimum value of 10 mol.%.
(Optimum reaction time and temperature were at 1.5 h and 100 °C, respectively)
[b] Yields are based on GC analysis.
[c] Isolated Yields.

As for mineral acid-catalyzed conversion of D-fructose into HMF in nitromethane solvent, the optimal conditions were determined as 100 °C and 90 min reaction time. As tabulated in Table 2, under the optimized conditions, HCl and $H_2SO_4$ led to the highest yields of about 72% and 70%, respectively. Moreover, they resulted in the formation of only small traces of levulinic acid, formic acids, and other by-products during the reaction. To achieve further improvements in HMF yield, different combinations of metal chlorides, and also combinations of metal chlorides and mineral acids were also studied. More specifically, the catalysts which highest yield and selectivity were considered in the combination experiments. But no significant increase in yields of HMF was observed (Table 3).

**Table 3.**
HMF synthesis from D-Fructose (10 wt %) using various combination patterns of metal chlorides catalysts as well as metal chorides/mineral acids as catalyst in Nitromethane.

| No.[a] | catalyst (1:1) | HMF yield (%)[a] |
|---|---|---|
| 1 | $ZrCl_4/CrCl_3$ | 70 |
| 2 | $ZrCl_4/CeCl_3$ | 74 |
| 3 | $CrCl_3/CeCl_3$ | 65 |
| 4 | $ZrCl_4.H_2SO_4$ | 76 |
| 5 | $CeCl_3.H_2SO_4$ | 68 |

[a] Catalyst concentration was set at the optimum value of 10 mol.%
(Optimum reaction time and temperature were at 3 h and 100 °C, respectively)

Having achieved efficient synthesis of HMF from D-fructose, different carbohydrates which could be obtained from biomass i.e. starch, sucrose, inulin, glucose, and cellulose were also investigated for HMF production.

Since, the combination of $ZrCl_4$ catalyst in nitromethane as solvent was proven to result in the maximum yield and selectivity in HMF formation from D-fructose, hence, the other carbohydrates were also reacted under the same conditions with $ZrCl_4$ as catalyst in nitromethane as solvent (Table 4). Inulin which is a constituent of biomass led to the highest HMF yield of about 85% (74% after isolation). Sucrose ranked second on the list producing about 14% HMF after isolation. Glucose and cellulose resulted in only traces of HMF formation. This could be ascribed to the fact that the conversion of cellulose into HMF requires the initial hydrolysis of the β–1,4 linkages present in cellulose to form glucose units. Moreover further glucose isomerisation to fructose is also needed. The hydrolysis and isomerisation processes were not found feasible using acid catalysts in nitromethane as solvent.

**Table 4.** HMF Synthesis from different sources of carbohydrates (10 wt %) using $ZrCl_4$ Catalyst in Nitromethane.

| No.[a] | Carbohydrate | Time (h) | HMF yield (%)[a] | HMF yield (%)[b] |
|---|---|---|---|---|
| 1 | Starch | 1.5 | 0 | - |
| 2 | Sucrose | 1.5 | 20 | 14 |
| 3 | Inulin | 1.5 | 85 | 74 |
| 4 | Glucose | 4 | $<1$ | - |
| 5 | Cellulose | 4 | $<1$ | - |

[a] Catalyst concentration was set at the optimum value of 10 mol.%
(Optimum reaction temperature was 100 C)

## 4. Conclusion

In conclusion, we have demonstrated that $ZrCl_4$ and nitromethane is an excellent catalyst/solvent combination for conversion of D-fructose and inulin into HMF with yields standing at 73% and 74%, respectively, and with almost 100% selectivity. Finally, the innexpensive catalyst and solvent system, efficient and eco-friendly reaction conditions, and the easy procedure offered in the present study for the formation of HMF, seems like a promising strategy for the production of HMF as a key precursor in the formation of alternative fuel 2,5-dimethylfuran (DMF) and other value added chemicals from biomass.

## 5. Acknowledgements

The authors would like to thank DST, New Delhi for the INSPIRE Faculty Award and Fast Track Grant. The authors would also like to thank Dr. Sourav Pal, Director, NCL for providing all the infra-structural facilities required during the course of this study.

## References

Alonso, D.M., Gallo, J.M.R., Mellmer, M.A., Wettstein, S.G., Dumesic, J.A., 2013. Direct conversion of cellulose to levulinic acid and gamma-valerolactone using solid acid catalysts. Catal. Sci. Technol. 3, 927-931.

Arias, K.S., Al-Resayes, S.I., Climent, M.J., Corma, A., Iborra, S., 2013. From biomass to chemicals: Synthesis of precursors to biodegradable surfactants from 5-hydroxymethylfurfural. ChemSusChem. 6, 123-131.

Binder, J.B., Raines, R.T., 2009. Simple chemical transformation of lignocellulosic biomass into furans for fuels and chemicals. J. Am. Chem. Soc. 131, 1979-1985.

Buntara, T., Noel, S., Phua, P.H., Melian-Cabrera, I., de Vries, J.G., Heeres, H.J., 2011. Caprolactum from renewable resources: Catalytic conversion of 5-hydroxymethylfurfural into caprolactone. Angew. Chem. 123, 7221-7225.

Caes, B.R., Raines, R.T., 2011. Conversion of fructose into 5-(Hydroxymethyl)furfural in sulfolane. ChemSusChem. 4, 353-356.

Deshpande, M.N., Cain, M.H., Patel, S.R., Singam, P.R., Brown, D., Gupta, A., Barkalow, J., Callen, G., Patel, K., Koops, R., Chorghade, M., Foote, H.,Pariza, R., 1998. A scalable process for the novel antidepressent ABT-200. Org. Process Res. Dev. 2, 351-356.

Du, Z., Ma, J., Wang, F., Liu, J., Xu, J., 2011. Oxidation of 5-hydroxymethylfurfural to *maleic anhydride* with molecular oxygen. Green Chem. 13, 554-557.

Gallo, J.M.R., Alonso, D.M., Mellmer, M.A., Dumesic, J.A., 2013. Production and upgrading of 5-hydroxymethylfurfural using heterogeneous catalysts and biomass-derived solvents. Green Chem. 15, 85-90.

Girisuta, B., Janssen, L.P.B.M., Heeres, H.J., 2006. A kinetic study on decomposition of 5-hydroxymethylfurfural into levulinic acid. Green Chem. 8, 701-709.

Huber, G.W., Chheda, J.N., Barrett, C.J., Dumesic, J.A., 2005. Production of liquid alkanes by liquid phase processing of biomass-derived carbohydrates. Science 308, 1446-1450.

Karimi, B., Mirzaei, H.M., 2013. The influence of hydrophobic/hydrophilic balance of the mesoporous solid acid catalysts in the selective dehydration of fructose into HMF. RSC Adv. 3, 20655-20661.

Kumari, N., Olesen, J. K., Pedersen, C.M., Bols, M., 2011. Synthesis of 5-bromomethylfurfural from cellulose as a potential intermediate for biofuel. Eur. J. Org. Chem. 7, 1266-1270.

Mascal, M., Nikitin, E.B., 2008. Direct, high-yield conversion of cellulose in biofuel. Angew. Chem. 120, 8042-8044.

Pagan-Torres, Y.J., Wang, T., Gallo, J.M.R., Shanks, B.H., Dumesic, J.A., 2012. Production of 5-hydroxymethylfurfural from glucose using a combination of lewis and bronsted acid catalysts in water in a biphasic reactor with an alkylphenol solvent. ACS Catalysis 2, 930-934.

Pan, T., Deng, J., Xu, Q., Zuo, Y., Guo. Q.X., Fu, Y., 2013. Catalytic conversion of furfural into a 2,5-furandicarboxylic acid-based polyester with total carbon utilization. ChemSusChem. 6, 47-50.

Román-Leshkov, Y., Chheda, J.N., Dumesic, J.A., 2006. Phase modifiers promote efficient production of hydroxymethylfurfural from fructose. Science. 312, 1933-1937.

Román-Leshkov, Y., Barrett, C.J., Liu, Z.Y., Dumesic, J.A., 2007. Production of dimethylfuran for liquid fuels from biomass-derived carbohydrates. Nature. 447, 982-986.

Roper, H., 2002. Renewable Raw Materials in Europe-Industrial Utilisation of Starch and Sugar. Starch-Stärke. 54, 89-99.

Rosatella, A.A., Simeonov, S.P., Frade, R.F.M., Afonso, C.A.M., 2011. 5-Hydroxymethylfurfural (HMF) as a building block platform: Biological properties, synthesis and synthetic applications. Green Chem. 13, 754-793.

Sutton, A.D., Waldie, F.D., Wu, R., Schlaf, M., Pete Silks III, L.A., Gordon, J.C., 2013. The hydrodeoxygenation of bioderived furans into alkanes. Nature. 5, 428-432.

Van Putten, R.J., Vander Waal, J.C., De Jong, E., Rasrendra, C.B., Heeres, H.J., deVries, J.G., 2013. Hydroxymethylfurfural, a versatile platform chemical made from renewable sources. Chem. Rev. 113 (3), 1499-1597.

Yang, F., Liu, Q., Yue, M., Bai, X., Du, Y., 2011. Tantalum compounds as heterogeneous catalysts for saccharide dehydration to 5-hydroxymethylfurfural. Chem. Comm. 47, 4469-4471.

Zhao, H., Holladay, J.E., Brown, H., Zhang, Z.C., 2007. Metal chlorides in ionic liquids solvents convert sugars to 5-hydroxymethylfurfural. Science. 316, 1597-1600.

# 17

# Fungal biomass and ethanol from lignocelluloses using *Rhizopus* pellets under simultaneous saccharification, filtration, and fermentation (SSFF)

Somayeh FazeliNejad, Jorge A. Ferreira*, Tomas Brandberg, Patrik R. Lennartsson, Mohammad J. Taherzadeh

*Swedish Centre for Resource Recovery, University of Borås, SE 501 90, Borås, Sweden.*

## HIGHLIGHTS

- Economically viable production of 2nd generation bioethanol cannot rely on a single product.
- SSFF can be used for production of ethanol and biomass from wheat straw.
- Glucose present in the feed controlled the assimilation of xylose and acetic acid.
- The fungal growth rate was found not to be influenced by the feed composition.
- *Rhizopus* biomass yields of up 0.34 g/g and ethanol yields of 0.40 g/g were obtained.

## GRAPHICAL ABSTRACT

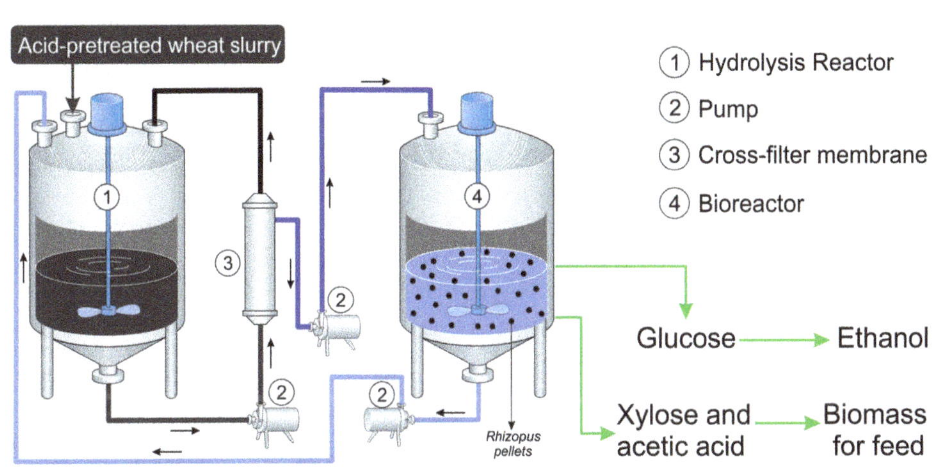


## ABSTRACT

The economic viability of the 2nd generation bioethanol production process cannot rely on a single product but on a biorefinery built around it. In this work, ethanol and fungal biomass (animal feed) were produced from acid-pretreated wheat straw slurry under an innovative simultaneous saccharification, fermentation, and filtration (SSFF) strategy. A membrane unit separated the solids from the liquid and the latter was converted to biomass or to both biomass and ethanol in the fermentation reactor containing *Rhizopus* sp. pellets. Biomass yields of up to 0.34 g/g based on the consumed monomeric sugars and acetic acid were achieved. A surplus of glucose in the feed resulted in ethanol production and reduced the biomass yield, whereas limiting glucose concentrations resulted in higher consumption of xylose and acetic acid. The specific growth rate, in the range of 0.013-0.015/h, did not appear to be influenced by the composition of the carbon source. Under anaerobic conditions, an ethanol yield of 0.40 g/g was obtained. The present strategy benefits from the easier separation of the biomass from the medium and the fungus ability to assimilate carbon residuals in comparison with when yeast is used. More specifically, it allows *in-situ* separation of insoluble solids leading to the production of pure fungal biomass as a value-added product.

**Keywords:**
Cellulosic ethanol
Animal feed
*Rhizopus* sp.
SSFF
Wheat straw

* Corresponding author
E-mail address: Jorge.Ferreira@hb.se

## 1. Introduction

Nowadays, the production of 1st generation bioethanol from agricultural sugar- or starch-rich crops, as a replacement to gasoline, is well established at commercial scale. The leading ethanol-producing countries, USA and Brazil, use corn and sugarcane as main feedstocks, respectively (RFA, 2014). However, ethical issues related to the use of sugar- or starch-rich feedstocks for fuel production instead of being directed to human consumption have put pressure on finding alternative feedstocks (Cherubini, 2010).

The production of ethanol from lignocellulosic materials has been considered for several decades (Leonard and Hajny, 1945). Nevertheless, due to their recalcitrant structure, a feasible commercial facility producing the so-called 2nd generation bioethanol is presently inexistent and is only limited to some pilot plants (Pandey et al., 2015). Constraints include the cost-intensive pretreatment needed to open up the lignocellulosic structure, the cost of enzymes needed in the post-pretreatment stage, the lack of robust microorganisms that can cope with inhibitors, and robust cultivation strategies that can meet all requirements for feasible 2nd generation bioethanol prodution (Ishola, 2014). Another conclusion drawn by the intensive studies conducted over years is that a facility using lignocelluloses as feedstock cannot rely on a single product (i.e., ethanol) for achieving an economically-viable operation (Pandey et al., 2015).

The most commonly used strategies for production of ethanol include simultaneous saccharification and fermentation (SSF) and separate hydrolysis and fermentation (SHF). Running a SSF instead of a SHF circumvents the product inhibition of cellulase enzymes due to glucose accumulation (Olofsson et al., 2008). However, SSF disadvantageously requires the use of new microorganisms at each batch since it is difficult to separate them from the medium (Wingren et al., 2003). Therefore, a new cultivation strategy, i.e., simultaneous saccharification, filtration, and fermentation (SSFF) was developed by Ishola et al. (2013). This new concept consists of a membrane unit (cross-flow membrane) connecting a hydrolysis reactor to a fermentation reactor. The enzyme-slurry mixture from the hydrolysis reactor is filtered and the sugar-rich permeate is continuously supplied to the fermentation reactor while the residues are returned to the hydrolysis reactor. The fermented medium in the fermentation reactor is also pumped back to the hydrolysis reactor for volume balance (**Fig. 1**).

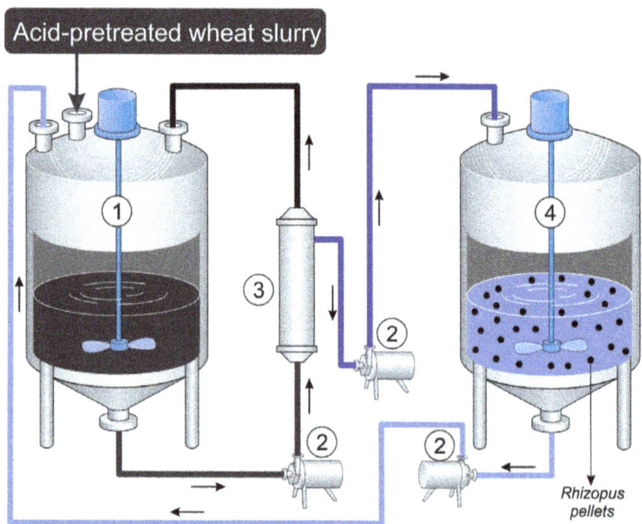


**Fig.1.** Schematic representation of SSFF for aerobic biomass (*Rhizopus* sp.) production.

Therefore, SSFF combines the advantages of both SSF and SHF by simultaneously solving their limitations, i.e., both hydrolysis and fermentation can be carried out at optimal conditions, the end-product inhibition is avoided, and there is also the possibility to reuse the fermenting cells. The fact that the fermenting cells is free from solid substrates (i.e., lignocellulosic materials) in the fermentation reactor under the SSFF, opens up the opportunity for the production of a second value-added product (i.e., biomass as animal feed) from the residual glucose, pentose sugars (such as xylose), and other components (such as acetic acid) contained in the pretreated wheat straw slurry. This would be a similar situation as exists for the 1st generation bioethanol plants from starch grains where both ethanol and animal feed products are generated (Kim et al., 2008).

Edible filamentous fungi have been previously used for production of protein-rich biomass (animal feed) from various types of substrates (Ferreira et al., 2013). For instance, tempe-isolated zygomycete *Rhizopus* sp. has been found to cope with inhibitors contained in lignocellulosic hydrolysates (FazeliNejad et al., 2013). Moreover, these microorganisms form pellets, and therefore, could be easily separated from the medium (Nyman et al., 2013). They also consume pentose sugars contrary to baker's yeast *Saccharomyces cerevisiae* (Wikandari et al., 2012). These attributes mark these fungi as appropriate candidates for application in the SSFF. Among the different lignocellulosic materials, wheat straw is an economic agricultural by-product available at huge amounts with potentials for production of ethanol in view of its cellulose (33%) and hemicellulose (33%) dry weight contents (Canilha et al., 2006). It is worth quoting that various studies have been carried out on the pretreatment and hydrolysis of wheat straw (Talebnia et al., 2010; Baboukani et al., 2012; Peng et al., 2012).

In the present work, SSFF was used for ethanol production from glucose using *S. cerevisiae* as well as biomass (animal feed) production from residual carbon sources using *Rhizopus* sp. in pellet form. The effects of the enzyme addition to the hydrolysis reactor, temperature in the hydrolysis and fermentation reactors, and aeration for assimilation of carbon sources in the filtered permeate for production of biomass (animal feed) were investigated. This is the first work reporting the use of SSFF with filamentous fungi *Rhizopus* sp. for the production of two value-added products, i.e., bioethanol and animal feed from lignocellulosic materials.

## 2. Materials and methods

### 2.1. Microorganisms

*Rhizopus* sp. CCUG 61147 (Culture Collection University of Gothenburg, Sweden) isolated from Indonesian leaves traditionally used for preparation of tempe, was used in this work. The strain was identified as RM4 in a previous publication (Wikandari et al., 2012). The fungus was kept in Potato Dextrose Agar (PDA) plates and its incubation and preparation for inoculation were carried out according to FazeliNejad et al. (2013). A strain of *S. cerevisiae*, Ethanol Red, kindly provided by Fermentis (France) in dry form was also used.

### 2.2. Pretreated wheat straw slurry

Slurry of wheat straw delivered by SEKAB E-Technology (Örnsköldsvik, Sweden) was produced by continuous treatment of wheat straw at 22 bar for 5-7 min. The resulting slurry, a liquid fraction with fine particles, had 14.6% suspended solids (SS) and 23.8% total solids (TS). The liquid had a pH value of 2.0 and contained (in g/L): glucose, 7.2; xylose, 22.1; galactose, 2.3; arabinose, 4.6; acetic acid, 5.9; 5-hydroxy methyl-furfural (HMF), 2.1; and furfural, 4.2. The solid fraction contained 34.7% (w/w) glucan and 4.6% (w/w) xylan.

### 2.3. Enzyme cocktail

Cellic® CTec2, kindly provided by Novozymes (Denmark), was used in the experiments for enzymatic treatment. The product had an activity of 168 filter paper units (FPU)/mL.

### 2.4. Inoculant preparation and cultivations in shake-flasks

A complex medium containing (in g/L): xylose, 20; potato extract, 4; soybean peptone, 6; and $CaCO_3$, 6 was used for preparation of *Rhizopus* sp. pellets as inoculant. The medium (50 mL) was transferred to 250 mL cotton-plugged Erlenmeyer flasks followed by sterilization in an autoclave at 121 °C for 20 min. It should be noted that xylose was autoclaved separately. After mixing and inoculation with $1.0 \times 10^5$ spores/mL of *Rhizopus* sp., the flasks were kept in a water-bath at 30 °C and 150 rpm for 72 h. The produced *Rhizopus* sp. pellets were transferred to new cultivations to a cell concentration of 1.65 ± 0.10 g/L (dry weight ± 1 SD). The new cultivations were carried out in 250-mL Erlenmeyer flasks containing 100 mL of the same medium which also consisted of (in g/L): $(NH_4)_2SO_4$, 7.5; $KH_2PO_4$, 3.5; $CaCl_2 \cdot 2H_2O$, 1; $MgSO_4 \cdot 7H_2O$, 0.75; and one of the following carbon sources namely acetic acid, 5.0; ethanol, 10; glucose, 10; lactic acid, 10; and xylose, 10. The cultivations were kept in a water-bath at 30 °C while being shaken at 150 rpm.

With similar preparation of the initial inoculant, 2.20 ± 0.12 g/L pellets (dry weight, ± 1 SD) were transferred to a new medium and cultivated as described above but a combination of the carbon sources at 3.5 g/L was used. This experiment was carried out in duplicate.

Liquid samples were withdrawn and stored at -20 °C for subsequent analysis. At the end of the cultivation, the pellets were harvested using a sieve, washed with distilled water, and dried in an oven at 70 °C to constant weight for 24 h. The cultivations using single-carbon sources or their combination were performed in quadruplicate and duplicate, respectively.

### *2.5. Cultivations under SSFF*

The SSFF as previously described by Ishola et al. (2013) was employed. The lignocellulosic feedstock was hydrolysed enzymatically in a separate vessel (hydrolysis reactor) and the resulting sugar-rich liquid was circulated through the fermentation reactor, where the fungal biomass production took place. The solid fraction was separated from the sugar-rich stream by a cross-flow membrane. However, a cell retention system was not needed in this work since 5 mm spherical pellets of *Rhizopus* sp. were used and they maintained this morphology throughout the cultivations.

For the SSFF trials, pellets were prepared as described above and transferred to a 750 mL fermentor (Ant, Belach Bioteknik AB, Sweden) containing sterilized salt solution as described above and 0.1 g/L antifoam. The transferred *Rhizopus* sp. pellets had an initial dry weight within the range 1.5-2.1 g and the volume was adjusted to a total volume of 500 mL. Wheat straw slurry was transferred to a parallel hydrolysis reactor (Memma, Belach Bioteknik AB, Sweden) and diluted with deionized water to 5.0% SS to a total volume of 3.5 L. The salt and antifoam content was the same as in the fermentation reactor. The cross-flow membrane unit was set up according to Ishola et al. (2013). After integration (**Fig. 1**), the flow of the filtrate through the fermentation bioreactor was 40 mL/h.

In a first experiment, the integration of the SSFF system was preceded by 24 h enzymatic decomposition by addition of Cellic®CTec2, corresponding to 10 FPU/g SS. The pH was initially adjusted to 5.5 in both reactors and regulated to 5.5 in the fermentation reactor by on-line addition of 2.0 M NaOH. The temperature was kept at 50 °C in the hydrolysis reactor and 35 °C in the fermentation reactor. The stirring was 350 rpm in the hydrolysis vessel and 100 rpm in the fermentation bioreactor, which was aerated at 1 vvm (volume of air per volume of liquid per minute). The experiment was carried out in duplicate where the integration phase lasted for 140 h and 168 h. Samples were withdrawn directly from the tubes channeling medium in and out of the fermentation vessel. The final cell (biomass) content was analysed by weighing it after drying at 70 °C for 24 h. The experiment was then repeated with the same parameters but with the following differences; no enzyme was added and the integration phase lasted for 72 h. This was intended to investigate the impact of enzyme addition into the hydrolysis reactor.

In a similarly initiated experiment, the effect of aeration in the fermentation reactor was investigated by switching off the air supply after 72 h. Moreover, 10 FPU/g SS enzyme was simultaneously added into the hydrolysis reactor. This anaerobic fermentation phase lasted for 94 h.

Another SSFF trial was also carried out where the temperature was adjusted to 35 °C in both reactors. Enzyme (10 FPU/g SS) and 15 g of dry baker's yeast were added to the hydrolysis reactor. The experiment was carried out in duplicate where the integration phase lasted for 96 and 120 h.

In a different set-up, the similar cultivation as of above (i.e., 35 °C, 10 FPU/g SS, and 15 g dry yeast) was initially performed without any integration for 54 h. The resulting fermented slurry was then distilled using a rotary evaporator (Labinett, Sweden) at 140 °C (oil bath), and 30 rpm rotation speed at atmospheric pressure. The water content lost during distillation was re-adjusted by the addition of sterile ultrapure water. The resulting slurry, now without ethanol, was used for integration with the SSFF and aerobic production of *Rhizopus* sp. biomass as described above during 96 h.

### *2.6. Analytical methods*

The measurements of glucose, metabolites, and inhibitors concentrations as well as spore counting were performed according to FazeliNejad et al. (2013). The SS was determined by filtration with Munktell filters, Grade 3 (5-8 μm) while the TS were determined by drying the samples to a constant weight at 105 °C overnight. The solid fraction of the wheat straw slurry and the enzyme activity were analyzed according to the NREL protocols (Adney and Baker, 2008; Sluiter et al., 2011).

## 3. Results and discussion

### *3.1. SSFF of wheat straw slurry with Rhizopus sp. pellets*

Production of additional products in a biorefinery concept has been proposed to improve the process economy of ethanol production from cellulosic raw materials (Wheals et al., 1999; Gnansounou and Dauriat, 2010). Animal feed in the form of *Rhizopus* sp. biomass has been suggested as a valuable co-product for ligno-ethanol in the present study. Implementing SSFF for aerobic production of *Rhizopus* sp. biomass entails the application of a cross-flow membrane to separate available sugars and other organic compounds from a pretreated lignocellulosic slurry (**Fig. 1**). The filtrate is supplied to an aerated fermentor, where carbon sources are consumed by *Rhizopus* sp. pellets in order to produce biomass (animal feed).

The pellet morphology is useful in order to prevent leakage of biomass when liquid is pumped back to the hydrolysis reactor. This reflux is necessary in order to maintain the liquid balance between the vessels and to prevent increasing the dry matter content of the slurry. Besides, the glucose concentration, which would increase as a result of enzymatic decomposition of cellulose and could inhibit the enzymes, can be controlled in this way.

On the other hand, the filtration of the slurry is in itself a very important operation since the solid fraction must not be mixed with the biomass, which would result in a downstream separation problem. In addition to biomass, the *Rhizopus* sp. used in this work is also a potential producer of ethanol (Wikandari et al., 2012).

The implementation of SSFF for production of ethanol and biomass includes the use of continuous cross-flow membrane as described earlier. The results obtained revealed that the filtration unit was used for up to 168 h without regeneration of the membrane and without any fouling. In a similar experiment, involving the slurry of pretreated spruce, the same operation was performed during 28 d without interruption, regeneration, or fouling (Ishola et al., 2013).

### *3.2. Specific growth rate and biomass yield*

Various SSFF experiments with *Rhizopus* sp. production from wheat straw slurry were carried out in order to validate this concept. The main difference between the different trials was the composition of the substrate, notably its glucose content. Enzymatic decomposition of cellulose in the solid fraction prior to integration with SSFF (**Fig. 2**) produced a relatively high initial glucose concentration in contrast to a similar experiment without enzyme addition (**Table 1**).

The addition of baker's yeast to the hydrolysis vessel in a different experiment nearly eliminated the glucose in the inflow to the fermentation reactor. Furthermore, an experiment was carried out where the amount of glucose was reduced by the addition of baker's yeast and the produced ethanol was also removed by distillation. The resulting mix was used for *Rhizopus* sp. production by SSFF (**Table 1**).

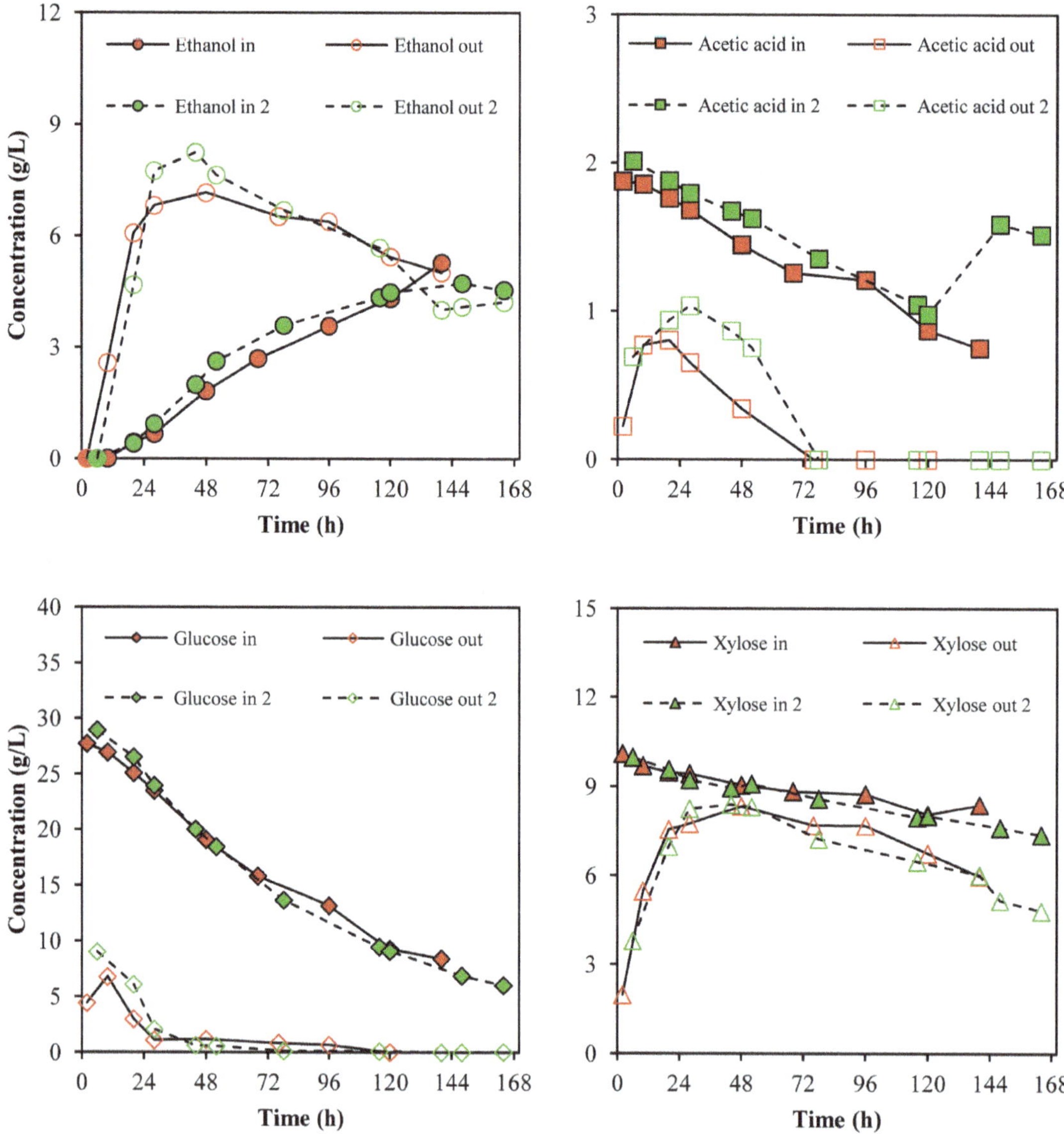


**Fig.2.** Concentrations of ethanol, acetic acid, glucose, and xylose in SSFF for aerobic production of Rhizopus sp. biomass and ethanol. The integration (connection between the hydrolysis reactor and bioreactor by a filtration unit) was preceded by 24 h of enzymatic hydrolysis. Closed symbols denote concentrations in the ingoing feed to the bioreactor (number 4 in Fig. 1), open symbols denote concentrations in the recirculation feed from the bioreactor to the hydrolysis reactor. The expressions "in" and "out" represent the medium going from the filtration unit into the bioreactor and the medium that leaves the bioreactor to the hydrolysis reactor, respectively, whereas "2" stands for the replicate 2 of the experiment.

The specific growth rate, μ (/h), was calculated according to the following equation (**Eq. 1**):

$$\mu = \frac{\ln(\frac{x}{x_0})}{t} \quad \text{Eq. 1}$$

Where $x_0$ denotes the initial biomass concentration and x is the biomass concentration after the elapsed time $t$. Assuming a constant μ is debatable because of the dynamic conditions such as substrate concentrations and the fact that the results are sensitive to the accuracy of wet weight measurements of the initial biomass. Growth in the form of pellets is also known to be different from that of free cells (Metz and Kossen, 1977), but considering the small size of the used pellets, this effect can be assumed to be relatively low. However, the results tabulated in **Table 1** show that μ as measured, hardly changed corresponding to the substrate composition, i.e., $0.013/h < \mu < 0.015/h$ under aerobic conditions. It could be concluded that the growth rate showed no tendency to be affected by glucose concentrations as long as alternative carbon sources such as acetic acid, ethanol, and xylose were present. However, more efficient aeration could have resulted in a higher growth rate.

The biomass yields reported in **Table 1** ranged between 0.24 and 0.34 g/g consumed monomeric sugars (arabinose, galactose, glucose, and xylose), acetic acid, and ethanol, except for the case with enzymatic hydrolysis (i.e., high initial glucose concentration, **Fig. 2**) where the biomass yield dropped due to ethanol production. These biomass yields were in harmony with the yield obtained in separate batch experiments with synthetic medium containing individual carbon sources (10 g/L of each compound except for acetic acid; 5 g/L). The measured yields of biomass for acetic acid, ethanol, and xylose in these trials were 0.30 g/g, 0.30 g/g, and 0.29 g/g, respectively, after 140 h batch cultivation (96 h for acetic acid, data not shown). The corresponding consumption of glucose was faster (less than 42 h), but the resulting biomass yield was as low as

**Table 1.**
Overview of SSFF trials including: (**1**) integration preceded by 24 h of enzymatic hydrolysis; (**2**) no addition of enzyme to the hydrolysis reactor; (**3**) integration preceded by 54 h of enzymatic hydrolysis and ethanol production with yeast in the hydrolysis reactor followed by evaporation of the ethanol; (**4**) Enzyme and yeast were added to the hydrolysis reactor with no ethanol evaporation before integration; (**5**) The air supply to the bioreactor was switched off at t = 72 h with concomitant addition of enzymes to the hydrolysis reactor. The glucose column refers to the concentration of glucose in the bioreactor and how it developed (the uptake of arabinose and galactose are not reported).

| SSFF trial | Glucose (g/L) | Cultivation time (h) | $Y_{X/S}$ (g/g) | $Y_{E/S}$ (g/g) | μ (/h) | Distribution of uptake (%) | | | |
|---|---|---|---|---|---|---|---|---|---|
| | | | | | | Glucose | Xylose | Ethanol | Acetic acid |
| (**1**) Enzyme addition | ~ 27 decreases to ~ 6 | 140 | 0.11 [a] | 0.21 [a] | 0.013 | 86 | 9 | - | 6 |
| | | 168 | 0.14 [a] | 0.14 [a] | 0.014 | 80 | 12 | - | 7 |
| (**2**) No enzyme addition | 2.2 decreases to 0.2 | 72 | 0.32 [a] | - | 0.015 | 46 | 43 | - | 12 |
| (**3**) SSF & evaporation [b] | 2.2 to ND | 96 | 0.34 [a] | 0.10 [a] | 0.015 | 28 | 39 | - | 25 |
| (**4**) Yeast in hydrolysis [c] | <0.3 | 96 | 0.24 [a] | Cons. | 0.015 | <1 | 53 | 22 | 25 |
| | | 120 | 0.30 [a] | | 0.013 | 1 | 37 | 28 | 34 |
| (**5**) Anaerobic with enzyme addition | peaks at 17.5 | 94 | 0.034 [d] | 0.40 [d] | 0.002 | 95 | 5 | - | <1 |

$Y_{X/S}$ = yield (g of fungal biomass/g of consumed carbon source) $Y_{E/S}$ = yield (g of ethanol/g of consumed carbon source) "Cons." = consumed "ND" = not detected.
[a] Biomass and ethanol yields related to consumed amounts of acetic acid, arabinose, ethanol, glucose, galactose, and xylose.
[b] This treatment resulted in reduced amounts of glucose and ethanol prior to SSFF integration.
[c] This method sharply reduced the glucose content in the flux into the bioreactor.
[d] Biomass and ethanol yields related to consumed amounts of glucose and xylose.

0.11 g/g due to formation of ethanol and glycerol (data not shown), confirming overflow metabolism (Crabtree effect) for *Rhizopus* sp. (Millati et al., 2005; Lennartsson et al., 2009). The pooled standard deviation for the biomass yields was 0.042 (± 1 SD).

### *3.3. Steering the uptake of carbon sources*

In a separate experiment with synthetic medium, the uptake pattern was studied in a cultivation, where acetic acid, ethanol, glucose, lactic acid, and xylose were added to the same cultivation of *Rhizopus* sp. in aerobic shake-flasks. The results showed a relatively rapid consumption of glucose, followed by acetic acid, whereas xylose and ethanol with similar consumption trends were not totally consumed after 72 h of cultivation (**Fig. 3**). Lactic acid frequently occurs as an undesired metabolite produced by contaminants (Skinner and Leathers, 2004) and its uptake by other zygomycetes is documented (Ferreira et al., 2013). However, no measurable consumption of lactic acid by the *Rhizopus* sp. strain was confirmed in this experiment. It is observed that the preference of carbon source, among those examined, under the examined conditions can be ranked as follows: glucose > acetic acid > xylose & ethanol (**Fig. 3**). The measured specific growth rate, μ, was 0.013/h, i.e., similar to the level in the SSFF experiments with wheat straw hydrolysate (**Table 1**), but it is difficult to differentiate the effects of inhibitors and different aeration rates.

In the SSFF experiment with cellulase addition, it is clearly visible that the glucose uptake was relatively efficient but had no visible positive effects on the specific growth rate. Instead, ethanol was produced in a respire-fermentative pattern (**Fig. 2 and Table 1**). In a biorefinery context, it is probable that glucose, if available, would be used for other

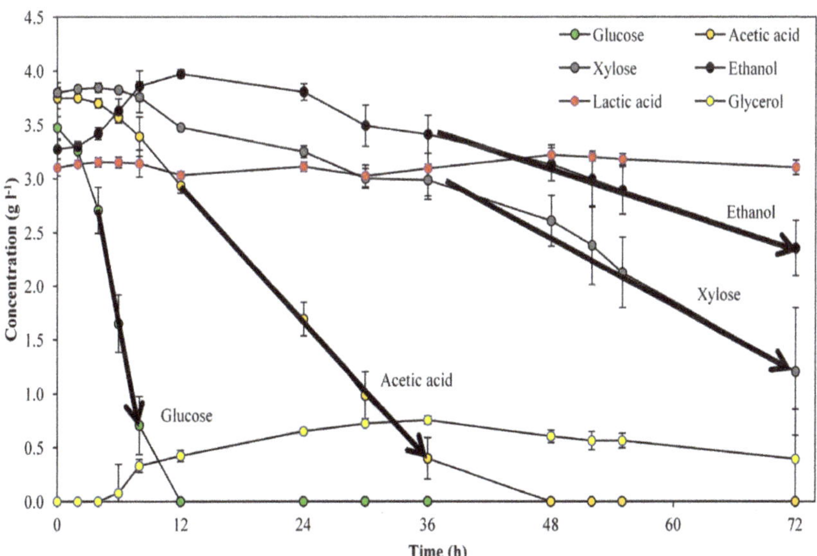


**Fig.3.** Concentration profiles of glucose, acetic acid, xylose, ethanol, lactic acid, and glycerol in an aerated shake-flask experiment inoculated with Rhizopus sp. pellets. At the beginning of the cultivation, the concentration of all mixed components except glycerol, which was produced during cultivation, was 3.5 g/L. Error bars denote ±1 SD.

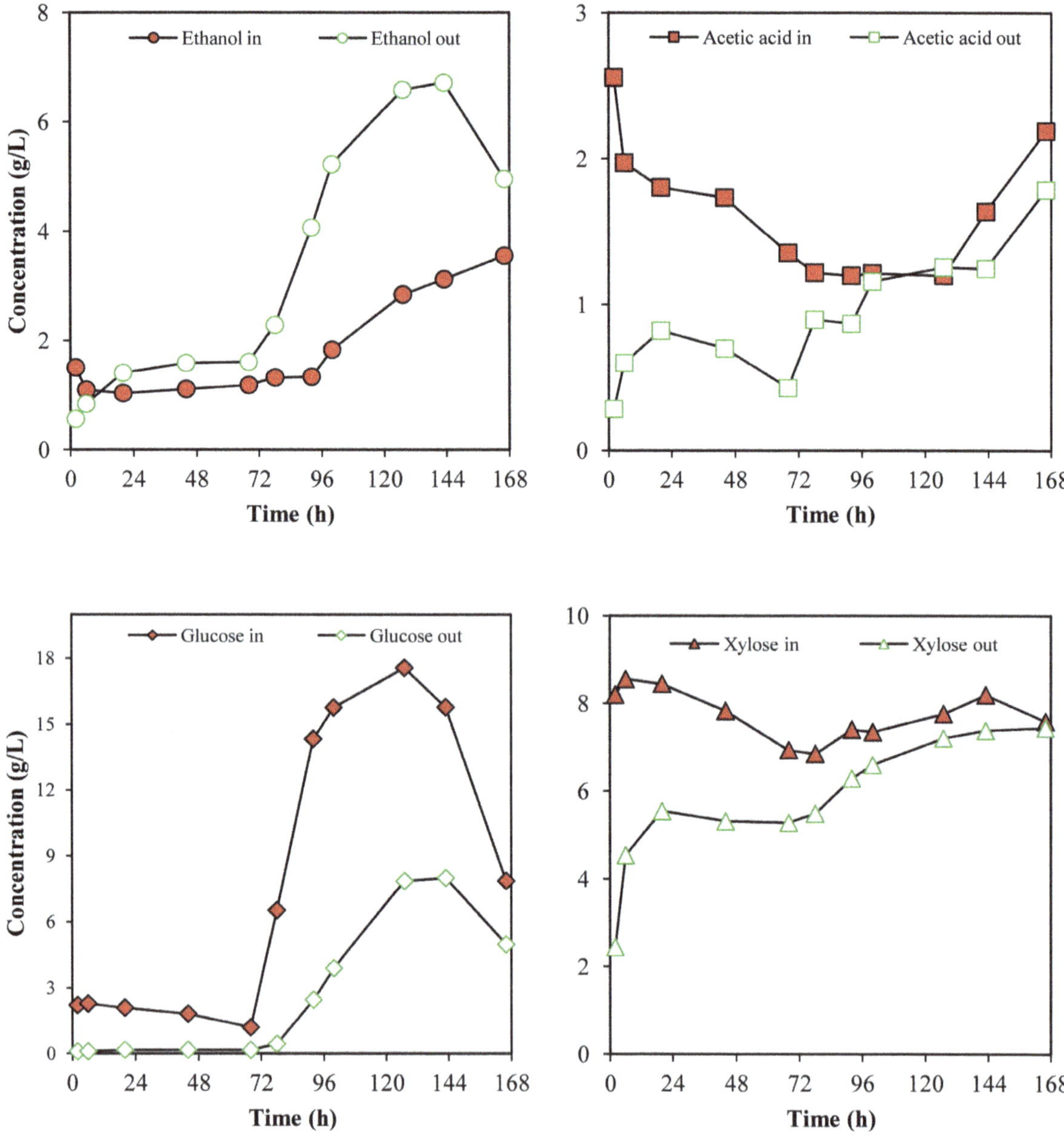


**Fig.4.** Concentrations of ethanol, acetic acid, glucose, and xylose in the SSFF for aerobic production of biomass followed by anaerobic fermentation by *Rhizopus* sp. The air supply to the bioreactor was switched off and enzymes were added into the hydrolysis reactor at t = 72 h. The expressions "in" and "out" represent the medium going from the filtration unit into the bioreactor and the medium that leaves the bioreactor to the hydrolysis reactor, respectively.

purposes, such as ethanol production by fermentation. Therefore, it may be advantageous to utilize other compounds than glucose for biomass formation. By excluding the enzymatic decomposition, the glucose concentration was sharply reduced, resulting in a higher uptake of xylose and acetic acid. Combining the cultivation of *Rhizopus* sp. by SSFF with the addition of *S. cerevisiae* in the hydrolysis reactor further reduced the glucose concentration in the feed and increased the consumption of xylose, acetic acid, and ethanol. The use of enzymatic hydrolysis and fermentation followed by distillation prior to SSFF cultivation produced a result remarkably similar to the case with the untreated slurry, i.e., without enzyme addition (**Table 1**).

In conclusion, reducing the glucose concentration in the present study steered the uptake by *Rhizopus sp.* to xylose and acetic acid, which both can be present as residual compounds in a biorefinery, without reducing either the biomass yield or the specific growth rate.

### *3.4. Fermentation by Rhizopus sp.*

*Rhizopus* sp. is also useful as a fermenting organism for ethanol production and the combined production of valuable biomass and ethanol is interesting in a biorefinery perspective. Performing a complete list of process possibilities is beyond the scope of this study, and only the production of ethanol and animal feed using *Rhizopus* sp. as producing organism was investigated herein.

Two experiments were carried out, where *Rhizopus* sp. was initially grown aerobically on straw hydrolysate in order to build up biomass. One of the trials was stopped after 72 h (referred to in **Table 1** as cultivation without enzyme addition), and the biomass was harvested and measured (4.4 g). The second experiment was initiated in a similar way, but after 72 h, cellulase (10 FPU/g SS) was added and the air supply to the fermentation reactor was switched off (**Fig. 4**).

During the subsequent 94 h, the amount of biomass increased to 6.4 g, suggesting a specific growth rate (μ) of 0.034/h during the anaerobic phase (**Table 1**).

In the time span from 72 to 166 h, at least 11.5 g of ethanol was produced (some may have evaporated), which would suggest an ethanol yield of 0.40 g/g consumed glucose and xylose, and an ethanol productivity of 0.023 g/g/h, based on the average biomass concentration. The volumetric productivity of ethanol, 0.24 g/L/h, was relatively low if compared with optimized fermentations with *S. cerevisiae* (Balat, 2011). The xylose uptake corresponded to only 5% of the consumed carbon source and the uptake should be a sign of leakage of oxygen into the fermentation vessel, considering that fungi normally do not consume xylose under anaerobic conditions. According to the measurements, the concentration of xylose in the hydrolysis vessel increased after 72 h, indicating the release of xylose from the solid fraction as a result of the enzymatic decomposition.

### *3.5. Impact of inhibitors*

After dilution to 5.0% SS, the concentrations of acetic acid, furfural, and HMF were 1.8 g/L, 0.65 g/L, and 1.3 g/L, respectively. Considering a previous study (FazeliNejad et al., 2013), these levels should not be very inhibiting by themselves. It is worth quoting that hydrolysates of lignocellulosic material usually contain other inhibitors as well, such as phenolic compounds, whose concentrations were not measured in the present study. It was observed in all the SSFF trials that ingoing furfural and HMF were completely converted (i.e., not detected in the outflow) after 6-8 h of the integrated fermentation, confirming *in situ* conversion of these compounds by *Rhizopus* sp. (FazeliNejad et al., 2013). Furthermore, the specific growth rate, μ, was similar in the hydrolysate-based SSFF experiments and the shake-flask experiments with synthetic medium. It could thus be assumed that the impact of inhibitors was limited.

## 4. Conclusions

*Rhizopus* sp. in pellet form was successfully used for aerobic production of biomass (animal feed) by SSFF from acid-pretreated wheat straw slurry with biomass yields of up to 0.34 g biomass/g consumed monomeric sugars and acetic acid. A surplus of glucose in the feed resulted in ethanol production and reduced the biomass yield, whereas limiting glucose concentrations resulted in higher consumption of xylose and acetic acid. The specific growth rate was in the range of 0.013/h and 0.015/h and did not appear to be influenced by the composition of the carbon source. Under anaerobic conditions, an ethanol yield of 0.40 g/g and an ethanol productivity of 0.023 g/g/h were obtained using the *Rhizopus* sp. pellets. Overall, the present strategy benefits from the easier separation of the biomass from the medium and the fungus ability to assimilate carbon residuals in comparison with when yeast is used. More specifically, it allows *in situ* separation of insoluble solids and hence, a two-stage cultivation system practiced for production of biomass and ethanol from whole stillage is not needed to be applied if biomass is desired as a separate value-added product.

## Acknowledgement

This work was financially supported by the Swedish Energy Agency.

## References

[1] Adney, B., Baker, J., 2008. Measurement of cellulase activities. National Renewable Energy Laboratory.

[2] Baboukani, B.S., Vossoughi, M., Alemzadeh, I., 2012. Optimisation of dilute-acid pretreatment conditions for enhancement sugar recovery and enzymatic hydrolysis of wheat straw. Biosys. Eng. 111, 166-174.

[3] Balat, M., 2011. Production of bioethanol from lignocellulosic materials via the biochemical pathway: A review. Energy Convers. Manage. 52, 858-875.

[4] Canilha, L., Carvalho, W., Batista, J., e Silva, A., 2006. Xylitol bioproduction from wheat straw: hemicellulose hydrolysis and hydrolyzate fermentation. J. Sci. Food Agr. 86, 1371-1376.

[5] Cherubini, F., 2010. The biorefinery concept: Using biomass instead of oil for producing energy and chemicals. Energy Convers. Manage. 51, 1412-1421.

[6] FazeliNejad, S., Brandberg, T., Lennartsson, P.R., Taherzadeh, M. J., 2013. Inhibitor Tolerance: A Comparison between *Rhizopus* sp. and *Saccharomyces cerevisiae*. Bioresources. 8, 5524-5535.

[7] Ferreira, J.A., Lennartsson, P.R., Edebo, L., Taherzadeh, M.J., 2013. Zygomycetes-based biorefinery: Present status and future prospects. Bioresour. Technol. 135, 523-532.

[8] Gnansounou, E., Dauriat, A., 2010. Techno-economic analysis of lignocellulosic ethanol: A review. Bioresour. Technol. 101, 4980-4991.

[9] Ishola, M.M., 2014. Novel application of membrane bioreactors in lignocellulosic ethanol production, PhD, University of Borås, Borås, Sweden.

[10] Ishola, M.M., Jahandideh, A., Haidarian, B., Brandberg, T., Taherzadeh, M.J., 2013. Simultaneous saccharification, filtration and fermentation (SSFF): A novel method for bioethanol production from lignocellulosic biomass. Bioresour. Technol. 133, 68-73.

[11] Kim, Y., Mosier, N. S., Hendrickson, R., Ezeji, T., Blaschek, H., Dien, B., Cotta, M., Dale, B., Ladisch, M.R., 2008. Composition of corn dry-grind ethanol by-products: DDGS, wet cake, and thin stillage. Bioresour. Technol. 99, 5165-5176.

[12] Lennartsson, P.R., Karimi, K., Edebo, L., Taherzadeh, M.J., 2009. Effects of different growth forms of Mucor indicus on cultivation on dilute-acid lignocellulosic hydrolyzate, inhibitor tolerance, and cell wall composition. J. Biotechnol. 143, 255-261.

[13] Leonard, R.H., Hajny, G.J., 1945. Fermentation of wood sugars to ethyl alcohol. Ind. Eng. Chem. 37, 390-395.

[14] Metz, B., Kossen, N.W.F., 1977. The growth of molds in the form of pellets - a literature review. Biotechnol. Bioeng. 19, 781-799.

[15] Millati, R., Edebo, L., Taherzadeh, M.J., 2005. Performance of *Rhizopus*, *Rhizomucor*, and *Mucor* in ethanol production from glucose, xylose, and wood hydrolyzates. Enzyme Microb. Technol. 36, 294-300.

[16] Nyman, J., Lacintra, M.G., Westman, J.O., Berglin, M., Lundin, M., Lennartsson, P.R., Taherzadeh, M.J., 2013. Pellet formation of zygomycetes and immobilization of yeast. New Biotechnol. 30, 516-522.

[17] Olofsson, K., Bertilsson, M., Lidén, G., 2008. A short review on SSF-an interesting process option for ethanol production from lignocellulosic feedstocks. Biotechnol. Biofuels. 1, 1-14.

[18] Pandey, A., Höfer, R., Larroche, C., Taherzadeh, M.J., Nampoothiri, M., 2015. Industrial Biorefineries and White Biotechnology. Elsevier.

[19] Peng, F., Peng, P., Xu, F., Sun, R.C., 2012. Fractional purification and bioconversion of hemicelluloses. Biotechnol. Adv. 30, 879-903.

[20] RFA, 2014. Ethanol industry outlook. Renewable Fuels Association.

[21] Skinner, K., Leathers, T., 2004. Bacterial contaminants of fuel ethanol production. J. Ind. Microbiol. Biot. 31, 401-408.

[22] Sluiter, A., Hames, B., Ruiz, R., Scarlata, C., Sluiter, J., Templeton, D., Crocker, D., 2011. Determination of structural carbohydrates and lignin in biomass. National Renewable Energy Laboratory, Colorado, USA.

[23] Talebnia, F., Karakashev, D., Angelidaki, I., 2010. Production of bioethanol from wheat straw: An overview on pretreatment, hydrolysis and fermentation. Bioresour. Technol. 101, 4744-4753.

[24] Wheals, A.E., Basso, L.C., Alves, D.M.G., Amorim, H.V., 1999. Fuel ethanol after 25 years. Trends Biotechnol. 17, 482-487.

[25] Wikandari, R., Millati, R., Lennartsson, P.R., Harmayani, E., Taherzadeh, M.J., 2012. Isolation and characterization of Zygomycetes fungi from tempe for ethanol production and biomass applications. Appl. Biochem. Biotechnol. 167, 1501-1512.

[26] Wingren, A., Galbe, M., Zacchi, G., 2003. Techno-economic evaluation of producing ethanol from softwood: Comparison of SSF and SHF and identification of bottlenecks. Biotechnol. Progr. 19, 1109-1117.

# 18

# Microbial growth in *Acrocomia aculeata* pulp oil, *Jatropha curcas* oil, and their respective biodiesels under simulated storage conditions

Juciana Clarice Cazarolli[1,*], Patrícia Dörr de Quadros[1], Francielle Bücker[1], Mariana Ruiz Frazão Santiago[2], Clarisse Maria Sartori Piatnicki[3], Maria do Carmo Ruaro Peralba[3], Eduardo Homem de Siqueira Cavalcanti[2], Fátima Menezes Bento[1]

[1] *Department of Microbiology, Federal University of Rio Grande do Sul. Rua Sarmento Leite, N° 500, 90050-170, Porto Alegre, RS, Brazil.*

[2] *Corrosion and Degradation Division, National Institute of Technology, Av. Venezuela, N° 82, Sala 608, 200081-312, Rio de Janeiro, RJ, Brazil.*

[3] *Department of Inorganic Chemistry, Federal University of Rio Grande do Sul. Av. Bento Gonçalves, N° 9500, 91501-970, Porto Alegre, RS, Brazil.*

## HIGHLIGHTS

➢ Microbial growth capacity of filamentous fungi in *Acrocomia aculeata* pulp oil and *Jatropha curcas* oil and their respective biodiesels investigated.
➢ Order of susceptibility to microbial growth was *A. aculeata* pulp Biodiesel > *J. curcas* Biodiesel > *A. aculeata* pulp oil > *J. curcas* oil.
➢ Esters contents of *A. aculeata* pulp and *J. curcas* biodiesels decreased by approx. 8 and 12% when inoculated by fungi.
➢The occurrence of biodiesel biodegradation even during a relatively short storage period of only 30 d was observed.

## GRAPHICAL ABSTRACT

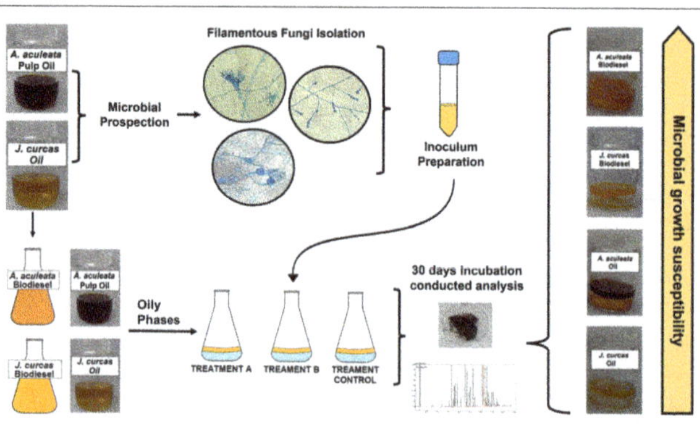


## ABSTRACT

With increasing demands for biodiesel in Brazil, diverse oil feedstocks have been investigated for their potentials for biodiesel production. Due to the high biodegradability of natural oils and their respective biodiesels, microbial growths and consequent deterioration of final product quality are generally observed during storage. This study was aimed at evaluating the susceptibility of *Acrocomia aculeata* pulp oil and *Jatropha curcas* oil as well as their respective biodiesels to biodeterioration during a simulated storage period. The experiment was conducted in microcosms containing oil/biodiesel and an aqueous phase over 30 d. The levels of microbial contamination included biodiesel and oil as received, inoculated with fungi, and sterile. Samples were collected every 7 d to measure pH, surface tension, acidity index, and microbial biomass. The initial and final ester contents of the biodiesels were also determined by gas chromatography. The major microbial biomass was detected in *A. aculeata* pulp and *J. curcas* biodiesels. Significant reductions in pH values were observed for treatments with *A. aculeata* pulp biodiesel as a carbon source ($p < 0.05$). The surface tension values decreased for all treatments ($p < 0.05$). Total ester contents were decreased in *A. aculeata* pulp and *J. curcas* biodiesels when inoculated by fungi by approximately 8 and 12%, respectively, indicating the occurrence of biodegradation during the relatively short storage period of only 30 d.

**Keywords:**
*Acrocomia aculeate* pulp oil
*Jatropha curcas* oil
Biodiesel
Storage
Degradation
Filamentous fungi

* Corresponding author
E-mail address: jucianacazarolli@gmail.com

## 1. Introduction

Petroleum-based fuels have been the world's main energy source. Recently, concerns about the environment and the depletion of non-renewable sources have been driving research activities aiming at introducing sustainable alternatives. The use of various vegetable oil and fatty animal waste feedstocks as raw material for the production of biofuels, such as biodiesel, has been made possible through such efforts. According to the Brazilian Law Nº 11.097 / 2005, the addition of biodiesel to petroleum diesel in Brazil is mandatory, and currently this ratio stands at 7% (B7) which has been predicted to increase to 10% by 2017.

As mentioned earlier, different vegetable oils have been used as substrate for biodiesel production, but their production costs are still limiting the production process (Meneghetti et al., 2013). Thus, the use of alternative and less expensive fatty acids is encouraged. Nowadays, soybean is the most used crop for biodiesel production in Brazil. According to the National Agency of Petroleum (ANP), in 2014 more than 76% of 3.42 million cubic meters of biodiesel produced were derived from the *Glycine max* seed oil (ANP, 2015; Souza et al., 2015).

Among potential oilseeds to be used in biodiesel production are Pinhão Manso (*Jatropha curcas*) and Macaúba (*AcrocomiA. aculeata*). *J. curcas*, a drought-resistant plant, is a shrub belonging to the family Euphorbia which is cultivated in Central and South America, Southeast Asia, India, and Africa. *J. curcas* is grown in soils with low fertility and in climates considered unfavorable for most traditional food crops (Pandey et al., 2012). In addition to being perennial and easy growing, *J. curcas* has a high annual oil productivity of 2 – 4 ton ha$^{-1}$, and therefore, it is considered as an attractive biodiesel feedstock (Mofijur et al., 2013).

Another potential feedstock to be used for biodiesel production is Macaúba (*Acrocomia aculeata*). It is an arborescent perennial palm native to tropical forests, which belongs to the Palmae family, whose fruits have great potentials for oil production. The mesocarp is rich in oil which is yellow in color due to the presence of carotenes, while the kernel oil is light yellow (Duarte et al., 2012). In Brazil, *A. aculeata* is mainly found in the southeast and northeast regions (Lorenzi, 2006; Amaral, 2007). Compared with soybean with an oil productivity of 400-800 kg ha$^{-1}$ on average, oil production capacity of Macaúba is much higher, reaching up to 6600 kg ha$^{-1}$ which is considered the highest productivity among Brazilian palm trees. Thus, this plant is highly recommended as a potential oleaginous feedstock for biofuel production in Brazil (Lorenzi, 2006).

The predominant technology used for biodiesel production is based on transesterification reactions. The quality of the resultant product obtained through this process is highly dependent on the raw materials used and their intrinsic chemical features. Different feedstocks could lead to the generation of esters of different chemical properties, which could conseqently influence biodiesel degradability potentials (Teixeira et al., 2010; Kaushik et al., 2015). In fact, being composed of esters of fatty acids, biodiesel is more susceptible to degradation than diesel (Mariano et al., 2008). This may be considered as an advantage in scenarios where environmental contamination is of prominent concern. However, from the storage point of view, it can be considered as a disadvantage, since microbial degradation of biodiesel could significantly compromise the quality of the final product. Therefore, in line with the widespread search for alternative oilseeds as raw material for biodiesel production, it is also essential to explore the susceptibility of these biodiesel feedstock candidates to microbial growth during storage.

The aim of the present study was to evaluate the susceptibility of *Acrocomia aculeata* (Macaúba) pulp oil and *Jatropha curcas* oil, and their respective biodiesels to microbial contamination. More specifically, the physical and chemical characteristics in response to microbial contamination were investigated during a 30-d storage period.

## 2. Materials and Methods

### *2.1. Oil and biodiesel samples*

*A. aculeata* pulp oil was produced by the Association of Small Rural Workers Riacho D'antas, Minas Gerais while *J. curcas* oil was obtained from the company Fabrica BRAZIL Ecoenergia. Biodiesels were produced by the National Institute of Technology (INT) using transesterification as production method according to Nascimento et al. (2016) and Silveira (2014).

### *2.2. Microbial prospection*

In order to investigate the cultivable native microbiota of *A. aculeata* pulp oil and *J. curcas* oil, these oils were placed in contact with an aqueous phase consisting of the mineral nutrients (Bushnell and Haas, 1941). More specifically, 20 ml of each oil sample was added into 20 ml of the mineral minimum medium and the suspension was incubated for 15 d in a greenhouse at 30 °C. At sampling times of 0, 7, and 15 d, 100 µl of the minimal mineral medium of each flask was taken and added to Petri dishes containing malt extract agar medium (for fungi isolation) and triptone soy agar medium (TSA: for bacteria isolation). Moreover, at the end of this experiment, i.e., day 15, the biomass grown in the oil-water interface by the native population of the oils was filtered by using filter paper discs; weight 80, thickness 205 µm, and porosity 14 mM (J. Prolab, Curitiba, Brazil), with the addition of hexane (Grade PA) as solvent to remove possible oil residues. Subsequently, the biomass was placed in a drying oven for 3 d and was then weighed on a precision scale.

#### *2.2.1. Isolation and identification of microorganisms*

After seeding by the aqueous phase, the culture plates were incubated at 30 °C. On days 5 and 7 after cultivation, microbial isolation was performed. After obtaining pure cultures, only the identification of filamentous fungal cultures was performed. The DNA of the isolates were obtained by DNA extraction according to the method described by Grattapaglia and Ferreira (1996). Identification was performed by microculture followed by Sanger sequencing through targeting the ITS1-5.8S rRNA-ITS2 region for filamentous fungi using ITS-1 and ITS-4 primers (Covino et al., 2015). The amplicons were sequenced by automated Amersham Mega BACE 1000 sequencing system using the standard protocols of the Brazilian Genome Network. The sequences generated were blasted against to the GenBank database (available at http://blast.be-md.ncbi.nlm.nih.gov/).

### *2.3. Determination of the susceptibility of A. aculeata pulp oil, J. curcas oil, and their respective biodiesels to microbial contamination during a 30-d simulated storage period*

A flowchart presenting the methodology developed in this work for assessing the susceptibility of *A. aculeata* pulp oil, *J. curcas* oil, and their respective biodiesels to microbial contamination is shown in **Figure 1**. The experiments were carried out in glass bottles containing 10 ml of oil or biodiesel phase and 20 ml of the mineral minimum medium (Bushnell and Haas, 1941) as the aqueous phase. The flasks were covered to prevent photo-oxidation. The experiments were performed in triplicate, assembled under destructive repetitions.

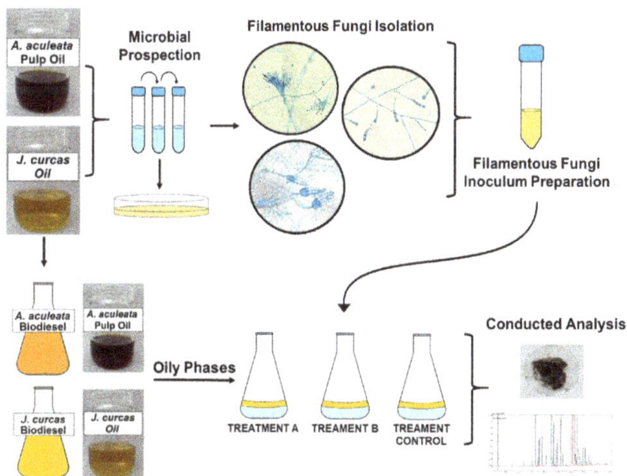


**Fig.1.** The flowchart presenting the methodology used to determine the susceptibility of *A. aculeata* pulp oil, *J. curcas* oil, and their respective biodiesels to microbial growth.

*2.3.1. Microbial contamination of the samples*

The conditions evaluated in the present study included oil or biodiesel as RECEIVED (low contamination - Treatment A); oil or biodiesel inoculated with fungi (INOCULATED - Treatment B); and sterile oil or biodiesel (CONTROL). The inoculum was prepared using a consortium of the filamentous fungi isolated from the oils previously, i.e., two isolates of fungi obtained from *J. curcas* and 6 fungal isolates obtained from *A. aculeata* pulp oil samples. The spore suspensions of the fungi were prepared from 7-d old cultures on inclined malt extract agar. In order to facilitate spore dispersion, 2 ml of sterile saline (NaCl 8.5 $gL^{-1}$) and 2 ml of a surfactant (Tween 80, prepared at a concentration of 0.01%) were added to the tubes containing the fungal growth. The resultant solution was added to 10 ml of the mineral medium in an Erlenmeyer flask and was kept under stirring at 120 rpm at 30 °C for 24 h. Counting of the spores of both inoculums was carried out in a Neubauer chamber to obtain a suspension containing of $10^7$ spores $mL^{-1}$. An aliquot of this suspension was added to the flasks corresponding to Treatment B (with inoculum). All flasks were incubated at 30 °C without stirring.

*2.3.2. Conducted analyses*

Sampling was conducted on 0, 7, 14, 21, and 30 d after incubation. At each sample time, the contents of the vials were added to a 250 ml separation funnel, where the separation of the oil phase (*A. aculeata* pulp oil, *J. curcas* oil, and their respective biodiesels), aqueous phase (mineral minimal medium), and the biomass formed in oil-water interface was performed.

Subsequently, the analyses carried out on the aqueous phase at every sampling time included determination of pH and surface tension. To measure the surface tension, the aqueous phase was used in the absence of biomass. The surface tension values were determined on a digital surface tension meter (Gilbertini, Milan, Italy) using the Wilhelmy plate method. For this measurement, about 10 ml of the aqueous phase were used. Liquid distilled water (72.0 mN $m^{-1}$) and ethanol (24.0 mN $m^{-1}$) were used as the standards for instrument calibration.

The biomass formed in the oil / water interface was also evaluated at each time after phase separation, when it was filtered in a vacuum system by using filter paper discs; weight 80, thickness 205 μm, and porosity 14 mM (J. Prolab, Curitiba, Brazil), with the addition of hexane (Grade PA). After oven drying (to constant weight), the biomass was weighed on a precision balance.

At the sampling times of 0 and 30 d, *A. aculeata* pulp oil, *J. curcas* oil, and their respective biodiesel samples were evaluated in terms of their constituent chemical structures using gas chromatography with a flame ionization detector (GC-FID) conducted in the Analytical Chemistry Laboratory of UFRGS, according to the EN 14103: 2011 method. The acid value of the oil and biodiesel samples was also determined according to the ABNT NBR 14448 (2005) method.

*2.4. Statistical analysis*

The data related to the biomass, surface tension, and pH measurements were statistically analyzed using Statistica 10.0. The variance analysis and the Tukey's test were performed at 5% significance level to verify the differences among the treatments against time.

## 3. Results and Discussion

*3.1. Microbial prospection*

The isolation of cultivable native microbiota was performed from both oil samples and the microorganisms isolated were divided according to their taxonomic group,i.e., fungi or bacteria. Overall, a total of 23 and 13 bacteria were isolated from the *A. aculeata* pulp oil and the *J. curcas* oil, respectively. The isolated filamentous fungi from *J. curcas* oil included *Monascus ruber* and *Penicillium chrysogenum* while *Monascus ruber*, *Penicillium chermesinum*, *Monascus pilosus*, *Pestalotiopsis vismiae*, *Penicillium rubens*, and *Penicillium chrysogenum* were isolated from the *A. aculeata* pulp oil.

*3.2. Susceptibility to microbial contamination*

*3.2.1. Microbial growth*

*A. aculeata* pulp oil and *J. curcas* oil as candidates for biodiesel production have been well explored by the previous studies (Raspe et al., 2014; Souza et al., 2016), but little has been reported on how their stability under storage condition is affected by microbial growth. It should be mentioned that some other biodiesel feedstocks such as soybean oil, olive oil, linseed oil, and beef tallow have been investigated in terms of their susceptibility to microbial contamination during storage (Cazarolli et al., 2012; Cazarolli et al., 2014; Soriano et al., 2015). However, based on their physicochemical properties, the biodiesel produced may have different characteristics which can directly influence the degradation reactions of abiotic or biotic nature during storage (Ramos et al., 2009). In line with this, the susceptibility of *A. aculeata* pulp oil, *J. curcas* oil, and their respective biodiesels to microbial contamination during simulated storage period is reported herein for the first time.

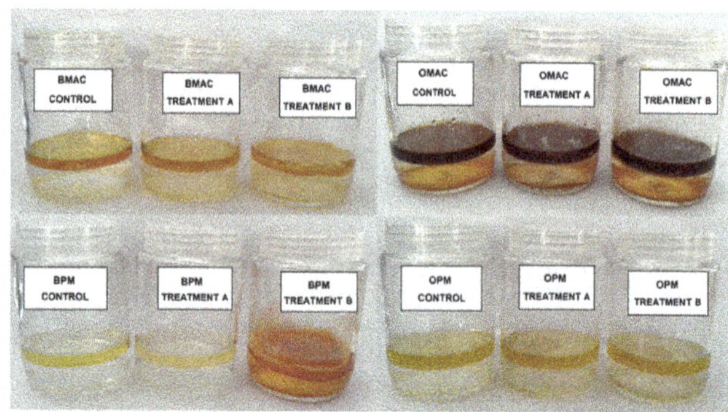


**Fig.2.** The appearance of the bottles used in the experiment after 30 d of incubation. (Captions: BMAC: *A. aculeata* pulp Biodiesel; BPM: *J. curcas* Biodiesel; OMAC: *A. aculeata* pulp oil; OPM: *J. curcas* oil; A: condition as received; B: inoculated with fungi, and CTE: sterile condition, control).

During the 30 days of growth, it was observed that all carbon sources (oils and biodiesels) well supported the growth of the microorganisms in all studied treatments. The values and characteristics of the total biomass formed in the oil / water interfaces (**Fig. 2**) are shown in **Table 1**. More biomass was generally formed in Treatment B; more specifically, in the samples containing *A. aculeata* pulp and *J. curcas* biodiesels (315.0 ± 5.1 and 235.3 ± 5.3 mg, respectively), indicating that the microorganisms which use the oil for growth can also, in case of possible contaminations during the transesterification process, utilize biodiesel as a source of carbon.

It is worth highlighting that the survival and growth of microorganisms in oil and/or biodiesel is related to their metabolic competence in accessing the carbon present in the structure of these substrates. In fact, the presence of fatty acids and esters in the environment would favor the growth of a portion of the indigenous microbial community with different capabilities in biodegrading such compounds as a carbon source, while promoting co-metabolism which could result in the persistence of microorganisms (Pasqualino et al., 2006; Kostka et al., 2011; Cruz et al, 2014).

The results obtained herein showed that in Treatment A, which included only native microbiota, i.e., microorganisms naturally present in the samples, the highest biomass values were found in *A. aculeata* pulp and *J. curcas* oil samples (128.3 ± 9.4 mg and 86.7 ± 7.7 mg, respectively), indicating that the native microbiota has a substantial deteriogenic capacity using the oils as a carbon source. It should also be noted that a microbiota capable of developing at the expense of fatty acid esters might also be added to the oil feedstock or the resultant biodiesel during shipping and handling.

Moreover, the results observed for biodiesel samples in Treatment A revealed that the transesterification process may have contributed to a reduction in the microbial population/growth of oil samples leading to lower microbial presence in biodiesel samples (**Table 1**). After the 30-d experiment, the final biomass values between biodiesels were not significantly different (p>0,05).

**Table 1.**
Comparison of different the oil and biodiesel samples in terms of pH, surface tension, and total biomass formed in the oil-water interface after 30 d of incubation. In each column, values with the same superscript symbol are not statistically different by Tukey test (p> 0.05).

| Oil/biodiesel sample | Treatment | pH * | Surface Tension (mN m$^{-1}$) ** | Total Dry Weight (mg) |
|---|---|---|---|---|
| | Control | 6,96±0,01$^{\#}$ | 19,5±0,86$^{‡}$ | - |
| *Acrocomia aculeata* **pulp Biodiesel** | Treatment A | 6,25±0,69$^{\#}$ | 21,2±0,60$^{‡}$ | 78,8±8,6$^{‡}$ |
| | Treatment B | 5,76±0,15$^{£}$ | 25,2±0,96$^{£}$ | 315,0±5,1$^{£}$ |
| | Control | 6,92±0,01$^{\#}$ | 25,6±0,79$^{§}$ | - |
| *Jatropha curcas* **Biodiesel** | Treatment A | 6,71±0,18$^{§}$ | 24,7±0,52$^{§}$ | 71,0±4,9$^{‡}$ |
| | Treatment B | 6,32±0,03$^{§}$ | 25,8±0,83$^{§}$ | 235,3±5,3$^{•}$ |
| | Control | 5,94±0,04$^{•}$ | 23,9±1,00$^{•□}$ | - |
| *Acrocomia aculeata* **pulp Oil** | Treatment A | 5,88±0,01$^{▪}$ | 25,6±0,95$^{•}$ | 128,3±9,4$^{\#}$ |
| | Treatment B | 5,66±0,05$^{□}$ | 23,1±0,49$^{□}$ | 218,8±6,2$^{•}$ |
| | Control | 6,78±0,12$^{▼}$ | 22,0±0,95$^{▼}$ | - |
| *Jatropha curcas* **Oil** | Treatment A | 6,73±0,06$^{▼}$ | 22,2±0,41$^{▼}$ | 86,7±7,7$^{‡}$ |
| | Treatment B | 6,71±0,04$^{▼}$ | 19,5±0,87$^{€}$ | 98,8±6,6$^{‡}$ |

* pH Initial value: 7.2
** The initial surface tension value of the mineral medium: 54.2 mN m$^{-1}$.

### 3.2.2. pH variations

The pH values measured during the storage simulation period are shown in **Table 1**. As presented, in the growth assays using *A. aculeata* pulp biodiesel, a significant reduction in pH was only observed in Treatment B, when compared with the initial pH value of the aqueous phase (i.e., 7.2) (p<0.05). In this treatment, a fungal inoculum was added, which was likely to have contributed to the observed results. No significant changes were observed in other treatments with *A. aculeata* pulp biodiesel as the oily phase (p> 0.05).

In case of the *J. curcas* biodiesel, a significant decrease in pH was also observed in both Treatments A and B (p <0.05), whereas the pH value of the control sample did not change significantly over the 30-d storage period (p> 0, 05), suggesting that in this case the decrease in pH may be ascribed to the observed microbial growth. The pH values measured for the *A. aculeata* pulp oil were significantly reduced compared with the baseline value (p <0.05). Since reductions were observed in all treatments, it could be concluded that these could be related to the *A. aculeata* pulp oil itself.

During the microbial degradation of hydrocarbons various compounds are produced, including organic acids (Bento et al., 2005; Callaghan, 2013). The mineral minimum medium developed by Bushnell and Haas (1941), which includes phosphate in its formulation, is generally used as a nutrient source in biodiesel biodegradation experiments (Bücker et al., 2011; Cazarolli et al., 2014). Phosphate is a limiting factor to microbial growth while it is also considered as an important buffering agent, an effect which is advantageous for maintaining the action of certain enzymes. However, possible changes in pH may not be detected due to low production of acidic metabolites by the microorganisms under study. These changes might have been missed during the short incubation period used in the present investigation. Polar organic molecules that might have partitioned into the aqueous phase are also likely to have functioned as buffers. Overall, the changes in pH in the aqueous phase could indicate the production of metabolites of acidic nature in quantities sufficient to overcome the buffering capacity of the mineral minimum medium. Beside the metabolic products formed during the fungal development, a reduction in the pH as a result of the treatments performed could be related to cellular lysis, polymeric products, and production of organic acids (Bento et al., 2005; Goldberg et al., 2006). Thus, one could conclude that by prolonging the cultivation period and without any limitations on oxygen and nutrients, the fungal isolates used in this work could have produced more metabolites which could have in turn led to a greater reduction in pH.

The surface tension values of the aqueous phase decreased shortly after 7 d (p <0.05), and these values remained unchanged until the end of the incubation period. This reduction was also observed in the control treatment, indicating that it might have been caused by the nature of the carbon chains present in the oils and biodiesels. In fact, due to their chemical structure, which is similar to the structure of some types of surfactants, the compounds present in the samples (i.e., esters and fatty acids) can act as surfactants in the presence of water and reduce the surface tension (Bücker et al., 2011; Cazarolli et al., 2012). Cazarolli et al. (2014) confirmed the presence of long-chain fatty acid esters by chromatography in a mineral minimum medium used as the aqueous phase in their experiment with biodiesel. Their results showed that the biodiesel components could migrate into the aqueous phase of the experiment. This well explains the reductions observed herein in the surface tension both in the control as well as in the other treatments (**Table 1**).

In a study by Kaczorek et al. (2011), they isolated *Pseudomonas stutzeri* from soils contaminated with crude oil and incubated the isolated with diesel oil for 14 d. After the incubation period, a degradation rate of 88% was observed in the treatment which received 120 mg of surfactant, in contrast to the treatment where no surfactant was added (54% degradation). In the present study, it was not possible to detect the production of surfactants by microorganisms but biodiesel may have had a surfactant effect on the mineral medium used. Considering the fact that biodiesel can perform a surfactant function which can spontaneously increase its bioavailability, it could be concluded that this attribute is a determining factor in its greater degradability when compared with diesel oil.

### 3.2.3. Acid value variations

The variations in acid values in response to prolonged storage and the treatments performed are shown in **Table 2**. Overall, the initial acid value of the oils ($T_0$) were considerably different from those reported in the literature for some other oil feedstocks. For instance, in a study aimed at biodiesel production from soybean oil, Cunha (2008) determined the acidity of the oil at 0.104 mg KOH g$^{-1}$. After the 30 d experiment ($T_{30}$), the acid values decreased for both oils with native microbiota (Treatment A). The results observed in biodiesel samples were statistically indistinguishable among the samples at the beginning and after the incubation period (p>0.05). It should also be noted that it was not possible to determine the acid values for all the samples and for all the treatments.

**Table 2.**
Acid value (mg KOH g$^{-1}$) of the oil and biodiesel samples at the beginning of the experiment ($T_0$) and after 30 d of incubation ($T_{30}$). In each row, values with the same superscript symbol are not statistically different by Tukey test (p> 0.05).

| Oil/biodiesel sample | $T_0$ Values (mg KOH g$^{-1}$) | $T_{30}$ Values (mg KOH g$^{-1}$) | | |
|---|---|---|---|---|
| | | Control | Treatment A | Treatment B |
| *Acrocomia aculeata* **pulp Biodiesel** | 0.455±0.004$^{▼♯}$ | 0.316±0.14$^{▼♯▪}$ | 0.354±0.16$^{▼♯▪}$ | 0.523±0.06$^{▼}$ |
| *Jatropha curcas* **Biodiesel** | 0.182±0.003$^{▪}$ | 0.360±0.09$^{▼♯▪}$ | -$^{a}$ | 0.360±0.14$^{▼♯▪}$ |
| *Acrocomia aculeata* **pulp Oil** | 2.582±0.062$^{•}$ | 0.375$^{b▼♯▪}$ | -$^{a}$ | 0.275$^{b♯▪}$ |
| *Jatropha curcas* **Oil** | 11.375±0.019* | 0.305±0.05$^{▼♯▪}$ | 0.225$^{b▪}$ | -$^{a}$ |

[a] Results were not conclusive; the turning point was not found.
[b] Results were not conclusive; the turning point was not found for all replicates.

According to Silveira (2014), biodiesel properties such as kinematic viscosity, acidity, triglycerides, diglycerides and monoglycerides content, free and total glycerol contents, and ester content are dependent on the transesterification process. Other parameters such as oxidation stability, iodine index, carbon residue, and cold filter plugging point depend on the nature of the oil used for the production of biodiesel. The acid value refers to the amount of free fatty acids in a given sample and is linked to aging and degradation when stored. Furthermore, the possibility of corrosion is indicated as a result of high acid values (Shalaby, 2015). Raw materials with high acidity values require a pretreatment before transesterification with basic catalysts because they can affect the yield and reaction time as well as the results of the biodiesel purification step (Tiwari et al., 2007).

Silveira (2014) and Nascimento et al. (2016) also produced *A. aculeata* and *J. curcas* biodiesels in different studies using the alkaline transesterification method, respectively, and achieved 97.8% of esters for *A. aculeata* biodiesel and 98.5% of esters for *J. curcas* biodiesel. Accordingly, the *J. curcas* and *A. aculeata* pulp oils were found to have different characteristics, such as fatty acid composition and acidity index, and the alkaline transesterification reaction was found effective in framing the biodiesels with the standard parameters required by the Resolution N° 14 of the National Agency of petroleum, natural gas, and biofuels (ANP) (ANP, 2012; Silveira, 2014; Nascimento et al., 2016).

**Table 3.**
Ester contents of *J. curcas* Biodiesel at $T_0$ and $T_{30}$. In each column, values with the same superscript symbol are not statistically different by Tukey test ($p > 0.05$).

| Esters (%) | *Jatropha curcas* Biodiesel | | | |
|---|---|---|---|---|
| | $T_0$ | $T_{30}$ | | |
| | | Control | Treatment A | Treatment B |
| $C_{12}$ | - | - | - | - |
| $C_{13}$ | - | - | - | - |
| $C_{14:1}$ | - | - | - | - |
| $C_{16}$ | 0,04±0,00* | 0,03±0,02* | 0,00±0,00$^{\#}$ | 0,00±0,00$^{\#}$ |
| $C_{16:1}$ | - | - | - | - |
| $C_{17}$ | - | - | - | - |
| $C_{18}$ | - | - | - | - |
| $C_{18:1}$ | 13,46±0,03* | 13,32±0,03* | 13,88±6,25* | 11,37±0,72* |
| $C_{18:2}$ | 0,65±0,00* | 0,65±0,01* | 0,66±0,30* | 0,51±0,04* |
| $C_{20}$ | 6,12±0,03* | 5,80±0,04* | 6,19±2,94* | 6,33±0,14* |
| $C_{20:1}$ | 44,43±0,08* | 44,37±0,10* | 45,65±4,38* | 45,03±0,29* |
| $C_{18:3}$ | 34,70±0,07* | 34,68±0,17* | 32,88±1,14* | 35,62±0,47* |
| $C_{22}$ | 0,25±0,08* | 0,53±0,08* | 0,50±0,28* | 0,51±0,07* |
| $C_{22:1}$ | 0,19±0,00* | 0,19±0,00* | 0,20±0,09* | 0,22±0,01* |
| $C_{23}$ | 0,06±0,00* | 0,06±0,00* | 0,06±0,03* | 0,06±0,00* |
| $C_{24:1}$ | 0,08±0,09* | 0,350,08* | 0,32±0,32* | 0,35±0,07* |
| Esters Total | 95,77±0,17*$^{\#}$ | 99,82±8,63* | 98,78±6,85* | 83,45±3,28$^{\#}$ |

### 3.2.4. Biodiesel biodegradation

The results obtained on the biodegradation of biodiesel samples during the storage period are tabulated in **Tables 3** and **4**. According to the total esters values, none of the biodiesels met the requirements set by the Resolution N° 14 of the ANP (ANP, 2012) even before they were subjected to biodeteriogenic activity. For *J. curcas* biodiesel, the total esters at $T_0$ was close to the standard total esters limit of 96.5%, but could still not be released for consumption. Similarly, based on the total esters value measured by gas chromatography analysis for *A. aculeata* pulp biodiesel, its commercialization would not be allowed according to Resolution N° 14 of the ANP (ANP, 2012).

**Table 4.**
Ester contents of *A. aculeata* pulp Biodiesel at $T_0$ and $T_{30}$. In each column, values with the same superscript symbol are not statistically different by Tukey test ($p > 0.05$).

| Esters (%) | *Acrocomia aculeata* pulp Biodiesel | | | |
|---|---|---|---|---|
| | $T_0$ | $T_{30}$ | | |
| | | Control | Treatment A | Treatment B |
| $C_{12}$ | 0,14±0,00* | 0,14±0,01* | 0,17±0,02* | 0,07±0,06$^{\#}$ |
| $C_{13}$ | 0,27±0,01* | 0,29±0,02* | 0,32±0,04* | 0,12±0.02$^{\#}$ |
| $C_{14:1}$ | 2,03±0,04* | 2,09±0,15* | 2,59±0,55* | 1,82±0,17* |
| $C_{16}$ | 1,77±0,03*$^{\#}$ | 1,50±0,10* | 2,28±0,50$^{\#}$ | 1,81±0,09*$^{\#}$ |
| $C_{16:1}$ | 0,14±0,00*$^{\#}$ | 0,10±0,00* | 0,17±0,04$^{\#}$ | 0,13±0,01*$^{\#}$ |
| $C_{17}$ | 0,15±0,00* | 0,11±0,01* | 0,21±0,08* | 0,16±0,00* |
| $C_{18}$ | 0,23±0,01* | 0,17±0,01$^{\#}$ | 0,26±0,03* | 0,25±0,01* |
| $C_{18:1}$ | 23,64±0,25* | 24,53±1,59* | 22,93±2,98* | 23,23±0,65* |
| $C_{18:2}$ | 3,12±0,08* | 2,88±0,41* | 3,42±0,43* | 3,30±0,24* |
| $C_{20}$ | 9,92±0,14* | 9,56±2,72* | 12,70±1,70* | 11,18±0,41* |
| $C_{20:1}$ | 52,40±0,50* | 51,51±1,61* | 49,90±1,23* | 50,86±0,91* |
| $C_{18:3}$ | 4,42±0,07* | 5,17±0,36* | 2,42±1,98* | 4,54±0,17* |
| $C_{22}$ | 0,37±0,47* | 0,95±0,03*$^{\#}$ | 1,42±0,34$^{\#\&}$ | 1,97±0,08$^{\&}$ |
| $C_{22:1}$ | 0,27±0,02* | 0,28±0,01* | 0,37±0,11* | 0,32±0,01* |
| $C_{23}$ | 0,60±0,02* | 0,66±0,04* | 0,77±0,18* | 0,00±0,00$^{\#}$ |
| $C_{24:1}$ | 0,51±0,44* | 0,04±0,03* | 0,04±0,03* | 0,22±0,38* |
| Esters Total | 81,44±0,89* | 81,38±2,17* | 82,69±0,67* | 73,90±2,98$^{\#}$ |

After the transesterification of the *A. aculeata* pulp oil in the present study, the esters percentages of methyl oleate (C18.1), methyl linoleate (C18:2), and methyl palmitate (C16) were different (23.64±0,25%; 3.12±0.08%, and 1.77±0,03%, respectively) from those reported by Amaral et al. (2011), who also analyzed the fatty acid composition of *A. aculeata* pulp oil (69.07% for oleic acid (C18.1), 6,77% for linoleic acid (C18.2), and 12.13% for palmitic acid (C16)). Similar observations indicating the predominance of the C18.1 esters have been reported by other studies conducted on *A. aculeata* pulp oil (Duarte et al., 2012; Melo, 2012). In the *A. aculeata* pulp biodiesel evaluated in this study, some esters of smaller chains (i.e., C12, C13, etc.) were also detected. The most predominant ester found in the sample, however, was the eicosenoate methyl ester (C20.1) (52,40±0,50%), in contrast to the literature (Doná, 2012). Studies using the *J. curcas* oil as raw material found oleic and linoleic acids as the major constituents of the oil making up between 70-77% of the oil composition (Ribeiro et al., 2012). In the present study, the major esters produced from *J. curcas* oil were linolenate methyl ester (C18.3) and eicosenoate methyl ester (C20.1), totaling 75% of the samples' esters.

It should be noted that the oil content and fatty acid composition of oil seeds depend on the plant species and cooul be altered by their maturation time, the cultivar, and their growth conditions (Atabani et al., 2013; Schulte et al., 2013). The oil samples used for the production of biodiesel in the present investigation remained in storage prior to the study for four months at temperatures ranging from 26 - 28 °C, which may have contributed to the differences observed in the esters content of these samples (Silveira, 2014).

The total value of the esters detected in Treatment B decreased significantly after 30 d for *J. curcas* biodiesel ($p<0.05$). A reduction of approximately 12±2.2% in the total ester content was observed compared with the content measured at the beginning of the experiment. No significant changes between the initial and ending time esters contents were detected for the Control treatment and biodiesel as received (Treatment A) ($p>0.05$). These findings indicated that the fungi inoculum could be considered responsible for the reductions observed in the total esters contents. More specifically, in Treatment B, the *J. curcas* biodiesel promoted the second largest biomass formed, which was consistent with the total biodegradation percentage observed. When analyzing each ester individually, only palmitic ester showed a significant reduction, although in small percentages ($p<0,05$). For *A. aculeata* pulp biodiesel, total esters percentage showed a reduction of 8±2.1% after 30 d in Treatment B. Moreover, a statistically-significant reduction was observed in the contents of the esters with lower carbon chains (C12 to C13), minor constituents of the sample though ($p<0.05$). The impurities (non-esters components) of the *A. aculeata* pulp and *J. curcas* biodiesels investigated herein stood at 18,55±0,88% and 4,15±0,15% of, respectively. Therefore, it could be concluded that these impurities, such as glycerol, or mono, di- and triacylglycerides (resulted from low transesterification yield and/or a poor purification process), might have been consumed by fungi and thus could have contributed to the increased susceptibility of biodiesel to microbial contamination (Bücker et al., 2011; Cazarolli et al., 2014).

Vieira et al. (2006) observed the degradation of palm oil by an unidentified microbial isolate and obtained the following degradation profile of methyl esters; myristic acid by 31.67%, palmitic acid by 51.28%, stearic acid by 34.11%, oleic acid by 33.31%, and linoleic acid by 25.16%. These results suggested a bacterial specificity for different methyl esters constituents of palm oil biodiesel. On the contrary, in the present work, it was not possible to observe such degradation ratios for each ester, confirming that other biofuel elements may have been consumed, or the time of incubation experiment was too short to observe such results.

After the 30 d experiment, the highest microbial biomass was obtained in both oil samples in Treatment A. This suggested that the carbon source present as fatty acids and other oil components were more easily assimilated by the native microbiota, including the non-cultivable microorganisms.

It has been reported that several processes including oxidation under aerobic conditions, thermal decomposition by excessive heat, as well as contaminations and impurities can modify fuels properties, significantly decreasing their stability and, therefore, their durability during storage (Yaakob et al., 2014). Furthermore, when water is present in storage systems, hydrolysis reactions may also occur, resulting in the release of free fatty acids, which in turn increases the acidity leading to the instability of biodiesel (Vieira et al., 2006).

It should also be noted that oil feedstocks with high acidity values which also have high unsaturation degrees are less stable during storage periods, and therefore, would be more susceptible to oxidative reactions induced chemically or biologically (Pullen and Saeed, 2012). In the present work, the *A. aculeata* pulp biodiesel had a higher percentage of saturated fatty acid esters compared with *J. curcas* biodiesel (13.45% *vs.* 6.47%, respectively), and this could explain the difference observed in the biodegradation rates (8±2.1% *vs.* 12±2.2%, respectively).

Several vegetable oil feedstocks, extremely adapted to the Brazilian climate and soil, have been considered for biodiesel production in Brazil including palm oil, coconut oil, *Orbignya* sp., peanut, canola, turnip, *J. curcas*, *A. aculeata*, and almonds. As previously mentioned, diversification of raw materials could make the production process less onerous and could further encourage the use of biofuels such as biodiesel. Among these resources, *A. aculeata* pulp oil and *J. curcas* oil, due to their desirable characteristics suitable for biodiesel production, are of great importance. Moreover, when compared with the other oleaginous, these plants have high productivities and high oil yields, and the production of biodiesel from them is already well established (Nascimento et al., 2014; Cesar et al., 2015). Nevertheless. studies on the chemical and biological stability of the biodiesel produced during storage were scarce and to the best of our knowledge, this was the first report shedding light on the storage stability of these oil feedstocks and their respective biodiesels. Future investigations are required to further compare the stability of these oil feedstocks with the other resources such as soybean oil and beef tallow, the major constituents of biodiesel production in Brazil today.

## 4. Conclusions

The *J. curcas* and *A. aculeata* pulp oil samples were evaluated for their cultivable microbial populations. According to the fungal growth results, the stored oils and biodiesels were potentially susceptible to microbial degradation. At the end of 30 days of simulated storage, the biomass formed (mg) in the samples indicated a predisposition order to microbial growth under the conditions of this study as follows: *A. aculeata* pulp biodiesel> *J. curcas* biodiesel> *A. aculeata* pulp Oil> *J. curcas* Oil. When *A. aculeata* pulp biodiesel was inoculated with the fungal consortium, the biomass formed was measured at 315 ± 5.1 mg and a 8±2.1% reduction in the total esters was observed. Moreover, *J. curcas* biodiesel showed a 12±2.2% degradation in total esters in response to the fungal inoculation where a considerable microbial biomass was formed (235.3 ± 5.3 mg). The results obtained herein revealed that the investigated biodiesels deteriorated during storage in the presence of microbial communities even over the short period of 30 d. Longer test times at pilot scale could help to better understand the susceptibility of these biodiesel feedstock candidates and to develop microbial growth control methods in order to increase their shelf life during storage.

## Acknowledgements

The authors would like to thank LAB-BIO/UFRGS, CECOM/UFRGS, INT (CNPq Grant N° 558896/2010-3), and CNPq (Process N ° 381691/2013-6) for the funds provided during the course of this study.

## References

[1] ABNT NBR 14448, 2005. Available at http://www.abntcatalogo.com.br (accessed on 11 November 2016).

[2] Amaral, F.P.D., Broetto, F., Batistela, C.B., Jorge, S.M.A., 2011. Extração e caracterização qualitativa do óleo da polpa e amêndoas de frutos de macaúba [*Acrocomia aculeata* (Jacq) Lodd. ex Mart] coletada na região de Botucatu-SP. Energia na Agricultura. 12-20.

[3] Amaral, F.P., 2007. Study of physico-chemical characteristics of almond oil and macaúba pulp [*Acrocomia aculeata* (Jacq.) Lodd. ex Mart]. Master of Science Dissertation, Faculdade de Ciências Agronômicas da Unesp, Brasil.

[4] ANP Resolutions, National Agency of Petroleum, 2012. Available at http://www.anp.gov.br (accessed on 11 November 2016).

[5] ANP, National Agency of Petroleum, Statistical Yearbook, 2015. Available at http://www.anp.gov.br (accessed on 11 November 2016).

[6] Atabani, A.E., Silitonga, A.S., Ong, H.C., Mahlia, T.M.I., Masjuki, H.H., Badruddin, I.A., Fayaz, H., 2013. Non-edible vegetable oils: a critical evaluation of oil extraction, fatty acid compositions, biodiesel production, characteristics, engine performance and emissions production. Renew. Sust. Energy Rev.18, 211-245.

[7] Bento, F.M., Beech, I.B., Gaylarde, C.C., Englert, G.E., Muller, I.L., 2005. Degradation and corrosive activities of fungi in a diesel-mild steel-aqueous system. World J. Microbiol. Biotechnol. 21(2), 135-142.

[8] Bücker, F., Santestevan, N.A., Roesch, L.F., Jacques, R.J.S., Peralba, M.D.C.R., Camargo, F.A., Bento, F.M., 2011. Impact of biodiesel on biodeterioration of stored Brazilian diesel oil. Int. Biodeterior. Biodegrad. 65(1), 172-178.

[9] Bushnell, L. D., Haas, H. F., 1941. The utilization of certain hydrocarbons by microorganisms. J. Bacteriol. 41, 653-673.

[10] Callaghan, A.V., 2013. Metabolomic investigations of anaerobic hydrocarbon-impacted environments. Curr. Opin. Biotechnol. 24(3), 506-515.

[11] Cazarolli, J.C., Bücker, F., Manique, M.C., Krause, L.C., Maciel, G.P.D.S., Onorevoli, B., Caramão, E.B., Cavalcanti, E.H.D.S., Samios, D., Peralba, M.D.C.R., Bento, F.M., 2012. Suscetibilidade do biodiesel de sebo bovino à biodegradação por Pseudallescheria boydii. Rev. Bras. de Bioc. 10(3), 251-257.

[12] Cazarolli, J.C., Guzatto, R., Samios, D., Peralba, M.D.C.R., Cavalcanti, E.H.D.S., Bento, F.M., 2014. Susceptibility of linseed,

soybean, and olive biodiesel to growth of the deteriogenic fungus *Pseudallescheria boydii*. Int. Biodeterior. Biodegrad. 95, 364-372.

[13] da Silva César, A., de Azedias Almeida, F., de Souza, R.P., Silva, G.C., Atabani, A.E., 2015. The prospects of using *Acrocomia aculeata (macaúba)* a non-edible biodiesel feedstock in Brazil. Renew. Sust. Energy Reviews. 49, 1213-1220.

[14] Covino, S., D'Annibale, A., Stazi, S.R., Cajthaml, T., Čvančarová, M., Stella, T., Petruccioli, M., 2015. Assessment of degradation potential of aliphatic hydrocarbons by autochthonous filamentous fungi from a historically polluted clay soil. Sci. Total Environ. 505, 545-554.

[15] Cruz, J.M., Tamada, I.S., Lopes, P.R.M., Montagnolli, R.N., Bidoia, E.D., 2014. Biodegradation and phytotoxicity of biodiesel, diesel, and petroleum in soil. Water Air Soil Pollut. 225(5), 1-9.

[16] Cunha, E. M., 2008. Characterization of biodiesel produced from mixtures of raw materials: bovine tallow, chicken oil and soybean oil. Master of Science Dissertation, Universidade Federal do Rio Grande do Sul - UFRGS, Brasil.

[17] Doná, G., 2012. Production of methyl esters with ethyl acetate supercritical tubular reactor. Master of Science Dissertation, Universidade Federal do Paraná, Brasil.

[18] Duarte, I. D., Rogério, J. B., Licurgo, F. M. S., Back, G. R., Santos, M. C. S., Antoniassi, R., Faria-Machado, A. F., Bizzo, H. R., Junqueira, N. T. V., 2012. Efeito da maturação de frutos de macaúba no rendimento de óleo e na composição em ácidos graxos. Resumos do 5° Congresso da Rede Brasileira de Tecnologia de Biodiesel e 8° Congresso Brasileiro de Plantas Oleaginosas, Óleos, Gorduras e Biodiesel, Salvador, Brasil, 1, 253-254.

[19] Ferreira, M.E., Grattapaglia, D., 1996. Introdução ao uso de marcadores moleculares em análise genética. EMBRAPA-CENARGEN.

[20] Goldberg, I., Rokem, J.S., Pines, O., 2006. Organic acids: old metabolites, new themes. J. Chem. Technol. Biotechnol. 81(10), 1601-1611.

[21] Kaczorek, E., Olszanowski, A., 2011. Uptake of hydrocarbon by *Pseudomonas fluorescens* (P1) and *Pseudomonas putida* (K1) strains in the presence of surfactants: a cell surface modification. Water Air Soil Pollut. 214(1), 451-459.

[22] Kaushik, S., Kumar, M., Thakur, S., Chhabra, M., Aggarwal, K.M., Tyagi, R.K., 2015. Biodiesel production and its performance characteristics measurement: a review and analysis. J. Chem. Pharm. Res. 7(5), 1075-1082.

[23] Kostka, J.E., Prakash, O., Overholt, W.A., Green, S.J., Freyer, G., Canion, A., Delgardio, J., Norton, N., Hazen, T.C., Huettel, M., 2011. Hydrocarbon-degrading bacteria and the bacterial community response in gulf of mexico beach sands impacted by the deepwater horizon oil spill. Appl. Environ. Microbiol. 77(22), 7962-7974.

[24] Lorenzi, G.M.A.C., 2006. *Acrocomia aculeata* (Jacq.) Lodd. Mart. ex-Arecaceae: foundations for sustainable harvesting. Doctoral thesis, Universidade Federal do Paraná, Brasil.

[25] Mariano, A.P., Tomasella, R.C., Oliveira, L.M., Contiero, J., Angelis, D.F., 2008. Biodegradability of diesel and biodiesel blends. Afr. J. Biotechnol. 7(9). 1323-1328.

[26] Melo, P.G., 2012. Production and characterization of biodiesel obtained from the oilseed Macaúba (*Acrocomia aculeata*). Doctoral thesis, Universidade Federal de Uberlândia, Brasil.

[27] Meneghetti, S.M.P., Meneghetti, M.R., Brito, Y.C., 2013. A Reação de Transesterificação, Algumas Aplicações e Obtenção de Biodiesel. Rev. Virt. Quím. 5(1), 63-73.

[28] Mofijur, M., Masjuki, H.H., Kalam, M.A., Atabani, A.E., 2013. Evaluation of biodiesel blending, engine performance and emissions characteristics of *Jatropha curcas* methyl ester: Malaysian perspective. Energy. 55, 879-887.

[29] Nascimento, M. R. F., 2014. Production and shelf life of biodiesel from Pinhão Manso oil (*Jatropha curcas*) obtained by esterification followed transesterification. Masters Dissertation, Universidade Federal do Rio de Janeiro, Brasil.

[30] Pandey, V.C., Singh, K., Singh, J.S., Kumar, A., Singh, B., Singh, R.P., 2012. *Jatropha curcas*: A potential biofuel plant for sustainable environmental development. Renew. Sust. Energy Rev. 16(5), 2870-2883.

[31] Pasqualino, J.C., Montane, D., Salvado, J., 2006. Synergic effects of biodiesel in the biodegradability of fossil-derived fuels. Biomass Bioenergy. 30(10), 874-879.

[32] Pullen, J., Saeed, K., 2012. An overview of biodiesel oxidation stability. Renew. Sust. Energy Rev. 16(8), 5924-5950.

[33] Ramos, M.J., Fernández, C.M., Casas, A., Rodríguez, L., Pérez, Á., 2009. Influence of fatty acid composition of raw materials on biodiesel properties. Bioresour. Technol. 100(1), 261-268.

[34] Raspe, D., Mello, B., Silva, P.J., da Silva, C., 2014. Esterificação Homogênea dos ácidos graxos livres do óleo da polpa de Macaúba (*Acrocomia aculeata*). E-xacta. 7(1), 45-54.

[35] Ribeiro, R. A., Queiroz, M. G. M. N., Alves, V. L., Starling, M. F. R., Cardoso, C. A., 2012. Perfil de ácidos graxos de óleo bruto de pinhão-manso (*Jartropha curcas*) submetido ao armazenamento. Resumos do 5° Congresso da Rede Brasileira de Tecnologia de Biodiesel e 8° Congresso Brasileiro de Plantas Oleaginosas, Óleos, Gorduras e Biodiesel, Salvador, Brasil. 305-306.

[36] Schulte, L.R., Ballard, T., Samarakoon, T., Yao, L., Vadlani, P., Staggenborg, S., Rezac, M., 2013. Increased growing temperature reduces content of polyunsaturated fatty acids in four oilseed crops. Ind. Crops Prod. 51, 212-219.

[37] Shalaby, E.A., 2015. A review of selected non-edible biomass sources as feedstock for biodiesel production. Biofuels - Status and Perspective, InTech, pp. 3-20.

[38] Silveira, S.D., 2014. Production of Biodiesel from Macaúba oil (*Acrocomia aculeata*) via esterification followed methyl transesterification with basic catalyst. Master of Science Dissertation, Universidade Federal do Rio de Janeiro, Brasil.

[39] Soriano, A.U., Martins, L.F., de Assumpção Ventura, E.S., de Landa, F.H.T.G., de Araújo Valoni, É., Faria, F.R.D., Ferreira, R.F., Faller, M.C.K., Valério, R.R., de Assis Leite, D.C., do Carmo, F.L., 2015. Microbiological aspects of biodiesel and biodiesel/diesel blends biodeterioration. Int. Biodeterior. Biodegrad. 99, 102-114.

[40] Souza, G.K., Scheufele, F.B., Pasa, T.L.B., Arroyo, P.A., Pereira, N.C., 2016. Synthesis of ethyl esters from crude macauba oil (*Acrocomia aculeata*) for biodiesel production. Fuel. 165, 360-366.

[41] Souza, V.H.A., Dos Santos, L.T., Campos, A.F., Carolino, J., 2015. Análise do Programa Nacional de Produção e Uso do Biodiesel (PNPB): Resultados e Críticas. Rev. Adm. Geral. 1(1), 23-41.

[42] Teixeira, G.A.A., Queiroz, N., Souza, A. L., Garcia, I. M., Maia, A. S., Souza, A. G., 2010. Avaliação dos parâmetros de qualidade para o biodiesel metílico obtido de misturas de óleos vegetais e gordura animal durante armazenamento. Resumos do 4° Congresso da Rede Brasileira de Tecnologia de Biodiesel, Belo Horizonte, Brasil. 3, 1287-1288.

[43] Tiwari, A.K., Kumar, A., Raheman, H., 2007. Biodiesel production from Jatropha oil (*Jatropha curcas*) with high free fatty acids: an optimized process. Biomass bioenergy. 31(8), 569-575.

[44] Vieira, T.M., Silva, E. P., Antoniosi Filho, N. R., Vieira, J. D. G., 2006. Determinação e quantificação da degradação bacteriana de biodiesel de óleo de palma. Resumos do 1° Congresso da Rede Brasileira de Tecnologia do Biodiesel, Brasília, Brasil. 218-223.

[45] Yaakob, Z., Narayanan, B.N., Padikkaparambil, S., 2006. A review on the oxidation stability of biodiesel. Renew. Sust. Energy Rev. 35, 136-153.

# 19

# Recent updates on biogas production

Ilona Sárvári Horváth[1,*], Meisam Tabatabaei[2,3], Keikhosro Karimi[4,5], Rajeev Kumar[6]

[1] *Swedish Centre for Resource Recovery, University of Borås, 501 90 Borås, Sweden.*

[2] *Microbial Biotechnology Department, Agricultural Biotechnology Research Institute of Iran (ABRII), AREEO, Karaj, Iran.*

[3] *Biofuel Research Team (BRTeam), Karaj, Iran.*

[4] *Department of Chemical Engineering, Isfahan University of Technology, Isfahan 84156-83111, Iran.*

[5] *Microbial Industrial Biotechnology Group, Institute of Biotechnology and Bioengineering, Isfahan University of Technology, Isfahan 84156-83111, Iran.*

[6] *Center for Environmental Research and Technology (CE-CERT), Bourns College of Engineering, University of California, Riverside, California, USA.*

## HIGHLIGHTS

- Biogas; a promising renewable alternative for natural gas with similar applications.
- Biogas can be produced from different types of organic wastes.
- AD process is accompanied with several environmental advantages compared with incineration, landfilling, and composting.
- Besides energy, AD process generated a nutrient-rich biological fertilizer.
- Recent developments in metagenomics techniques have provided valuable tools to achieve improved AD process.

## GRAPHICAL ABSTRACT

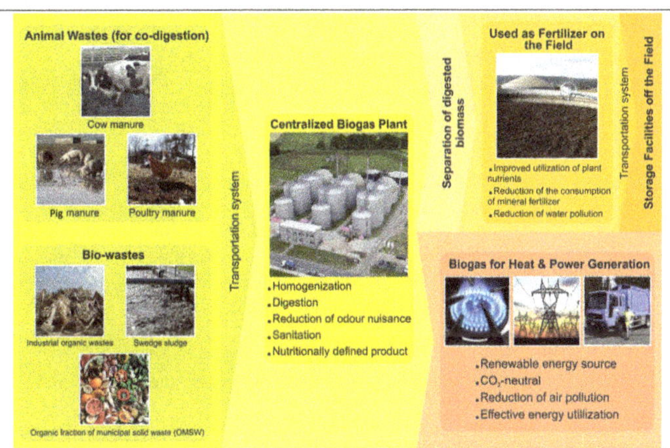


## ABSTRACT

One of the greatest challenges facing the societies now and in the future is the reduction of green house gas emissions and thus preventing the climate change. It is therefore important to replace fossil fuels with renewable sources, such as biogas. Biogas can be produced from various organic waste streams or as a byproduct from industrial processes. Beside energy production, the degradation of organic waste through anaerobic digestion offers other advantages, such as the prevention of odor release and the decrease of pathogens. Moreover, the nutrient rich digested residues can be utilized as fertilizer for recycling the nutrients back to the fields. However, the amount of organic materials currently available for biogas production is limited and new substrates as well as new effective technologies are therefore needed to facilitate the growth of the biogas industry all over the world. Hence, major developments have been made during the last decades regarding the utilization of lignocellulosic biomass, the development of high rate systems, and the application of membrane technologies within the anaerobic digestion process in order to overcome the shortcomings encountered. The degradation of organic material requires a synchronized action of different groups of microorganisms with different metabolic capacities. Recent developments in molecular biology techniques have provided the research community with a valuable tool for improved understanding of this complex microbiological system, which in turn could help optimize and control the process in an effective way in the future.

**Keywords:**
Biogas plants
Anaerobic digestion (AD)
Anaerobic membrane reactor
Microbial community analysis
Metagenomics

* Corresponding author
E-mail address: ilona.horvath@hb.se

**Contents**



## 1. Introduction

Biogas production through anaerobic digestion (AD) is an environmental friendly process utilizing the increasing amounts of organic waste produced worldwide. A wide range of waste streams, including industrial and municipal waste waters, agricultural, municipal, and food industrial wastes, as well as plant residues, can be treated with this technology. It offers significant advantages over many other waste treatment processes. The main product of this treatment, *i.e.*, the biogas, is a renewable energy resource, while the by-product, *i.e.*, the digester residue, can be utilized as fertilizer because of its high nutrient content available to plants (Ward et al., 2008). The performance of the AD process is highly dependent on the characteristics of feedstock as well as on the activity of the microorganisms involved in different degradation steps (Batstone et al., 2002). The conversion of organic matters into biogas can be divided in three stages: hydrolysis, acid formation, and methane production. In these different stages which are however carried out in parallel, different groups of bacteria collaborate by forming an anaerobic food chain where the products of one group will be the substrates of another group. The process proceeds efficiently if the degradation rates of the different stages are in balance (Yong et al., 2015).

This review presents an overview of the biogas industry worldwide and discusses some new technologies aiming at utilizing new substrates and enhancing the efficiency of the process.

## 2. Biogas, driving forces and the biogas industry

There is an increasing interest in bioenergy production across the world for environmental as well as economic and social reasons. The production of biogas contributes to the production of renewable and sustainable energy since biogas works as a flexible and predictable alternative for fossil fuels. The main political driving forces linked to the biogas system has a country-specific variation (Huttunen et al., 2014). Within the European Union, well-developed biogas industry can be found in Germany, Denmark, Austria, and Sweden followed by the Netherlands, France, Spain, Italy, the United Kingdom, and Belgium. In these countries, with a strong agro-sector, reduction of nutrient emissions and renewable energy production are equally strong driving forces supporting biogas production. In other countries, like Portugal, Greece, and Ireland, as well as in many of the new East-European member states, the biogas sector is currently under development, due to the identified large potential for biomass utilization there.

The biogas plants in Europe are classified based on the type of digested substrates, the technology applied, or the size of the plant. In this sense, they are usually considered as (1) large scale, joint co-digestion plants or (2) farm scale plants. Nevertheless, there are no major differences between these two categories regarding the technology used.

### *2.1. Joint co-digestion plants*

Simultaneous digestion of a mixture of two or more substrates is called co-digestion. The coexistence of different types of residues in the same geographic area enables integrated management, offering considerable environmental benefits, like energy savings, recycling of nutrients back to the agricultural land, and reduction of $CO_2$ emissions (Kacprzak et al., 2010).

Due to the different characteristics of waste streams treated together, co-digestion may enhance the performance of the AD process owing to a positive synergism established in the digestion medium by providing a balanced nutrient supply and sometimes by suitably increasing the moisture content required in the digester (Mata-Alvarez et al., 2000).

Joint biogas plants are referred to large scale plants, with digester capacities ranging from few hundreds $m^3$ up to several thousands $m^3$. Different organic waste streams are collected and transported to the plant and co-digested there. The process is running either at thermophilic or mesophilic conditions, using hydraulic retention times (HRT) of 12–25 d. HRT is normally inversely proportional to the process temperature. Generally, the substrates and in particular animal by-products, which are to be sent to the digester, first go through a controlled pre-sanitation phase, to inactivate pathogens and to break their propagation cycles. After the AD process, the digested residue is transferred to storage tanks, which are typically covered with a gas proof membrane for the recovery of the remaining gas and to prevent methane leakage to the atmosphere. The digested residue has a high nutrient content, and therefore, it can be recycled to the fields as fertilizer. The produced biogas is utilized as a renewable energy source.

In Europe, biogas is mainly used for generating heat and electricity. Some of the produced heat is utilized within the biogas plant as process heating and the remaining heat is distributed through districts` heating systems to consumers. The produced power is sold to the grid. In some countries, like Sweden, the produced biogas is upgraded to bio-methane which is utilized as vehicle fuel (Nielsen et al., 2002; Persson et al., 2006). **Figure 1** shows the biogas production cycle within an integrated system.

Recently, co-digestion has taken much attention since it is one of the interesting ways of improving the yield of AD. Most of the investigations on co-digestion were carried out in batch operation mode and many researchers have pointed out the influence of synergy, due to a balanced mixture composition, on methane yield (Misi and Forster, 2001; Pagés Díaz et al., 2011; Esposito et al., 2012; Wang et al., 2012; Pagés-Díaz et al., 2014). Pagés-Díaz et al. (2011) reported that it was possible to relate synergetic effects with up to 43% enhancement in methane yield ($Y_{CH4}$) compared with the expected $Y_{CH4}$ calculated on basis of methane potentials obtained for the individual substrates. The substrates investigated were

Animal Wastes (for co-digestion)

Cow manure

Pig manure

Poultry manure

Bio-wastes

Industrial organic wastes

Swedge sludge

Organic fraction of municipal solid waste (OMSW)

Transportation system

Centralized Biogas Plant

- Homogenization
- Digestion
- Reduction of odour nuisance
- Sanitation
- Nutritionally defined product

Separation of digested biomass

Used as Fertilizer on the Field

- Improved utilization of plant nutrients
- Reduction of the consumption of mineral fertilizer
- Reduction of water pollution

Transportation system

Storage Facilities off the Field

Biogas for Heat & Power Generation

- Renewable energy source
- $CO_2$-neutral
- Reduction of air pollution
- Effective energy utillization

**Fig.1**. The main streams of the integrated concept of a centralized biogas plant (adapted from Holm-Nielsen et al., 2004).

four different waste streams, such as slaughterhouse waste, various crop residues, manure, and the organic fractions of municipal solid waste (OMSW). A successful co-digestion is not simply a digestion of several waste streams treated at the same time. In fact, biogas production and the stability of the process are highly dependent on waste composition, process conditions, and the activity of microbial community in the system. In that sense, for certain mixing ratios, co-digestion may also lead to antagonistic interactions, resulting in methane yields lower than expected (Pagés-Díaz et al., 2014 and 2015).

### *2.2. The farm scale biogas plants*

It has been reported that more than 4,000 farm scale biogas digesters were in operation in Germany; followed by about 350 in Austria, 72 in Switzerland, 65 in the United Kingdom, 35 in Denmark, and 12 in Sweden (Raven and Gregersen, 2007; Wilkinson, 2011). The main substrate fractions, which are utilized in these farm scale biogas plants are animal manure and energy crops. One of the important aspects of biogas production for farmers is to reduce leaching of nutrients from agricultural lands to the aquatic environments (Bojesen et al., 2014). Hence, farm scale plants are usually established at large pig farms, aiming at solving the problems caused by the excessive slurry production. **Figure 2** presents the closed cycle of organic waste AD and the main steps involved in the quality management process. The most common and recent digester type that is used in farm scale applications is a vertical tank generally made of concrete and equipped with a flexible membrane and light roof making it possible to be used as digester and gas-storage tank simultaneously. The average digester size here is typically from a couple of hundreds to one thousand $m^3$ (Garcia, 2005).

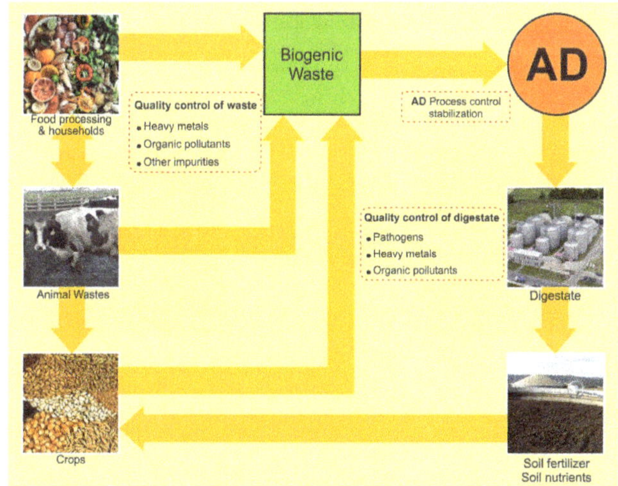


**Fig.2**. Schematic representation of the closed cycle of anaerobic digestion of organic waste and the main steps involved in the quality management process (adapted from Al Seadi (2002)).

*2.3. Domestic biogas technologies in developing countries*

Domestic biogas digesters are abundant in developing countries, especially Asian countries, such as Nepal or Vietnam. Prior to the development of domestic biogas projects, it is important to check the current biogas diffusion in a given country in order to realize the maturity of the sector. The definition of national diffusion targets (*i.e.,* a targeted amount of biogas units that should be built within a specified time frame) by the governments also provides information about the actual diffusion levels. In many countries already promoting domestic biogas production, the governments have implemented national programs aiming at establishing a proper biogas sector. Such programs typically include financing schemes, as well as training campaigns for local workforce, and providing technical support to project developers. These programs involve different players including non-profit organizations cooperating together with the local public institutions and the private sector in order to benefit potential synergies. The German GIZ (Society for International Cooperation, formerly GTZ) and the Dutch SNV are the two main international organizations acting worldwide for domestic biogas advancement, delivering technical service and documentation on this issue.

Some countries like India, Nepal, and China host much more domestic biogas plants than others. It has been reported that about 250,000 domestic plants were installed within the past 20 years in Nepal and 125,000 in Vietnam. Furthermore, 12,500 domestic biogas units are planned to be installed by the end of 2016 in Rwanda, 8,000 in Kenya, and 12,000 plants in Tanzania (Rakotojaona, 2013; TDBP, 2013; Cheng et al., 2014). The domestic biogas plant development is only at an earlier stage in Peru compared with the other Latin American countries. In 2013, the Dutch development organization in cooperation with the Peruvian, planned to set up a national program to construct 10,000 domestic biogas plants within the next 5 years (Rakotojaona, 2013).

## 3. Current biogas process technologies

The production of biogas through AD offers major advantages over other forms of bioenergy production. In fact, it has been defined as one of the most energy-efficient and environmentally beneficial technology for bioenergy production (Deublein and Steinhauser, 2011). The degradation process can be divided into four phases: hydrolysis, acidogenesis, acetogenesis, and methanogenesis; and in each individual phase, different groups of facultative or obligatory anaerobic microorganisms are involved as shown in **Figure 3** (Merlin Christy et al., 2014; Chasnyk et al., 2015; Abdeshahian et al., 2016).

Beside energy production, the degradation of organic waste also offers some other advantages including the reduction of odour release and decreased level of pathogens. Moreover, the nutrient rich digested residue could be used as organic fertilizer for arable land instead of mineral fertilizer, as well as an organic substrate for green house cultivation (De Vries et al., 2012; Abdeshahian et al., 2016). Among the raw substances, organic materials obtained from farm and animal waste streams, as well as from industrial and household activities are pivotal sources for biogas production.

*3.1 Substrates traditionally used*

Through human activities, a huge amount of organic solid waste is generated, which as discussed earlier can be used as feedstock for biogas production. Based on the origin, the different waste streams can be classified as municipal solid waste (MSW), agricultural residues, and wastes from industrial activities. According to a 2012 world bank report, 1.3 billion tons of MSW was generated per year by 3 billion urban residents all over the word, which will increase to 2.2 billion tons by 2025 (Hoornweg and Bhada-Tata, 2012). MSW mainly consists of food waste, paper and paperboard, yard trimmings, wood, plastic, metal, and glass. However, its composition differs depending on regions and countries in which it is collected. To be able to utilize this fraction for biogas production, all the inert material, including plastic, metal, and glass should be removed prior to AD. Moreover, around 15 billion tons of waste, like crops residues and animal manure, is generated worldwide annually from the agricultural sector (Donkin et al., 2013).

Food processing industries also generate waste, however the estimation of its amount is excessively difficult, since it greatly depends on the industry and technology applied. As an example, in the juice producing industry up to 50% of the processed fruit will end up as waste. Moreover, 30% of the weight of a chicken is not suitable for human consumption, and it is therefore removed as waste during slaughtering and other processing steps (Salminen and Rintala, 2002; Forgács et al., 2012).

Although all these different waste fractions are suitable for biogas production, their biogas potential varies significantly. The biogas yield mainly depends on the composition and the biodegradability (under anaerobic conditions) of the waste. Theoretically, the highest biogas yield can be achieved from lipids (1.01 $Nm^3$ $CH_4$/ kg VS), followed by proteins (0.50 $Nm^3$ $CH_4$/ kg VS), and carbohydrates (0.42 $Nm^3$ $CH_4$/ kg VS) (Møller et al., 2004). On the other hand, biodegradability defines how much of a given material is actually utilized during the process. Some compounds like sugars degrade fast and completely, while the degradation of some other materials take longer times, as for example, lignocellulose-rich biomass degrades at very low rates.

*3.2. Pretreatment for enhanced biogas production*

The growing global energy demand together with the limited availability of fossil fuels, unstable energy prices, and environmental problems necessitate the use of renewable energies. The currently used feedstocks for AD are limited, and therefore, it is important to explore new substrates for their utilization in AD to reserve the growing needs. The abundance and availability of lignocellulosic biomasses worldwide as well as their high carbohydrate content make these materials an attractive feedstock for biofuel production. Lignocelluloses have been accounted for approximately 50% of the biomass in the world and the production of lignocelluloses can count up to about 200 billion tons per year (Claassen et al., 1999; Zhang, 2008). Currently, the utilization of lignocellulosic residues as feedstock for methane production is not widespread (Lehtomäki, 2006; Seppälä et al., 2007) due to their recalcitrant structure, which is the main challenge (Hendriks and Zeeman, 2009).

During the first step of AD, *i.e.,* in the hydrolysis step, the hydrolytic bacteria convert the insoluble complex organic matters into monomers and soluble oligomers such as fatty acids, amino acids, and sugars (**Fig. 3**). The enzymes involved in this process are cellulases, hemicellulases, lipases, amylases, and proteases (Taherzadeh and Karimi, 2008). Therefore, in biogas processes, almost all kinds of substrates can be hydrolyzed. However, the rate of the hydrolysis step is highly dependent

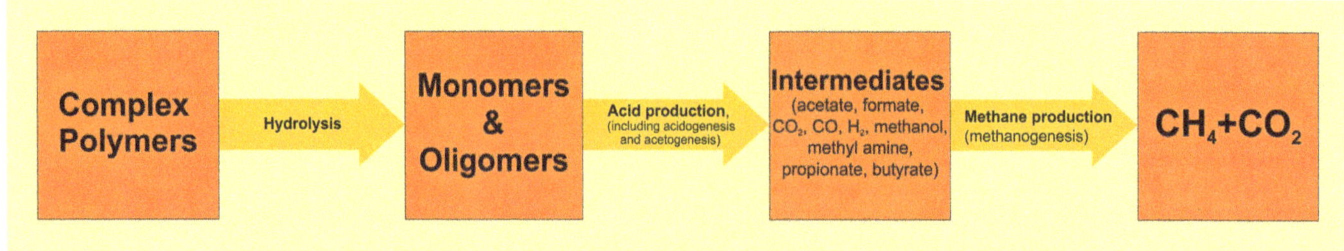


**Fig.3.** The degradation process taking place during AD, i.e., hydrolysis, acidogenesis, acetogenesis, and methanogenesis.

on the characteristics of a given substrate. Hydrolysis can proceed relatively fast if the necessary enzymes are produced by microorganisms and suitable surface area for physical contact between the enzymes and the substrate is provided (Taherzadeh and Karimi, 2008). Nevertheless, substrates with more recalcitrant structure, like cellulose, need longer period to be degraded, and the degradation is usually not complete (Deublein and Steinhauser, 2011). Hence, the hydrolysis step is often considered as the rate-limiting step when utilizing these kinds of substrates (Vavilin et al., 1996; Taherzadeh and Karimi, 2008).

Therefore, an initial pretreatment step, which converts raw materials to a form that is amenable to microbial and enzymatic degradation is needed (Zhang, 2008). A suitable pretreatment by the disruption of the secondary cell walls structure will reduce biomass recalcitrance and thus facilitate downstream processes. Optimally, a pretreatment should also be cost-effective and yield a polysaccharidic-rich substrate with limited amounts of inhibitory by-products.

A numbers of pretreatment methods have been suggested for enhancing biogas production from lignocellulosic biomass, which can be classified as, physical, physicochemical, chemical, and biological pretreatments (Chandra et al., 2007; Taherzadeh and Karimi, 2008; Yang and Wyman, 2008; Hendriks and Zeeman, 2009). Milling, among the physical pretreatments was proven to be effective by shearing, increasing the specific surface area, and reducing the degree of polymerization (DP), thus improving the hydrolysis yield by 5–25%. Degree of such improvement depends on type of biomass, as well as the duration and type of milling (Jin and Chen, 2006; Zeng et al., 2007). Overall, it has been repeatedly shown that smaller particle sizes result in higher yields (Jin and Chen, 2006; Monavari et al., 2009; Lennartsson et al., 2011; Teghammar et al., 2012). That is why the physical pretreatment is often carried out in combination with other pretreatment methods. However, in some cases, the chemical agent used for the pretreatment can act as a potential inhibitor for the microbial community involved in the AD. In a recent study, it was found that the remaining solvent affected the digestion process negatively when forest residues was pretreated with an organic solvent, N-methylmorpholine-N-oxide, even at concentrations as low as 0.008% (Kabir et al., 2013). Besides, the pretreatment process itself might lead to the production of inhibitory products; and despite optimization of pretreatment conditions, some inhibitors will still occur in the pretreated slurry. These may be either degradation products, such as furans through dilute-acid hydrolysis and steam explosion pretreatments, and furfural through alkaline pretreatments, or biomass constituents of varying molecular weights and concentrations (Ahring et al., 1996; Taherzadeh and Karimi, 2008).

Recently, it was shown that using alcohols or weak organic acid for the pretreatment of lignocelluloses seems to be an interesting method. Since they are intermediary products during the anaerobic degradation process, the above-mentioned inhibitory problems can be avoided and moreover, the remaining traces of these solvents after the pretreatment can be consumed for additional methane production. In a recent study, Kabir et al. (2015) applied ethanol, methanol, or acetic acid for the pretreatment of forest residues prior to AD. It was found that although according to the batch experimental results, treatments with ethanol or acetic acid resulted in higher methane yields; the techno-economic calculations showed that treatment with methanol was economically more feasible due to the lower price of methanol and the lower costs for its recovery after the treatment.

### *3.3. Challenges of the current processes*

In general and as mentioned earlier, the AD of organic material requires combined activity of several different groups of microorganisms with different metabolic capacities (Himmel et al., 1994). To obtain a stable biogas process, all the conversion steps involved in the degradation of organic matters and the microorganisms carrying out these steps must work in a synchronised manner. Methanogens have longer duplication times (of up to 30 d) and are generally considered as the most sensitive group to process disturbances (Griffin et al., 1998). It is therefore important to prevent these groups of microorganisms from being washed out from the system, by decoupling the solid retention time (SRT) and the HRT. Major developments have been therefore made during the last decades with regard to development of high rate systems, lowering the effects of toxic compounds, integrating the biological process with membrane separation techniques, as well as better understanding of anaerobic metabolism, and interactions among different microbial species.

## 4. Novel anaerobic digestion technologies

AD systems have undergone several modifications in the last decades to increase the efficiency of the process. In this sense and aiming at overcoming the methanogenesis as the rate-limiting step, efficient retention of the slow-growing methanogenic biomass has been the most important challenge. An important milestone was the development of a new reactor design, *i.e.*, the up flow anaerobic sludge blanket (UASB) reactor, containing a well-settleable methanogenic sludge due to the formation of a dense sludge bed. Another technology making possible to retain active biomass within the system was the application of membrane bioreactors (MBRs). Besides separating cells, the membrane can also be used for the separation of inhibitory compounds, which otherwise would negatively affect the biological process, or for *in situ* recovery of the product could result in decreased cost of down stream processing. Additionally, the development of molecular biology techniques provided researchers with a valuable tool to understand the complex microbiological system involved in anaerobic degradation of organic matters. By the application of these techniques, it would be possible to regulate and control the process and discover disturbances much earlier then using traditional process parameters for monitoring the process.

### *4.1. High rate anaerobic reactors*

The UASB reactor, which was developed by Dr. Gatze Lettinga in the Netherlands during the early 70s, is probably the most popular high-rate reactor system applied for anaerobic biological treatment of "wastewater", as more than 1000 UASB reactors are in operation throughout the World. This process is attractive because of its compactness, high loading rates, relatively low retention times for anaerobic treatment, low operational cost, low sludge production, and high methane production rates. The granular or flocculated sludge is the main prominent characteristic of this type of reactors as compared with other anaerobic technologies. In an UASB reactor, anaerobic microorganisms can form granules through self-immobilization of the cells, and the performance of the system is strongly dependent upon the granulation process together with the characteristics of a particular wastewater treated (Schmidt and Ahring, 1996). Thus, changing the waste type will also affect the sludge quality and thereby the efficiency of the process. Moreover, substrates with a high fraction of particulate organic material are not suitable to be treated with this technology. A modified reactor configuration was therefore proposed recently aiming at separating the hydrolysis and acid formation steps from the methanogenesis step when treating MSW using a two-stage process including a continuously stirred tank reactor (CSTR) and an UASB reactor (Aslanzadeh, 2014). Comparing the performance of this two-stage system with that of a traditional one stage digestion, it was found that using this novel technology, organic loading rate (OLR) of 10 gVS/L/d could be achieved while the HRT could be reduced to 3 d.

### *4.2. Anaerobic membrane bioreactors (AnMBR)*

In membrane bioreactors (MBRs), the membrane forms a selective barrier allowing certain components to pass while retaining others, thereby the biological system can be protected. The application of MBRs provides both increased SRT by avoiding the wash out of the cells and decreasing inhibitor concentrations by the separation of inhibitors (Visvanathan and Abeynayaka, 2012).

Today, there are two different designs for membrane bioreactors applied. The membrane can be placed either in an external loop or submerged within the reactor (**Fig. 4**).

The submerged system requires less space and energy, since compared with the external loop system, energy input is not required to maintain a continuous flow through the membrane. However, it could be problematic to operate this system at high particulate and/or cell concentrations, due to fouling (Judd, 2010). Membrane technologies developed and applied in waste water treatment processes can also be used for biogas production processes. Different studies on membrane technologies in biogas systems

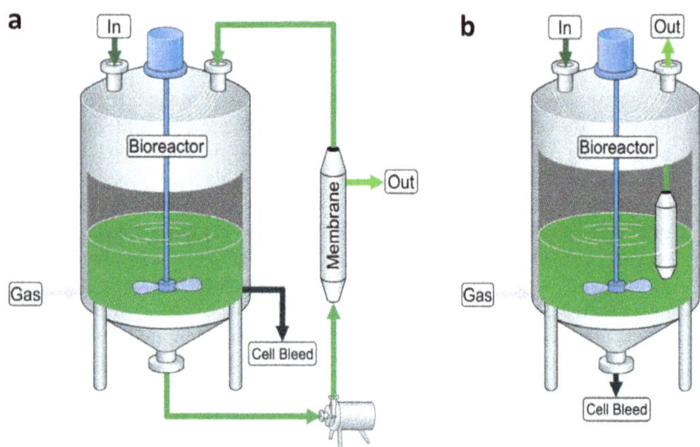


**Fig.4.** Membrane bioreactor designs; a) external loop, b) submerged (adapted from Ylitervo et al. (2013)).

reported yields comparable with those obtained with high rate systems, *i.e.*, UASB systems (Lin et al., 2011; Wijekoon et al., 2011).

Encapsulation of methane-producing bacteria was carried out to test the viability of this technique in biogas processes. One-step liquid-droplet-forming method was used to form spherical capsules of alginate. Chitosan or $Ca^{2+}$ was used as counter-ions together with the addition of carboxymethyl cellulose. Furthermore, a synthetic Durapore® membrane (hydrophilic polyvinyldifluorid (PVDF)) was also tested by making encapsulating sachets with dimension of 3×3 or 3×6 $cm^2$ for holding the bacteria. The results indicated that these membranes allowed the penetration of nutrients into the cells while the gas produced could escape out of the capsules by diffusion. Hence, encapsulation can be a promising method, keeping high density of microorganisms in the system (Youngsukkasem et al., 2012). This theory was further investigated by comparing the ability of encapsulated cells with free cells to handle limonene containing synthetic media during AD. Limonene naturally occurs in citrus waste, making the utilization of this waste stream in biogas processes difficult, due to its inhibitory effects on the biogas producing microorganisms. The results showed the protective effect of the PVDF membrane resulting in faster biogas production by the encased bacteria compared to the free cells (Youngsukkasem et al., 2013).

Furthermore, a novel *An*MBR configuration was investigated later, where both free cells and encased cells worked simultaneously in a single reactor treating a model substrate, Avicel, with limonene addition (Wikandari et al., 2014). The experiments were carried out at thermophilic conditions under semi-continuous operation at OLR of 1 gVS/L/d and HRT of 30 d. Generally, citrus waste contains 8 g/L limonene, and it was found that this reactor configuration could overcome the inhibitor problem with the addition of up to 5 g/L limonene. Thus, this technique has a potential to be applied for anaerobic digestion of fruit wastes containing certain inhibitory compounds.

As it was mentioned earlier, the recalcitrant structure makes the utilization of lignocellulosic biomass in biogas processes difficult. Besides the introduction of different pretreatment technologies prior to AD with an aim to open up their structure, another approach was recently introduced by processing the lignocellulosic biomass thermochemically instead, aiming at obtaining intermediary gases, called syngas. Syngas primarily contains carbon monoxide (CO) hydrogen ($H_2$), and carbon dioxide ($CO_2$). Hence, this gas mixture can be utilized by the anaerobic microorganisms, using the CO and/or $CO_2$ as carbon source and $H_2$ as energy source, to produce methane. In order to increase the productivity and the efficiency of the conversion, a reverse MBR (RMBR) was applied retaining the cells inside the reactor (Youngsukkasem et al., 2015). Using anaerobic sludge encased in PVDF membranes, the conversion of syngas to methane could be carried out at a retention time of 1 d. Furthermore, co-digestion of syngas with a synthetic organic medium was also successful by allowing the diffusion of both gas and liquid through the surface of the membrane.

### *4.3. Integration of membranes and high rate systems*

The combination of anaerobic membrane technology and high rate systems is increasingly being investigated. These integrated systems have several advantages such as improved methane production and less fouling problems and are especially suitable to treat high strength industrial and municipal wastewaters aiming at achieving solids free effluents with a high degree of pathogen removal.

Kraft evaporator condensate was treated at mesophilic conditions with a submerged combined UASB-MBR system achieving a methane yield of 0.35 L $CH_4$/gCOD$_{removed}$ which was very close to the theoretical yield of 0.397 L $CH_4$/gCOD at 37°C (Xie et al., 2010). However, seeding the UASB reactor with non-granule sludge required a long start up period (up to 3-4 months) to be able to achieve the formation of granules and hence, a stable biogas production. In that sense, the presence of a membrane in the reactor could eliminate the hydraulic pressure and negatively affect the granular sludge properties (Ozgun et al., 2013). Further investigations are therefore needed to determine the most optimal process configurations, *i.e.*, the reactor type and the way of coupling it with the membrane module.

## 5. Microbial community analysis and biogas process control

As mentioned earlier, AD involves different degradation steps, *i.e.*, hydrolysis, acidogenesis, acetogenesis, and methanogenesis that are facilitated by various groups of microorganisms (**Fig. 3**). These microorganisms can be divided into three functional groups: hydrolysing and fermenting bacteria, obligate hydrogen-producing acetogenic bacteria, and methanogenic archaea (Ahring, 2003). Hydrolytic acidogenic bacteria (HABs) hydrolyze complex organic polymers into simple compounds during the first step of the degradation. During the acidogenesis process, volatile fatty acids (VFA), alcohols, $H_2$, and $CO_2$ are produced. Similarly, acetic acid, $H_2$, and $CO_2$ are produced in the acetogenesis step by the obligate $H_2$-producing acetogens. *Syntrophobacter* (PUAs: propionate-utilizing acetogens) and *Syntrophomonas* (BUAs: butyrate-utilizing acetogens) represent the major part of acetogens. A key factor in the degradation is that anaerobic oxidation of butyrate and propionate occurs only in syntrophic association with $H_2$-utilizing methanogens (HUMs), consuming $H_2$ and $CO_2$ for methane ($CH_4$) production, preventing the accumulation of increasing $H_2$ pressure in the digester. Another way of methane formation is the conversion of acetate to $CH_4$ and $CO_2$ by the action of acetate-utilizing methanogens (AUMs) (Climent et al., 2007; Zahedi et al., 2013; Ennouri et al., 2016). In general, the operational parameters as well as substrate characteristics will influence the composition of the anaerobic microbial consortium present in a digester.

Molecular biology techniques provide valuable tools for improved understanding of microbial communities and their function in connection with different aspects of AD, which in turn may help optimize the biogas production process more efficiently. A broad range of studies was published recently on investigations on microbial community structures in biogas reactors. The methodologies applied included analysis of total bacteria and archaeal community by targeting 16S rRNA using 454 next generation sequencing (NGS) technique (Zakrzewski et al., 2012) or terminal restriction fragment length polymorphism (T-RFLP) (Wang et al., 2010); as well as detection and quantification of methanogenic Archaea by quantitative real time polymerase chain reaction (qPCR). qPCR is a commonly used method in microbial community studies to detect and quantify a targeted DNA sequence. The principle of qPCR is very similar to that of conventional PCR. The target gene is amplified over a number of cycles. However, the conventional PCR allows only end point detection, whereas using a fluorescent dye or probe, the concentration of the target gene can be monitored after each cycle in qPCR. The detected change in fluorescence intensity reflects the concentration of the amplified gene in real time (VanGuilder et al., 2008).

Among the first studies aiming at understanding the relationship between biodiversity, operating conditions, and process performance, the prokaryotic community of seven digesters treating sewage sludge was examined by constructing and analyzing a total of 9890 16S rRNA gene clones. The results showed that the bacterial community could be divided in three components: one-third of the phylotypes could be found in most of the digesters, one-third were phylotypes shared among a few digesters, and the rest were specific phylotypes found under certain conditions (Riviere et al., 2009).

*5.1. Metagenomics approaches*

The traditional molecular biology technologies help with identifying only the most abundant microbial populations present in the reactor. Due to their high sequencing depth, the newly developed sequencing techniques make the determination of both the most abundant and also the minor populations possible. The NGS-based metagenomic approach enables following up changes in the microbial community structure starting from the very initial stage to souring of the digester. Coding gene sequences (mRNA) especially those representing critical steps of specific metabolic pathways can be mapped to assess the functional profiles of microbial communities. The high throughput sequencing-based metagenomic characterization of various microbial communities involved in biomethanation of a range of substrates has been elucidated with the help of 454 pyrosequencing and SOLiD NGS methods (Kovács et al., 2013; Sundberg et al., 2013; Pore et al., 2016). For example, the Ion Torrent PGM technique, which was launched in 2011, provided the highest throughput compared with that of 454 NGS and it was recently used for microbial composition analysis in several studies (Luo et al., 2013; Wang et al., 2013). Investigations on the microbial community in 21 full scale anaerobic digestion plants using 454 pyrosequencing of 16S rRNA gene sequences showed that the bacterial community was always more abundant and more diverse than the archaeal community in all reactors. Moreover, it was found that while acetoclastic methanogens or AUMs were detected in plants digesting sewage sludge, they were absent in co-digestions plants. Hence, methane is generated from acetate mainly *via* syntrophic acetate oxidation in the co-digestion plants (Sundberg et al., 2013). To date, most studies have strived to investigate the microbial community inside the reactors without taking into account the whole biogas process chain.

Using Ion Torrent PGM technique, investigations on bacterial composition analysis and the presence of bacterial pathogens were performed recently by Luo and Angelidaki (2014) within the whole biogas producing system including the influent, the biogas reactor, and the post-digesters. They found that bacterial community composition of the influent was changed after AD. More specifically, the richness and relative abundance of bacterial pathogens reduced during AD, however, an increase in the relative abundance of pathogens was observed after prolonged post digestion times of 30 d. The authors pointed out that special attention should be therefore paid to the post digestion step aiming at avoiding the re-growth of bacterial pathogens, which otherwise will limit the disposal of the digested residue as bio-fertilizer. Similarly, the denaturing gradient gel electrophoresis (DGGE) technique is still among the promising methods to perform a preliminary analysis of the microbial community profile and to monitor the various experimental stages during the biogas production process. In a recent study, Dias et al. (2016) compared the sequences from DGGE bands with NCBI and RDP databases and reported the significant presence of *Proteobacteria* (6 from 7 sequences), specifically *Gammaproteobacteria* in the biogas system from vinasse methanisation.

In another study, the microbial community structure in a solid-state anaerobic digester (SS-AD) treating lignocellulosic residues, *i.e.*, waste from palm oil mill industry or wheat straw was investigated. The samples were analyzed by 16S rRNA gene (*rrs*) sequence analysis combined with PCR-DGGE. The bacterial community in SS-AD was comprised of *Ruminococcus sp., Thiomargarita sp., Clostridium sp., Anaerobacter sp., Bacillus sp., Sporobacterium sp., Saccharofermentans sp., Oscillibacter sp., Sporobacter sp., Lachnospiraceae sp., etc.* (Heeg et al., 2014; Suksong et al., 2016).

Moreover, the high-throughput Illumina Miseq approach is also widely considered as a promising culture-independent method to perform microbial community analysis of AD systems. By the application of this method, the specific syntrophic relationships between acetogens and methanogens could be better understood, especially in terms of how it can be related to disturbances occurring in the biogas production process. Anaerobic digesters treating lipid-extracted microalgae residue at various inoculum-to-substrate ratios were investigated using Illumina Miseq analysis. Differences in the phylum distribution of the bacterial community were detected in accordance with the changes in inoculum to substrate ratios. The different levels of long chain fatty acids (LCFAs) affected each functional microbial group. Although methanogens were the most sensitive group to LCFA inhibition, the LCFA inhibition factor for hydrolytic bacteria was more highly affected by the inoculum to substrate ratios. Syntrophic acetogens showed a decreased abundance in case of high LCFA concentrations (Ma et al., 2015; Aydin, 2016).

## 6. Concluding remarks

The increasing demand for renewable energy compels the exploration of new substrates and the development of new technologies for biogas production. Regarding raw materials for AD, it is preferable to utilize waste streams since in this way, the process addresses both waste reduction and energy production. Lignocellulosic residues are readily available; however, further development of novel pretreatment technologies are needed to achieve economically viable processes. Anaerobic degradation of organic material requires a well functioning microbial consortium, and methanogenic microorganisms, responsible for methane production within the final step of the digestion process, are known to be the most sensitive ones to process disturbances. This together with their slow growing rate made it necessary to develop novel process configurations aiming at preventing their wash out from the system. In this sense, the development of UASB reactor was an important milestone. In UASB system the formation of a dense well-settleable granular sludge makes an efficient decoupling of SRT and HRT possible. In better words, a crucial factor for a successful anaerobic high-rate treatment is the retention of all slow-growing microorganisms. Hence, when sludge granulation is hindered or lacking, membranes can be applied for biomass separation and recycling back into the reactor.

Therefore, the interest in using different membrane configurations is driven by the requirement for increasing productivity. However, with high particulate and/or cell concentrations, the operation of these kinds of systems can be problematic due to fouling. Thus, full-scale implementation of the AnMBR technology will be highly dependent on flux levels achieved during long-term operation. Finally, since AD is a complex microbial process, a broad range of studies have recently aimed at understanding the relationship between the microbial community structure, operating conditions, and process performance. By using novel newly-developed molecular biology tools, it would be possible to control and regulate the process in an effective way. To date, these techniques were mainly applied for the digestion step itself, however, it is necessary to pay attention to the whole biogas production system, including storage and feeding together with the post digestion step in the future as well.

## References


[1] Abdeshahian, P., Lim, J.S., Ho, W.S., Hashim, H., Lee, C.T., 2016. Potential of biogas production from farm animal waste in Malaysia. Renew. Sust. Energy Rev. 60, 714-723.

[2] Ahring, B.K., Jensen, K., Nielsen, P., Bjerre, A.B., Schmidt, A.S., 1996. Pretreatment of wheat straw and conversion of xylose and xylan to ethanol by thermophilic anaerobic bacteria. Bioresour. Technol. 58(2), 107-113.

[3] Ahring, B.K., 2003. Perspectives for anaerobic digestion, in: In Biomethanation I. Advances in biochemical engineering/biotechnology. 81, 1-30.

[4] Al Seadi, T., 2002. Quality management of AD residues from biogas production. Proc. IEA Bioenergy, Task 24 - Energy from Biological Conversion of Organic Waste.

[5] Aslanzadeh, S., 2014. Pretreatement of cellulosic waste and high rate biogas production. University of Borås, Borås, Sweden.

[6] Aydin, S., 2016. Enhancement of microbial diversity and methane yield by bacterial bioaugmentation through the anaerobic digestion of *Haematococcus pluvialis*. Appl. Microbiol. Biotechnol.100, 5631-5637.

[7] Batstone, D.J., Keller, J., Angelidaki, I., Kalyuzhnyi, S.V., Pavlostathis, S.G., Rozzi, A., Sanders, W.T.M., Siegrist, H., Vavilin, V.A., 2002. The IWA Anaerobic Digestion Model No 1 (ADM 1). Water Sci. Technol. 45(10), 65-73.

[8] TDBP, 2013. Tanzania Domestic Biogas Programme.

[9] Chandra, R.P., Bura, R., Mabee, W.E., Berlin, A., Pan, X., Saddler, J.N., 2007. Substrate pretreatment: The key to effective enzymatic

hydrolysis of lignocellulosics?, In: Olsson L.(Ed.), Biofuels. Springer Berlin Heidelberg, pp. 67-93.
[10] Chasnyk, O., Sołowski, G., Shkarupa, O., 2015. Historical, technical and economic aspects of biogas development: Case of Poland and Ukraine. Renew. Sust. Energy Rev. 52, 227-239.
[11] Cheng, S., Li, Z., Mang, H.P., Huba, E.M., Gao, R., Wang, X., 2014. Development and application of prefabricated biogas digesters in developing countries. Renew. Sust. Energy Rev. 34, 387-400.
[12] Claassen, P.A.M., Van Lier, J.B., Contreras, A.M.L., Van Niel, E.W.J., Sijtsma, L et al., 1999. Utilisation of biomass for the supply of energy carriers. Appl. Microbiol. Biotechnol. 52(6), 741-755.
[13] Climent, M., Ferrer, I., Baeza, M.d.M., Artola, A., Vázquez, F., Font, X., 2007. Effects of thermal and mechanical pretreatments of secondary sludge on biogas production under thermophilic conditions. Chem. Eng. J. 133(1), 335-342.
[14] Deublein, D., Steinhauser, A., 2011. Biogas from waste and renewable resources: an introduction. Mörlenbach, Germany: Wiley-VCH Verlag GmbH & Co. KGaA.
[15] de Vries, J.W., Vinken, T.M.W.J., Hamelin, L., De Boer, I.J.M., 2012. Comparing environmental consequences of anaerobic mono- and co-digestion of pig manure to produce bio-energy - A life cycle perspective. Bioresour. Technol. 125, 239-248.
[16] Dias, M.F., Colturato, L.F., de Oliveira, J.P., Leite, L.R., Oliveira, G., Chernicharo, C.A., de Araújo, J.C., 2016. Metagenomic analysis of a desulphurisation system used to treat biogas from vinasse methanisation. Bioresour. Technol. 205, 58-66.
[17] Donkin, S.S., Doane, P.H., Cecava, M.J., 2013. Expanding the role of crop residues and biofuel co-products as ruminant feedstuffs. Anim. Front. 3(2), 54-60.
[18] Ennouri, H., Miladi, B., Diaz, S.Z., Güelfo, L.A.F., Solera, R., Hamdi, M., Bouallagui, H., 2016. Effect of thermal pretreatment on the biogas production and microbial communities balance during anaerobic digestion of urban and industrial waste activated sludge. Bioresour. Technol. 214, 184-191.
[19] Esposito, G., Frunzo, L., Panico, A., Pirozzi, F., 2012. Enhanced bio-methane production from co-digestion of different organic wastes. Environ. Technol. 33(24), 2733-2740.
[20] Forgács, G., Pourbafrani, M., Niklasson, C., Taherzadeh, M.J, Sárvári Horváth, I., 2012. Methane production from citrus wastes: Process development and cost estimation. J. Chem. Technol. Biotechnol. 87(2), 250-255.
[21] Garcia, S.G., 2005. Farm scale anaerobic digestion integrated in an organic farming system. JTI (Institutet för Jordbruks-och Miljöteknik). Report, 34.
[22] Griffin, M.E., McMahon, K.D., Mackie, R.I., Raskin, L., 1998. Methanogenic population dynamics during start-up of anaerobic digesters treating municipal solid waste and biosolids. Biotechnol. Bioeng. 57(3), 342-355.
[23] Heeg, K., Pohl, M., Sontag, M., Mumme, J., Klocke, M., Nettmann, E., 2014. Microbial communities involved in biogas production from wheat straw as the sole substrate within a two-phase solid-state anaerobic digestion. Syst. Appl. Microbiol. 37(8), 590-600.
[24] Hendriks, A.T.W.M., Zeeman, G., 2009. Pretreatments to enhance the digestibility of lignocellulosic biomass. Bioresour. Technol. 100(1), 10-18.
[25] Himmel, M.E., Baker, J.O., Overend, R.P., 1994. Enzymatic Conversion of Biomass for Fuels Production. American Chemical Society, Washington DC.
[26] Holm-Nielsen, J.B., Al-Seadi, T., Lens, P., Hamelers, B., Hoitink, H., Bidlingmaier, W., 2004. Manure-based biogas systems-Danish experience, in: Lens, P., Hamelers, B., Hoitink, H., Bidlingmaier, W. (Eds.), Resource recovery and reuse in organic solid waste management. IWA Publishing, London, UK, pp. 377-394.
[27] Hoornweg, D., Bhada-Tata, P., 2012. What a waste: a global review of solid waste management. World Bank.
[28] Huttunen, S., Kivimaa, P., Virkamäki, V., 2014. The need for policy coherence to trigger a transition to biogas production. Environ. Innovat. Soc. Transit. 12, 14-30.
[29] Jin, S., Chen, H., 2006. Superfine grinding of steam-exploded rice straw and its enzymatic hydrolysis. Biochem. Eng. J. 30(3), 225-230.
[30] Judd, S., 2010. The MBR book, principles and applications of membrane bioreactors for water and wastewater treatment. Elsevier, Oxford.
[31] Kabir, M.M., del Pilar Castillo, M., Taherzadeh, M.J., Sárvári Horváth, I., 2013. Enhanced methane production from forest residues by N-methylmorpholine-N-oxide (NMMO) pretreatment. BioResources. 8(4), 5409-5423.
[32] Kabir, M.M., Rajendran, K., Taherzadeh, M.J., Sarvari Horvath, I., 2015. Experimental and economical evaluation of bioconversion of forest residues to biogas using organosolv pretreatment. Bioresour. Technol. 178, 201-208.
[33] Kacprzak, A., Krzystek, L., Ledakowicz, S., 2010. Co-digestion of agricultural and industrial wastes. Chem. Pap. 64(2), 127-131.
[34] Kovács, E., Wirth, R., Maróti, G., Bagi, Z., Rákhely, G., Kovács, K.L., 2013. Biogas production from protein-rich biomass: fed-batch anaerobic fermentation of casein and of pig blood and associated Changes in microbial community composition. PLoS One. 8(10), e77265.
[35] Lehtomäki, A., 2006. Biogas production from energy crops and crop residues. University of Jyväskylä.
[36] Lennartsson, P.R., Niklasson, C., Taherzadeh, M.J., 2011. A pilot study on lignocelluloses to ethanol and fish feed using NMMO pretreatment and cultivation with zygomycetes in an air-lift reactor. Bioresour. Technol. 102(6), 4425-4432.
[37] Lin, H., Liao, B-Q., Chen, J., Gao, W., Wang, L., et al., 2011. New insights into membrane fouling in a submerged anaerobic membrane bioreactor based on characterization of cake sludge and bulk sludge. Bioresour. Technol. 102(3), 2373-2379.
[38] Luo, G., Angelidaki, I., 2014. Analysis of bacterial communities and bacterial pathogens in a biogas plant by the combination of ethidium monoazide, PCR and Ion Torrent sequencing. Water Res. 60, 156-163.
[39] Luo, G., Wang, W., Angelidaki, I., 2013. Anaerobic Digestion for Simultaneous Sewage Sludge Treatment and CO Biomethanation: Process Performance and Microbial Ecology. Environ. Sci. Technol. 47(18), 10685-10693.
[40] Ma, J., Zhao, Q-B., Laurens, L.L.M., Jarvis, E.E., Nagle, N.J., et al., 2015. Mechanism, kinetics and microbiology of inhibition caused by long-chain fatty acids in anaerobic digestion of algal biomass. Biotechnol. Biofuels. 8(1), 1-12.
[41] Mata-Alvarez, J., Mace, S., Llabres, P., 2000. Anaerobic digestion of organic solid wastes. An overview of research achievements and perspectives. Bioresour. Technol. 74(1), 3-16.
[42] Merlin Christy, P., Gopinath, L.R., Divya, D., 2014. A review on anaerobic decomposition and enhancement of biogas production through enzymes and microorganisms. Renew. Sust. Energy Rev. 34, 167-173.
[43] Misi, S.N., Forster, C.F., 2001. Batch co-digestion of multi-component agro-wastes. Bioresour. Technol. 80(1), 19-28.
[44] Møller, H.B., Sommer, S.G., Ahring, B.K., 2004. Methane productivity of manure, straw and solid fractions of manure. Biomass Bioenergy. 26(5), 485-495.
[45] Monavari, S., Galbe, M., Zacchi, G., 2009. Impact of impregnation time and chip size on sugar yield in pretreatment of softwood for ethanol production. Bioresour. Technol. 100(24), 6312-6316.
[46] Nielsen, L.H., Hjort-Gregersen, K., Thygesen, P., Christensen, J., 2002. Samfundsøkonomiske analyser af biogasf llesanlg. Fødevareøkonomisk Institut, Rapport, 136.
[47] Ozgun, H., Dereli, R.K., Ersahin, M.E., Kinaci, C., Spanjers, H., van Lier J.B., 2013. A review of anaerobic membrane bioreactors for municipal wastewater treatment: Integration options, limitations and expectations. Sep. Purif. Technol. 118, 89-104.
[48] Pagés Díaz, J., Pereda Reyes, I., Lundin, M., Sárvári Horváth, I., 2011. Co-digestion of different waste mixtures from agro-industrial activities: Kinetic evaluation and synergetic effects. Bioresour. Technol. 102(23), 10834-10840.
[49] Pagés-Díaz, J., Pereda-Reyes, I., Taherzadeh, M.J., Sárvári-Horváth, I., Lundin, M., 2014. Anaerobic co-digestion of solid slaughterhouse wastes with agro-residues: Synergistic and antagonistic interactions determined in batch digestion assays. Chem. Eng. J. 245, 89-98.

[50] Pagés-Díaz, J., Westman, J., Taherzadeh, M.J., Pereda-Reyes, I., Sárvári Horváth, I., 2015. Semi-continuous co-digestion of solid cattle slaughterhouse wastes with other waste streams: Interactions within the mixtures and methanogenic community structure. Chem. Eng. J. 273, 28-36.
[51] Rakotojaona, L., 2013. Domestic biogas development in developing countries ENEA Consulting.
[52] Persson, M., Jönsson, O., Wellinger, A., 2006. Biogas upgrading to vehicle fuel standards and grid injection. IEA Bioenergy, Task, 37.
[53] Pore, S.D., Shetty, D., Arora, P., Maheshwari, S., Dhakephalkar, P.K., 2016. Metagenome changes in the biogas producing community during anaerobic digestion of rice straw. Bioresour. Technol. doi:10.1016/j.biortech.2016.03.045
[54] Raven, R., Gregersen, K.H., 2007. Biogas plants in Denmark: successes and setbacks. Renew. Sust. Energy Rev. 11(1), 116-132.
[55] Riviere, D., Desvignes, V., Pelletier, E., Chaussonnerie, S., Guermazi, S., et al., 2009. Towards the definition of a core of microorganisms involved in anaerobic digestion of sludge. ISME J. 3(6), 700-714.
[56] Salminen, E., Rintala, J., 2002. Anaerobic digestion of organic solid poultry slaughterhouse waste-a review. Bioresour. Technol. 83(1), 13-26.
[57] Seppälä, M., Paavola, T., Rintala, J., 2007. Methane yields of different grass species on the second and third harvest in boreal conditions. Proc. 11th IWA World Congress on Anaerobic Digestion, Brisbane, Australia. pp. 23-27.
[58] Schmidt, J.E., Ahring, B.K., 1996. Granular sludge formation in upflow anaerobic sludge blanket (UASB) reactors. Biotechnol. Bioeng. 49(3), 229-246.
[59] Bojesen, M., Boerboom, L., Skov-Petersen, H., 2014. Towards a sustainable capacity expansion of the Danish biogas sector. Land Use Policy 42, 264-277.
[60] Suksong, W., Kongjan, P., Prasertsan, P., Imai, T., O-Thong, S., 2016. Optimization and microbial community analysis for production of biogas from solid waste residues of palm oil mill industry by solid-state anaerobic digestion. Bioresour. Technol. 214, 166-174.
[61] Taherzadeh, M., Karimi, K., 2008. Pretreatment of lignocellulosic wastes to improve ethanol and biogas production: a review. Int. J. Mol. Sci. 9(9), 1621-1651.
[62] Teghammar, A., Karimi, K., Sárvári Horváth, I., Taherzadeh, M.J., 2012. Enhanced biogas production from rice straw, triticale straw and softwood spruce by NMMO pretreatment. Biomass Bioenergy. 36, 116-120.
[63] VanGuilder, H.D., Vrana, K.E., Freeman, W.M., 2008. Twenty-five years of quantitative PCR for gene expression analysis. Biotechniques. 44(5), 619- 626.
[64] Visvanathan, C., Abeynayaka, A., 2012. Developments and future potentials of anaerobic membrane bioreactors (AnMBRs). Membr. Water. Treat. 3, 1-23.
[65] Vavilin, V.A., Rytov, S.V., Lokshina, L.Y., 1996. A description of hydrolysis kinetics in anaerobic degradation of particulate organic matter. Bioresour. Technol. 56(2), 229-237.
[66] Wang, H., Vuorela, M., Keränen, A-L., Lehtinen, T.M., Lensu, A., Lehtomäki, A., Rintala, J., 2010. Development of microbial populations in the anaerobic hydrolysis of grass silage for methane production. FEMS Microbiol. Ecol. 72(3), 496-506.
[67] Wang, L.H., Wang, Q., Cai, W., Sun, X., 2012. Influence of mixing proportion on the solid-state anaerobic co-digestion of distiller's grains and food waste. Biosyst. Eng. 112(2), 130-137.
[68] Wang, W., Xie, L., Luo, G., Zhou, Q., Angelidaki, I., 2013. Performance and microbial community analysis of the anaerobic reactor with coke oven gas biomethanation and in situ biogas upgrading. Bioresour. Technol. 146, 234-239.
[69] Ward, A.J., Hobbs, P.J., Holliman, P.J., Jones, D.L., 2008. Optimisation of the anaerobic digestion of agricultural resources. Bioresour. Technol. 99(17), 7928-7940.
[70] Wijekoon, K.C., Visvanathan, C., Abeynayaka, A., 2011. Effect of organic loading rate on VFA production, organic matter removal and microbial activity of a two-stage thermophilic anaerobic membrane bioreactor. Bioresour. Technol. 102(9), 5353-5360.
[71] Wikandari, R., Youngsukkasem, S., Millati, R., Taherzadeh, M.J., 2014. Performance of semi-continuous membrane bioreactor in biogas production from toxic feedstock containing D-Limonene. Bioresour. Technol. 170, 350-355.
[72] Wilkinson, K.G., 2011. A comparison of the drivers influencing adoption of on-farm anaerobic digestion in Germany and Australia. Biomass Bioenergy. 35(5), 1613-1622.
[73] Xie, K., Lin, H.J., Mahendran, B., Bagley, D.M., Leung, K.T., Liss, S.N., Liao, B.Q., 2010. Performance and fouling characteristics of a submerged anaerobic membrane bioreactor for kraft evaporator condensate treatment. Environ. Technol. 31(5), 511-521.
[74] Yang, B., Wyman, C.E., 2008. Pretreatment: the key to unlocking low-cost cellulosic ethanol. Biofuels, Bioprod. Biorefin. 2(1), 26-40.
[75] Ylitervo, P., Akinbomia, J., Taherzadeha, M.J., 2013. Membrane bioreactors' potential for ethanol and biogas production: a review. Environ. Technol. 34(13-14), 1711-1723.
[76] Yong, Z., Dong, Y., Zhang, X., Tan, T., 2015. Anaerobic co-digestion of food waste and straw for biogas production. Renew. Energ. 78, 527-530.
[77] Youngsukkasem, S., Akinbomi, J., Rakshit, S.K., Taherzadeh, M.J., 2013. Biogas production by encased bacteria in synthetic membranes: protective effects in toxic media and high loading rates. Environ. Technol. 34(13-16), 2077-2084.
[78] Youngsukkasem, S., Chandolias, K., Taherzadeh, M.J., 2015. Rapid bio-methanation of syngas in a reverse membrane bioreactor: membrane encased microorganisms. Bioresour. Technol. 178, 334-340.
[79] Youngsukkasem, S., Rakshit, S.K., Taherzadeh, M.J., 2012. Biogas production by encapsulated methane-producing bacteria. BioResources. 7(1), 56-65.
[80] Zahedi, S., Sales, D., Romero, L.I., Solera, R., 2013. Optimisation of single-phase dry-thermophilic anaerobic digestion under high organic loading rates of industrial municipal solid waste: population dynamics. Bioresour. Technol. 146, 109-117.
[81] Zakrzewski, M., Goesmann, A., Jaenicke, S., Jünemann, S., Eikmeyer, F., Szczepanowski, R., Al-Soud, W.A., Sørensen, S., Pühler, A., Schlüter, A., 2012. Profiling of the metabolically active community from a production-scale biogas plant by means of high-throughput metatranscriptome sequencing. J. Biotechnol. 158(4), 248-258.
[82] Zeng, M., Mosier, N.S., Huang, C.P., Sherman, D.M., Ladisch, M.R., 2007. Microscopic examination of changes of plant cell structure in corn stover due to hot water pretreatment and enzymatic hydrolysis. Biotechnol. Bioeng. 97(2), 265-278.
[83] Zhang, Y.H.P., 2008. Reviving the carbohydrate economy via multi-product lignocellulose biorefineries. J. Ind. Microbiol. Biotechnol. 35(5), 367-375.

# 20

# A critical review on biomass gasification, co-gasification, and their environmental assessments

Somayeh Farzad*, Mohsen Ali Mandegari, Johann F. Görgens

*Department of Process Engineering, University of Stellenbosch, Private Bag X1, Matieland, 7602, South Africa.*

## HIGHLIGHTS

➢ Conventional and new gasification technologies were compared.
➢ Studies dealing with co-gasification of different feedstocks were summarized.
➢ Life cycle assessments of biomass gasification and co-gasification were studied.

## GRAPHICAL ABSTRACT

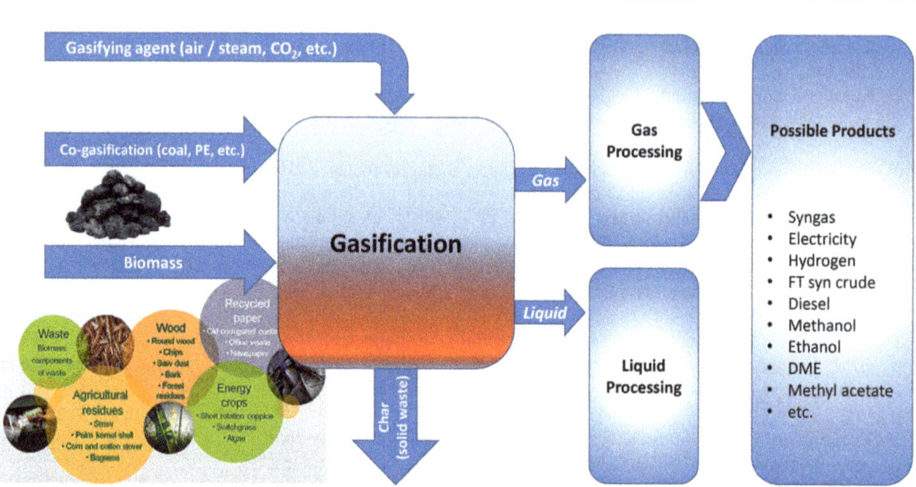


## ABSTRACT

Gasification is an efficient process to obtain valuable products from biomass with several potential applications, which has received increasing attention over the last decades. Further development of gasification technology requires innovative and economical gasification methods with high efficiencies. Various conventional mechanisms of biomass gasification as well as new technologies are discussed in this paper. Furthermore, co-gasification of biomass and coal as an efficient method to protect the environment by reduction of greenhouse gas (GHG) emissions has been comparatively discussed. In fact, the increasing attention to renewable resources is driven by the climate change due to GHG emissions caused by the widespread utilization of conventional fossil fuels, while biomass gasification is considered as a potentially sustainable and environmentally-friendly technology. Nevertheless, social and environmental aspects should also be taken into account when designing such facilities, to guarantee the sustainable use of biomass. This paper also reviews the life cycle assessment (LCA) studies conducted on biomass gasification, considering different technologies and various feedstocks.

**Keywords:**
Biomass gasification
Plasma gasification
Supercritical water gasification
Co-gasification
Life Cycle Assessment (LCA)

* Corresponding author
E-mail address: sfarzad@sun.ac.za

**Contents**



**Abbreviations**

| | |
|---|---|
| AC | Alternating current |
| AER | Absorption enhanced reforming |
| BIG-GT | Biomass integrated gasification/gas turbine |
| BFD | Bubbling fluidized bed |
| CFD | Circulating fluidized bed |
| CHP | Combined heat and power |
| CLC | Chemical loop combustion |
| CSCWG | Catalytic supercritical water gasification |
| DC | Direct current |
| DFBG | Dual fluidized-bed biomass gasifiers |
| DME | Dimethylether |
| ECN | Energy Research Center of the Netherlands |
| FBG | Fluidized bed gasifier |
| F-T | Fischer-Tropsch |
| GHG | Greenhouse gas |
| HHV | High heating value |
| IEA | International energy agency |
| ISO | International Organization for Standardization |
| IGCC | Integrated gasification combined cycle |
| LCA | Life cycle assessment |
| LHV | Lower heating value |
| ORC | Organic Rankine cycle |
| PSA | Pressure swing adsorption |
| PSI | Paul-Scherrer Institute |
| RF | Radio frequency |
| RPM | Random pore model |
| SCWG | Supercritical water gasification |
| SNG | Synthetic natural gas |
| S/B ratio | Steam-to-biomass (S/B) ratio |
| WGSR | Water-gas shift reaction |

## 1. Introduction

Climate change phenomenon or the global temperature rise caused by the emissions of $CO_2$, $NO_x$, and $SO_x$ pose a serious threat to mankind and the other species. According to the international energy outlook (www.eia.gov), world energy related $CO_2$ emissions will increase from 30.2 (in 2008) to 43.2 billion metric tons in 2035. Since greenhouse gas (GHG) emissions from burning fossil fuels for power generation is a major contributor to climate change, a switch from conventional to renewable power resources, i.e., biomass, solar, wind, and hydroelectric energy generation, is vital (Sikarwar et al., 2016).

Biomass has an advantage over the other renewable sources as it is more evenly distributed over the earth and is also abundantly available (Akia et al., 2014; Din and Zainal, 2016; Gottumukkala et al., 2016). In fact, biomass is the fourth-most important source of energy after coal, petroleum, and natural gas, and currently provides more than 10% of the global energy (Saidur et al., 2011). It is estimated that biomass and waste will contribute a quarter or third of global primary energy supply by 2050 (Bauen et al., 2009).

The first confirmed application of gasification for electricity production was reported in 1792. However, the first successful gasifier unit was installed in 1861 by Siemens, while the fluidized bed gasifier (FBG) was only developed in 1926, leading to the establishment of the first commercial coal gasification plant at Wabash River in the USA in 1999. As a consequence of unstable oil prices and concerns over climate change, biomass gasification has increasingly received interest since 2001 (Basu, 2010).

Biomass gasification is a thermochemical partial oxidation process that converts biomass into gas in the presence of gasifying agents, i.e., air, steam, oxygen, carbon dioxide, or a mixture of these (Ruiz et al., 2013). The syngas product is a mixture of CO, $H_2$, $CH_4$, and $CO_2$, as well as light hydrocarbons, i.e., ethane and propane, and heavier hydrocarbons such as tars. The quality of produced gas is affected by the feedstock material, gasifying agent, design of the reactor, the presence of catalyst, and operational conditions of the reactor (Parthasarathy and Narayanan, 2014). The lower heating value (LHV) of the syngas ranges from 4 to 13 MJ/Nm$^3$, as a function of feedstock, the gasification technology, and the operational conditions (Basu, 2013). The produced char is a mixture of unconverted organic fraction and ash (as a function of the treated biomass). The LHV of the char lies in the range of 25 to 30 MJ/kg depending on the amount of unconverted organic fraction (Molino et al., 2016). Biomass can be utilized as a substitute for fossil fuels in generating syngas, hydrogen, electricity, and heat, while syngas can be further processed into methanol, dimethyl ether, Fischer Tropsch (F-T) syncrude, or other chemicals (Leibbrandt et al., 2013; Petersen et al., 2015). Biomass gasification and subsequent conversions lead to several potential benefits such as sustainability, regional economic development, social and agricultural development, and reduction in GHG emissions (Demirbas and Demirbas, 2007). The gasification process still requires optimization to enhance the energy efficiency of the process by overcoming the main challenges such as tar production and moisture content of the biomass. New technologies have been developed as effective ways to utilize even toxic and wet biomass for power generation.

Environmental performance of gasification should be investigated for better design of the process. Life cycle Assessment (LCA) is a cradle-to-grave approach formalized by the International Organization for Standardization (ISO, 2006), which has been regarded as a valuable environmental assessment tool for the chemical industries (Khoo et al., 2016). LCA has been widely applied to the assessment of gasification technologies (Renó et al., 2014), but the majority of the studies focused on the GHGs and energy balance with less attention paid to the wider range of environmental impact categories.

Recently some review papers have been published on gasification processes in general. Ahmad et al. (2016) reviewed biomass gasification considering process conditions, simulation, optimization, and economic evaluation. Heidenreich and Foscolo (2015) and Sikarwar et al. (2016) conducted a comprehensive study about gasification fundamentals, advanced process, polygeneration strategies, and new gasification concepts. Furthermore, there are some review papers about specific aspects of gasification, i.e., dual fluidized bed gasifier (Corella et al., 2007), syngas production and clean up (Göransson et al., 2011; Abdoulmoumine et al., 2015; Samiran et al., 2016), modelling (Baruah and Baruah, 2014), electricity production (Ruiz et al., 2013), and hydrogen production (Parthasarathy and Narayanan, 2014; Udomsirichakorn and Salam, 2014). While this review has focused on biomass gasification to survey the latest progress on conventional and new gasification technologies, effective parameters, different products, and applications as well as its environmental performance. Moreover, co-gasification of different feedstocks (coal and wastes) as a new technique for process improvements and waste management,is reviewed based on the recent research activities carried out.

## 2. Gasification technologies

During the gasification process, biomass undergoes a combination of drying, pyrolysis, combustion, and gasification reactions. Biomass gasification has been developed as a waste valorisation method to obtain products such as syngas, $H_2$, $CH_4$, and chemical feedstocks. The conventional gasification technologies include fixed bed (updraft and downdraft), fluidized bed, and entrained flow reactors, as demonstrated in **Figure 1**. A wider variety of new gasification technologies have been further developed, including plasma gasification and gasification in supercritical water of wet biomass, to convert different feedstocks to gas products (Heidenreich and Foscolo, 2015; Sikarwar et al., 2016). Besides, process integrations and combinations aim to achieve higher process efficiencies, better gas quality and purity, with lower investment costs. Therefore, the so called "emerging technologies" have received increasing attention recently, such as integration of gasification and gas cleaning technologies, or pyrolysis combined with gasification and combustion. A summary of new technologies applied for biomass gasification is represented in **Table 1**.

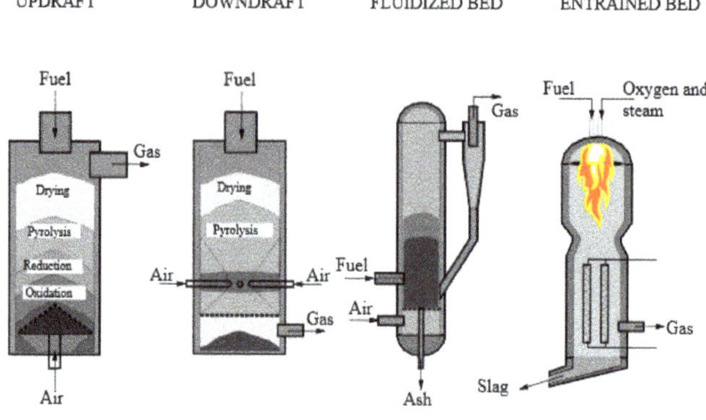


**Fig.1.** Conventional gasification technologies (With permission from www.biorootenergy.com).

### 2.1. *Fluidized bed gasifier*

Fluidized bed gasifiers are typically operated in the range of 800-1000 °C to avoid ash agglomeration, which is satisfactory for biomass utilization. Unlike other reactor types, a fluidized bed gasifier contains a bed of inert materials that serves as heat carrier and mixer, while the gasifying medium acts as the fluidizing gas. Typically, biomass particles are heated to bed temperature (as a result of contact with hot bed solids) and undergo rapid drying and pyrolysis, producing char and gases. The pyrolysis products break down into non-condensable gases after contact with hot solids. Bubbling fluidized bed (BFD) and circulating fluidized bed (CFD) are the most conventional types of fluidized bed gasifiers.

A BFD cannot achieve complete char conversion because of the back-mixing of solids. As a consequence of high degree of solid mixing, BFD gasifiers achieve temperature uniformity. An important drawback of BFD gasifiers is the slow diffusion of oxygen from the bubbles to the emulsion phase, which decreases gasification efficiency (the combustion occurs in the bubble phase) (Basu, 2013).

**Table 1.**
Summary of new technologies applied for biomass gasification (adopted from Heidenreich and Foscolo (2015) and Sikarwar et al. (2016)).

| Strategy employed | Advantages | Limitations |
|---|---|---|
| Combination of gasification and gas clean-up in one reactor | (i) Robust process design<br>(ii) Cost-effective | More research is needed for large-scale commercial applications |
| Multi-staged gasification concept | (i) High quality clean syngas<br>(ii) Improved process efficiency | Enhanced complexity |
| Distributed pyrolysis plants with central gasification plant | (i) Usage of distributed, low-grade biomass<br>(ii) Cost-effective transportation of char oil slurry | Gasoline and olefins production *via* this process is not economically viable |
| Plasma gasification | (i) Decomposition of any organic mattes<br>(ii) Treatment of hazardous waste | (i) High investment cost<br>(ii) High power requirement<br>(iii) Low efficiency |
| Supercritical water gasification (SCWG) | (i) Liquid and biomass with high moisture content are treated<br>(ii) No pre-treatment is required | (i) High energy requirement<br>(ii) High investment cost |
| Co-generation of thermal energy with power | Enhanced process efficiency | Only decentralized heat and power production is feasible as heat needs to be produced near consumers |
| Poly-generation of heat, power, and $H_2$/SNG | (i) Enhanced process efficiency<br>(ii) Generation of renewable $H_2$/renewable fuel for transportation | (i) Enhanced complexity in process design<br>(ii) Not economical in the absence of a natural gas distribution system |
| F-T process coupled with gasification | Production of clean, carbon neutral liquid biofuels | Enhanced complexity in process design |

In a CFD gasifier, gasification takes place in two stages; 1) combustion occurs in BFD to generate the necessary heat for gasification, and 2) pyrolysis and gasification takes place in the presence of high speed gas. The produced gas passes through a cyclone where product gas is separated from the bed materials which are re-circulated to the first stage.

Currently fluidized bed is the most promising technology in biomass gasification because of its potential to gasify a wide range of fuels (or mixture of fuels), high mixing capacity, high mass and heat transfer rate, and moreover, the possibility of using catalysts as part of the bed, which affects tar reforming (Kirnbauer et al., 2012; Gómez-Barea et al., 2013a; Udomsirichakorn et al., 2013).

### 2.2. Fixed bed gasifier

In a typical fixed bed (updraft) gasifier, fuel is fed from the top, while the pre-heated gasifying agent is fed through a grid at the bottom. As the gasifying medium enters the bottom of the bed, it meets hot ash and unconverted chars descending from the top and complete combustion takes place, producing $H_2O$ and $CO_2$ while also raising the temperature. The released heat will heat up the upward moving gas as well as descending solids. The combustion reaction rapidly consumes most of the available oxygen; further up partial oxidation occurs, releasing CO and moderate amounts of heat. The mixture of CO, $CO_2$, and gasifying medium from the combustion zone, moves up into the gasification zone where the char from upper bed is gasified. The residual heat of the rising hot gas pyrolyzes the dry biomass (Basu, 2010). Updraft gasifier is not appropriate for many advanced application, due to production of 10-20 wt.% tar in the produced gas (Ciferno and Marano, 2002).

In downdraft gasifiers, the reaction regions differ from the updraft gasifiers, as biomass fed from the top descends, while gasifying agent is fed into a lower section of the reactor. The hot gas then moves downward over the remaining hot char, where the gasification happens.

### 2.3. Entrained flow gasifier

Entrained flow gasifiers are highly efficient and useful for large scale gasification and are typically operated at high temperature (1300-1500 °C) and pressure values (20-70 bar), where the feed fine fuel (<75 µm) and the gasifying agent (commonly pure oxygen) are injected in co-current (**Fig. 1**). The high operating temperature (well above melting point of ash) results in complete destruction of tar; therefore, these gasifiers are advantageous for biomass gasification where tar is a serious issue. To facilitate feeding into the reactor, the fuel may be mixed with water to prepare a slurry, which will lead to additional reactor volume for evaporation of the large amount of water (Basu, 2013) and 20% higher oxygen consumption than that of dry-feed system (Higman and Van der Burgt, 2011). Utilization of biomass fine particles usually requires a torrefaction based pre-treatment (Couhert et al., 2009; Svoboda et al., 2009).

### 2.4. Supercritical water gasification (SCWG)

Conversion and gasification of organic hydrocarbons in supercritical water has been fundamentally investigated since 1970s (Heidenreich and Foscolo, 2015). Water above its critical point (T = 374.12 °C and P = 221.2 bar) is termed as supercritical, where the liquid and gas phases do not exist separately, and supercritical water shows distinctive reactivity and solvency characteristics. The properties of supercritical water lie between those of the liquid and gaseous phases and a drastic reduction of density causes a significant decrease in the static relative dielectric constant (Kruse, 2008; Sikarwar et al., 2016). Water is not only a reactant involved in the reaction, but also a catalyst with significant impacts on the supercritical water gasification (SCWG) reaction process. Using supercritical water for biomass gasification is attracting growing interest for $H_2$ and/or $CH_4$ production and much progress has been made in the technical aspects of the processes, because it is safe, non-toxic, readily available, inexpensive, and environmentally-benign (Kruse, 2008; Guo et al., 2010; Heidenreich and Foscolo, 2015). Furthermore, SCWG is applied to wet biomass without the need for pre-drying, which is a major advantage over conventional gasification techniques. Moreover, even liquid biomass such as olive mill water can also be utilized for production of low-tar $H_2$ gas using SCWG (Kruse, 2008; Sikarwar et al., 2016). A schematic process flow of a SCWG system is presented in **Figure 2**.

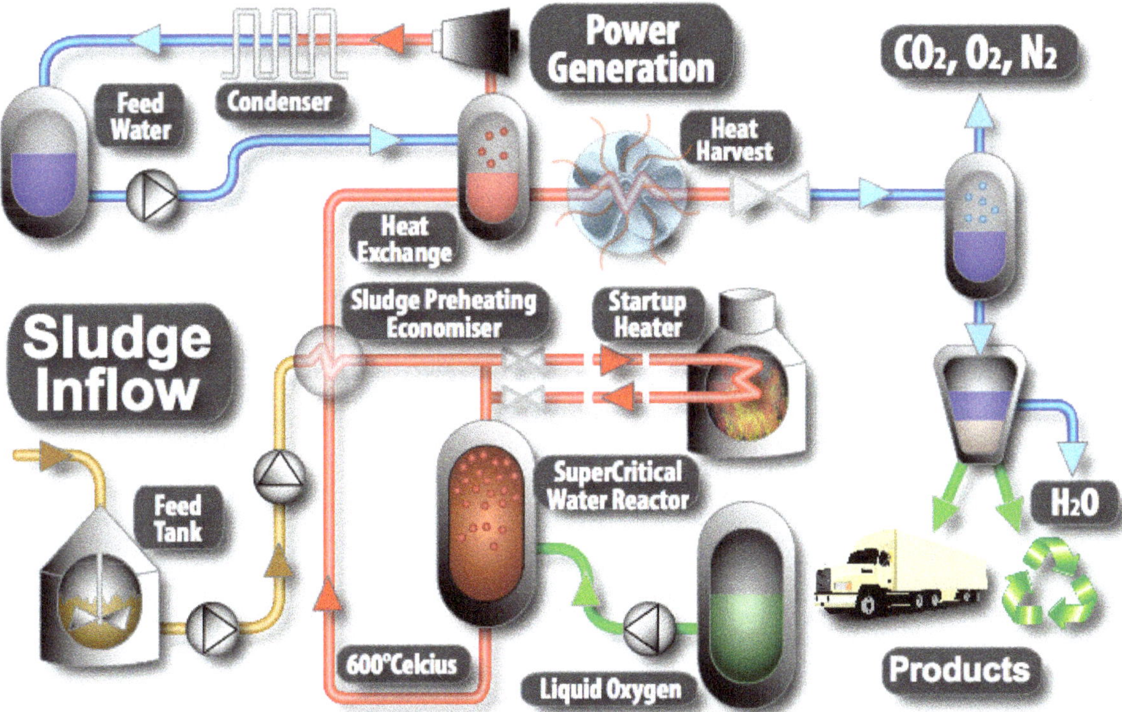


**Fig.2.** A schematic process flow for a SCWG system (Kamler and Andres, 2012). Copyright (2016), with permission from InTech.

There are two approaches for biomass gasification in supercritical water, i.e., high temperature and catalytic SCWG. High-temperature SCWG employs reaction temperatures ranging from 500 to 750 °C (Matsumura et al., 2005), leading to high operating cost, which is the biggest obstacle to the development of this technology. To overcome this bottleneck, many researchers have carried out intensive research work on the catalytic supercritical water gasification (CSCWG), which employs reaction temperatures ranging from 350 to 600 °C, and gasifies the feedstock with the aid of metal catalysts (Savage, 2009). At reaction temperatures below 450 °C, $CH_4$ is the main component in the produced gas, whereas at reaction temperatures above 600 °C hydrogen is dominant. At temperatures above 600 °C, water is a strong oxidant and reacts with the carbon and releases hydrogen (Guo et al., 2010; Heidenreich and Foscolo, 2015).

SCWG can be considered as the most promising method for hydrogen production from biomass, due to the relatively high process efficiency. Generally, the calculated energy efficiencies of the different approaches and process designs of SCWG vary between 44% and 65% and the exergy efficiencies lie in the range of 41–52% (Kruse, 2008; Lu et al., 2012). Although, SCWG has been significantly improved since its initial conception and presents a feasible technology especially for wet biomass, large-scale or commercial gasification requires further studies.

### *2.5. Plasma gasification*

Plasma is defined as the fourth state of the matter, which is highly reactive due to the free electrons, ions, and neutral particles in the gas (Saber et al., 2016). To generate a plasma, a direct current (DC) discharge, alternating current (AC) discharge, radio frequency (RF) induction discharge, or microwave discharge can be used. Plasmas are classified into two categories, including "thermal or equilibrium" (atmospheric pressure) and "cold or non-equilibrium" (vacuum pressure). Thermal plasmas are produced with gases such as argon, nitrogen, hydrogen, water vapour, or a gas mixture at 4700-20,000 °C (Pfender, 1999; Gomez et al., 2009; Heidenreich and Foscolo, 2015). Thermal plasmas have some advantages, i.e., high temperature, high intensity, non-ionising radiation, and high energy density, while its drawback especially from an economic perspective, is the use of electrical power as the energy source, which leads to high construction, operation, and maintenance costs. However, a complete comparative cost evaluation often demonstrates the economic viability of plasma-based technologies (Gomez et al., 2009; Sikarwar et al., 2016). Compared with thermal plasmas, cold plasmas have lower temperatures, degrees of ionisation, and energy densities, and therefore, are applied for applications such as tar removal, local surface modification, or surface activation (Gomez et al., 2009; Du et al., 2015).

Thermal Plasma treatment has been employed for pyrolysis, gasification, and compaction of waste materials as illustrated in **Figure 3** (Heberlein and Murphy, 2008). For the gasification process, plasma is applied: 1) as a heat source during gasification and 2) for tar cracking after standard gasification. Because of extremely high temperatures, thermal plasma is applicable for wet biomass, i.e., sewage sludge (Mountouris, Voutsas, and Tassios 2008) regardless of the particle size and biomass structure (Heidenreich and Foscolo, 2015).

Recently, thermal plasma gasification of biomass has been investigated by several researchers. Rutberg et al. (2011) evaluated experimentally high temperature air plasma gasification of wood for the production of syngas for combined heat and power (CHP) production. Experimental results of using AC plasma torches integrated with a thermodynamic model showed that the chemical energy in the produced syngas was 13.8-14.3 MJ/kg with a power input of 2.2-3.3 MJ/kg, while the LHV energy content of wood is 13.9 MJ/kg. Motycka (2013) studied an integrated plasma gasification (biomass-to-liquids) plant to determine the production cost of F-T syncrude. The results showed that, assuming zero cost for waste refuse feedstock, the products (i.e., F-T diesel and kerosene) would be cost-competitive with similar products obtained from a petroleum process. Hlina et al. (2014) experimentally studied a plasma torch with DC electric for high temperature (18000 °C) plasma gasification of wood, waste plastics, and pyrolysis oil. The ratio of net arc power to the mass flow rate of plasma was drastically higher than standard regime of arc, although

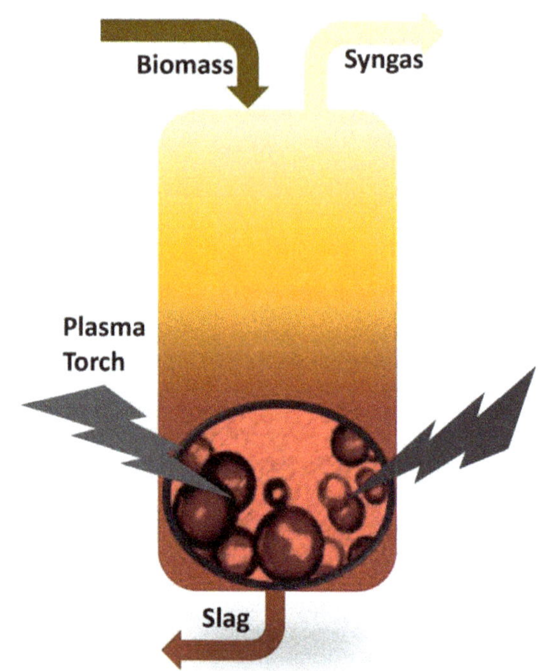


**Fig.3.** Schematic diagram of a plasma gasifier.

the LHV of produced syngas was also higher than normal. Furthermore, different angles of thermal plasma gasification of biomass have also been studied by previous studies (Brothier et al., 2007; Van Oost et al., 2008; Hrabovsky et al., 2009). The main reported benefits of this process are, 1) higher syngas yield with high $H_2$ and CO content, 2) improved heat content, 3) low $CO_2$ yield, and 4) low tar content (Sanlisoy and Carpinlioglu, 2016; Sikarwar et al., 2016). There are some thermal plasma facilities around the world of various capacities form 1 t/d to 300 t/d, with most in the range of 5-30 t/d (Li et al., 2016) and also there are some ongoing project with higher capacities of up to 910 t/d (Fabry et al., 2013). Air Products started to build a 49 MW waste gasification plant at Teesside in England, which could produce either electricity or hydrogen from wastes as the biggest of its kind in the world (Stockford et al., 2015). However, recently (in April 2016) this project has been dropped by the company because additional design and operational challenges would require significant time and cost to rectify the current design (www.airproducts.com).

### *2.6. Integration of gasification and gas cleaning*

Currently, in biomass gasification plants clean gas is produced at ambient temperature (after filtration and scrubbing), which limits its applications. Therefore, gas conditioning preceded by clean-up at elevated temperatures (i.e., hot gas clean) is necessary to ensure high efficiency in industrial applications, specifically for steam gasification. Recent developments in innovative catalysts, sorbents, and high temperature filtration media offer the opportunity to integrate biomass gasification and gas cleaning/conditioning in one reactor (Sikarwar et al., 2016). The strategy to unite biomass gasification with product gas clean-up followed by conditioning *so-called* UNIQUE concept gasifier, is currently in the Lab-scale testing (Heidenreich et al., 2013; Heidenreich and Foscolo, 2015).

### *2.7. Integration of gasification and pyrolysis*

Gasification process of carbonaceous materials into gas comprises several overlapping process steps, such as heating and drying, pyrolysis, oxidation, and gasification. The overlapping of these process steps makes it impossible to control and optimize the different steps separately. Modern, advanced gasification concepts separate the pyrolysis and the gasification steps in single controlled stages to produce high gas purity with low levels of tar to improve the process efficiency as well as environmental compliance (Malkow, 2004; Heidenreich and Foscolo, 2015).

There are two different applications of this technique. 1) To combine pyrolysis and gasification directly in a two or three stage gasification process (multi-stage gasification processes) to optimize operating conditions (Ahrenfeldt et al., 2013). Several gasification processes based on the multi-stage gasification processes concept have been developed recently, i.e., 75 kW Viking gasifier developed at the Danish Technical University (Henriksen et al., 2006) and a three-stage gasifier (FLETGAS) process developed at the University of Sevilla in Spain (Gómez-Barea et al., 2013a). Staged gasification is identified as a method capable of (i) maximizing energy utilization of the fuel (maximizing char conversion), (ii) minimizing secondary treatment of the gas (by avoiding complex tar cleaning), and (iii) being applied in small (0.5-10 MWe) biomass-to-electricity gasification plants (Henriksen et al., 2006; Gómez-Barea et al., 2013a; Heidenreich and Foscolo, 2015). 2) To perform pyrolysis and gasification at different locations to concentrate biomass at decentralized small pyrolysis plants for an economical transport of the biomass pyrolysis products (liquid and solid) to a centralized large gasification plant in order to produce biofuels (Dahmen et al., 2010).

### *2.8. Combination of gasification and combustion*

Combination of gasification with a combustion stage has been developed aiming at increasing the overall process efficiency, through combustion of unreacted char for additional heat production, and production of gas with a lower tar concentration (by conversion of tar through partial combustion). Biomass gasification with pure steam in a fluidized bed is connected to a fluidized-bed combustor to burn the generated char in the gasifier. This arrangement is called dual fluidized-bed biomass gasifier (DFBG) on which a significant progress in R&D and technology demonstration have made since 1975 (Corella et al., 2007; Göransson et al., 2011).

DFBGs have been employed for three purposes as illustrated in **Figure 4**; 1) to supply heat for gasification (the common), 2) to supply oxygen (the chemical loop combustion (CLC) process), and 3) to capture $CO_2$ (the absorption enhanced reforming (AER) process) (Göransson et al., 2011; Shrestha et al., 2016). Apart from DFBG, partial combustion has also attracted an increasing deal of interest in recent years as a method to achieve thermal tar conversion. Air/fuel ratio, hydrogen concentration, methane concentration, temperature, and free radicals produced during the combustion, influence the cracking or polymerization reactions of the tar components (Houben et al., 2005; Anis and Zainal, 2011; Gómez-Barea, et al., 2013b).

## 3. Co-Gasification

Co-asification is defined as gasification of a mixture of waste/biomass and oal which offers several opportunities, especially to utility companies and customers, to protect the environment by reducing GHG emissions from existing process equipment. In recent years, co-gasification of biomass and coal has been broadly investigated by researchers (Collot et al., 1999; Aigner et al., 2011; Taba et al., 2012), because it creates opportunities in industries such as forestry, agriculture, and food processing to manage large quantities of combustible agricultural and wood wastes. In addition, the cost of adapting an existing coal power plant to co-fire biomass is significantly lower than the cost of building new systems dedicated only to biomass power. The biomass rate in the range of 3–5% on energy basis

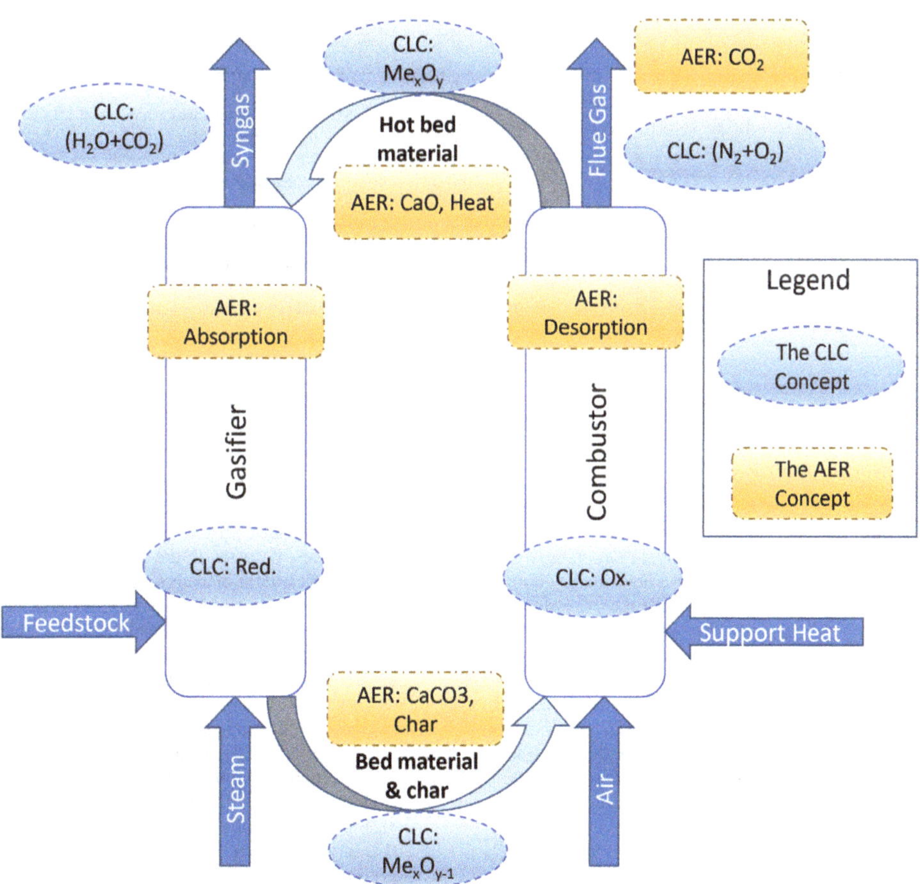


**Fig.4.** A Schematic view of dual fluidized-bed biomass gasifiers (DFBG) (adopted from Göransson et al., 2011).

**Table 2.**
Summary of research works on co-gasification of different feedstocks.

| Feedstock | Gasifier type | Gasification conditions | Concise results | Reference |
|---|---|---|---|---|
| ▪ Petroleum coke<br>▪ Pine pellets | Fluidized bed | (i) Gasification agent: steam<br>(ii) Biomass ratio: 50%, 80%, and 100%<br>(iii)Temperature: 800 and 900 °C,<br>(iv) Total gasification time: 2.5-3 h | (i) The activation energy decreased with increasing biomass ratio.<br>(ii) Higher gasification temperature and oxygen concentration led to higher petcoke conversion and decreased tar concentration. | Nemanova et al. (2014) |
| ▪ Shinhwa coal<br>▪ Pine sawdust | Fluidized bed | (i) Gasification agent: $CO_2$ 40%, and $N_2$ 60%<br>(ii) Biomass ratio: 0%, 25%, 75%, and 100%<br>(iii) Temperature: 900, 1000, and 1100 °C<br>(iv) The ratio of fuel/$CO_2$: 0.20, 0.21, 0.21, and 0.23 | (i) The reactivity of char was improved with an increasing amount of biomass.<br>(ii) The random pore model (RPM) could be used to interpret the carbon conversion data. | Jeong et al. (2014) |
| ▪ Plastics (PE)<br>▪ Wood pellets | Dual fluidized bed | (i) Gasification agent: steam<br>(ii) Biomass ratio: 0%, 25%, 75%, and 100%<br>(iii) Temperature: 850 °C<br>(iv) Steam-to-carbon mass ratio (SCR): 2.3<br>(v) Heterogeneous catalyst: olivine | (i) Co-gasification led to successful thermochemical conversion of plastics as opposed to mono-gasification.<br>(ii) Elevating the plastics content in feed resulted in increased fractions of ethane and ethylene and decreased $CO_2$ | Narobe et al. (2014) |
| ▪ Hard coal<br>▪ Energy crops | Fixed bed | (i) Gasification agent: steam<br>(ii) Biomass ratio: 0%-100% with 20% intervals<br>(iii) Temperature: 700, 800, and 900 °C | (i) The reactivity of char increased with temperature.<br>(ii) The reactivity for chars of fuel blends was higher than biomass chars irrespective of the temperature. | Howaniec and Smoliński (2013) |
| ▪ Bituminous coals<br>▪ Cedar bark. | Entrained flow | (i) Gasification agent: $CO_2$<br>(ii) Biomass ratio: 0%-30%<br>(iii) Temperature: 1200 and 1300 °C<br>(iv) Pressure: 0.5 MPa<br>(v) The ratio of fuel/$CO_2$: 0.20, 0.21, 0.21, and 0.23 | (i) The reactivity of mixture was higher than single coal at 1200 °C.<br>(ii) The reactivity was almost the same at 1400 °C.<br>(iii) Distinguished synergy to improve the gasification reactivity was not observed. | Kajitani et al. (2009) |
| ▪ Pine sawdust<br>▪ Plastic<br>▪ Coal | Fluidized bed | (i) Gasification agent: air; ER: 0.3-0.46<br>(ii) Feed blend: 60% coal, 20% pine, and 20% plastic<br>(iii) Temperature: 750- 880 °C<br>(iv) Catalyst: dolomite | (i) The optimal condition was: temperature, 850 °C and ER: 0.36 equivalent ratio.<br>(ii) Resulted gas contained medium hydrogen content (up to 15% dry basis) and low tar content. | Aznar et al. (2006) |
| ▪ Pine chips<br>▪ Black coal<br>▪ Sabero coal | Fluidized bed | (i) Gasification agent: air-steam<br>(ii) Biomass ratio: 0%, 25%,40%, and 100%<br>(iii) Temperature: 840–910°C | (i) CO increased<br>(ii) $H_2$ first increased up to 25% of biomass and then decreased.<br>(iii) Overall thermal efficiency increased (40% to 68%).<br>(iv) Carbon conversion efficiency increased (63% to 83.4%). | Pan et al. (2000) |

can be directly co-fired. However, this rate may rise to 20% when cyclone boilers are used (Savolainen, 2003; Agbor et al., 2014).

The produced syngas of co-gasification is hydrogen-rich and contains $CH_4$, which can be used for power plants. During the co-gasification process, the volatiles readily decompose and form free radicals which react with the organic matters of the coal, thus, the conversion rate increases while the $CO_2$, $SO_2$ and $NO_x$ emissions reduce.

Since different kinds of coal and biomass have different properties, it is possible to vary the contents and yield of gaseous products from the co-gasification process by changing the amounts and properties of the fuel mixture and temperature (Taba et al., 2012; Emami-Taba et al., 2013). The results of an experimental study of coal and biomass mixture (0–100%) showed linear relationship with changing fuel ratios and gas components, while high wood ratios led to a gas, more suitable for F-T synthesis and synthetic natural gas (SNG) production due to a higher $H_2$/CO ratio (Aigner et al., 2011). Pinto et al. (2009 and 2010) have evaluated the gas produced by co-gasification of coal and wastes blends (olive oil bagasse, pine, and polyethylene) in two catalytic fixed bed (dolomite and Ni based catalysts) reactors. Based on their results, it was possible to substitute one type of waste by another, without great changes to the gasifier but both the released tar and hydrocarbons were different. The presence of wastes in the feedstock led to higher concentrations of hydrocarbons and tar in the gas obtained.

Direct co-firing of biomass can result in several problems, due to high alkaline and chlorine contents of biomass. Main reported problems are corrosion, slagging, fouling in the boiler as well as heat exchanger and piping, poisoning of catalysts, and performance problems in electrostatic precipitators (Heidenreich and Foscolo, 2015). To overcome these problems, indirect and parallel co-firing have been introduced (Sami et al., 2001; Agbor et al., 2014), but the production cost of the plant (CAPEX and OPEX) is higher than the direct co-firing. In addition to biomass and coal co-gasification, co-gasification of biomass with plastic wastes (Pinto et al., 2002), petroleum coke (Nemanova et al., 2014), and tire (Lahijani et al., 2013) have also been studied. A summary of co-gasification studies have been presented in **Table 2**.

## 4. Products of biomass gasification

### *4.1. Syngas production*

Syngas is known as an important source for production of valuable chemicals, i.e., diesel or gasoline (*via* F-T synthesis), hydrogen (produced in refineries), fertilizers (through ammonia), and methanol (Diederichs et al., 2016; Leibbrandt et al., 2011). The syngas from a typical gasifier contains $H_2$, CO, $CO_2$, $CH_4$, $H_2O$, trace amount of higher hydrocarbons, possible inert gases present in the gasification agent, and various contaminants (Göransson et al., 2011). The composition of syngas is dependent on the gasifying medium and utilization of steam or oxygen (the most appropriate gasifying medium for syngas production) instead of air will lead to lower nitrogen content in the product gas (Yin et al., 2004). In low temperature gasification, heavier hydrocarbons are also produced along with CO and $H_2$, which are further cracked and separated from the products. For maximum syngas production with minimum tar formation, the reaction temperature should be increased (because of endothermic gasification reactions), while the volatile residence time should be extended (to increase tar cracking). In terms of biomass gasification, a high alkali content (influencing the softening temperature of the generated ash and consequently agglomeration problems) limits the maximum allowable gasification temperature (Corella et al., 2008). Since high ash content causes slagging, usually biomass with low ash content should be utilized for syngas production (Sikarwar et al., 2016). Considering the requirements of the downstream process, gasification is often followed by the shift reaction to adjust $H_2$/CO ratio.

### *4.2. Hydrogen enriched gas production*

Hydrogen plays a very important role in the development of hydrogen economy and many studies are conducted in this regard ( Lu et al., 2012; Ni et al., 2006; Guo et al., 2010; Sekoai and Daramola, 2015). Hydrogen is mostly produced from fossil fuels, i.e., natural gas, coal, and oil, while only 4% of hydrogen is produced from renewable resources (Parthasarathy and Narayanan, 2014). Amongst renewable sources of hydrogen (biomass, solar, and wind) only biomass can directly generate hydrogen, while other sources have to undertake electrolysis of water. A lot of initiatives have been undertaken to promote hydrogen production from biomass, i.e., the international energy agency's (IEA) program launched the project of $H_2$ production and utilization from carbon-coating materials. Hydrogen can be produced from biomass through thermochemical (pyrolysis, gasification, steam gasification, and SCWG) or biochemical routes, while thermochemical pathways deliver higher efficiency at a lower cost (Balat and Kırtay, 2010; Sekoai and Daramola, 2015). Although biological pathways are less energy-intensive and more environmentally friendly, their low rate of hydrogen production is the major challenge. Therefore, biological methods have not been considered in most scenarios of future hydrogen economy (Ni et al., 2006).

During the gasification process, water-gas shift reaction (WGSR) converts the reformed gas into hydrogen, while pressure swing adsorption (PSA) process is used for product purification. Steam gasification technology is a well-established method of producing renewable $H_2$ with highest yield of $H_2$ from biomass and minimal environmental impacts (Parthasarathy and Narayanan, 2014). It has been reported that steam gasification will increase the yield by three folds, compared with air gasification (Nipattummakul et al., 2010), because WGSR (which is necessary for $H_2$ production) will be enhanced (Wei et al., 2007). Utilisation of pure steam is proven to be more economical than the other conventional gasifying agents (Franco et al., 2003). The evolution of $H_2$ will be increased at higher temperatures, due to significantly faster gasification reaction at temperatures above 800 °C (Nipattummakul et al., 2010).

The SCWG is a method for hydrogen production that is particularly appropriate for high moisture content biomass, but it is more expensive than the current price of $H_2$ from steam methane reforming (Matsumura et al., 2006). Although this method is believed to deliver higher efficiencies, it is still under development and requires more research to make it proven.

An important factor in maximizing $H_2$ production is known to be utilisation of catalyst in the gasification process, where nickel-based catalysts have been efficient in tar reduction and $H_2$ production (Sutton et al., 2001; Wu et al., 2011; Ruoppolo et al., 2012).

A higher steam-to-biomass (S/B) ratio leads to higher steam partial pressure and enhances the shift reaction to $H_2$ production (Göransson et al., 2011). But, increasing the S/B ratio beyond the threshold limit, produces excess steam in the syngas which will lead to efficiency reduction (Sharma and Sheth, 2016). Production of hydrogen from biomass gasification is facing problems due to presence of tar in the gas product and low energy content (by volume) of hydrogen (Ahmed et al., 2012).

### *4.3. Electricity production*

Generating electricity is one potential application of biomass gasification that has been widely applied worldwide. Syngas carries particulate matters and light hydrocarbons which should be cleaned up before its combustion for electricity generation. The syngas cleaning (which is the less developed aspect) is a critical and costly step, which caused closures of some electricity production plants due to technical issues and ash problems (Negro et al., 2008; Ruiz et al., 2013).

### *4.4. Biomass gasification co-generation*

Co-generation is an approach to improve the economic and sustainability aspects of the biomass gasification. Co-generation refers to the combined production of two products or more (poly-generation) to maximize the transformation efficiency of the energy and material of the feedstock into products. As an additional advantage, co-generation offers flexibility regarding the changes of market demands. CHP production is a classic example for a co-generation process (Ahrenfeldt et al., 2013; Heidenreich and Foscolo, 2015).

CHP production units can provide heat and power to industrial, commercial, and residential buildings. CHP by biomass combustion is prevalent, however, gasification is better in terms of electrical efficiency and the acceptable range of biomass qualities (Berggren et al., 2008). The combination of biomass gasification and a gas engine for CHP is a logical choice in the small-scale range and with a biomass to power efficiency potential of 35-40%, which is high compared with conventional technology (Ahrenfeldt et al., 2013; Kumar et al., 2015). In order to reduce the technical problems, a small size (1–10 MW) of the plant could be attractive (Ahrenfeldt et al., 2013; Asadullah, 2014). Many researchers have investigated the CHP co-generation strategy to enhance electricity production. Some researchers have coupled an Organic Rankine cycle (ORC), which additionally transforms 10–15% of heat into electricity, and improved biomass power efficiency (Heidenreich and Foscolo, 2015; Sikarwar et al., 2016). Another approach is the integrated gasification combined cycle (IGCC) process, where a gas turbine and a steam turbine are combined to generate electricity. Since small steam turbines have a low electrical efficiency, an IGCC process is only interesting for larger scale applications (Corti and Lombardi, 2004; Sikarwar et al., 2016).

By using selective syngas conversion reactions with different catalysts various organic products can be manufactured, e.g., methanol, dimethylether (DME), olefins, methane, hydrogen, F-T diesel, etc. (Henrich et al., 2009).

Newer processes compared with CHP co-generation; aim to combine SNG or hydrogen and heat production, or biofuels, heat, and power production. SNG from biomass is considered as a renewable clean fuel substitute for fossil fuels in heating, CHP, and transportation systems. During the last 10 years, the production of SNG from biomass gasification syngas has gained increasing interest and has been investigated by some research groups such as the Energy Research Center of the Netherlands (ECN), the Paul-Scherrer Institute (PSI) in Switzerland, and Güssing in Germany (Dahmen et al., 2010; Heidenreich and Foscolo, 2015). Sweden is a pioneer country for bio-SNG production and GoBiGas project in Göteborg as a commercial plant with 20 MW has been fully operational since 2014 while the second phase of this plant with a capacity of 80 MW is currently under construction (Ahrenfeldt et al., 2013; Zinn and Thunman, 2016).

## 5. Life cycle assessment of biomass gasification

One of the main drivers for the intensified utilization of biomass to produce energy and other materials is its potential to reduce the environmental impacts of fossil fuels utilization. Various methodologies have been applied for examination of environmental impacts, while LCA is one of the most widely used methods. LCA first received attention in 1960s, but only in 1997 International Organization for Standardization (ISO) developed the LCA standard (Kalinci et al., 2012). Literature surveys have suggested that although there are various studies on energy analysis of biomass gasification, LCA of these systems has been rarely studied. Most of the research works have considered GHG emissions (Boerrigter and Rauch, 2006; Moreno and Dufour, 2013), but few have studied complete environmental impacts. A summary of environmental assessments on gasification, based on feedstock, technology, and product have been tabulated in **Table 3**.

Different studies have focused on GHG emissions of different plant sizes or feedstock types for electricity or heat generation from biomass. The GHG emissions of electricity production through biomass co-firing in coal plants have been analysed by different research groups (Zhang et al., 2009; Froese et al., 2010). Different sizes of co-generation plants have also been investigated by Upadhyay et al. (2012) in Canada.

Environmental studies on $H_2$ production have focused on different

**Table 3.**
Summary of LCA studies on biomass gasification.

| Purpose | Case | Scope | Reference |
|---|---|---|---|
| Evaluation of feedstocks | Evaluation of co-firing of biomass with coal<br>*Alternatives*:Wood pellets, coal, and natural gas | GHG emissions | Zhang et al. (2009) |
| | Evaluation of co-firing of biomass with coal for electricity production<br>*Alternatives*: Forestry residues, energy crops, and coal | GHG emissions | Froese et al. (2010) |
| | Heat production through gasification<br>*Alternatives*: Forestry residues, and recycled wood | GHG emissions | Puy et al. (2010) |
| | Heat production through gasification<br>*Alternatives*: Forestry residues, wood pellets, and natural gas | GHG emissions | Pa et al. (2011) |
| | Heat production through gasification<br>*Alternatives*: Forestry residues, woody energy crops, and natural gas | GHG emissions | Pucker et al. (2012) |
| | Production of $H_2$ through gasification<br>*Alternatives*: Vine and almond pruning, forest waste from pine, and eucalyptus plantation | FU*: production of 1 $Nm^3$ $H_2$ | Moreno and Dufour (2013) |
| Evaluation of technologies | Evaluation of integrated gasification combined cycle (IGCC)<br>*Alternatives*: IGCC with upstream $CO_2$ adsorption *vs.* chemical absorption of $CO_2$ | FU: produced energy unit | Corti and Lombardi (2004) |
| | Evaluation of $H_2$ production via biomass gasification<br>*Alternatives*: Gasification followed by syngas reforming *vs.* electricity generation | GHG emissions | Koroneos et al. (2008) |
| | Evaluation of production processes for ethanol production<br>*Alternatives*: Biochemical *vs.* thermochemical processes | GHG emissions | Bright and Strømman (2009) |
| | Evaluation of CHP plant with different sizes<br>*Alternatives*: 0.1, 1, and 50 MWe | GHG emissions | Guest et al. (2011) |
| | Evaluation of CHP plants for power and heat production in rural areas<br>*Alternatives*: Biomass fed CHP *vs.* fossil fuels in a large scale plant | FU: 1 year supply of heat and power to a modern village | Kimming et al. (2011) |
| | Comparison of different gasifiers for $H_2$ production<br>*Alternatives*: Downdraft gasifier and fluidized bed gasifier | GHG emissions | Kalinci et al. (2012) |
| | Evaluation of energy production systems<br>*Alternatives*: Electricity *via* gasification *vs.* bioethanol through enzymatic hydrolysis | FU: the use of biomass chips from 1 ha | González-García et al. (2012) |
| | Evaluation of potential future energy systems<br>*Alternatives*: F-T liquid through biomass gasification, rapeseed based biodiesel, and fossil fuels | FU: 1 energy unit of diesel fuel | Tonini and Astrup (2012) |
| | Evaluation of methanol production (*via* gasification) based on an autonomous distillery or sugar mill<br>*Alternatives*: Co-generation plant combined with methanol or biomass integrated gasification/gas turbine (BIG-GT) system | 1 MJ of each product | Renó et al. (2014) |
| | Evaluation of bioenergy generation alternatives using forest and wood residues<br>*Alternatives*: Combustion and gasification technologies with different capacities | GHG emissions | Cambero et al. (2015) |
| | Evaluation of $H_2$ production through biomass gasification<br>*Alternatives*: Bio-$H_2$ with/without $CO_2$ capturing *vs.* fossil based $H_2$ | FU: 1 kg $H_2$ produced | Susmozas et al. (2016) |
| Product alternatives | Evaluation of synthetic natural gas (SNG) production through gasification<br>*Alternatives*: SNG for heat, power, and transportation | GHG emissions | Steubing et al. (2011) |

* FU; Functional unit of LCA study

technologies or different feedstock sources, separately. The environmental feasibility of $H_2$ production through biomass gasification – by investigating several feedstocks - have been studied by Moreno and Dufour (2013). Their results indicated that main factors contributing to environmental performance of biomass gasification are yield to gas and requirements of fertilizers and pesticides in biomass growth (Moreno and Dufour, 2013). Their study also showed that recovery and use of valuable products such as non-converted methane improved the environmental performance of the process. Koroneos et al. (2008) studied the environmental aspects of $H_2$ production *via* different renewable sources including biomass.

A comparative LCA study of two different gasification systems (downdraft gasifier and CFB gasifier) for $H_2$ production proved that downdraft gasifier delivered better environmental performance over CFB gasifier (Kalinci et al., 2012). According to the LCA study of hydrogen production by Susmozas et al. (2016), direct emission to air, external electricity production, and biomass production are the key processes contributing to environmental impacts, while bio-hydrogen production with $CO_2$ capture delivers superior environmental performance over conventional processes.

Since biomass gasification is an economically interesting solution to produce syngas with low/medium heating value which can be transformed into electricity (González-García et al., 2012), LCA has been applied by different researchers to assess the environmental impacts of electricity generation from biomass. Environmental performance of different electricity production technologies has also been studied. Study of environmental impacts of electricity production in Denmark showed that GHG emissions can be significantly reduced (from 68 to 17 Gg $CO_2$-eq/PJ) by increased utilization of residual biomass (Tonini and Astrup, 2012 ). A comparative study revealed that electricity production from biomass delivered significantly lower $CO_2$ emissions (35-178 g-$CO_2$/kWh) than coal fired systems (975.3 g-$CO_2$/kWh) (Varun et al., 2009).

Environmental effects of electricity production *via* co-gasification of coal and biomass resulted in much lower $CO_2$ emission, in comparison with coal gasification (Hartmann and Kaltschmitt, 1999). IGCC of biomass - with upstream $CO_2$ adsorption - has been compared with IGCC with chemical absorption of $CO_2$ at the stack (Corti and Lombardi, 2004). The environmental performance of an IGCC with $CO_2$ removal - through chemical absorption – has also been studied on the basis of Eco-indicator 95 methodology and compared with similar energy conversion cycle fed by coal (Carpentieri et al., 2005). In a different investigation, the environmental assessment of three different CHP systems revealed that biomass-based scenarios reduced GHG emissions considerably, but delivered higher acidification impacts compared with fossil fuel-based scenarios (Kimming et al., 2011).

Life cycle analyses of GHG emissions of bioenergy systems including combustion and gasification technologies - with different capacities - in British Columbia were investigated by Cambero et al. (2015). Their results implied that in the community where all energy needs were satisfied with fossil fuels and biomass residues were disposed of by burning, net reduction of up to 40,909 t of $CO_2$ equivalent GHG emissions could be achieved. However, in the community where the current energy was mostly supplied from other renewable sources, the net achievable GHG emissions reduction was significantly lower.

## 6. Concluding remarks and future prospects

Gasification of biomass is a promising technology which converts biomass to valuable products such as $H_2$, electricity, and syngas (can be further processed to methanol, F-T syncrude, etc.). Gasification products are a function of applied technology, temperature, pressure, gasifying agent, and the fuel/gasifying medium ratio. More new technologies such as plasma gasification and SCWG deliver higher efficiencies and lower tar productions, while are capable of treating wider ranges of biomass and are mostly appropriate for wet biomass. Various co-generation approaches to produce heat and power along with other products, demonstrate more economically-viable scenarios. Furthermore, co-gasification of biomass and coal can be applied to reduce the consumption of fossil fuels and increase utilization of waste/biomass, leading to less unpleasant products, (i.e., tar), higher carbon conversion, and higher gas yield than coal/biomass gasification. Environmental studies of biomass gasification have proven the potential of reducing GHG emissions, but there is a need for more comprehensive LCAs, taking into account the whole environmental impact categories. Co-gasification and co-generation can be promising future renewable energy scenarios, which require further studies specifically by considering their environmental effects (LCA analysis).

## References


[1] Abdoulmoumine, N., Adhikari, S., Kulkarni, A., Chattanathan, S., 2015. A review on biomass gasification syngas cleanup. Appl. Energy. 155, 294-307.

[2] Agbor, E., Zhang, X., Kumar, A., 2014. A review of biomass co-firing in North America. Renew. Sust. Energy Rev. 40, 930-943.

[3] Ahmad, A.A., Zawawi, N.A., Kasim, F.H., Inayat, A., Khasri, A., 2016. Assessing the gasification performance of biomass: a review on biomass gasification process conditions, optimization and economic evaluation. Renew. Sust. Energy Rev. 53, 1333-1347.

[4] Ahmed, T.Y., Ahmad, M.M., Yusup, S., Inayat, A., Khan, Z., 2012. Mathematical and computational approaches for design of biomass gasification for hydrogen production: a review. Renew. Sust. Energy Rev. 16(4), 2304-2315.

[5] Ahrenfeldt, J., Thomsen, T.P., Henriksen, U., Clausen, L.R., 2013. Biomass gasification cogeneration: a review of state of the art technology and near future perspectives. Appl. Therm. Eng. 50(2), 1407-1417.

[6] Aigner, I., Pfeifer, C., Hofbauer, H., 2011. Co-gasification of coal and wood in a dual fluidized bed gasifier. Fuel. 90(7), 2404-2412.

[7] Akia, M., Yazdani, F., Motaee, E., Han, D., Arandiyan, H., 2014. A review on conversion of biomass to biofuel by nanocatalysts. Biofuel Res. J. 1(1), 16-25.

[8] Anis, S., Zainal, Z.A., 2011. Tar reduction in biomass producer gas via mechanical, catalytic and thermal methods: a review. Renew. Sust. Energy Rev. 15(5), 2355-2377.

[9] Asadullah, M., 2014. Biomass gasification gas cleaning for downstream applications: a comparative critical review. Renew. Sust. Energy Rev. 40, 118-132.

[10] Aznar, M.P., Caballero, M.A., Sancho, J.A., Francés, E., 2006. Plastic waste elimination by co-gasification with coal and biomass in fluidized bed with air in pilot plant. Fuel Process. Technol. 87(5), 409-420.

[11] Balat, H., Kırtay, E., 2010. Hydrogen from biomass-present scenario and future prospects. Int. J. Hydrogen Energy. 35(14), 7416-7426.

[12] Baruah, D., Baruah, D.C., 2014. Modeling of biomass gasification: a review. Renew. Sust. Energy Rev. 39, 806-815.

[13] Basu, P., 2010. Biomass gasification and pyrolysis: practical design and theory. Academic press.

[14] Bauen, A., Berndes, G., Junginger, M., Londo, M., Vuille, F., Ball, R., Bole, T., Chudziak, C., Faaij, A., Mozaffarian, H., 2009. Bioenergy-a sustainable and reliable energy source. a review of status and prospects. bioenergy: a sustainable and reliable energy source. a review of status and prospects. IEA Bioenergy.

[15] Berggren, M., Ljunggren, E., Johnsson, F., 2008. Biomass co-firing potentials for electricity generation in poland-matching supply and co-firing opportunities. Biomass Bioenergy. 32(9), 865-879.

[16] Bhat, I.K., Prakash, R., 2009. LCA of renewable energy for electricity generation systems-a review. Renew. Sust. Energy Rev. 13(5), 1067-1073.

[17] Knoef, H.A.M., 2005. Handbook biomass gasification. Meppel, The Nederlands: BTG Biomass Technology Group B.V.

[18] Bright, R.M., Strømman, A.H., 2009. Life cycle assessment of second generation bioethanols produced from Scandinavian boreal forest resources. J. Ind. Ecol. 13(4), 514-531.

[19] Brothier, M., Gramondi, P., Poletiko, C., Michon, U., Labrot, M., Hacala. A., 2007. Biofuel and hydrogen production from biomass gasification by use of thermal plasma. High Temperature Material Processes: An International Quarterly of High-Technology Plasma Processes. 11, 2.

[20] Cambero, C., Alexandre, H.M., Sowlati, T., 2015. Life cycle greenhouse gas analysis of bioenergy generation alternatives using forest and wood residues in remote locations: a case study in British Columbia, Canada. Resour. Conserv. Recycl. 105, 59-

[21] Carpentieri, M., Corti, A., Lombardi, L., 2005. Life cycle assessment (LCA) of an integrated biomass gasification combined cycle (IBGCC) with $CO_2$ removal. Energy Convers. Manage. 46(11-12), 1790-1808.

[22] Ciferno, J.P., Marano, J.J., 2002. Benchmarking biomass gasification technologies for fuels, chemicals and hydrogen production. US department of energy. US Department of Energy. National Energy Technology Laboratory.

[23] Collot, A.G., Zhuo, Y., Dugwell, D.R., Kandiyoti, R., 1999. Co-pyrolysis and co-gasification of coal and biomass in bench-scale fixed-bed and fluidised bed reactors. Fuel. 78(6), 667-679.

[24] Corella, J., Toledo, J.M., Molina, G., 2008. Biomass gasification with pure steam in fluidised bed: 12 variables that affect the effectiveness of the biomass gasifier. Int. J. Oil Gas Coal Technol. 1(1-2), 194-207.

[25] Corella, J., Toledo, J.M., Molina, G., 2007. A review on dual fluidized-bed biomass gasifiers. Ind. Eng. Chem. Res. 46(21), 6831-6839.

[26] Corti, A., Lombardi, L., 2004. Biomass integrated gasification combined cycle with reduced $CO_2$ emissions: performance analysis and life cycle assessment (LCA). Energy. 29(12-15), 2109-2124.

[27] Couhert, C., Salvador, S., Commandré, J.M., 2009. Impact of torrefaction on syngas production from wood. Fuel. 88(11), 2286-2290.

[28] Dahmen, N., Henrich, E., Kruse, A., Raffelt, K., 2010. Biomass liquefaction and gasification. biomass to biofuels: strategies for global industries. Wiley Blackwell Science, UK.

[29] Demirbas, A.H., Demirbas, I., 2007. Importance of rural bioenergy for developing countries. Energy Convers. Manage. 48(8), 2386-2398.

[30] Diederichs, G.W., Mandegari, M.A., Farzad, S., Görgens, J.F., 2016. Techno-economic comparison of biojet fuel production from lignocellulose, vegetable oil and sugar cane juice. Bioresour. Technol. 216, 331-339.

[31] Din, Z.U., Zainal, Z.A., 2016. Biomass integrated gasification-SOFC systems: technology overview. Renew. Sust. Energy Rev. 53 , 1356-1376.

[32] Du, C., Wu, J., Ma, D., Liu, Y., Qiu, P., Qiu, R., Liao, S., Gao, D., 2015. Gasification of corn cob using non-thermal arc plasma. Int. J. Hydrogen Energy. 40(37), 12634-12649.

[33] Emami-Taba, L., Irfan, M.F., Daud, W.M.A.W., Chakrabarti, M.H., 2013. Fuel blending effects on the co-gasification of coal and biomass-a review. Biomass Bioenergy. 57, 249-263.

[34] Fabry, F., Rehmet, C., Rohani, V., Fulcheri, L., 2013. Waste gasification by thermal plasma: a review. Waste Biomass Valorization. 4(3), 421-439.

[35] Franco, C., Pinto, F., Gulyurtlu, I., Cabrita, I., 2003. The study of reactions influencing the biomass steam gasification process. Fuel. 82(7), 835-842.

[36] Froese, R.E., Shonnard, D.R., Miller, C.A., Koers, K.P., Johnson, D.M., 2010. An evaluation of greenhouse gas mitigation options for coal-fired power plants in the US great lakes states. Biomass Bioenergy. 34(3), 251-262.

[37] Gómez-Barea, A., Leckner, B., Perales, A.V., Nilsson, S., Cano, D.F., 2013a. Improving the performance of fluidized bed biomass/waste gasifiers for distributed electricity: a new three-stage gasification system. Appl. Therm. Eng. 50(2), 1453-1462.

[38] Gómez-Barea, A., Ollero, P., Leckner, B., 2013b. Optimization of char and tar conversion in fluidized bed biomass gasifiers. Fuel. 103, 42-52.

[39] Gomez, E., Rani, D.A., Cheeseman, C.R., Deegan, D., Wise, M., Boccaccini, A.R., 2009. Thermal plasma technology for the treatment of wastes: a critical review. J. Hazard. Mater. 161(2-3), 614-626.

[40] González-García, S., Iribarren, D., Susmozas, A., Dufour, J., Murphy, R.J., 2012. Life cycle assessment of two alternative bioenergy systems involving *salix* spp. biomass: bioethanol production and power generation. Appl. Energy. 95, 111-122.

[41] Göransson, K., Söderlind, U., He, J., Zhang, W., 2011. Review of syngas production via biomass DFBGs. Renew. Sust. Energy Rev. 15(1), 482-492.

[42] Gottumukkala, L.D., Haigh, K., Collard, F.X., van Rensburg, E., Görgens, J., 2016. Opportunities and prospects of biorefinery-based valorisation of pulp and paper sludge. Bioresour. Technol. 215, 37-49.

[43] Guest, G., Bright, R.M., Cherubini, F., Michelsen, O., Strømman, A.H., 2011. Life cycle assessment of biomass-based combined heat and power plants. J. Ind. Ecol. 15(6), 908-921.

[44] Guo, Y., Wang, S.Z., Xu, D.H., Gong, Y.M., Ma, H.H., Tang, X.Y., 2010. Review of catalytic supercritical water gasification for hydrogen production from biomass. Renew. Sust. Energy Rev. 14(1), 334-343.

[45] Hartmann, D., Kaltschmitt, M., 1999. Electricity generation from solid biomass via co-combustion with coal: energy and emission balances from a German case study. Biomass Bioenergy. 16(6), 397-406.

[46] Heberlein, J., Murphy, A.B., 2008. Thermal plasma waste treatment. J. Phys. D: Appl. Phys. 41(5), 053001.

[47] Heidenreich, S., Foscolo, P.U., 2015. New concepts in biomass gasification. Prog. Energy Combust. Sci. 46, 72-95.

[48] Heidenreich, S., Nacken, M., Foscolo, P.U., Rapagna, S., 2013. Gasification apparatus and method for generating syngas from gasifiable feedstock material. U.S. Patent 8,562,701.

[49] Henrich, E., Dahmen, N., Dinjus, E., 2009. Cost estimate for biosynfuel production via biosyncrude gasification. Biofuels, Bioprod. Biorefin. 3(1), 28-41.

[50] Henriksen, U., Ahrenfeldt, J., Jensen, T.K., Gøbel, B., Bentzen, J.D., Hindsgaul, C., Sørensen, L.H., 2006. The design, construction and operation of a 75 kW two-stage gasifier. Energy. 31(10-11), 1542-1553._

[51] Higman, C., Van der Burgt, M., 2011. Gasification. gulf professional publishing.

[52] Hlina, M., Hrabovsky, M., Kavka, T., Konrad, M., 2014. Production of high quality syngas from argon/water plasma gasification of biomass and waste. Waste Manage. 34(1), 63-66.

[53] Houben, M.P., De Lange, H.C., Van Steenhoven, A.A., 2005. Tar reduction through partial combustion of fuel gas. Fuel. 84(7-8), 817-824.

[54] Howaniec, N., Smoliński, A., 2013. Steam co-gasification of coal and biomass - Synergy in reactivity of fuel blends Chars. Int. J. Hydrogen Energy. 38(36), 16152-16160.

[55] Hrabovsky, M., Hlina, M., Konrad, M., Kopecky, V., Kavka, T., Chumak, O., Maslani, A., 2009. Thermal plasma gasification of biomass for fuel gas production. High Temp. Mater. Processes. 13(3-4), 229._

[56] ISO, I., 2006. 14040: Environmental management-life cycle assessment–principles and framework. London: British Standards Institution.

[57] Jeong, H.J., Park, S.S., Hwang, J., 2014. Co-gasification of coal-biomass blended char with $CO_2$ at temperatures of 900-1100 °C. Fuel. 116, 465-470.

[58] Kajitani, S., Zhang, Y., Umemoto, S., Ashizawa, M., Hara, S., 2009. Co-gasification reactivity of coal and woody biomass in high-temperature gasification. Energy Fuels. 24(1), 145-151.

[59] Kalinci, Y., Hepbasli, A., Dincer, I., 2012. Life cycle assessment of hydrogen production from biomass gasification systems. Int. J. Hydrogen Energy. 37(19), 14026-14039.

[60] Kamler, J., Andres, J., 2012. Supercritical water gasification of municipal sludge: a novel approach to waste treatment and energy recovery. InTech.

[61] Khoo, H.H., Ee, W.L., Isoni, V., 2016. Bio-chemicals from lignocellulose feedstock: sustainability, LCA and the green conundrum. Green Chem. 18(7), 1912-1922.

[62] Kimming, M., Sundberg, C., Nordberg, A., Baky, A., Bernesson, S., Norén, O., Hansson, P.A., 2011. Biomass from agriculture in small-scale combined heat and power plants-a comparative life cycle assessment. Biomass Bioenergy. 35(4), 1572-1581.

[63] Kirnbauer, F., Wilk, V., Kitzler, H., Kern, S., Hofbauer, H., 2012. The positive effects of bed material coating on tar reduction in a dual fluidized bed gasifier. Fuel. 95, 553-562.

[64] Koroneos, C., Dompros, A., Roumbas, G., 2008. Hydrogen production via biomass gasification-a life cycle assessment approach. Chem. Eng. Process. Process Intensif. 47(8), 1261-1268.

[65] Kruse, A., 2008. Supercritical water gasification. Biofuels, Bioprod. Biorefin. 2(5), 415-437.

[66] Kumar, A., Kumar, N., Baredar, P., Shukla, A., 2015. A review on biomass energy resources, potential, conversion and policy in India. Renew. Sust.Energy Rev. 45, 530-539.

[67] Lahijani, P., Zainal, Z.A., Mohamed, A.R., Mohammadi, M., 2013. Co-gasification of tire and biomass for enhancement of tire-char reactivity in $CO_2$ gasification process. Bioresour. Technol. 138, 124-130.

[68] Leibbrandt, N.H., Aboyade, A.O., Knoetze, J.H., Görgens, J.F., 2013. Process efficiency of biofuel production via gasification and fischer-tropsch synthesis. Fuel. 109, 484-492.

[69] Leibbrandt, N.H., Knoetze, J.H., Görgens, J.F., 2011. Comparing biological and thermochemical processing of sugarcane bagasse: an energy balance perspective. Biomass Bioenergy. 35(5), 2117-2126.

[70] Li, J., Liu, K., Yan, S., Li, Y., Han, D., 2016. Application of thermal plasma technology for the treatment of solid wastes in China: an overview. Waste Manage. 58, 260-269.

[71] Lu, Y., Guo, L., Zhang, X., Ji, C., 2012. Hydrogen production by supercritical water gasification of biomass: explore the way to maximum hydrogen yield and high carbon gasification efficiency. Int. J. Hydrogen Energy. 37(4), 3177-3185.

[72] Malkow, T., 2004. Novel and innovative pyrolysis and gasification technologies for energy efficient and environmentally sound MSW disposal. Waste Manage. 24(1), 53-79.

[73] Matsumura, Y., Sasaki, M., Okuda, K., Takami, S., Ohara, S., Umetsu, M., Adschiri, T., 2006. Supercritical water treatment of biomass for energy and material recovery. Combust. Sci. Technol. 178(1-3), 509-536.

[74] Matsumura, Y., Minowa, T., Potic, B., Kersten, S.R., Prins, W., van Swaaij, W.P., van de Beld, B., Elliott, D.C., Neuenschwander, G.G., Kruse, A., Antal Jr, M.J., 2005. Biomass gasification in near-and super-critical water: status and prospects. Biomass Bioenergy. 29(4), 269-292.

[75] Molino, A., Chianese, S., Musmarra, D., 2016. Biomass gasification technology: the state of the art overview. J. Energy Chem. 25(1), 10-25.

[76] Moreno, J., Dufour, J., 2013. Life cycle assessment of hydrogen production from biomass gasification. Evaluation of different Spanish feedstocks. Int. J. Hydrogen Energy. 38(18), 7616-7622.

[77] Motycka, S.A., 2013. Techno economic analysis of a plasma gasification biomass to liquids plant. Doctoral dissertation, George Washington University.

[78] Mountouris, A., Voutsas, E., Tassios, D., 2008. Plasma gasification of sewage sludge: process development and energy optimization. Energy Convers. Manageme. 49(8), 2264-2271.

[79] Narobe, M., Golob, J., Klinar, D., Francetič, V., Likozar, B., 2014. Co-gasification of biomass and plastics: pyrolysis kinetics studies, experiments on 100 kW dual fluidized bed pilot plant and development of thermodynamic equilibrium model and balances. Bioresour. Technol. 162, 21-29.

[80] Negro, S.O., Suurs, R.A., Hekkert, M.P., 2008. The bumpy road of biomass gasification in the Netherlands: explaining the rise and fall of an emerging innovation system. Technol. Forecasting Social Change. 75(1), 57-77.

[81] Nemanova, V., Abedini, A., Liliedahl, T., Engvall, K., 2014. Co-gasification of petroleum coke and biomass. Fuel. 117, 870-875.

[82] Ni, M., Leung, D.Y., Leung, M.K., Sumathy, K., 2006. An overview of hydrogen production from biomass. Fuel Process. Technol. 87(5), 461-472.

[83] Nipattummakul, N., Ahmed, I.I., Kerdsuwan, S., Gupta, A.K., 2010. Hydrogen and syngas production from sewage sludge via steam gasification. Int. J. Hydrogen Energy. 35(21), 11738-11745.

[84] Pa, A., Bi, X.T., Sokhansanj, S., 2011. A life cycle evaluation of wood pellet gasification for district heating in British Columbia. Bioresour. Technol. 102(10), 6167-6177.

[85] Pan, Y.G., Velo, E., Roca, X., Manya, J.J., Puigjaner, L., 2000. Fluidized bed co-gasification of residual biomass/poor coal blends for fuel gas production. Fuel. 79(11), 1317-1326.

[86] Parthasarathy, P., Narayanan, K.S., 2014. Hydrogen production from steam gasification of biomass: influence of process parameters on hydrogen yield-a review. Renew. Energ. 66, 570-579.

[87] Petersen, A.M., Farzad, S., Görgens, J.F., 2015. Techno-economic assessment of integrating methanol or Fischer-Tropsch synthesis in a South African sugar mill. Bioresour. Technol. 183, 141-152.

[88] Pfender, E., 1999. Thermal plasma technology: where do we stand and where are we going?. Plasma Chem. Plasma Process. 19(1), 1-31.

[89] Pinto, F., André, R.N., Franco, C., Lopes, H., Carolino, C., Costa, R., Gulyurtlu, I., 2010. Co-gasification of coal and wastes in a pilot-scale installation. 2: Effect of catalysts in syngas treatment to achieve sulphur and nitrogen compounds abatement. Fuel. 89(11), 3340-3351.

[90] Pinto, F., André, R.N., Franco, C., Lopes, H., Gulyurtlu, I., Cabrita, I., 2009. Co-gasification of coal and wastes in a pilot-scale installation. 1: Effect of catalysts in syngas treatment to achieve tar abatement. Fuel. 88 (12), 2392-2402.

[91] Pinto, F., Franco, C., André, R.N., Miranda, M., Gulyurtlu, I., Cabrita, I., 2002. Co-gasification study of biomass mixed with plastic wastes. Fuel. 81(3), 291-297.

[92] Pucker, J., Zwart, R., Jungmeier, G., 2012. Greenhouse gas and energy analysis of substitute natural gas from biomass for space heat. Biomass Bioenergy. 38, 95-101.

[93] Puy, N., Rieradevall, J., Bartrolí, J., 2010. Environmental assessment of post-consumer wood and forest residues gasification: the case study of Barcelona metropolitan area. Biomass Bioenergy. 34(10), 1457-1465.

[94] Renó, M.L.G., del Olmo, O.A., Palacio, J.C.E., Lora, E.E.S., Venturini, O.J., 2014. Sugarcane biorefineries: case studies applied to the Brazilian sugar-alcohol industry. Energy Convers. Manage. 86, 981-991.

[95] Ruiz, J.A., Juárez, M.C., Morales, M.P., Muñoz, P., Mendívil, M.A., 2013. Biomass gasification for electricity generation: review of current technology barriers. Renew. Sust. Energy Rev. 18, 174-183.

[96] Ruoppolo, G., Ammendola, P., Chirone, R., Miccio, F., 2012. $H_2$-rich syngas production by fluidized bed gasification of biomass and plastic fuel. Waste Manage. 32(4),724-732.

[97] Rutberg, P.G., Bratsev, A.N., Kuznetsov, V.A., Popov, V.E., Ufimtsev, A.A., 2011. On efficiency of plasma gasification of wood residues. Biomass Bioenergy. 35(1), 495-504.

[98] Saber, E.M., Tham, K.W., Leibundgut, H., 2016. A review of high temperature cooling systems in tropical buildings. Build. Environ. 96, 237-249.

[99] Saidur, R., Abdelaziz, E.A., Demirbas, A., Hossain, M.S., Mekhilef, S., 2011. A review on biomass as a fuel for boilers. Renew. Sust. Energy Rev. 15(5), 2262-2289.

[100] Sami, M., Annamalai, K., Wooldridge, M., 2001. Co-firing of coal and biomass fuel blends. Prog. Energy Combust. Sci. 27(2), 171-214.

[101] Samiran, N.A., Jaafar, M.N.M., Ng, J.N., Lam, S.S., Chong, C.T., 2016. Progress in biomass gasification technique - with focus on Malaysian palm biomass for syngas production. Renew. Sust. Energy Rev. 62, 1047-1062.

[102] Sanlisoy, A., Carpinlioglu, M.O., 2016. A review on plasma gasification for solid waste disposal. Int. J. Hydrogen Energy. DOI: 10.1016/j.ijhydene.2016.06.008.

[103] Savage, P.E., 2009. A perspective on catalysis in sub- and supercritical water. J. Supercrit. Fluids. 47(3), 407-414.

[104] Savolainen, K., 2003. Co-firing of biomass in coal-fired utility boilers. Appl. Energy. 74(3-4), 369-381.

[105] Sekoai, P.T., Daramola, M.O., 2015. Biohydrogen production as a potential energy fuel in South Africa. Biofuel Res. J. 2(2), 223-226.

[106] Sharma, S., Sheth, P.N., 2016. Air-steam biomass gasification: experiments, modeling and simulation. Energy Convers. Manage. 110,307-318.

[107] Shrestha, S., Ali, B.S., Hamid, M.B.D., 2016. Cold flow model of dual fluidized bed: a review. Renew. Sust. Energy Rev. 53, 1529-1548.

[108] Sikarwar, V.S., Zhao, M., Clough, P., Yao, J., Zhong, X., Memon, M.Z., Shah, N., Anthony, E., Fennell, P., 2016. An overview of advances in biomass gasification. Energy Environ. Sci. 9, 2939-2977.

[109] Steubing, B., Zah, R., Ludwig, C., 2011. Life cycle assessment of SNG from wood for heating, electricity, and transportation. Biomass Bioenergy. 35(7), 2950-2960.

[110] Stockford, C., Brandon, N., Irvine, J., Mays, T., Metcalfe, I., Book, D., Ekins, P., Kucernak, A., Molkov, V., Steinberger-Wilckens, R., Shah, N., 2015. H2FC SUPERGEN: an overview of the hydrogen and fuel cell research across the UK. Int. J. Hydrogen Energy. 40(15), 5534-5543.

[111] Susmozas, A., Iribarren, D., Zapp, P., Linβen, J., Dufour, J., 2016. Life-cycle performance of hydrogen production via indirect biomass gasification with $CO_2$ capture. Int. J. Hydrogen Energy 41(42),

19484-19491.

[112]Sutton, D., Kelleher, B., Ross, J.R., 2001. Review of literature on catalysts for biomass gasification. Fuel Process. Technol. 73(3), 155-173.

[113]Svoboda, K., Pohořelý, M., Hartman, M., Martinec, J., 2009. Pretreatment and feeding of biomass for pressurized entrained flow gasification. Fuel Process. Technol. 90(5), 629-635.

[114]Taba, L.E., Irfan, M.F., Daud, W.A.M.W., Chakrabarti, M.H., 2012. The effect of temperature on various parameters in coal, biomass and CO-gasification: a review. Renew. Sust. Energy Rev. 16(8), 5584-5596.

[115]Tonini, D., Astrup, T., 2012. LCA of biomass-based energy systems: a case study for Denmark. Appl. Energy. 99, 234-246.

[116]Udomsirichakorn, J., Basu, P., Salam, P.A., Acharya, B., 2013. Effect of CaO on tar reforming to hydrogen-enriched gas with in-process $CO_2$ capture in a bubbling fluidized bed biomass steam gasifier. Int. J. Hydrogen Energy. 38(34), 14495-14504.

[117]Udomsirichakorn, J., Salam, P.A., 2014. Review of hydrogen-enriched gas production from steam gasification of biomass: the prospect of CaO-based chemical looping gasification. Renew. Sust. Energy Rev. 30, 565-579.

[118]Upadhyay, T.P., Shahi, C., Leitch, M., Pulkki, R., 2012. Economic feasibility of biomass gasification for power generation in three selected communities of northwestern Ontario, Canada. Energy Policy. 44, 235-244.

[119]Van Oost, G., Hrabovsky, M., Kopecky, M., Konrad, M., Hlina, M., Kavka, T., 2008. Pyrolysis/gasification of biomass for synthetic fuel production using a hybrid gas-water stabilized plasma torch. Vacuum. 83(1), 209-212.

[120]Wei, L., Xu, S., Zhang, L., Liu, C., Zhu, H., Liu, S., 2007. Steam gasification of biomass for hydrogen-rich gas in a free-fall reactor. Int. J. Hydrogen Energy. 32(1), 24-31.

[121]Wu, C., Wang, L., Williams, P.T., Shi, J., Huang, J., 2011. Hydrogen production from biomass gasification with Ni/MCM-41 catalysts: influence of Ni content. Appl. Catal., B. 108, 6-13.

[122]Yin, X., Chang, J., Wang, J., Fu, Y., Wu, C., Leung, D.Y.C., 2004. Exploration of possibility and technical route for methanol synthesis via biomass gasification in China. Coal Convers. *27*(3), 17-22.

[123]Zhang, Y., McKechnie, J., Cormier, D., Lyng, R., Mabee, W., Ogino, A., Maclean, H.L., 2009. Life cycle emissions and cost of producing electricity from coal, natural gas, and wood pellets in Ontario, Canada. Environ. Sci. Technol. 44(1), 538-544.

[124]Zinn, E., Thunman, H., 2016. Göteborg Energi: Vehicle fuel from organic waste. Industrial Biorenewables: Practical Viewpoint, A, pp. 255-266.

# 21

# Advanced nanocomposite membranes for fuel cell applications

Kolsoum Pourzare[1], Yaghoub Mansourpanah[1,2,*], Saeed Farhadi[1]

[1]*Membrane Research Laboratory, Lorestan University, Khorramabad, P.O. Box 68137-17133, Iran.*

[2]*Membrane Separation Technology (MST) Group, Biofuel Research Team (BRTeam), Karaj, Iran.*

## HIGHLIGHTS

- Nanocomposite proton exchange membranes based on different fillers have been comprehensively discussed.
- Analytical methods used for proton exchange membranes properties have been reviewed.
- Properties of polymer composites based on a variety of nanoparticles have been scrutinized.

## GRAPHICAL ABSTRACT

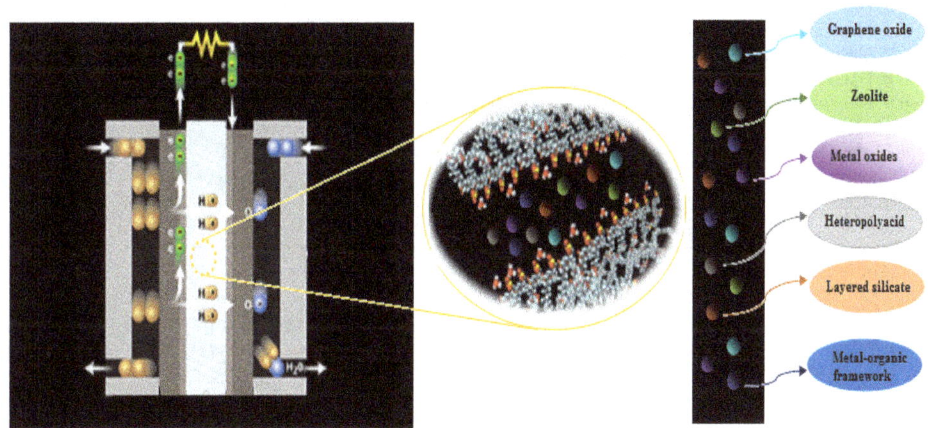


## ABSTRACT

Combination of inorganic fillers into organic polymer membranes (organic–inorganic hybrid membranes) has drawn a significant deal of attention over the last few decades. This is because of the incorporated influence of the organic and inorganic phases towards proton conductivity and membrane stability, in addition to cost decline, improved water retention property, and also suppressing fuel crossover by increasing the transport pathway tortuousness. The preparation methods of the composite membranes and the intrinsic characteristics of the used particles as filler, such as size, type, surface acidity, shape, and their interactions with the polymer matrix can significantly affect the properties of the resultant matrix. The membranes currently used in proton exchange membrane fuel cells (PEMFCs) are perfluorinated polymers containing sulfonic acid, such as Nafion®. Although these membranes possess superior properties, such as high proton conductivity and acceptable chemical, mechanical, and thermal stability, they suffer from several disadvantages such as water management, CO poisoning, and fuel crossover. Organic-inorganic nanocomposite PEMs offer excellent potentials for overcoming these shortcomings in order to achieve improved FC performance. Various inorganic fillers for the fabrication of composite membranes have been comprehensively reviewed in the present article. Moreover, the properties of polymer composites containing different nanoparticles have been thoroughly discussed.

**Keywords:**
Organic-inorganic nanocomposite
Proton exchange membrane
Inorganic fillers
Fuel cell

* Corresponding author
E-mail address: mansourpanah.y@lu.ac.ir , mansourpanah.y@gmail.com

**Contents**



**Abbreviations**

| | | | |
|---|---|---|---|
| AFC | Alkaline fuel cell | PAMPS | Poly(2-acrylamido-2-methylpropane sulfonic acid) |
| ATMP | Amino trimethylene phosphonic acid | PBI | Poly(benzimidazole) |
| CCVD | Catalytic chemical vapour deposition | PDASA | 1,4-phenylenediamine-2-sulfonic acid |
| CLs | Catalyst layers | PDDA | Poly(diallyldimethylammoniumchloride) |
| CNTs | Carbon nanotubes | PDHC | 1,4-phenyldiamine hydrochloride |
| CPs | Coordination polymers | PEEK | poly(ether ether ketone) |
| CTS-SPIONs | Chitosan coated superparamagnetic iron oxide nanoparticles | PEM | Proton exchange membrane |
| CTS | Chitosan | PEMFC | Proton exchange membrane fuel cell |
| CV | Cyclic voltammograms | PES | poly(ether sulfone) |
| CWs | Cellulose whiskers | POMs | Polyoxometalates |
| DI water | Deionized water | PPO | poly(2,6-dimethyl-1,4-phenylene oxide) |
| DMA | Dynamical mechanical analysis | PTA | Phosphotungstic acid |
| DMFC | Direct methanol fuel cell | PVA | Poly(vinyl alcohol) |
| DSC | Differential scanning calorimetry | PVP | Polyvinyl pyrrolidone |
| EFC | Enzymatic fuel cell | RH | Relative humidity |
| EIS | Electrochemical impedance spectroscopy | SBUs | Secondary building units |
| FC | Fuel Cell | SHNTs | Sulfonated halloysite nanotubes |
| GC | Gas chromatography | SGO | Sulfonated graphene oxide |
| GDLs | Gas diffusion layers | SNP | Sodium nitroprusside |
| GO | Graphene oxide | SOFC | Solid oxide fuel cell |
| GPTMS | (3-glycidoxypropyl)-methyldiethoxysilane | SPEEK | Sulfonated poly(ether ether ketone) |
| HR-TEM | High resolution transmission electron microscopy | SPAES | Sulfonated poly(arylene ether sulfone |
| IEC | Ion exchange capacity | SPAEK–COOH | Sulfonated poly(arylene ether ketone) |
| LBL | Layer-by-layer | SPS | Sulfonated poly(styrene) |
| MCFC | Molten carbonate fuel cell | SPSU | Sulfonated polysulfone |
| MeOEGMA | Monomethoxy oligoethylene glycol methacrylate | SW | Membrane swelling |
| MFC | Microbial fuel cell | SWy | Smectite clays |
| MOFs | Metal–organic frameworks | TEOS | Tetra ethyl orthosilicate |
| MPTMS | 3-mercaptopropyltrimethoxysilane | TGA | Thermogravimetric analyzer |
| MRI | Magnetic resonance imaging | TS | Sulfated nanotitania |
| MWNTs | Multi-walled carbon nanotubes | WU | Water uptake |
| NAFB | Nafion®/acid functionalized zeolite beta | ZIF-8 | Zeolitic imidazolate framework-8 |
| NMPA | Nitrilotri(methyl triphosphonic acid) | ZrNT | Zirconium oxide nanotube |
| OCV | Open circuit voltage | ZrP | Zirconium phosphate |
| PAFC | Phosphoric acid fuel cell | | |

## 1. Introduction

Global energy consumption is projected to increase by 56% by the year 2040 as a result of various factors such as rapid urbanization and population growth. In response to this growing energy demand, several alternative energy technologies have been proposed, among which fuel cell (FC) technology has attracted considerable attention (Ramaswamy et al., 2014). FC technology is identified as promising electric energy generators that could furnish clean and efficient energy for stationary applications, transportation, and portable power applications in the 21st century (Carrette et al., 2001; Steele and Heinzel, 2001; Jacobson et al., 2005). Technically, an FC is an electrochemical device that continuously converts chemical energy fuel (such as hydrogen, natural gas, methanol, ethanol, etc.) into electric energy (and some heat) as long as fuel and oxidant are supplied (Zhang et al., 2012). The utilisation of liquid fuels such as ethanol and methanol in FCs would reduce the requirements of establishing totally new infrastructure as required for hydrogen as a fuel source (Zakaria et al., 2016). Furthermore, both methanol and ethanol are considered renewable fuels (biofuels) when generated from biomass resources. It is worth quoting that currently bioethanol dominates the global biofuel production capacity and can be produced in bulk from biomass feedstocks *via* a fermentation process (Badwal et al., 2015; Zakaria et al., 2016).

FCs diminish power loses by avoiding the intermediate steps required in similar diesel-powered generators (Mishra et al., 2012). FCs are commonly named and classified based on the nature of the electrolyte used in the cell as follows: proton exchange membrane fuel cell (PEMFC), direct methanol fuel cell (DMFC), phosphoric acid fuel cell (PAFC), molten carbonate fuel cell (MCFC), solid oxide fuel cell (SOFC), alkaline fuel cell (AFC), microbial fuel cell (MFC), and enzymatic fuel cell (EFC), and in general, two main classes can be defined: i) low temperature FCs operating at temperatures lower than 250 °C (such as proton exchange membrane fuel cells (PEMCs)) and ii) high temperature FCs operating at temperatures ranging from 600 °C up to 1100 °C (such as solid oxide fuel cells (SOFCs)) (Laberty-Robert et al., 2011). **Table 1** summarizes the operating properties of the main types of FCs (Sharaf and Orhan, 2014).

**Table1.**
Typical electrolyte and operating properties of main fuel cell technologies (modified from (Sharaf and Orhan, 2014)).

| Fuel cell type | Typical electrolyte | Typical fuel | Operation temperature (°C) | Electrical efficiency (%) | Charge carrier |
|---|---|---|---|---|---|
| Proton exchange membrane fuel cell (PEMFC) | Perfluorosulfonic acid | $H_2$ | 50-80 | 60 | $H^+$ |
| Phosphoric acid fuel cell (PAFC) | Phosphoric acid ($H_3PO_4$) ins ilicon carbide (SiC) | $H_2$ | 160-220 | 40 | $H^+$ |
| Alkaline fuel cell (AFC) | aqueous solution of potassium hydroxide (KOH) | $H_2$ | 50-200 | 60 | $OH^-$ |
| Molten carbonate fuel cell (MCFC) | Solution of ($Li_2CO_3$, $Na_2CO_3$, $K_2CO_3$) in Lithium aluminate ($LiAlO_2$) | $CH_4$ | 600-700 | 45-50 | $CO_3^{2-}$ |
| Solid oxide fuel cell (SOFC | Yttria- stabilized zirconia (YSZ) | $H_2$ | 800-1000 | 60 | $O^{2-}$ |
| Direct methanol fuel cell (DMFC) | Perfluorosulfonic acid | $CH_3OH/H_2O$ | 90-120 | 60 | $H^+$ |
| Microbial fuel cell (MFC) | Ion exchange membrane | Wastewater glucose, acetate | 20-60 | 65 | $H^+$ |

Among the various kinds of FCs, PEMFCs have attracted a great deal of attention and have provided the strongest motivation for technological enlargement owing to their unique features such as ambient temperature working conditions, fast start-up, high specific power density, and easy portability (Steele and Heinzel, 2001; Narayanamoorthy et al., 2012). PEMFCs themselves can also be classified into two groups: LT-PEMFC (operation temperature between 60-80 °C) and HT-PEMFC (operation temperature between 100-200 °C). There are some intrinsic problems with operating PEMFCs at low temperatures (around 80 °C) such as heat and water management and CO poisoning while the HT-PEMFCs are well capable of overcoming these drawbacks (Chandan et al., 2013; Authayanun et al., 2015). As schematically shown in **Figure 1a**, a PEMFC is composed of the anode and cathode flow field plates, gas diffusion layers (GDLs), catalyst layers (CLs), and proton exchange membrane (PEM) (Wang et al., 2011; Ye and Zhan, 2013). A single FC is only able to produce a certain voltage and current. In order to obtain a higher voltage and current or power, FCs are connected in either series or parallel, called stacks (Wang et al., 2011), as shown in **Figure 1b**. The important properties of PEM as central component in hydrogen and methanol FC systems include high proton conductivity, low electronic conductivity, high mechanical and thermal stability, good oxidative and hydrolytic stability, low fuel and oxidant permeability, low cost, good dimensional, and morphological stability (Kraytsberg and Ein-Eli, 2014).

Nafion® as the most common and commercially available PEM for PEMFC, DMFC, and BFC possess wonderful properties, such as high proton conductivity and good chemical, mechanical, and thermal stability.

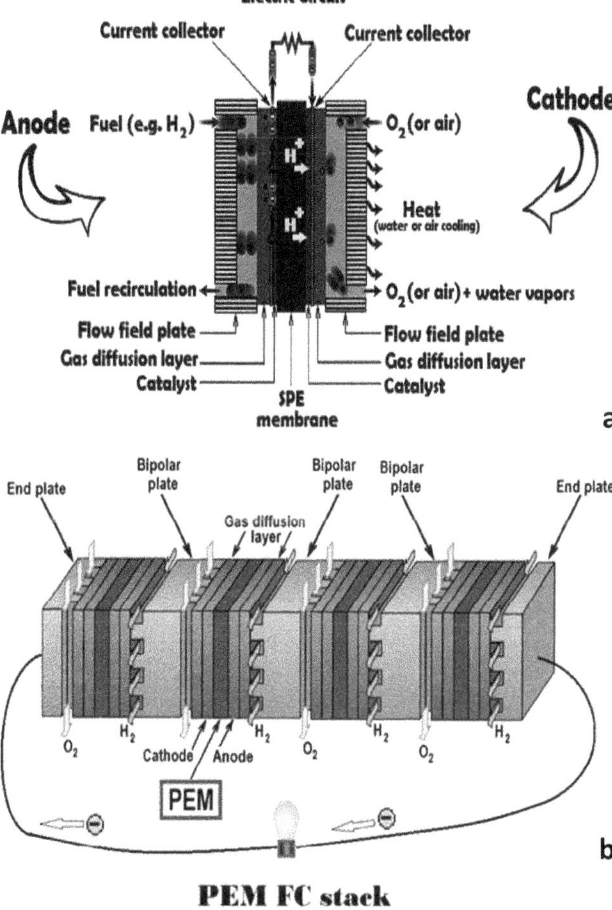


**Fig. 1.** (a) Schematic representation of a PEMFC design and (b) PEMFC stack (Kraytsberg and Ein-Eli, 2014). Copyright (2016), reprinted with permission from ACS.

Despite their favourable features, these membranes have several disadvantages, i.e., high cost, fuel permeability, and diminished performance accompanying with dehydration at temperatures above 80 °C (Hickner et al., 2004). In recent years, many studies have been focused on developing proton conducting membranes for operation at higher temperatures under lower humidification conditions to overcome the above-mentioned drawbacks aiming at generating higher FC performance compared with perfluorinated ionomers (Jalani et al., 2005; Ramani et al., 2005a; Jung et al., 2006; Ren et al., 2006; Sacca et al., 2006; Zhai et al.,

2006; Zeng et al., 2007; Park et al., 2008). Basically, to achieve this goal, there are two strategies: 1) fabrication of polymeric membranes based on hydrocarbon polymers such as poly(benzimidazole) (PBI), poly(arylene ether sulfone) (PSU), poly(styrene) (PS), poly(ether ether ketone) (PEEK), poly(ether sulfone) (PES), poly(2,6-dimethyl-1,4-phenylene oxide) (PPO), etc. (Rikukawa and Sanui, 2000; Gil et al., 2004; Xu et al., 2008; Ahmad et al., 2010), and 2) preparation of composite membranes through the application of inorganic micro-/nanoparticle (Di et al., 2015). The use of organic–inorganic composite membranes by incorporating inorganic fillers, in the ionomer matrix can significantly affect the properties of the matrix. The composite membranes exhibit suppressed fuel crossover, improved thermal, mechanical, dimensional, and oxidative stability. The properties of polymer composites depend on the type of nanoparticles that are incorporated, their size and shape, as well as their concentration and their interactions with the polymer matrix. Dispersion of inorganic nanoparticles in a polymer matrix usually leads to nanoparticles agglomerate due to their specific surface area and volume effects while it can also result in a reduction in proton conductivity due to decreased number of sulfonate groups per unit volume. These problems can be overcome by modification of the surface of the inorganic particles (Laberty-Robert et al., 2011; Kango et al., 2013; Liu et al., 2015).

### *1.1. Scope of the review*

Various inorganic fillers have been used in many polymer membranes developed for PEMFC. The objective of this review is to highlight these fillers and their impacts on proton conductivity, methanol permeability, water uptake, mechanical and thermal properties and cell performance of the resulting nanocomposite membranes.

## 2. Category of the materials used in PEMs

Generally, the materials used in synthesizing the polymer electrolyte membranes, also known as PEMs, can be categorized into five different groups: perfluorinated ionomers, partially-fluorinated polymers, non-fluorinated hydrocarbons, non-fluorinated membranes with aromatic backbones, and acid-base complexes (**Table 2**). Several reviews are available describing the materials used in the synthesis of the polymer electrolyte membranes (Peighambardoust et al., 2010; Dupuis, 2011; Awang et al., 2015).

## 3. Fabrication methods of nanocomposite membranes

Among a variety of approaches applied to incorporate inorganic fillers into an ionomer matrix, blending, infiltration (also called *"in situ* method"), and sol-gel approach (Tripathi and Shahi, 2011; Li et al., 2013), have received a great deal of attention owing to their wide availability which are described in the following sections.

### *3.1. Blending method*

The simplest method for the fabrication of polymer/inorganic nanocomposites is direct mixing of the nanoparticles into the polymer matrix. The mixing can be done by melt blending or solution blending. Filler agglomeration is the main difficulty faced in the blending method which can be overcome by the modification of the surface of the inorganic particles (**Fig. 2a**).

### *3.2. Sol-gel method*

The sol-gel method is a low temperature synthesis method and has been extensively used to synthesize organic-inorganic nanocomposites since the 1980s (**Fig. 2b**). This process is generally done by hydrolysis and condensation reactions of metal alkoxides, M $(OR)_n$ (M = Si, Ti, Zr, VO, Zn, Al, Sn, Ce, Mo, W, etc. and R = Me, Et, . . .) inside a polymer dissolved in non-aqueous or aqueous solutions. These reactions are as follows (**Eqs. 1 and 2**) (Pomogailo, 2005):

$$M(OR)_4 + 4\ H_2O \rightarrow M(OH)_4 + 4\ ROH \qquad \text{Eq. 1}$$

**Table 2.**
The chemical structure of some of the most commonly-used polymers in PEMs

| Name | Structure | Refference |
|---|---|---|
| Nafion® | 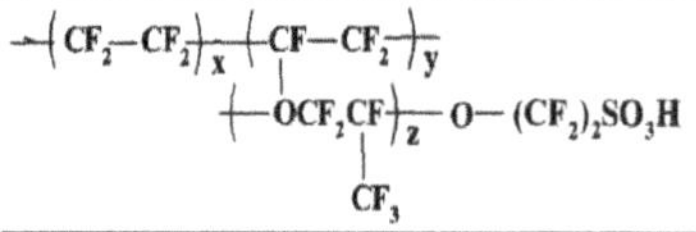 | Peighambardoust et al. (2010) |
| SPSU | 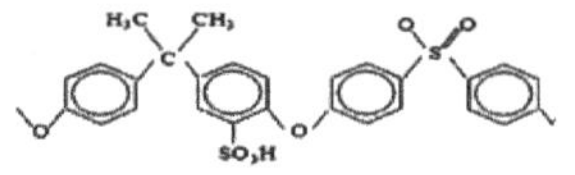 | Awang et al. (2015) |
| SPS | 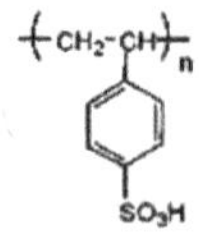 | Li et al. (2013) |
| SPEEK | 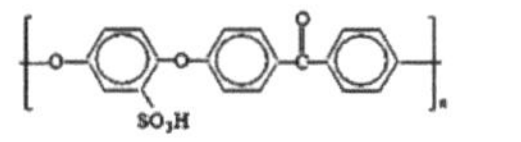 | Dupuis (2011) |
| PBI | 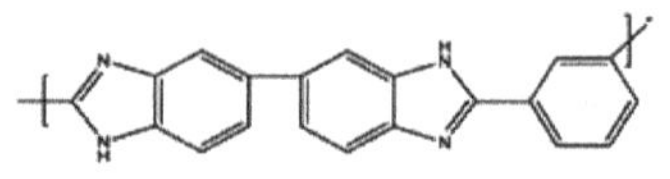 | Sharaf and Orhan (2014) |
| SPES | 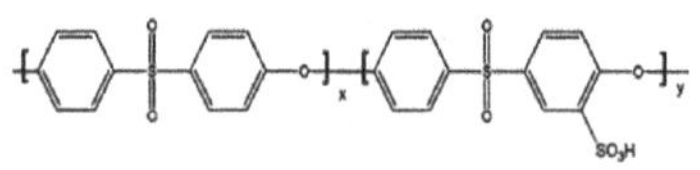 | Tripathi and Shahi (2011) |
| SPPBP | 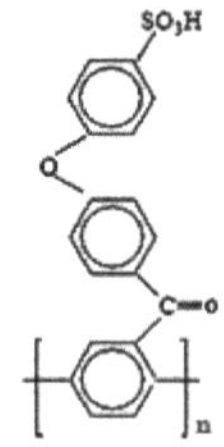 | Dupuis (2011) |
| SPPO | 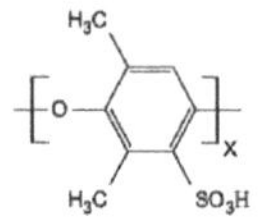 | Pomogailo (2005) |

$$mM(OH)_4 \rightarrow (MO_2)m + 2\ mH_2O \qquad \text{Eq. 2}$$

Silicon alkoxides are not very sensitive to hydrolysis and gelation may take place within several days when pure water is added. Therefore, hydrolysis and condensation proceed without catalysts for non-silicate metal alkoxides, whereas acid or base catalysts are required for silicon alkoxides (Livage, 2004). Four factors affect the kinetics reactions and consequently the final structure and properties of the product, including: the molar ratio of water/silane, the tape catalyst, temperature, and the solvent nature which were briefly described previously by Bounor-Legaré and Cassagnau (2014).

### *3.3. Infiltration method*

*In-situ* or infiltration methods have also been used to prepare organic-inorganic nanocomposites, in which the precursors of inorganic fillers

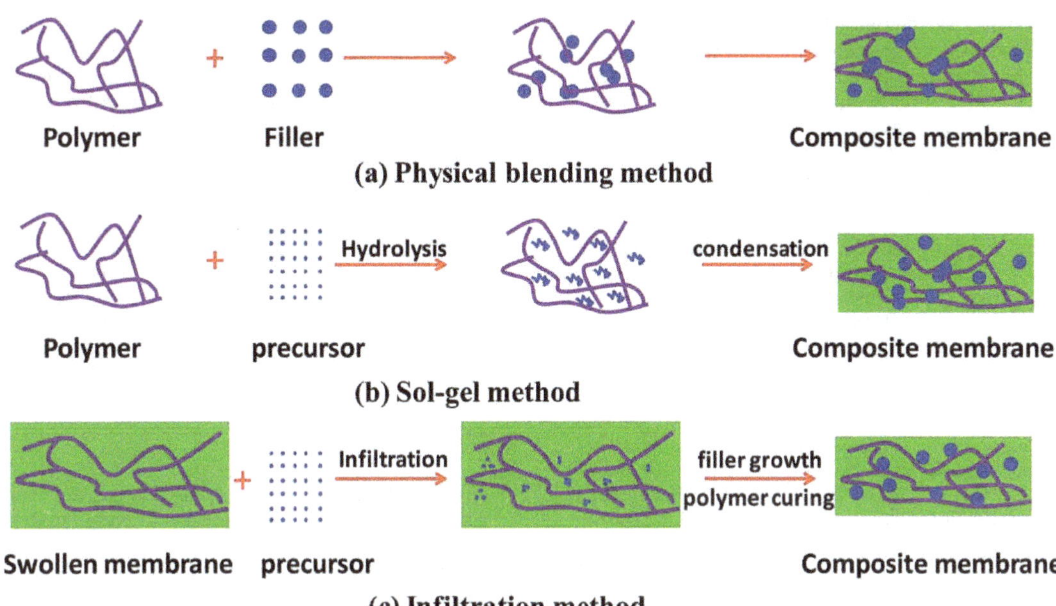


**Fig. 2.** Three typical methods for fabricating composite membranes: (a) physical blending method; (b) sol-gel method; and (c) infiltration method (Li et al., 2013). Copyright (2016), reprinted with permission from RSC.

infiltrate into a swollen or hydrogel-like polymer matrix (to increase the pore or void volume before infiltration), and then the nanocomposite membranes are obtained through filler growth, removing the impurities, and polymer curing. Meanwhile, the isolation effect caused by the polymer network can hinder the undesirable agglomeration of nanoparticles and simultaneously lead to controled particles size and uniform distribution (**Fig. 2c**).

## 4. Proton conduction in PEMs

The high proton conductivity of Nafion® at low temperatures and its lower conductivity at high temperatures have prompted many researchers to investigate its proton conduction mechanism (Mishra et al., 2012). Proton conductivity is a major parameter to evaluate the membrane performance. It has been reported that proton conductivity depends on the degree of sulfonation, pre-treatment of the membrane, water uptake, as well as ambient relative humidity (RH) and temperature (Jaafar et al., 2009). Proton conduction through membranes follows two types of mechanisms: Vehicle mechanism and Grotthuss mechanism. In the Grotthuss mechanism, the protons jump from one ionic site ($H_3O^+.SO_3^-$) to another while in the vehicle mechanism, protons attach to free water molecules and diffuse. Proton conduction is accompanied with activation energies; i.e., 0.1 <Eact< 0.4 eV by the Grotthuss mechanism and Eact > 0.5 eV by vehicle mechanism (Ren et al., 2013). A schematic representation of the two mechanisms is shown in **Figure 3**.

**Fig. 3.** Illustration of proton conduction models the Grotthuss mechanism (top) and the vehicle mechanism (bottom). Adopted from Ren et al. (2013).

## 5. Measurement methods of the PEMs properties

### *5.1. Pre-treatment of PEMs*

In order to measure the PEMs properties, they should be in H-form and free from impurities. Therefore, PEMs are activated by chemical treatments prior to analyses. Commonly, membranes are first soaked into a $H_2O_2$ solution, followed by treatment with a mixture of water and $H_2SO_4$ under continuous stirring. The membranes are then washed with deionized water (DI) water (Neelakandan et al., 2014; Dutta et al., 2015). However, Nair's group suggested that the introduction of harsh conditions such as $H_2O_2$ and acid treatment may destroy or disrupt the structure of the layered inorganic materials in the polymer matrix (Hudiono et al., 2009).

### *5.2. Membrane proton conductivity measurement*

The proton conductivity of membranes is calculated by the electrochemical impedance spectroscopy (EIS) and can be measured in two directions; i.e., in-plane or through-plane (Tang et al., 2012), by either four-probe or two-probe method. Conductivity ($\sigma$) is calculated using the following equation (**Eq. 3**):

$$\sigma = \frac{L}{RA} \qquad \text{Eq. 3}$$

For the in-plane test, L is the distance between the two electrodes and A is the cross-sectional area of the membrane; for the through-plane test, L is the thickness of the membrane and A is the overlap area of the two electrodes. R is the impedance of the membrane in Nyquist plot, which is determined from the intercept of the impedance curve with the real axis for through-plane test (Silva et al., 2004; Luan et al., 2008) and from the frequency at the minimum imaginary response for the in-plane experiment (Choi et al., 2008).

### *5.3. Methanol permeability measurement*

Methanol permeability is the product of the diffusion coefficient and the sorption coefficient in which the diffusion coefficient reflects the effect of a surrounding environment on the molecular motion of the permeate and the sorption coefficient correlates with the concentration of a component in the fluid phase (Marx et al., 2002; Kumar et al., 2009). A proton conducting membrane with low methanol permeability is required for DMFC. Because methanol transport through the membranes causes loss of fuel, reduced fuel efficiency, reduced cathode voltage and cell performance, and excess thermal load in the cell which can be controlled by factors such as the hydrophilic channel size, water uptake, membrane compaction, and other operating conditions (Won et al., 2003; Li and Yang, 2009). Methanol permeability is determined by a side-by-side cell, one side is filled with a methanol solution in DI water (side A) while pure water is placed in the other side (side B). The solutions in both compartments are continuously stirred with a magnetic stirrer during the experiment to ensure homogeneity (see **Fig. 4**). Several spectroscopic techniques have been used to measure methanol concentration in the water compartment. For example Chang's and Javanbakht' groups recorded methanol concentration in the side B using a density meter (Chien et al., 2013; Beydaghi et al., 2015). Recently, the crossover methanol (from side A to side B) has been examined by UV–vis spectroscopic technique using sodium nitroprusside (SNP) as chromogenic reagent. In fact, SNP is a mixture of sodium nitroprusside, potassium ferrocyanide, NaOH, and water in a certain proportion (Das et al., 2014; Dutta et al., 2015). The concentration of the methanol in side B can also be measured by gas chromatography (GC) method (Shahi, 2007; Hasanabadi et al., 2011). Methanol permeability using GC is determined as follows (**Eq. 4**):

$$C_B(t) = \frac{AP}{LV_B} C_A(t - t_0) \quad \text{Eq. 4}$$

where, $C_B(t)$ and $C_A$ are the concentrations of methanol in side A and B (mol $L^{-1}$), respectively, P is methanol permeability ($cm^2\ s^{-1}$), $V_B$ is the volume of DI water in side B ($cm^3$), L is the thickness of the membrane (cm), and A is the membrane area ($cm^2$). In other studies, cyclic voltammograms (CV) method (Escudero-Cid et al., 2015) and refractometry (Neelakandan et al., 2014) have also been used for measuring methanol concentration.

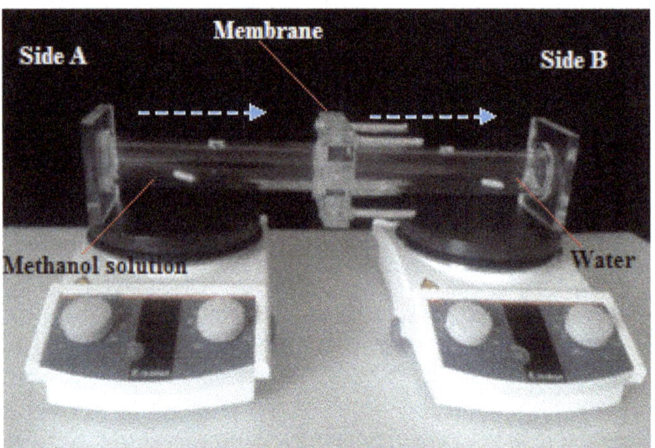


**Fig. 4.** A schematic illustration of the diffusion cell utilized for analyzing methanol crossover through membranes.

*5.4. Water uptake measurement*

The water uptake of the membranes is defined as mass ratio of the absorbed water to that of the dry membrane (Xi et al., 2007). Before measurements, membrane samples are dried at 80 °C for 12 h, weighed, and are then soaked in DI water at room temperature for 24 h. The samples whose surface water is fully hydrated are blotted by tissue papers and weighted immediately. The water uptake of the membranes is calculated using **Equation 5**:

$$WU\ (\%) = \frac{M_{wet} - M_{dry}}{M_{dry}} \times 100 \quad \text{Eq. 5}$$

where, $M_{wet}$ and $M_{dry}$ are the weights of the wet and dry membranes, respectively. A high water content in a membrane generally guarantees excellent proton conductivity. However, high water uptake can cause undesired side effects such as low mechanical strength, poor hydrolytic stability, low dimensional stability, and high methanol permeability, especially in DMFC applications (Kim et al., 2006). The membrane swelling (SW) is determined according to following equation (**Eq. 6**):

$$SW(\%) = \frac{L_{wet} - L_{dry}}{L_{dry}} \times 100 \quad \text{Eq. 6}$$

where $L_{wet}$ and $L_{dry}$ are the thicknesses of the wet and dry membranes, respectively. The membrane water content parameter (λ) is the ratio of the mole number of water molecules to the sulfonate groups, and is calculated using the following equation (**Eq. 7**):

$$\lambda = \frac{WU}{IEC \times M_{water}} \quad \text{Eq. 7}$$

where WU, IEC, and $M_{water}$ are the water uptake, ion exchange capacity, and molecular weight of water (18 g $mol^{-1}$), respectively.

*5.5. Ion exchange capacity measurement*

The ion exchange capacity (IEC) indicates the number of milli-equivalents of ions in 1 g of the dry polymer (Smitha et al., 2003). IEC of membranes is measured with the back-titration method. Membranes in acid form are immersed in a sodium chloride (NaCl) solution for about 24 h to replace $H^+$ ions with $Na^+$ ions. Then, the released $H^+$ is titrated with sodium hydroxide (NaOH) using phenolphthalein as indicator. The IEC value is calculated by using **Equation 8**:

$$IEC = \frac{V_{NaOH} - C_{NaOH}}{W_{dry}} \quad \text{Eq. 8}$$

where $V_{NaOH}$ is the NaOH solution volume used to neutralize (ml) the $H^+$, $C_{NaOH}$ is the concentration of NaOH (mol $L^{-1}$) and $W_{dry}$ is the mass of the dry membrane (g).

*5.6. Durability/stability*

*5.6.1. Oxidative stability*

The $H_2O_2$ generation at the cathode (due to the electrochemical two-electron reduction of oxygen) and at the anode (due to the chemical combination of crossover oxygen and hydrogen) and its decomposition to intermediate products, such as $HO^•$ and $HO_2^•$, could cause membrane chemical/electrochemical degradation (Bose et al., 2011). The oxidative stability of membranes is investigated by immersing of a membrane sample in Fenton's reagent (3 wt. % $H_2O_2$ + 2 ppm $FeSO_4$) in a shaking bath at 80 °C. The oxidative stability is evaluated in terms of the weight loss of the membranes when they start to break into pieces (due to the attack of radical species; $HO^•$ and $HOO^•$) for an elapsed time (Amirinejad et al., 2011).

*5.6.2. Thermal stability*

Thermal decomposition of membranes is investigated by using thermogravimetric analyzer (TGA). The decomposition profiles of membranes measured by TGA depend on many factors, including heating rate, gas (nitrogen or oxygen) flow rate, and sample preparation technique (Mishra et al., 2012). There are generally three stages in the decomposition process of membranes: (1) the release of free and bound water molecules; (2) the decomposition of functional groups; and (3) the separation of core series of the membrane (Zakaria et al., 2016). Meanwhile, thermal transition behaviour (glass transition) of PEMs is obtained by differential scanning calorimetry (DSC) study (Tripathi and Shahi, 2011).

*5.6.3. Mechanical stability*

In view of the potential practical applications in lithium batteries, fuel cells, electrochromic windows, and other electrochemical devices, it is essential for polymer electrolyte membranes to retain good mechanical strength (Wang and Kim, 2007). The membrane must have good mechanical resistance to stretching and shear in hydrated and dry states (Laberty-Robert et al., 2011). Dynamical mechanical analysis (DMA) technique has been used to determine mechanical stability of membranes. Briefly, samples are submitted to a periodic mechanical strain or stress, while the temperature is changed with a constant rate. The storage modulus E' (elastic) and loss modulus E" (viscous) of polymers are measured as a function of temperature. The E"/E' ratio, also named tan $\delta$, is related to mechanical damping (Sgreccia et al., 2010).

**6. Hybrid organic–inorganic membranes (nanocomposite membranes)**

Inorganic–organic composite membranes can be classified into two main categories: 1) membranes composed of proton conductive polymers and less-proton conductive inorganic particles and 2) membranes composed of proton conductive particles and less-proton conductive organic polymers (Zhang et al., 2012). Incorporation of inorganic nanoparticles into a polymer matrix strongly influence the original characteristic of the polymers, due to the interface interaction with the polymer matrix. The inclusion of inorganic fillers improves the mechanical properties and the membrane water management, while also suppresses fuel crossover by increasing the transport pathway tortuousness (Peighambardoust et al., 2010). The preparation method of composite membranes containing inorganic and inorgano–organic proton-conducting particles is very important as it will influence the microstructure of the membrane. The intrinsic characteristics of the particles such as size or the specific surface area, type, surface acidity, shape, and their interactions with the polymer matrix are particularly important and can induce large variations in membrane performance (Laberty-Robert et al., 2011). However, enrichment of the available fillers, especially nano-sized fillers, has been one of the most investigated subjects (Li et al., 2013). The fillers can be classified into two types, 1) solid nonporous filler, such as $SiO_2$, and $TiO_2$ nanoparticles, and 2) solid porous filler, such as Zeolites, porous metal oxides, metal–organic frameworks (MOFs), and carbon nanotubes (CNT). Countless organic–inorganic hybrid nanocomposite membranes have been reported in the literature by using inorganic material such as: silicates, titanium dioxide, zirconium dioxide, iron oxide, yttrium oxide, zirconium phosphate, heteropolyacids, and CNT. The nanocomposite membranes based on different fillers will be described in the following sections.

*6.1. Nanocomposite PEMs with graphene oxide*

Graphene oxide (GO) is an amphiphilic material with a two-dimensional laminated structure and contains epoxy and hydroxyl groups on the basal plane, and carboxylic acid groups along the sheet edge. The presence of these oxygen-containing functional groups facilitate the hydration of GO and helps to hold more water and improve the proton conductivity. Owing to this feature and also large surface area, as well as intrinsic mechanical (due to large Young's modulus) and chemical stability and fuel crossover barrier, GO is one of the best nanofillers to be used in PEMFCs (Chien et al., 2013; Bayer et al., 2014; Lue et al., 2015).

Beydaghi et al. (2015) prepared the nanocomposite blend membranes based on SPEEK/PVA blend polymers and SGO/$Fe_3O_4$ nanosheets using solution casting method for DMFC applications. The methanol permeability of SPEEK/PVA blend membrane decreased from $1.78 \times 10^{-6}$ cm$^2$s$^{-1}$ to $8.83 \times 10^{-7}$ cm$^2$s$^{-1}$ with the addition of 5 wt.% SGO/$Fe_3O_4$ nanosheets due to the fuel barrier properties of graphene-based nanosheets. Moreover, the proton conductivity (0.084 S cm$^{-1}$ at 25 °C) of the nanocomposite membranes increased due to the interactions between the sulfonic acid groups of SGO, and the surface hydroxyl groups of $Fe_3O_4$ nanoparticles with free water molecules by Grotthus and Vehicle mechanism, respectively.

In a study, Kumar et al. (2014) synthesized an SGO/SPEEK composite membrane, and the effects of SGO on the properties of SPEEK such as proton conductivity and fuel cell performance were studied. The composite membranes developed exhibited a very good conductivity (0.055 S cm$^{-1}$ at 80 °C and 30% RH) and reasonable PEMFC performance (378 mW cm$^{-2}$ at 80 °C and 30% RH), which were higher than that of recast SPEEK.

Among the extensive efforts devoted to the development of PEM with adequate conductivity operating under elevated temperatures and anhydrous (low humidity) conditions, acid–base composites have attracted a great deal of attention owing to their unique transport manner. Within these composites, proton donor (acid group) and acceptor (base group) are closely linked, and protons can transport between donor and acceptor *via* the Grotthuss mechanism without water.

An acid–base-paired nanocomposite membranes was fabricated by He et al. (2014), using polydopamine-modified GO into SPEEK matrix. Nanocomposite membrane with 10 wt.% of fillers displayed an anhydrous proton conductivity of 0.498 mS cm$^{-1}$ in comparison with SPEEK control membrane (0.387 mS cm$^{-1}$). This study showed a 47% increase in maximum current density (698.6 mA cm$^{-2}$) coupled with a 38% increase in maximum power density (192.1 mW cm$^{-2}$) with 5 wt.% fillers.

Zarrin et al. (2011) also investigated the effects of functionalized GO with 3-mercaptopropyltrimethoxysilane (MPTMS) and subsequent oxidation of the thiol groups to sulfonic acid groups, as inorganic fillers in a Nafion® composite membrane for high temperature PEMFCs. Proton conductivity and single cell test results exhibited significant improvements for functionalized GO/Nafion® membranes (4 times) over recast Nafion® at 120 °C with 25% humidity. In a recent study, He et al. (2016) fabricated Nafion®/GO composite membranes *via* spin-coating method using 1,4-phenylenediamine-2-sulfonic acid (PDASA) as crosslinker (see **Figure 5**). The authors claimed that methanol permeability decreased by 93% while retaining the high proton conductivity of Nafion®, due to the synergistic optimization of methanol-transport and proton-transport channels within the GO film.

The layer-by-layer (LBL) self-assembly is a simple and remarkably adaptable method introduced by Decher (1997) and Decher et al. (1998), and has been applied to deposit a thin multilayer film on a substrate by electrostatic association between alternately deposited, oppositely charged components. LBL technology has been used successfully in the area of both PEMs and DMFC electrodes and electrolytes (Kim et al., 2004; Farhat and Hammond, 2006; Jiang et al., 2006).

Recently, Nafion®/GO composite membranes was fabricated *via* the LBL procedure using 1,4-phenyldiamine hydrochloride (PDHC) as cross-linker and was used as a PEM for DMFC (Wang et al., 2015). The composite membranes exhibited higher selectivity and lower methanol permeability than Nafion® 117. Yuan et al. (2014) also demonstrated that the GO multilayer films on the Nafion® membranes not only reduced methanol crossover by 67% but also enhanced membrane strength. In their work, poly (diallyldimethylammoniumchloride) (PDDA) was used as cross-linker.

*6.2. Nanocomposite PEMs with heteropolyacids*

Polyoxometalates (POMs) are Brønsted acids with nano-size metal–oxygen clusters which have received much attention in recent years owing to their thermal stability and intrinsic proton conductivity (0.18 S cm$^{-1}$ in the hydrated crystalline form (phosphotungstic acid; PTA·29 $H_2O$)) (Malers et al., 2007). POMs incorporation in the polymer matrix enhance the proton conductivity and the FC performance due to hygroscopic properties and strong acid strength of POMs cluster compounds by providing a preferential pathway for proton hopping and sustaining a large number of water molecules in their hydration sphere. However, the major obstacle faced in the utilization of POMs in PEMs is their high solubility in water and the low surface area of PTA ((5-8) m$^2$ g$^{-1}$) which limits the accessibility to acid sites (Kourasi et al., 2014). In order to increase the stability and the surface area, two approaches have been suggested. The first involves fixation of the POMs on support oxides such as $SiO_2$, $Al_2O_3$, $ZrO_2$, and $TiO_2$ (Saccà et al., 2008). The second option is the use of heteropoly salts by substituting the protons ($H^+$) with large cations such as $Cs^+$, NH4$^+$, $Rb^+$ and $Tl^+$ (Ramani et al., 2005b). This second approach was investigated by Amirinejad et al. (2012). They dispersed $Cs_{2.5}H_{0.5}PW_{12}O_{40}$ (CsPW) and $H_3PW_{12}O_{40}/SiO_2$ (PWS) throughout the sulfonated fluorinated bi-phenol (ESF-BP) copolymer separately and the performances of the fabricated membranes were investigated. Both POMs used provided additional surface functional sites throughout the composite

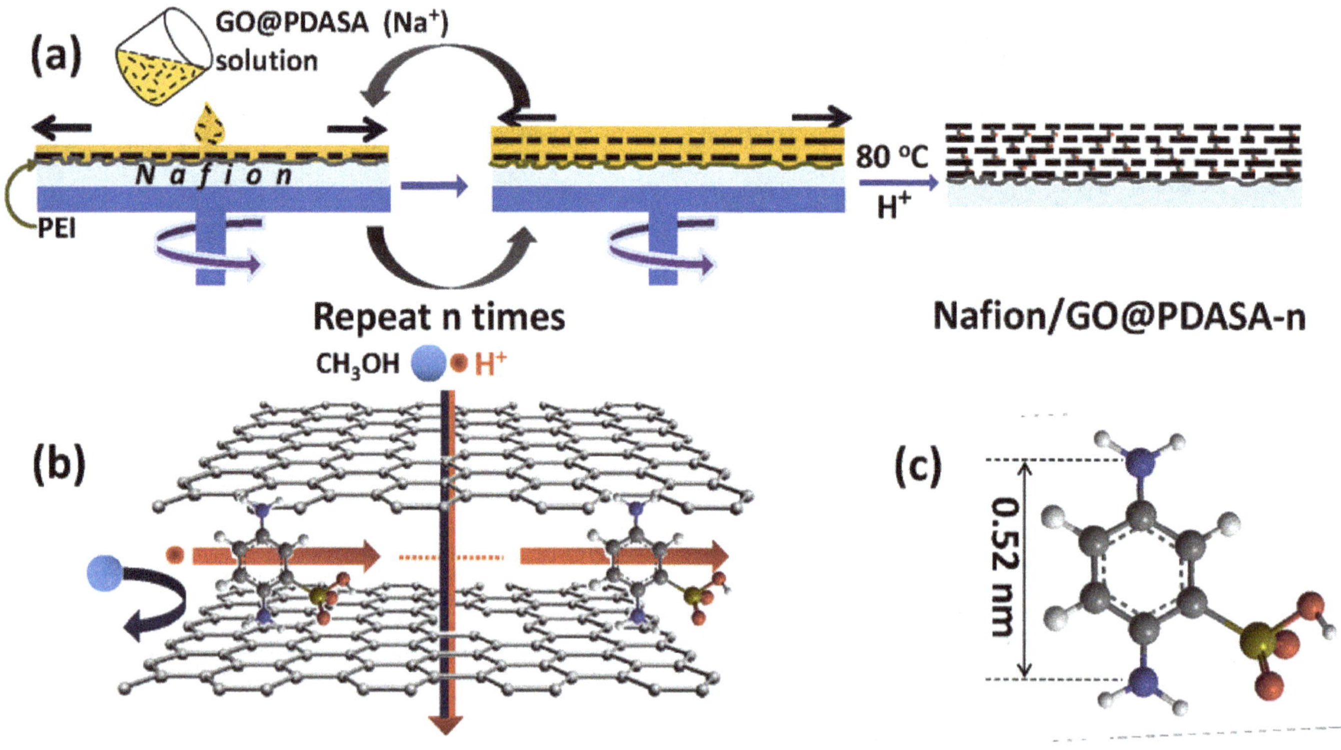


**Fig. 5.** (a) Fabrication of PDASA crosslinked GO film (GO@PDASA) by a spin-coating method; (b) transport mechanism of proton and methanol through the GO@PDASA film; and (c) structure of 1, 4-phenylenediamine-2-sulfonic acid (PDASA) (He et al., 2016). Copyright (2016), reprinted with permission from RSC.

membranes and facilitated proton transport. However, the conductivity and consequently the performance of the ESF-BP/7PWS membrane was reportedly higher than that of the ESF-BP/7CsPW membrane due to the higher number of acid sites and surface area.

Kim et al. (2015b) also studied polymer electrolyte FCs operating at elevated temperatures and low RH by utilizing a polyoxometalate modified GO–Nafion® membrane. The Nafion®/PW-mGO membrane exhibited a proton conductivity and a maximum power density of 10.4 mS $cm^{-1}$ and 841 mW $cm^{-2}$ at 20% RH at 80 °C, respectively.

In a similar attempt, water uptake, proton conductivity, and FC performance of Nafion® membrane was increased by the incorporation of $SiO_2$-PTA (Mahreni et al., 2009). The Nafion®/$SiO_2$/PTA composite membrane exhibited a current density of 82 mA $cm^{-2}$ at 0.6 V as compared with the Nafion® membrane (30 mA $cm^{-2}$ at 0.2 V). Recently, Hasani-Sadrabadi et al. (2016) reported modified Nafion® membranes with PTA-filled CNT nanostructures. The fabricated Nafion®/nanopeapod membranes demonstrated a proton conductivity and a power density of 0.202 S $cm^{-1}$ (at 90 °C) and 302 mW $cm^{-2}$ (at 40% RH and 120 °C), respectively, in comparison with 0.132 S $cm^{-1}$ and 84 mW $cm^{-2}$ for the recast Nafion® membrane.

Zhao et al. (2009) demonstrated that the multilayer films placed onto the surface of sulfonated poly(arylene ether ketone) (SPAEK–COOH) membrane by the LBL self-assembly of polycation chitosan (CTS) and negatively-charged inorganic particle PTA reduced methanol permeability by 2 orders of magnitude compared with Nafion® 117 while maintaining a high proton conductance. **Figure 6** illustrates the fabrication of the $(CTS/PTA)_n$ multilayer films.

### *6.3. Nanocomposite PEMs with NCTs*

Among inorganic fillers, CNTs, which are cylindrical graphene tubes with a nano-sized diameter consist of single or several graphene layers, have attracted considerable attention owing to their high aspect ratios of 100–1000, high specific surface areas, low densities, and remarkable mechanical properties. Because of these unique properties, recently, CNTs have been widely investigated as candidate PEM materials. Furthermore, the properties and performances of nanotubes can be significantly intensified through coordination effects by modification with different functionalized groups such as carboxylic acid-functionalized CNTs (Thomassin et al., 2007), sulfonated CNTs (Yun et al., 2011; Zhou et al., 2011; Yu et al., 2013), phosphonated CNTs (Kannan et al., 2010; Kannan et al., 2011), and Nafion®- and polybenzimidazole functionalized CNTs (Chang et al., 2011; Hasani-Sadrabadi et al., 2013).

Recently Cui et al. (2015) reported the incorporation of silica-coated CNTs as a new additive into the SPEEK matrix for DMFC applications. The composite membranes with a $SiO_2$@CNT loading of 5 wt.% showed one order of magnitude decrease in methanol crossover in comparison with pristine membranes, while the proton conductivity remained above 10-2 S $cm^{-1}$ at room temperature.

Zhang et al. (2014) synthesized sulfonated halloysite nanotubes (SHNTs) *via* a facile distillation precipitation polymerization, and found that incorporating 10% SHNTs enhanced the conductivity of SPEEK from 0.0152 to 0.0245 S $cm^{-1}$. The embedded SHNTs not only interconnected the ionic channels in SPEEK matrix and donated more continuous ionic networks which could serve as proton pathways allowing efficient proton transfer with low resistance, but also increased the thermal and mechanical stabilities of the resultant nanocomposite membranes by interfering with SPEEK chain motion and packing.

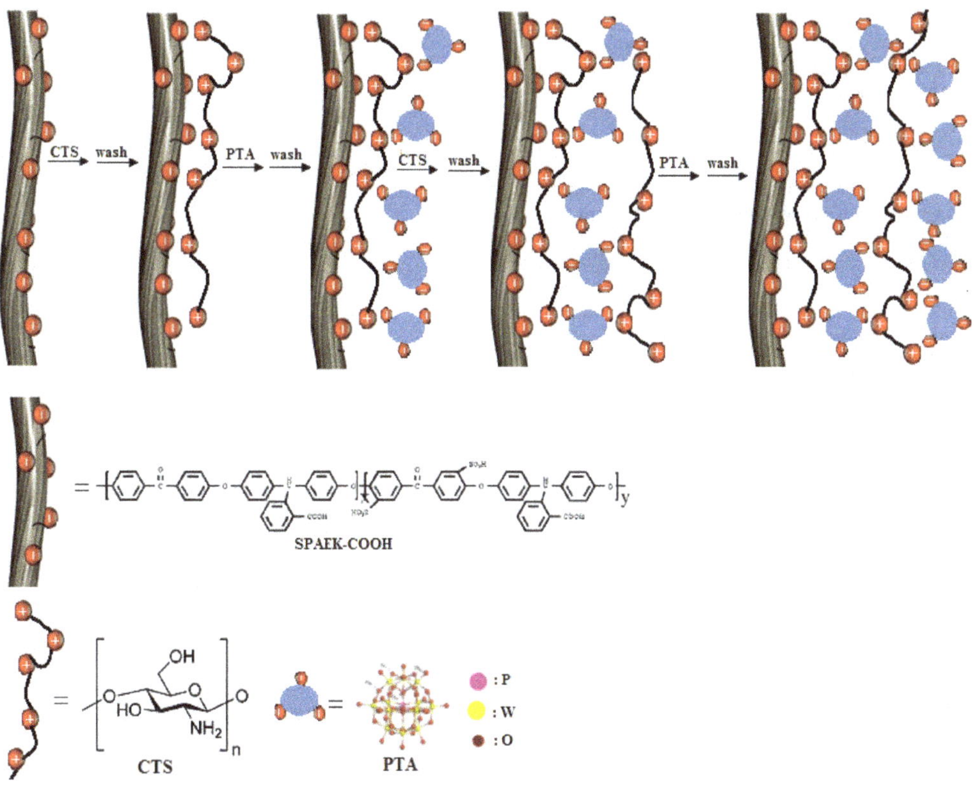


**Fig. 6.** Schematic representation of the fabrication of $(CTS/PTA)_n$ multilayer films. Adopted from Zhao et al. (2009).

By incorporating imidazole groups on the surface of CNTs into Nafion®, significant improvements in the power density, methanol permeability, and proton conductivity was achieved by Asgari et al. (2013). They investigated FC performance of the Nafion® doped with 0.5 wt.% CNTs and Im-CNT at two different methanol concentrations, i.e., 1 M and 5 M at 70 °C. The results obtained using the Nafion®/Im-CNT-0.5 wt.% showed current and power densities of 500 and 300 mA $cm^{-2}$ and 86.53, 74.22 mW $cm^{-2}$ at 1 and 5 M methanol, respectively. The open circuit voltage (OCV) of the Nafion®/Im-CNT-0.5 wt.% membranes was found higher than those of the other membranes. i.e., recast Nafion® and Nafion®/CNT-0.5 wt.%. This could be ascribed to the strong interactions between imidazole groups and Nafion® molecules which in turn decreased the size of channels and subsequently notably hindered methanol diffusion. Yun et al. (2011 and 2012) developed a sulfonated multi-walled CNT/sulfonated PES (s-MWCNT-s-PES) and sulfonated PVA/sulfonated multi-walled CNT (s-MWNTs/s-PVA) nanocomposite membranes. In both of these studies, s-MWCNT was added in order to act as filler for DMFC. The composite membranes demonstrated excellent proton conductivity and low methanol permeability.

### *6.4. Nanocomposite PEMs with $SiO_2$*

Silicate-based nanoparticles have been extensively studied because of their lower cost, inferior electrical conductivity, and better water retention properties compared with those of the other nanoparticles. By modifying the silicate surface using different modifiers phase inconsistency between organic polymer membranes and inorganic silicate could be avoided. In general, silica is synthesized through the hydrolysis and polycondensation of alkoxy silanes in an acidic or basic medium, using different precursors such as alkoxy silanes (like tetraethyl orthosilicate (TEOS)), sodium metasilicate, and fumed silica (Mishra et al., 2012). The hydrophilic nature of silicates helps in developing the proton conductivity of nanocomposite membranes. Moreover, addition of sulfonated and phosphonated silicates along with a high degree of dispersion of the nanomaterials in the polymer matrix can notably upgrade the conductivity of the nanocomposites.

Yoon et al. (2009) compared the performance of sulfonated poly(arylene ether sulfone (SPAES)/$SiO_2$ membranes prepared by wet-type milling method and sonication method. Comparatively, wet-type milling method remarkably improved the dispersion of $SiO_2$ in the SPAES matrix, due to the intensive impact of collisions between milling beads and nanoparticles. This enhancement in nanoparticle dispersion improved proton conductivity as well as methanol permeability and selectivity in the composite membranes.

The incorporation of $SiO_2$ nanoparticles in sulfonated polyimide containing triazole groups (SPI-8) remarkably improved the FC performances during low humidity operation at 53% RH and 80 °C (Sakamoto et al., 2014). Influence of the size and shape of silica nanoparticles on the properties and degradation of PBI-based membranes was investigated by Ossiander et al. (2014). PBI-based membranes with 40%, 80%, and 120% of the inorganic silica precursor TEOS were fabricated using *in situ* sol–gel reaction using (3-glycidoxypropyl)-methyldiethoxysilane (GPTMS) as crosslinker. The results obtained showed higher performance and mechanical stability in the composite membranes with 40% TEOS content.

Sulfonated GO–silica (S-GO–$SiO_2$)/Nafion® PEMs with enhanced transport properties were prepared *via* solution casting by Feng et al. (2014). The composite membranes showed an obvious reduction in methanol permeability (due to the increased tortuosity of the transport channels in the membrane matrix) as well as higher proton conductivity, and 2-fold increased selectivity (the ratio of proton conductivity to methanol permeability) compared with the recast Nafion® membrane. Farrukh et al. (2015) modified surface silica nanoparticles with poly(monomethoxy oligoethylene glycol methacrylate), poly(MeOEGMA), and employed them as conductivity enhancing additives for the fabrication of Nafion® nanocomposite membranes. The modified membranes containing 1% additives showed ~11 times higher proton conductivity at 20% RH and 25 °C, whereas at the same temperature and 80% RH, the proton conductivity of the nanocomposite membrane was ~4 times higher than that of the Nafion®.

## 6.5. Nanocomposite PEMs with $TiO_2$

$TiO_2$ is a hygroscopic metal oxide which improves the cell performance, in terms of higher operating temperature, easier water management, and thermo mechanical stability (Bose et al., 2011). Cozzi et al. (2014) synthesized and investigated propylsulfonic functionalized titania ($TiO_2$-$RSO_3H$) as inorganic fillers in a Nafion® composite membrane. The composite membrane containing 10 wt.% of fillers displayed the highest conductivity value ($\sigma$ = 0.08 S $cm^{-1}$ at 140 °C) and the best DMFC performance, with 64 mW $cm^{-2}$ power density (about 40% higher than the Nafion® cast membrane). SPEEK/phosphonic acid-functionalized titania nanohybrid membranes were fabricated by an *in situ* method using titanium tetrachloride ($TiCl_4$) as inorganic precursor and amino trimethylene phosphonic acid (ATMP) as modifier (Wu et al., 2015). The nanohybrid membranes demonstrated remarkably enhanced proton conduction (25%), a 23% decrement in methanol permeability, and also better thermal and mechanical stabilities. Gandhi et al. (2012) also showed that the addition of titanium dioxide had a very dramatic and positive effect on proton conductivity of pure polystyrene porous membranes. Aslan and Bozkurt (2014) synthesized proton conducting nanocomposite membranes *via* ternary mixtures comprising sulfated nanotitania (TS), sulfonated polysulfone (SPSU), and nitrilotri(methyl triphosphonic acid) (NMPA). These membranes displayed a maximum proton conductivity of 0.002 S $cm^{-1}$ at 150 °C.

Amjadi et al. (2010) prepared Nafion®-$TiO_2$ nanocomposite membranes by sol–gel and casting methods. Their results revealed that sol–gel method was better than casting due to the formation of fine particles and good distribution of $TiO_2$ particles. Water uptake (up to 3 wt.%) and thermal properties of these membranes were improved with increasing $TiO_2$ content. PEMFC performance at 110 °C was improved, in spite of a slight reduction in proton conductivity. To improve the interfacial compatibility between polymeric resin and inorganic materials and to enhance proton conductivity, Li et al. (2012) modified the Nafion® by of amine-tailored titanate nanotubes. The composite membranes developed showed about 4-5 times higher proton conductivity in comparison with pristine Nafion® and 3 times higher proton conductivity compared with that of composite membrane impregnated with unmodified titanate nanotubes. Wu et al. (2014) synthesized a series of amino acid functionalized titania submicrospheres (~200 nm) and incorporated them into SPEEK. All the as-prepared hybrid membranes exhibited improved methanol resistance compared with pristine SPEEK membrane. This could be rationalized by the size reduction of the ionic channels, which was unfavourable for methanol crossover. Moreover, the incorporation of this filler could introduce acid-base pairs as proton donors and acceptors into polymer, which helped to form continuous pathways for proton hopping, thus leading to increased proton conductivity.

## 6.6. Nanocomposite PEMs with perovskite-type oxides

Protonic conductors with perovskite structures have been considered attractive owing to their high chemical stability, excellent thermal and mechanical stability, relatively low cost, and high applicability in electrochemical devices for energy generation (An et al., 2012). Recently, nanocomposite membranes based on perovskite-type oxides for high temperature PEMFCs have been studied by Hooshyari et al. (2015) and Shabanikia et al. (2015). Accordingly, polybenzimidazole- nanocomposite membranes based on $BaZrO_3$ and $SrCeO_3$ reportedly displayed higher water uptake and proton conductivity compared with virgin PBI membranes. This improvement was attributed to the hygroscopic nature of $BaZrO_3$ and $SrCeO_3$ nanoparticles. Also they investigated the effects of variations in the percentages of the nanoparticles and the solvent used (water, ethanol, and water/ethanol (1:1 v/v)) for the dispersion of nanoparticles within of Nafion®/$Fe_2TiO_5$ nanocomposite membranes on the proton conductivity, water uptake, and also the thermal stability of the membranes (Hooshyari et al., 2014). The results displayed that 2 wt.% of the nanoparticles *vs.* Nafion® membrane in 10 mL water as solvent had the highest proton conductivity (226 ± 7 mS $cm^{-1}$) compared with the other membranes at 25 °C and 95% RH due to the existence of water as more polar solvent and large affinity of the $Fe_2TiO_5$ nanoparticles with water.

## 6.7. Nanocomposite PEMs with zeolite

Zeolites are a class of crystalline aluminosilicates, which form a framework of $SiO_2$ and $AlO_4$ tetrahedra and contain exchangeable cations on the extra-framework to maintain the electrical neutrality (Dyer, 1988). Zeolites are highly hydrophilic solids and have a high water sorption capacity because of the charged anionic framework and the extra-framework cations (Kornatowski, 2005; Ng and Mintova, 2008) in addition to their open structure, high pore volume, and vast surface area. Most zeolites reportedly used in the composite membranes for FCs are micrometer-sized particles with low proton conductivity (Libby et al., 2003; Tricoli and Nannetti, 2003). Nanometer-sized inorganic additives have been proven to be crucial to the compatibility between the inorganic filler and Nafion®, which has significant effects on the proton conductivity and methanol permeability of the composite membranes (Zimmerman et al., 1997). Different types of zeolite nanocrystals (A, NaX, NaY, Beta, etc.) have been prepared successfully by template and template-free methods (Wang et al., 2002; Wang et al., 2003; Chen et al., 2005; Holmberg et al., 2005). To minimize the loss of proton conductivity caused by the fillers while reducing the methanol permeability, acid functionalized (–$SO_3H$) zeolites have been prepared (Jones et al., 1998) and used successfully as an inorganic filler in composite membranes (Holmberg et al., 2005). The acid functionalized zeolites were selected for their favourable proton conductivity (~ 0.02 S $cm^{-1}$), excellent acid stability, and hydrophilic nature.

Recently, Devrim and Albostan (2015) modified Nafion® membranes with zeolit by solvent casting procedure; the modified membranes demonstrated increasing water uptake and proton conductivity owing to the water retention properties of the zeolite and interaction between the Nafion® polymer and zeolite particles. In another study, Auimviriyavat et al. (2011) reported ferrierite zeolite as inorganic filler in SPEEK membrane. In general, ferrierite zeolite in known to possess favourable acid and thermal stabilities because of its relatively high $SiO_2/Al_2O_3$ molar ratio (Göğebakan et al., 2007). Auimviriyavat et al. (2011) claimed that ferrierite zeolite improved the mechanical strength and water retention of the membrane. These membranes demonstrated better proton conductivity, low methanol permeability, and the highest selectivity of $7.148 \times 10^{-3}$ S. cm $s^{-1}$. Nafion®/acid functionalized zeolite beta (NAFB) nanocomposite membranes were also prepared *via in situ* hydrothermal crystallization by Chen et al. (2006), with showed a slightly lower proton conductivity but a markedly lower methanol permeability (40% less) in comparison with Nafion® membranes (**Fig. 7**). These membranes were also shown to offer higher OCV (by 3%) and higher maximum power density (by 21%) than Nafion®.

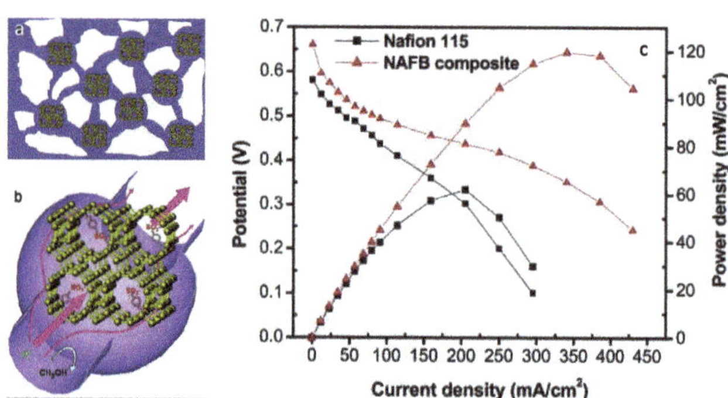


**Fig. 7.** Schematics of NAFB composite membrane and its proton and methanol transport: a) white region represents the hydrophobic section of the Nafion® membrane and the coloured region represents the hydrophilic section of the Nafion® membrane; b) proton and methanol transport in the NAFB composite membrane; and (c) DMFC single cell performance at 70 °C, fed with 5 M methanol (Chen et al., 2006). Copyright (2016), reprinted with permission from ACS.

### *6.8. Nanocomposite PEMs with iron oxide*

Iron oxide nanoparticles have received a great deal of attention owing to their easy and controlled synthesis, low toxicity, magnetic and catalytic properties, and as a result, their potential applications in different fields such as magnetic resonance imaging (MRI) (Sun et al., 2008), drug targeting (Chomoucka et al., 2010), catalysts (Maleki et al., 2014; Safari and Javadian, 2014), and hyperthermia treatments (Laurent et al., 2011). The fabrication of anisotropic membranes orientated in the desired direction in a matrix is an attractive idea to enhance ionic conductivity. Two methods to design anisotropic nanocomposite membranes are to align nanofillers in the matrix under electric field (Oren et al., 2004) or magnetic field (Brijmohan and Shaw, 2007). In a study, SGO/$Fe_3O_4$ nanosheets were aligned under magnetic field in PVP matrix by Beydaghi and Javanbakht (2015). With the orientation of the SGO/$Fe_3O_4$ nanosheets, the water transferring channels in the membrane became wide and the empty spaces accommodating water molecules in the membranes increased, hence, water uptake and swelling of membranes increased. Moreover, these membranes showed higher thermal stability, methanol permeability, and selectivity with a maximum power density of 25.57 mW $cm^{-1}$ at 30 °C compared with a nonaligned membrane (Beydaghi and Javanbakht, 2015). Hasani-Sadrabadi et al. (2014a) aligned CTS-coated superparamagnetic iron oxide nanoparticles (CTS-SPIONs) in Nafion® matrix. **Figure 8** presents a cross section of a high resolution transmission electron microscopy (HR-TEM) image of the aligned nanocomposite which clearly confirms a chain-like assembly of the CTS-SPIONs because of the magnetic field (Hasani-Sadrabadi et al., 2014a). The modified membranes displayed a power output over five times higher than that of the unmodified Nafion® at 120 °C and 40% RH. In an investigation, the surface of γ-$Fe_2O_3$ nanoparticles was modified by MPTMS as sulfonic acid functional group precursor and subsequently, a magnetic field was applied during solvent casting and evaporation to align the nanoparticles in Nafion matrix (Hasanabadi et al., 2013). The aligned nanocomposite membranes showed higher ionic conduction, drastic reduction in methanol permeability and activation energy for proton migration, and also significant higher selectivity as compared with randomly-distributed nanocomposite membranes.

### *6.9. Nanocomposite PEMs with MOFs*

MOFs, or coordination polymers (CPs), are open networks consisting of metal-centered secondary building units (SBUs) joined together by organic linkers to form large one-dimensional (1-D), two-dimensional (2-D), or three-dimensional (3-D) networks (Ren et al., 2013). Structural features of MOFs such as their crystallinity, regular arrangement of voids, tailorable porosity, and dynamic behaviour are especially attractive for their use as proton conductors (Ramaswamy et al., 2014). MOFs exhibit proton conductivity; protons can be passed through the coordination skeleton of a MOF (Ohkoshi et al., 2010) or through carriers, such as imidazole (Bureekaew et al., 2009), 1, 2, 4-triazole (Hurd et al., 2009), or water (Duan et al., 2009), loaded in the pores. Liang et al. (2013) incorporated a 2-D MOF containing protonated tertiary amines as proton carriers into a polyvinyl pyrrolidone (PVP) matrix. Conductivity measurements at 53% RH and 333 K indicated that the conductivity increased from 1.4 $\times 10^{-8}$ (for pure PVP) to 3.2 $\times$ $10^{-4}$ S $cm^{-1}$ for MOF–PVP composite. Wu et al. (2013) reported the fabrication and characterization of a composite membrane with a high proton conductivity by combining sulfonated PPO with Fe-MIL-101-$NH_2$ or [$Fe_3(O)(BDC\text{-}NH_2)_3(OH)(H_2O)_2$].$nH_2O$ (where BDC-$NH_2$ is 2-aminoterephthalate) *via* the Hinsberg reaction, wherein a sulfonyl chloride reacts with an amine to form a sulfonamide salt (see **Fig. 9**).

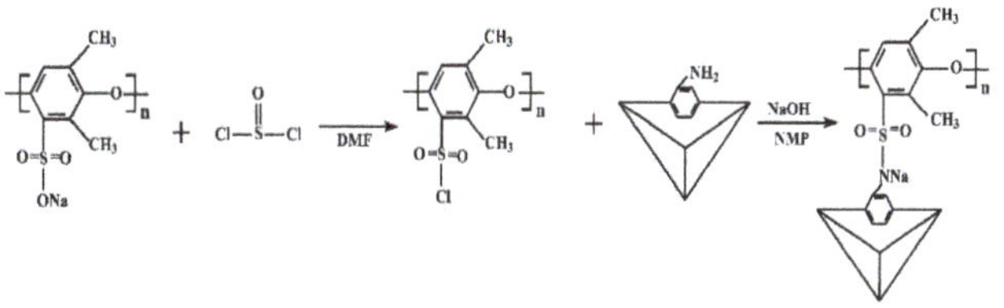

**Fig. 9.** The synthetic procedure of MOFs-PPO-$SO_2Cl$ membranes (Wu et al., 2013). Copyright (2016), reprinted with permission from RSC.

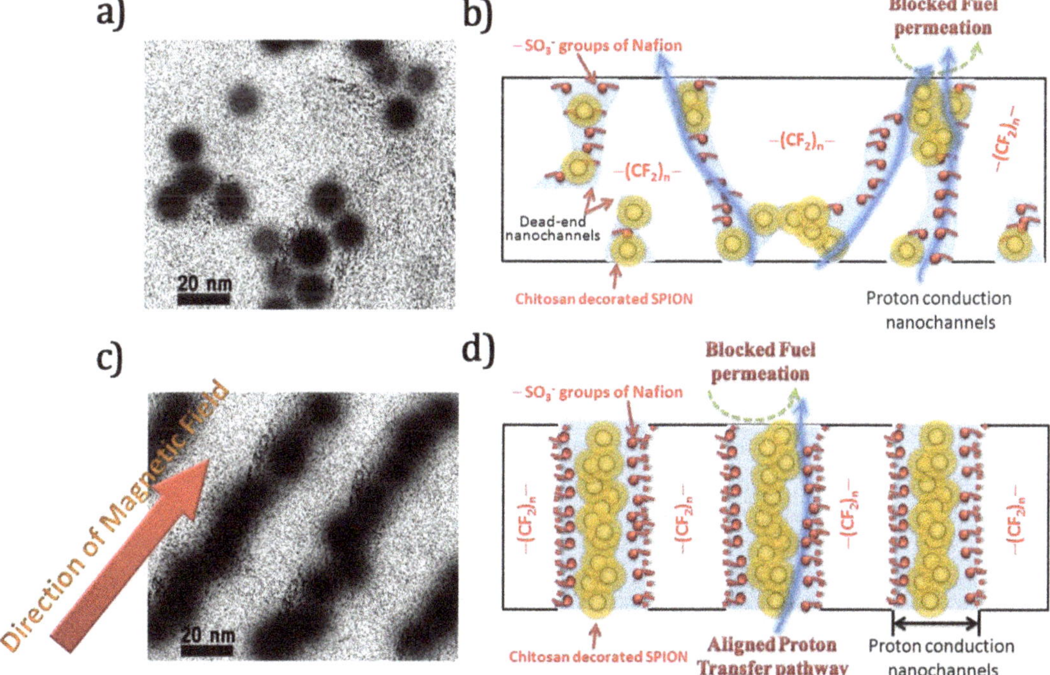


**Fig. 8.** High resolution transmission electron microscopy (HR-TEM) shows the dispersion of chitosan-coated magnetic nanoparticles (CTS-SPIONs) inside Nafion® matrix; a) without external magnetic field and c) in the presence of an applied magnetic field (c). b) presents a schematic representation of the proposed microstructure for randomly-dispersed nanoparticles and d) shows the unidirectional orientation of ion conduction nanochannels (Hasani-Sadrabadi et al., 2014a). Copyright (2016), reprinted with permission from ACS.

They claimed that the proton conductivity of the membranes was as high as 0.10 S $cm^{-1}$ at room temperature and 0.25 S $cm^{-1}$ at 90 °C.

Recently, a novel ternary composite membrane consisting of PVA, poly(2-acrylamido-2-methylpropane sulfonic acid) (PAMPS), and zeolitic imidazolate framework-8 (ZIF-8), was prepared by physical blending and casting methods (Erkartal et al., 2016). This study showed that ZIF-8 nanoparticles (40-60 nm) not only assisted with the water management because of their hydrophobic nature, but also contributed to the proton conductivity by forming hydrogen bonds with the polymer network. These membranes displayed 0.134 S $cm^{-1}$ proton conductivity under fully hydrated state at 80 °C.

In another study, novel Nafion®-based composite membranes (PEM-1 and PEM-2) using two 1-D channel microporous MOFs as fillers; i.e., CPO-27(Mg) and MIL-53(Al) were investigated for PEMFC applications (Tsai et al., 2014). The results obtained showed improved water uptake and proton conductivity by 1.7 times and 2.1 times in magnitude, respectively, as compared with the recast Nafion® membrane. CPO-27(Mg)-Nafion composite membrane exhibited maximum power density values of 818 mW $cm^{-2}$ and 591 mW $cm^{-2}$, at 50 °C and 80 °C, respectively.

Enhanced proton conductivity by sulfonated MIL101(Cr) into SPEEK matrix was also reported by Li et al. (2014). In general, MOFs, with low cost and various properties, can act as electrolytes, electrode catalysts, and catalyst precursors by adjusting the structures to obtain the optimized materials for FCs (Li and Xu, 2013).

### *6.10. Nanocomposite PEMs with layered silicate materials*

Clay (layered silicates), including montmorillonite and laponite, are one class of the most widely used inorganic fillers for the preparation of polymer-clay nanocomposite membranes for PEMFCs owing to high aspect ratio and excellent barrier properties (Mishra et al., 2012). The thickness of the layer is generally around 1 nm, and the lateral dimensions of these layers vary from 30 nm to several μm or larger depending on the particular layered silicate (Laberty-Robert et al., 2011). In order to improve the compatibility between the inorganic clay and the organic polymer, the surfaces of the nanoclays are modified *via* ionic (using alkyl ammonium ions), covalent (using alkoxy silanes), or plasma (using modifiers containing vinyl groups) modification techniques (Yen et al., 2006; Zhang, 2007; Buquet et al., 2010). The monovalent ions located between the clay layers are used for the modification allowing the absorption of polar solvents such as water (Kim et al., 2015a). Recently, organo-functionalization of CNT was grafted on smectite clays (SWy) by catalytic chemical vapour deposition (CCVD) method (Simari et al., 2016) (**Fig. 10**), and their composite membranes with Nafion® matrix exhibited $7 \times 10^{-2}$ S $cm^{-1}$ proton conductivity at 120 °C and 30% RH. Bentonite clay was also modified by grafting the organo sulfonic acid groups on the surface through silane condensation, and was subsequently dispersed in SPEEK to form a composite electrolyte for its use in DMFCs (Sasikala et al., 2014). The composite membranes showed 140 mW $cm^{-2}$ power density in comparison with 71 mW $cm^{-2}$ for pristine SPEEK membrane at 70 °C as well as high proton conduction and methanol restricting behaviour. Jana et al. (2015) prepared Poly(vinylidene fluoride) nanohybrid with organically modified 2-D layered silicate. Nanohybrids were functionalized by sulfonation using chlorosulfonic acid under controlled condition to fabricate FC membranes. Power density of the sulfonated nanohybrid membrane exhibited significantly higher value of 33 mW $cm^{-2}$, against the value of 11 mW $cm^{-2}$ measured for standard Nafion® at similar current density. He et al. (2015) investigated the effects of three types of clays; i.e., sodium montmorillonite (IC), hydrophobic organo-clay with long alkyl chains (OC), and organo-clay with carboxylic acid end groups (HC), on the structure and properties of SPEEK/clay nanocomposites. They found that the SPEEK/HC hybrid membranes achieved the best clay dispersion, higher proton conductivity (due to the interaction between the carboxylic acid groups in HC and sulfonic acid groups in SPEEK) and selectivity at low filler loading (<10 wt.%).

In a study, a Nafion® nanocomposite membrane, with 2 wt.% CTS-functionalized montmorillonite was prepared *via* a solvent casting method and indicated surprisingly 23-times higher membrane selectivity (Hasani-Sadrabadi et al., 2010).

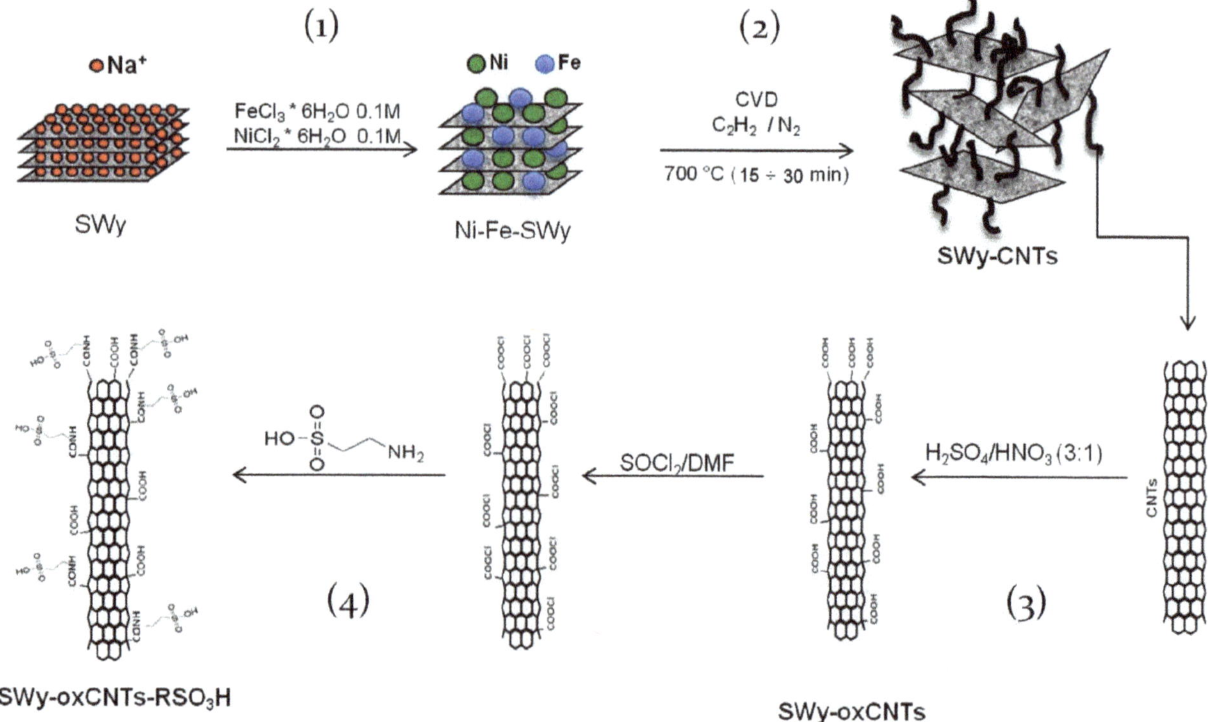


**Fig. 10.** Synthetic procedure for the production of clay–CNTs hybrid materials used as nanoadditives in Nafion® polymer (Simari et al., 2016). Copyright (2016), reprinted with permission from ACS.

*6.11. Nanocomposite PEMs with other miscellaneous fillers*

Cellulose whiskers (CWs), has also been used as a filler to alter the transport properties of Nafion®. These 1D nanoparticles with their unique properties such as high aspect ratio, excellent dispersibility in aqueous solvents owing to high surface charges, as well as high capacity to absorb and retain water, can improve proton conductivity, power density, and reduce methanol crossover in Nafion®-CW nanocomposite membranes. These improvements are ascribed to the formation of long-range oriented conduction pathways in the vicinity of 1D cellulosic nanostructures as shown schematically in **Figure 11** (Hasani-Sadrabadi et al., 2014b).

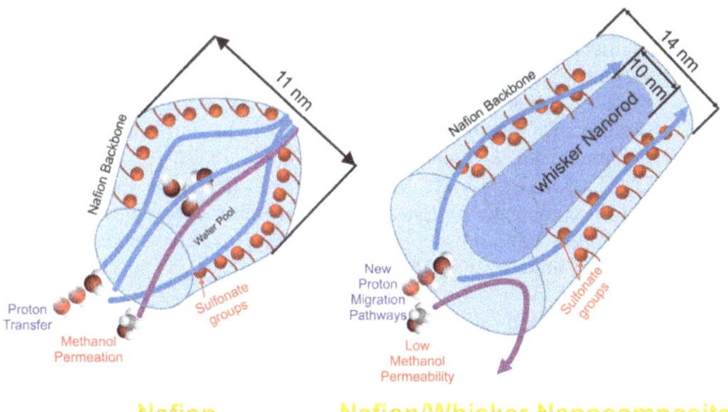


**Fig. 11.** Schematic representation of the proposed mechanism for enhancement in proton migration as well as blocking methanol diffusion pathways in the presence of cellulose whiskers (Hasani-Sadrabadi et al., 2014b). Copyright (2016), reprinted with permission from RSC.

Incorporation of $ZrO_2$ as filler has also been investigated for PEMs. For instance, Zheng and Mathe (2011) found that a poly(2,5-benzimidazole) membrane filled with $ZrO_2$ exhibited both higher proton conductivity (0.069 S $cm^{-1}$ at 100 °C) as well as higher thermal and mechanical stability, compared with an unfilled membrane. Incorporating a porous hygroscopic filler with tubular morphology into a Nafion® membrane could also be an effective approach to enhance the performance of Nafion® membranes operated under fully humid and low RH conditions (Matos et al., 2007; Jun et al., 2011). In line with that, Ketpang et al. (2014) incorporated porous $ZrO_2$ nanotube (ZrNT) into a Nafion® matrix, and the resultant nanocomposite membrane exhibited 2.7 and 1.2 times higher power density at 0.6 V, under 50% and 100% RH at 80 °C, respectively.

Graphitic carbon nitride (g-$C_3N_4$) is also a promising 2-D soft nanomaterial with a similar stacked 2- D structure as graphene (Zheng et al., 2012; Xu et al., 2013), which is becoming increasingly popular in the fields of photocatalysis, heterogeneous catalysis, etc. (Shinde et al., 2015; Wen et al., 2015; Li et al., 2016). g-$C_3N_4$ can be employed as potential filler for PEMs to enhance their performance by forming acid-base pairs between the amino (–$NH_2$) and imino (–NH) groups of g-$C_3N_4$ and by interacting with the acid groups (sulfonate groups) in the polymer matrix.

Recently, Gang et al. (2016) explored g-$C_3N_4$/SPEEK nanocomposite membrane for DMFC applications and achieved a 39% increase in maximum power density and 68% increase in ultimate tensile strength. Among various inorganic fillers, zirconium phosphate ($Zr(HPO_4)2.H_2O$) (ZrP), an acidic filler containing phosphonic acids ($HPO_4$), has the ability to donate protons and is thermally stable at temperatures above 180 °C (Truffier-Boutry et al., 2007). Moreover, it can increase the mobility of protons on its surface and thus the conductivity (Truffier-Boutry et al., 2007), and therefore, has been used as a doping material to improve Nafion® membrane properties (Bauer and Willert-Porada, 2004; Alberti et al., 2007). In fact, zirconium phosphate (ZrP) represent a series of layered structures of two forms, α-ZrP and γ-ZrP. Both structures can be exfoliated to produce suspensions of lamella in a variety of solvents suitable for inclusion into composite membranes (Alberti and Casciola, 2003). Pica et al. (2012) studied mechanical properties and proton conductivity of short-side-chain perfluorosulfonic acid membranes filled with different amounts of ZrP nanoparticles. Their results showed a significant increase in the Young's modulus (up to 80%) and in the yield stress (up to 124%) not only under ambient conditions but also at 80 °C and 80% RH in comparison with the neat polymer. S40(sulfonated poly(styrene-block-(ethylene-ran-butylene)-block-styrene) with 40% degree of sulfonation)-ZrP composite membranes were prepared *via* precursor infiltration method by Liu et al. (2015). The composite membrane containing 3 wt.% of ZrP indicated a remarkable 16 fold increment in selectivity as compared with the parent S40 membrane. Moreover, these membranes also demonstrated a power density of 66 mW $cm^{-2}$ and a maximum current density of 450 mA $cm^{-2}$ without cathode flooding (Liu et al., 2015).

## 7. Conclusions

As stated, the Nafion® membrane due to its disadvantages is the main obstacle to further improvements of FCs. Hybrid organic-inorganic membranes consisting of a polymeric material combined with an inorganic filler, could be of significant help to overcome these shortcomings. The inorganic fillers mainly used include hygroscopic oxides, ZrP, silane-based fillers, GO, MOFs, clays, CNTs, zeolites, and PTA. In fact, these membranes combine the intrinsic physical and chemical properties of both the inorganic and organic segments. In hybrid membranes, the inorganic segment provides high mechanical and thermal stability, while also suppress fuel crossover. On the other hand, the organic segment provides flexibility. The properties of polymer composites depend on the type of nanoparticles incorporated, their size and shape, their concentration, and their interactions with the polymer matrix.

## References

[1] Ahmad, H., Kamarudin, S.K., Hasran, U.A., Daud, W.R.W., 2010. Overview of hybrid membranes for direct-methanol fuel-cell applications. Int. J. Hydrogen Energy. 35(5), 2160-2175.

[2] Alberti, G., Casciola, M., 2003. Composite membranes for medium-temperature PEM fuel cells. Annu. Rev. Mater. Res. 33(1),129-154.

[3] Alberti, G., Casciola, M., Capitani, D., Donnadio, A., Narducci, R., Pica, M., Sganappa, M., 2007. Novel Nafion-zirconium phosphate nanocomposite membranes with enhanced stability of proton conductivity at medium temperature and high relative humidity. Electrochim. Acta. 52(28), 8125-8132.

[4] Amirinejad, M., Madaeni, S.S., Lee, K.S., Ko, U., Rafiee, E., Lee, J.S., 2012. Sulfonated poly(arylene ether)/heteropolyacids nanocomposite membranes for proton exchange membrane fuel cells. Electrochim. Acta. 62, 227-233.

[5] Amirinejad, M., Madaeni, S.S., Rafiee, E., Amirinejad, S., 2011. Cesium hydrogen salt of heteropolyacids/Nafion nanocomposite membranes for proton exchange membrane fuel cells. J. Membr. Sci. 377(1-2), 89-98.

[6] Amjadi, M., Rowshanzamir, S., Peighambardoust, S.J., Hosseini, M.G., Eikani, M.H., 2010. Investigation of physical properties and cell performance of Nafion/$TiO_2$ nanocomposite membranes for high temperature PEM fuel cells. Int. J. Hydrogen Energy. 35(17), 9252-9260.

[7] An, L., Zhao, T.S., Wu, Q.X., Zeng, L., 2012. Comparison of different types of membrane in alkaline direct ethanol fuel cells. Int. J. Hydrogen Energy. 37(19), 14536-14542.

[8] Asgari, M.S., Nikazar, M., Molla-Abbasi, P., Hasani-Sadrabadi, M.M., 2013. Nafion®/histidine functionalized carbon nanotube: High-performance fuel cell membranes. Int. J. Hydrogen Energy. 38(14), 5894-5902.

[9] Aslan, A., Bozkurt, A., 2014. Nanocomposite membranes based on sulfonated polysulfone and sulfated nano-titania/NMPA for proton exchange membrane fuel cells. Solid State Ionics. 255, 89-95.

[10] Auimviriyavat, J., Changkhamchom, S., Sirivat, A., 2011. Development of poly (ether ether ketone) (Peek) with inorganic filler for direct methanol fuel cells (DMFCS). Ind. Eng. Chem. Res. 50(22), 12527-12533.

[11] Authayanun, S., Im-orb, K., Arpornwichanop, A., 2015. A review of the development of high temperature proton exchange membrane fuel cells. Chin. J. Catal. 36(4), 473-483.

[12] Awang, N., Ismail, A.F., Jaafar, J., Matsuura, T., Junoh, H., Othman, M.H.D., Rahman, M.A., 2015. Functionalization of polymeric materials as a high performance membrane for direct methanol fuel cell: a review. React. Funct. Polym. 86, 248-258.

[13] Badwal, S.P.S., Giddey, S., Kulkarni, A., Goel, J., Basu, S., 2015. Direct ethanol fuel cells for transport and stationary applications-a comprehensive review. Appl. Energy. 145, 80-103.

[14] Bauer, F., Willert-Porada, M., 2004. Microstructural characterization of Zr-phosphate-Nafion® membranes for direct methanol fuel cell (DMFC) applications. J. Membr. Sci. 233(1-2), 141-149.

[15] Bayer, T., Bishop, S.R., Nishihara, M., Sasaki, K., Lyth, S.M., 2014. Characterization of a graphene oxide membrane fuel cell. J. Power Sources. 272, 239-247.

[16] Beydaghi, H., Javanbakht, M., Bagheri, A., Salarizadeh, P., Ghafarian-Zahmatkesh, H., Kashefi, S., Kowsari, E., 2015. Novel nanocomposite membranes based on blended sulfonated poly (ether ether ketone)/poly (vinyl alcohol) containing sulfonated graphene oxide/$Fe_3O_4$ nanosheets for DMFC applications. RSC Adv. 5(90), 74054-74064.

[17] Beydaghi, H., Javanbakht, M., 2015. Aligned nanocomposite membranes containing sulfonated graphene oxide with superior ionic conductivity for direct methanol fuel cell application. Ind. Eng. Chem. Res. 54(28) 7028-7037.

[18] Bose, S., Kuila, T., Nguyen, T.X.H., Kim, N.H., Lau, K.T., Lee, J.H., 2011. Polymer membranes for high temperature proton exchange membrane fuel cell: recent advances and challenges. Prog. Polym. Sci. 36(6), 813-843.

[19] Brijmohan, S.B., Shaw, M.T., 2007. Magnetic ion-exchange nanoparticles and their application in proton exchange membranes. J. Membr. Sci. 303(1-2), 64-71.

[20] Bounor-Legaré, V., Cassagnau, P., 2014. *In situ* synthesis of organic–inorganic hybrids or nanocomposites from sol-gel chemistry in molten polymers. Prog. Polym. Sci. 39(8), 1473-1497.

[21] Bureekaew, S., Horike, S., Higuchi, M., Mizuno, M., Kawamura, T., Tanaka, D., Yanai, N., Kitagawa, S., 2009. One-dimensional imidazole aggregate in aluminium porous coordination polymers with high proton conductivity. Nat. Mater. 8(10), 831-836.

[22] Buquet, C.L., Fatyeyeva, K., Poncin-Epaillard, F., Schaetzel, P., Dargent, E., Langevin, D., Nguyen, Q.T., Marais, S., 2010. New hybrid membranes for fuel cells: plasma treated laponite based sulfonated polysulfone. J. Membr. Sci. 351(1-2), 1-10.

[23] Carrette, L., Friedrich, K.L., Stimming, U., 2001. Fuel cells-fundamentals and applications. Fuel cells. 1(1), 5-39.

[24] Chandan, A., Hattenberger, M., El-Kharouf, A., Du, S., Dhir, A., Self, V., Pollet, B.G., Ingram, A., Bujalski, W., 2013. High temperature (HT) polymer electrolyte membrane fuel cells (PEMFC)-a review. J. Power Sources. 231, 264-278.

[25] Chang, C.M., Liu, Y.L., Lee, Y.M., 2011. Polybenzimidazole membranes modified with polyelectrolyte-functionalized multiwalled carbon nanotubes for proton exchange membrane fuel cells. J. Mater. Chem. 21(20), 7480-7486.

[26] Chen, Z., Holmberg, B., Li, W., Wang, X., Deng, W., Munoz, R., Yan, Y., 2006. Nafion/zeolite nanocomposite membrane by *in situ* crystallization for a direct methanol fuel cell. Chem. Mater. 18(24), 5669-5675.

[27] Chen, Z., Li, S., Yan, Y., 2005. Synthesis of template-free zeolite nanocrystals by reverse microemulsion-microwave method. Chem. Mater. 17(9), 2262-2266.

[28] Chien, H.C., Tsai, L.D., Huang, C.P., Kang, C.Y., Lin, J.N., Chang, F.C., 2013. Sulfonated graphene oxide/Nafion composite membranes for high-performance direct methanol fuel cells. Int. J. hydrogen energy. 38(31), 13792-13801.

[29] Choi, J., Kim, D.H., Kim, H.K., Shin, C., Kim, S.C., 2008. Polymer blend membranes of sulfonated poly (arylene ether ketone) for direct methanol fuel cell. J. Membr. Sci. 310(1), 384-392.

[30] Chomoucka, J., Drbohlavova, J., Huska, D., Adam, V., Kizek, R., Hubalek, J., 2010. Magnetic nanoparticles and targeted drug delivering. Pharmacol. Res. 62(2), 144-149.

[31] Cozzi, D., de Bonis, C., D'Epifanio, A., Mecheri, B., Tavares, A.C., Licoccia, S., 2014. Organically functionalized titanium oxide/Nafion composite proton exchange membranes for fuel cells applications. J. Power Sources. 248, 1127-1132.

[32] Cui, L., Geng, Q., Gong, C., Liu, H., Zheng, G., Wang, G., Liu, Q., Wen, S., 2015. Novel sulfonated poly (ether ether ketone)/silica coated carbon nanotubes high-performance composite membranes for direct methanol fuel cell. Polym. Adv. Technol. 26(5), 457-464.

[33] Das, S., Kumar, P., Dutta, K., Kundu, P.P., 2014. Partial sulfonation of PVdF-*co*-HFP: a preliminary study and characterization for application in direct methanol fuel cell. Appl. Energy. 113, 169-177.

[34] Decher, G., Eckle, M., Schmitt, J., Struth, B., 1998. Layer-by-layer assembled multicomposite films. Curr. Opin. Colloid Interface Sci. 3(1), 32-39.

[35] Decher, G., 1997. Fuzzy nanoassemblies: toward layered polymeric multicomposites. Science. 277(5330), 1232-1237.

[36] Devrim, Y., Albostan, A., 2015. Enhancement of PEM fuel cell performance at higher temperatures and lower humidities by high performance membrane electrode assembly based on Nafion/zeolite membrane. Int. J. Hydrogen Energy. 40(44), 15328-15335.

[37] Di, Z., Xie, Q., Li, H., Mao, D., Li, M., Zhou, D., Li, L., 2015. Novel composite proton-exchange membrane based on proton-conductive glass powders and sulfonated poly (ether ether ketone). J. Power Sources. 273, 688-696.

[38] Duan, C., Wei, M., Guo, D., He, C., Meng, Q., 2009. Crystal structures and properties of large protonated water clusters encapsulated by metal-organic frameworks. J. Am. Chem. Soc. 132(10), 3321-3330.

[39] Dupuis, A.C., 2011. Proton exchange membranes for fuel cells operated at medium temperatures: materials and experimental techniques. Prog. Mater. Sci. 56(3), 289-327.

[40] Dutta, K., Das, S., Kundu, P.P., 2015. Partially sulfonated polyaniline induced high ion-exchange capacity and selectivity of Nafion membrane for application in direct methanol fuel cells. J. Membr. Sci. 473, 94-101.

[41] Dyer, A., 1988. An introduction to zeolite molecular sieves. John Wiley & Sons, Chichester, ISBN 0471919810.

[42] Erkartal, M., Usta, H., Citir, M., Sen, U., 2016. Proton conducting poly (vinyl alcohol) (PVA)/poly (2-acrylamido-2-methylpropane sulfonic acid) (PAMPS)/zeolitic imidazolate framework (ZIF) ternary composite membrane. J. Membr. Sci. 499, 156-163.

[43] Escudero-Cid, R., Montiel, M., Sotomayor, L., Loureiro, B., Fatás, E., Ocón, P., 2015. Evaluation of polyaniline-Nafion® composite membranes for direct methanol fuel cells durability tests. Int. J. Hydrogen Energy. 40(25), 8182-8192.

[44] Farhat, T.R., Hammond, P.T., 2006. Engineering ionic and electronic conductivity in polymer catalytic electrodes using the layer-by-layer technique. Chem. Mater. 18(1), 41-49.

[45] Farrukh, A., Ashraf, F., Kaltbeitzel, A., Ling, X., Wagner, M., Duran, H., Ghaffar, A., ur Rehman, H., Parekh, S.H., Domke, K.F., Yameen, B., 2015. Polymer brush functionalized $SiO_2$ nanoparticle based Nafion nanocomposites: a novel avenue to low-humidity proton conducting membranes. Polym. Chem. 6(31), 5782-5789.

[46] Feng, K., Tang, B., Wu, P., 2014. Sulfonated graphene oxide-silica for highly selective Nafion-based proton exchange membranes. J. Mater. Chem. A. 2(38), 16083-16092.

[47] Gandhi, K., Dixit, B.K., Dixit, D.K., 2012. Effect of addition of zirconium tungstate, lead tungstate and titanium dioxide on the proton conductivity of polystyrene porous membrane. Int. J. Hydrogen Energy. 37(4), 3922-3930.

[48] Gang, M., He, G., Li, Z., Cao, K., Li, Z., Yin, Y., Wu, H., Jiang, Z., 2016. Graphitic carbon nitride nanosheets/sulfonated poly (ether ether ketone) nanocomposite membrane for direct methanol fuel cell application. J. Membr. Sci. 507, 1-11.

[49] Gil, M., Ji, X., Li, X., Na, H., Hampsey, J.E., Lu, Y., 2004. Direct synthesis of sulfonated aromatic poly (ether ether ketone) proton exchange membranes for fuel cell applications. J. Membr. Sci. 234(1-2), 75-81.

[50] Gögebakan, Z., Yücel, H., Culfaz, A., 2007. Crystallization field and rate study for the synthesis of Ferrierite. Ind. Eng. Chem. Res. 46(7), 2006-2012.

[51] Hasanabadi, N., Ghaffarian, S.R., Hasani-Sadrabadi, M.M., 2011. Magnetic field aligned nanocomposite proton exchange membranes based on sulfonated poly (ether sulfone) and $Fe_2O_3$ nanoparticles for direct methanol fuel cell application. Int. J. Hydrogen Energy. 36(23), 15323-15332.

[52] Hasanabadi, N., Ghaffarian, S.R., Hasani-Sadrabadi, M.M., 2013. Nafion-based magnetically aligned nanocomposite proton exchange membranes for direct methanol fuel cells. Solid State Ionics. 232, 58-67.

[53] Hasani-Sadrabadi, M.M., Dashtimoghadam, E., Majedi, F.S., Kabiri, K., Mokarram, N., Solati-Hashjin, M., Moaddel, H., 2010. Novel high-performance nanohybrid polyelectrolyte membranes based on bio-functionalized montmorillonite for fuel cell applications. Chem. Commun. 46(35), 6500-6502.

[54] Hasani-Sadrabadi, M.M., Dashtimoghadam, E., Majedi, F.S., Moaddel, H., Bertsch, A., Renaud, P., 2013. Superacid-doped polybenzimidazole-decorated carbon nanotubes: a novel high-performance proton exchange nanocomposite membrane. Nanoscale. 5(23), 11710-11717.

[55] Hasani-Sadrabadi, M.M., Majedi, F.S., Coullerez, G., Dashtimoghadam, E., VanDersarl, J.J., Bertsch, A., Moaddel, H., Jacob, K.I., Renaud, P., 2014a. Magnetically aligned nanodomains: application in high-performance ion conductive membranes. ACS Appl. Mater. Interfaces. 6(10), 7099-7107.

[56] Hasani-Sadrabadi, M.M., Dashtimoghadam, E., Nasseri, R., Karkhaneh, A., Majedi, F.S., Mokarram, N., Renaud, P., Jacob, K.I., 2014b. Cellulose nanowhiskers to regulate the microstructure of perfluorosulfonate ionomers for high-performance fuel cells. J. Mater. Chem. A. 2(29), 11334-11340.

[57] Hasani-Sadrabadi, M.M., Dashtimoghadam, E., Majedi, F.S., VanDersarl, J.J., Bertsch, A., Renaud, P., Jacob, K.I., 2016. Ionic nanopeapods: next-generation proton conducting membranes based on phosphotungstic acid filled carbon nanotube. Nano Energy. 23, 114-121.

[58] He, G., He, X., Wang, X., Chang, C., Zhao, J., Li, Z., Wu, H., Jiang, Z., 2016. A highly proton-conducting, methanol-blocking Nafion composite membrane enabled by surface-coating crosslinked sulfonated graphene oxide. Chem. Commun. 52(10), 2173-2176.

[59] He, S., Jia, H., Lin, Y., Qian, H., Lin, J., 2015. Effect of clay modification on the structure and properties of sulfonated poly (ether ether ketone)/clay nanocomposites. Polym. Compos. 37(9), 2632-2638.

[60] He, Y., Wang, J., Zhang, H., Zhang, T., Zhang, B., Cao, S., Liu, J., 2014. Polydopamine-modified graphene oxide nanocomposite membrane for proton exchange membrane fuel cell under anhydrous conditions. J. Mater. Chem. A. 2(25), 9548-9558.

[61] Hickner, M.A., Ghassemi, H., Kim, Y.S., Einsla, B.R., McGrath, J.E., 2004. Alternative polymer systems for proton exchange membranes (PEMs). Chem. Rev. 104(10), 4587-4612.

[62] Holmberg, B.A., Hwang, S.J., Davis, M.E., Yan, Y., 2005. Synthesis and proton conductivity of sulfonic acid functionalized zeolite BEA nanocrystals. Microporous Mesoporous Mater. 80(1-3), 347-356.

[63] Hooshyari, K., Javanbakht, M., Shabanikia, A., Enhessari, M., 2015. Fabrication $BaZrO_3$/PBI-based nanocomposite as a new proton conducting membrane for high temperature proton exchange membrane fuel cells. J. Power Sources. 276, 62-72.

[64] Hooshyari, K., Javanbakht, M., Naji, L., Enhessari, M., 2014. Nanocomposite proton exchange membranes based on Nafion containing $Fe_2TiO_5$ nanoparticles in water and alcohol environments for PEMFC. J. Membr. Sci. 454, 74-81.

[65] Hudiono, Y., Choi, S., Shu, S., Koros, W.J., Tsapatsis, M., Nair, S., 2009. Porous layered oxide/Nafion® nanocomposite membranes for direct methanol fuel cell applications. Microporous Mesoporous Mater. 118(1-3), 427-434.

[66] Hurd, J.A., Vaidhyanathan, R., Thangadurai, V., Ratcliffe, C.I., Moudrakovski, I.L., Shimizu, G.K., 2009. Anhydrous proton conduction at 150C in a crystalline metal-organic framework. Nat. Chem. 1(9), 705-710.

[67] Jaafar, J., Ismail, A.F., Matsuura, T., 2009. Preparation and barrier properties of SPEEK/Cloisite 15A®/TAP nanocomposite membrane for DMFC application. J. Membr. Sci. 345(1), 119-127.

[68] Jacobson, M.Z., Colella, W.G., Golden, D.M., 2005. Cleaning the air and improving health with hydrogen fuel-cell vehicles. Science. 308(5730), 1901-1905.

[69] Jalani, N.H., Dunn, K., Datta, R., 2005. Synthesis and characterization of Nafion®-$MO_2$ (M=Zr, Si, Ti) nanocomposite membranes for higher temperature PEM fuel cells. Electrochim. Acta. 51(3), 553-560.

[70] Jana, K.K., Charan, C., Shahi, V.K., Mitra, K., Ray, B., Rana, D., Maiti, P., 2015. Functionalized poly (vinylidene fluoride) nanohybrid for superior fuel cell membrane. J. Membr. Sci. 481, 124-136.

[71] Jiang, S.P., Liu, Z., Tian, Z.Q., 2006. Layer-by-layer self-assembly of composite polyelectrolyte-Nafion membranes for direct methanol fuel cells. Adv. Mater. 18(8), 1068-1072.

[72] Jones, C.W., Tsuji, K., Davis, M.E., 1998. Organic-functionalized molecular sieves as shape-selective catalysts. Nature. 393(6680), 52-54.

[73] Jun, Y., Zarrin, H., Fowler, M., Chen, Z., 2011. Functionalized titania nanotube composite membranes for high temperature proton exchange membrane fuel cells. Int. J. Hydrogen Energy. 36(10), 6073-6081.

[74] Jung, U.H., Park, K.T., Park, E.H., Kim, S.H., 2006. Improvement of low-humidity performance of PEMFC by addition of hydrophilic $SiO_2$ particles to catalyst layer. J. Power Sources. 159(1), 529-532.

[75] Kango, S., Kalia, S., Celli, A., Njuguna, J., Habibi, Y., Kumar, R., 2013. Surface modification of inorganic nanoparticles for development of organic-inorganic nanocomposites-a review. Prog. Polym. Sci. 38(8), 1232-1261.

[76] Kannan, R., Aher, P.P., Palaniselvam, T., Kurungot, S., Kharul, U.K., Pillai, V.K., 2010. Artificially designed membranes using phosphonated multiwall carbon nanotube-polybenzimidazole composites for polymer electrolyte fuel cells. J. Phys. Chem. Lett. 1(14), 2109-2113.

[77] Kannan, R., Kagalwala, H.N., Chaudhari, H.D., Kharul, U.K., Kurungot, S., Pillai, V.K., 2011. Improved performance of phosphonated carbon nanotube-polybenzimidazole composite membranes in proton exchange membrane fuel cells. J. Mater. Chem. 21(20), 7223-7231.

[78] Ketpang, K., Lee, K., Shanmugam, S., 2014. Facile synthesis of porous metal oxide nanotubes and modified Nafion composite membranes for polymer electrolyte fuel cells operated under low relative humidity. ACS Appl. Mater. Interfaces. 6(19), 16734-16744.

[79] Kim, D.J., Jo, M.J., Nam, S.Y., 2015. A review of polymer-nanocomposite electrolyte membranes for fuel cell application. J. Ind. Eng. Chem. 21, 36-52.

[80] Kim, D.W., Choi, H.S., Lee, C., Blumstein, A., Kang, Y., 2004. Investigation on methanol permeability of Nafion modified by self-assembled clay-nanocomposite multilayers. Electrochim. Acta. 50(2-3), 659-662.

[81] Kim, J.Y., Mulmi, S., Lee, C.H., Park, H.B., Chung, Y.S., Lee, Y.M., 2006. Preparation of organic-inorganic nanocomposite membrane using a reactive polymeric dispersant and compatibilizer: proton and methanol transport with respect to nano-phase separated structure. J. Membr. Sci. 283(1-2), 172-181.

[82] Kim, Y., Ketpang, K., Jaritphun, S., Park, J.S., Shanmugam, S., 2015. A polyoxometalate coupled graphene oxide-Nafion composite membrane for fuel cells operating at low relative humidity. J. Mater. Chem. A. 3(15), 8148-8155.

[83] Kraytsberg, A., Ein-Eli, Y., 2014. Review of advanced materials for proton exchange membrane fuel cells. Energy Fuels. 28(12), 7303-7330.

[84] Kumar, G.G., Uthirakumar, P., Nahm, K.S., Elizabeth, R.N., 2009. Fabrication and electro chemical properties of poly vinyl alcohol/para toluene sulfonic acid membranes for the applications of DMFC. Solid State Ionics. 180(2), 282-287.

[85] Kumar, R., Mamlouk, M., Scott, K., 2014. Sulfonated polyether ether ketone-sulfonated graphene oxide composite membranes for polymer electrolyte fuel cells. RSC Adv. 4(2), 617-623.

[86] Kornatowski, J., 2005. Expressiveness of adsorption measurements for characterization of zeolitic materials-a review. Adsorption. 11(3), 275-293.

[87] Kourasi, M., Wills, R.G.A., Shah, A.A., Walsh, F.C., 2014. Heteropolyacids for fuel cell applications. Electrochim. Acta. 127, 454-466.

[88] Laberty-Robert, C., Valle, K., Pereira, F., Sanchez, C., 2011. Design and properties of functional hybrid organic-inorganic membranes for fuel cells. Chem. Soc. Rev. 40(2), 961-1005.

[89] Laurent, S., Dutz, S., Häfeli, U.O., Mahmoudi, M., 2011. Magnetic fluid hyperthermia: focus on superparamagnetic iron oxide nanoparticles. Adv. Colloid Interface Sci. 166(1-2), 8-23.

[90] Libby, B., Smyrl, W.H., Cussler, E.L., 2003. Polymer-zeolite composite membranes for direct methanol fuel cells. AIChE J. 49(4), 991-1001.

[91] Li, S.L., Xu, Q., 2013. Metal-organic frameworks as platforms for clean energy. Energy Environ. Sci. 6(6), 1656-1683.

[92] Li, T., Yang, Y., 2009. A novel inorganic/organic composite membrane tailored by various organic silane coupling agents for use in direct methanol fuel cells. J. Power Sources. 187(2), 332-340.

[93] Li, Q., Zhang, H., Tu, Z., Yu, J., Xiong, C., Pan, M., 2012. Impregnation of amine-tailored titanate nanotubes in polymer electrolyte membranes. J. Membr. Sci. 423-424, 284-292.

[94] Liu, K.L., Lee, H.C., Wang, B.Y., Lue, S.J., Lu, C.Y., Tsai, L.D., Fang, J., Chao, C.Y., 2015. Sulfonated poly (styrene-*block*-(ethylene-*ran*-butylene)-*block*-styrene (SSEBS)-zirconium phosphate (ZrP) composite membranes for direct methanol fuel cells. J. Membr. Sci. 495, 110-120.

[95] Livage, J., 2004. Basic principles of sol-gel chemistry, in Sol-Gel Technologies for Glass Producers and Users. Springer US, pp. 3-14.

[96] Li, Y., Jin, R., Fang, X., Yang, Y., Yang, M., Liu, X., Xing, Y., Song, S., 2016. *In situ* loading of $Ag_2WO_4$ on ultrathin g-$C_3N_4$ nanosheets with highly enhanced photocatalytic performance. J. Hazard. Mater. 313, 219-228.

[97] Li, Y., He, G., Wang, S., Yu, S., Pan, F., Wu, H., Jiang, Z., 2013. Recent advances in the fabrication of advanced composite membranes. J. Mater. Chem. A. 1(35), 10058-10077.

[98] Li, Z., He, G., Zhao, Y., Cao, Y., Wu, H., Li, Y., Jiang, Z., 2014. Enhanced proton conductivity of proton exchange membranes by incorporating sulfonated metal-organic frameworks. J. Power Sources. 262, 372-379.

[99] Liang, X., Zhang, F., Feng, W., Zou, X., Zhao, C., Na, H., Liu, C., Sun, F., Zhu, G., 2013. From metal-organic framework (MOF) to MOF-polymer composite membrane: enhancement of low-humidity proton conductivity. Chem. Sci. 4(3), 983-992.

[100] Luan, Y., Zhang, H., Zhang, Y., Li, L., Li, H., Liu, Y., 2008. Study on structural evolution of perfluorosulfonic ionomer from concentrated DMF-based solution to membranes. J. Membr. Sci. 319(1-2), 91-101.

[101] Lue, S.J., Pai, Y.L., Shih, C.M., Wu, M.C., Lai, S.M., 2015. Novel bilayer well-aligned Nafion/graphene oxide composite membranes prepared using spin coating method for direct liquid fuel cells. J. Membr. Sci. 493, 212-223.

[102] Mahreni, A., Mohamad, A.B., Kadhum, A.A.H., Daud, W.R.W., Iyuke, S.E., 2009. Nafion/silicon oxide/phosphotungstic acid nanocomposite membrane with enhanced proton conductivity. J. Membr. Sci. 327(1-2), 32-40.

[103] Maleki, A., Ghamari, N., Kamalzare, M., 2014. Chitosan-supported $Fe_3O_4$ nanoparticles: a magnetically recyclable heterogeneous nanocatalyst for the syntheses of multifunctional benzimidazoles and benzodiazepines. RSC Adv. 4(19), 9416-9423.

[104] Malers, J.L., Sweikart, M.A., Horan, J.L., Turner, J.A., Herring, A.M., 2007. Studies of heteropoly acid/polyvinylidenedifluoride-hexafluoroproylene composite membranes and implication for the use of heteropoly acids as the proton conducting component in a fuel cell membrane. J. Power Sources. 172(1), 83-88.

[105] Marx, S., van der Gryp, P., Neomagus, H., Everson, R., Keizer, K., 2002. Pervaporation separation of methanol from methanol/tert-amyl methyl ether mixtures with a commercial membrane. J. Membr. Sci. 209(2), 353-362.

[106] Matos, B.R., Santiago, E.I., Fonseca, F.C., Linardi, M., Lavayen, V., Lacerda, R.G., Ladeira, L.O., Ferlauto, A.S., 2007. Nafion-titanate nanotube composite membranes for PEMFC operating at high temperature. J. Electrochem. Soc. 154(12), B1358-B1361.

[107] Mishra, A.K., Bose, S., Kuila, T., Kim, N.H., Lee, J.H., 2012. Silicate-based polymer-nanocomposite membranes for polymer electrolyte membrane fuel cells. Prog. Polym. Sci. 37(6), 842-869.

[108] Narayanamoorthy, B., Datta, K.K.R., Eswaramoorthy, M., Balaji, S., 2012. Improved oxygen reduction reaction catalyzed by pt/clay/Nafion nanocomposite for PEM fuel cells. ACS Appl. Mater. Interfaces. 4(7), 3620-3626.

[109] Neelakandan, S., Rana, D., Matsuura, T., Muthumeenal, A., Kanagaraj, P., Nagendran, A., 2014. Fabrication and electrochemical properties of surface modified sulfonated poly (vinylidenefluoride-co-hexafluoropropylene) membranes for DMFC application. Solid State Ionics. 268, 35-41.

[110] Ng, E.P., Mintova, S., 2008. Nanoporous materials with enhanced hydrophilicity and high water sorption capacity. Microporous Mesoporous Mater. 114(1-3), 1-26.

[111] Ohkoshi, S.I., Nakagawa, K., Tomono, K., Imoto, K., Tsunobuchi, Y., Tokoro, H., 2010. High proton conductivity in prussian blue analogues and the interference effect by magnetic ordering. J. Am. Chem. Soc. 132(19), 6620-6621.

[112] Oren, Y., Freger, V., Linder, C., 2004. Highly conductive ordered heterogeneous ion-exchange membranes. J. Membr. Sci. 239(1),17-26.

[113] Ossiander, T., Heinzl, C., Gleich, S., Schönberger, F., Völk, P., Welsch, M., Scheu, C., 2014. Influence of the size and shape of silica nanoparticles on the properties and degradation of a PBI-based high temperature polymer electrolyte membrane. J. Membr. Sci. 454, 12-19.

[114] Park, K.T., Jung, U.H., Choi, D.W., Chun, K., Lee, H.M., Kim, S.H., 2008. $ZrO_2$-$SiO_2$/Nafion® composite membrane for polymer electrolyte membrane fuel cells operation at high temperature and low humidity. J. Power Sources. 177(2), 247-253.

[115] Peighambardoust, S., Rowshanzamir, S., Amjadi, M., 2010. Review of the proton exchange membranes for fuel cell applications. Int. J. Hydrogen Energy. 35(17), 9349-9384.

[116] Pica, M., Donnadio, A., Casciola, M., Cojocaru, P., Merlo, L., 2012. Short side chain perfluorosulfonic acid membranes and their composites with nanosized zirconium phosphate: hydration, mechanical properties and proton conductivity. J. Mater. Chem. 22(47), 24902-24908.

[117] Pomogailo, A.D., 2005. Polymer sol-gel synthesis of hybrid nanocomposites. Colloid J. 67(6), 658-677.

[118] Ramani, V., Kunz, H.R., Fenton, J.M., 2005. Stabilized composite membranes and membrane electrode assemblies for elevated temperature/low relative humidity PEFC operation. J. Power Sources. 152, 182-188.

[119] Ramani, V., Kunz, H.R., Fenton, J.M., 2005. Stabilized heteropolyacid/Nafion® composite membranes for elevated temperature/low relative humidity PEFC operation. Electrochim. Acta. 50(5), 1181-1187.

[120] Ramaswamy, P., Wong, N.E., Shimizu, G.K., 2014. MOFs as proton conductors-challenges and opportunities. Chem. Soc. Rev. 43(16), 5913-5932.

[121] Ren, S., Sun, G., Li, C., Song, S., Xin, Q., Yang, X., 2006. Sulfated zirconia-Nafion composite membranes for higher temperature direct methanol fuel cells. J. Power Sources. 157(2), 724-726.

[122] Ren, Y., Chia, G.H., Gao, Z., 2013. Metal-organic frameworks in fuel cell technologies. Nano Today. 8(6), 577-597.

[123] Rikukawa, M. Sanui, K., 2000. Proton-conducting polymer electrolyte membranes based on hydrocarbon polymers. Prog. Polym. Sci. 25(10), 1463-1502.

[124] Sacca, A., Gatto, I., Carbone, A., Pedicini, R., Passalacqua, E., 2006. $ZrO_2$-Nafion composite membranes for polymer electrolyte fuel cells (PEFCs) at intermediate temperature. J. Power Sources. 163(1), 47-51.

[125] Saccà, A., Carbone, A., Pedicini, R., Marrony, M., Barrera, R., Elomaa, M., Passalacqua, E., 2008. Phosphotungstic acid supported on a nanopowdered $ZrO_2$ as a filler in Nafion-based membranes for polymer electrolyte fuel cells. Fuel Cells. 8(3-4), 225-235.

[126]Safari, J. Javadian, L., 2014. Chitosan decorated $Fe_3O_4$ nanoparticles as a magnetic catalyst in the synthesis of phenytoin derivatives. RSC Adv. 4(90), 48973-48979.

[127]Sakamoto, M., Nohara, S., Miyatake, K., Uchida, M., Watanabe, M., Uchida, H., 2014. Effects of incorporation of $SiO_2$ nanoparticles into sulfonated polyimide electrolyte membranes on fuel cell performance under low humidity conditions. Electrochim. Acta. 137, 213-218.

[128]Sasikala, S., Meenakshi, S., Bhat, S.D., Sahu, A.K., 2014. Functionalized Bentonite clay-sPEEK based composite membranes for direct methanol fuel cells. Electrochim. Acta. 135, 232-241.

[129]Sgreccia, E., Chailan, J.F., Khadhraoui, M., Di Vona, M.L., Knauth, P., 2010. Mechanical properties of proton-conducting sulfonated aromatic polymer membranes: stress-strain tests and dynamical analysis. J. Power Sources. 195(23), 7770-7775.

[130]Shabanikia, A., Javanbakht, M., Amoli, H.S., Hooshyari, K., Enhessari, M., 2015. Polybenzimidazole/strontium cerate nanocomposites with enhanced proton conductivity for proton exchange membrane fuel cells operating at high temperature. Electrochim. Acta. 154, 370-378.

[131]Shahi, V.K., 2007. Highly charged proton-exchange membrane: sulfonated poly (ether sulfone)-silica polyelectrolyte composite membranes for fuel cells. Solid State Ionics. 177(39-40), 3395-3404.

[132]Sharaf, O.Z. Orhan, M.F., 2014. An overview of fuel cell technology: fundamentals and applications. Renew. Sust. Energy Rev. 32, 810-853.

[133]Shinde, S., Sami, A., Lee, J.H., 2015. Electrocatalytic hydrogen evolution using graphitic carbon nitride coupled with nanoporous graphene co-doped by S and Se. J. Mater. Chem. A. 3(24), 12810-12819.

[134]Silva, R.F., De Francesco, M., Pozio, A., 2004. Tangential and normal conductivities of Nafion® membranes used in polymer electrolyte fuel cells. J. Power Sources. 134(1), 18-26.

[135]Simari, C., Potsi, G., Policicchio, A., Perrotta, I., Nicotera, I., 2016. Clay-Carbon nanotubes hybrid materials for nanocomposite membranes: advantages of branched structure for proton transport under low humidity conditions in PEMFCs. J. Phys. Chem. C. 120(5), 2574-2584.

[136]Smitha, B., Sridhar, S., Khan, A.A., 2003. Synthesis and characterization of proton conducting polymer membranes for fuel cells. J. Membr. Sci. 225(1-2), 63-76.

[137]Steele, B.C., Heinzel, A., 2001. Materials for fuel-cell technologies. Nature. 414, 345-352.

[138]Sun, C., Lee, J.S., Zhang, M., 2008. Magnetic nanoparticles in MR imaging and drug delivery. Adv. Drug Delivery Rev. 60(11), 1252-1265.

[139]Tang, J., Yuan, W., Wang, J., Tang, J., Li, H., Zhang, Y., 2012. Perfluorosulfonate ionomer membranes with improved through-plane proton conductivity fabricated under magnetic field. J. Membr. Sci. 423-424, 267-274.

[140]Thomassin, J.M., Kollar, J., Caldarella, G., Germain, A., Jérôme, R., Detrembleur, C., 2007. Beneficial effect of carbon nanotubes on the performances of Nafion membranes in fuel cell applications. J. Membr. Sci. 303(1-2), 252-257.

[141]Tricoli, V., Nannetti, F., 2003. Zeolite-Nafion composites as ion conducting membrane materials. Electrochim. Acta. 48(18), 2625-2633.

[142]Tripathi, B.P. Shahi, V.K., 2011. Organic-inorganic nanocomposite polymer electrolyte membranes for fuel cell applications. Prog. Polym. Sci. 36(7), 945-979.

[143]Truffier-Boutry, D., De Geyer, A., Guetaz, L., Diat, O., Gebel, G., 2007. Structural study of zirconium phosphate-Nafion hybrid membranes for high-temperature proton exchange membrane fuel cell applications. Macromolecules. 40(23), 8259-8264.

[144]Tsai, C.H., Wang, C.C., Chang, C.Y., Lin, C.H., Chen-Yang, Y.W., 2014. Enhancing performance of Nafion®-based PEMFC by 1-D channel metal-organic frameworks as PEM filler. Int. J. Hydrogen Energy. 39(28), 15696-15705.

[145]Wang, Y.J., Kim, D., 2007. Crystallinity, morphology, mechanical properties and conductivity study of in situ formed $PVdF/LiClO_4/TiO_2$ nanocomposite polymer electrolytes. Electrochim. Acta. 52(9), 3181-3189.

[146]Wang, Y., Chen, K.S., Mishler, J., Cho, S.C., Adroher, X.C., 2011. A review of polymer electrolyte membrane fuel cells: technology, applications, and needs on fundamental research. Appl. Energy. 88(4), 981-1007.

[147]Wang, H., Huang, L., Holmberg, B.A., Yan, Y., 2002. Nanostructured zeolite 4A molecular sieving air separation membranes. Chem. Commun. (16), 1708-1709.

[148]Wang, H., Holmberg, B.A. Yan, Y., 2003. Synthesis of template-free zeolite nanocrystals by using in situ thermoreversible polymer hydrogels. J. Am. Chem. Soc. 125(33), 9928-9929.

[149]Wang, L.S., Lai, A.N., Lin, C.X., Zhang, Q.G., Zhu, A.M., Liu, Q.L., 2015. Orderly sandwich-shaped graphene oxide/Nafion composite membranes for direct methanol fuel cells. J. Membr. Sci. 492, 58-66.

[150]Wen, P., Gong, P., Sun, J., Wang, J., Yang, S., 2015. Design and synthesis of Ni-MOF/CNT composites and rGO/carbon nitride composites for an asymmetric supercapacitor with high energy and power density. J. Mater. Chem. A. 3(26), 13874-13883.

[151]Won, J., Park, H.H., Kim, Y.J., Choi, S.W., Ha, H.Y., Oh, I.H., Kim, H.S., Kang, Y.S., Ihn, K.J., 2003. Fixation of nanosized proton transport channels in membranes. Macromolecules. 36(9), 3228-3234.

[152]Wu, B., Lin, X., Ge, L., Wu, L., Xu, T., 2013. A novel route for preparing highly proton conductive membrane materials with metal-organic frameworks. Chem. Commun. 49(2), 143-145.

[153]Wu, H., Cao, Y., Shen, X., Li, Z., Xu, T., Jiang, Z., 2014. Preparation and performance of different amino acids functionalized titania-embedded sulfonated poly (ether ether ketone) hybrid membranes for direct methanol fuel cells. J. Membr. Sci. 463, 134-144.

[154]Wu, H., Cao, Y., Li, Z., He, G., Jiang, Z., 2015. Novel sulfonated poly (ether ether ketone)/phosphonic acid-functionalized titania nanohybrid membrane by an *in situ* method for direct methanol fuel cells. J. Power Sources. 273, 544-553.

[155]Xi, J., Wu, Z., Qiu, X., Chen, L., 2007. $Nafion/SiO_2$ hybrid membrane for vanadium redox flow battery. J. Power Sources. 166(2), 531-536.

[156]Xu, T., Wu, D., Wu, L., 2008. Poly (2,6-dimethyl-1,4-phenylene oxide) (PPO)-a versatile starting polymer for proton conductive membranes (PCMs). Prog. Polym. Sci. 33(9), 894-915.

[157]Xu, J., Li, Y., Peng, S., Lu, G., Li, S., 2013. Eosin Y-sensitized graphitic carbon nitride fabricated by heating urea for visible light photocatalytic hydrogen evolution: the effect of the pyrolysis temperature of urea. Phys. Chem. Chem. Phys. 15(20), 7657-7665.

[158]Ye, D.H., Zhan, Z.G., 2013. A review on the sealing structures of membrane electrode assembly of proton exchange membrane fuel cells. J. Power Sources. 231, 285-292.

[159]Yen, C.Y., Liao, S.H., Lin, Y.F., Hung, C.H., Lin, Y.Y., Ma, C.C.M., 2006. Preparation and properties of high performance nanocomposite bipolar plate for fuel cell. J. Power Sources. 162(1), 309-315.

[160]Yoon, K.S., Choi, J.H., Hong, Y.T., Hong, S.K., Lee, S.Y., 2009. Control of nanoparticle dispersion in $SPAES/SiO_2$ composite proton conductors and its influence on DMFC membrane performance. Electrochem. Commun. 11(7),1492-1495.

[161]Yuan, T., Pu, L., Huang, Q., Zhang, H., Li, X., Yang, H., 2014. An effective methanol-blocking membrane modified with graphene oxide nanosheets for passive direct methanol fuel cells. Electrochim. Acta. 117, 393-397.

[162]Yu, D.M., Sung, I.H., Yoon, Y.J., Kim, T.H., Lee, J.Y., Hong, Y.T., 2013. Properties of sulfonated poly (arylene ether sulfone)/functionalized carbon nanotube composite membrane for high temperature PEMFCs. Fuel Cells. 13(5), 843-850.

[163]Yun, S., Im, H., Heo, Y., Kim, J., 2011. Crosslinked sulfonated poly (vinyl alcohol)/sulfonated multi-walled carbon nanotubes nanocomposite membranes for direct methanol fuel cells. J. Membr. Sci. 380(1-2),208-215.

[164]Yun, S., Heo, Y., Im, H., Kim, J., 2012. Sulfonated multiwalled carbon nanotube/sulfonated poly (ether sulfone) composite membrane with low methanol permeability for direct methanol fuel cells. J. Appl. Polym. Sci. 126(S2).

[165]Zhao, C., Lin, H., Cui, Z., Li, X., Na, H., Xing, W., 2009. Highly conductive, methanol resistant fuel cell membranes fabricated by layer-by-layer self-assembly of inorganic heteropolyacid. J. Power Sources. 194(1), 168-174.

[166]Zhang, H., Ma, C., Wang, J., Wang, X., Bai, H., Liu, J., 2014. Enhancement of proton conductivity of polymer electrolyte membrane enabled by sulfonated nanotubes. Int. J. Hydrogen Energy. 39(2), 974-986.

[167]Zhou, W., Xiao, J., Chen, Y., Zeng, R., Xiao, S., Nie, H., Li, F., Song, C., 2011. Sulfonated carbon nanotubes/sulfonated poly (ether sulfone ether ketone ketone) composites for polymer electrolyte membranes. Polym. Adv. Technol. 22(12), 1747-1752.

[168]Zhang, L., Chae, S.R., Hendren, Z., Park, J.S., Wiesner, M.R., 2012. Recent advances in proton exchange membranes for fuel cell applications. Chem. Eng. J. 204-206, 87-97.

[169]Zhang, X., 2007. Porous organic-inorganic hybrid electrolytes for high-temperature proton exchange membrane fuel cells. J. Electrochem. Soc. 154(3), B322-B326.

[170]Zakaria, Z., Kamarudin, S.K., Timmiati, S.N., 2016. Membranes for direct ethanol fuel cells: an overview. Appl. Energy. 163, 334-342.

[171]Zarrin, H., Higgins, D., Jun, Y., Chen, Z., Fowler, M., 2011. Functionalized graphene oxide nanocomposite membrane for low humidity and high temperature proton exchange membrane fuel cells. J. Phys. Chem. C. 115(42), 20774-20781.

[172]Zhai, Y., Zhang, H., Hu, J., Yi, B., 2006. Preparation and characterization of sulfated zirconia ($SO_4^{2-}$/$ZrO_2$)/Nafion composite membranes for PEMFC operation at high temperature/low humidity. J. Membr. Sci. 280(1-2), 148-155.

[173]Zheng, Y., Liu, J., Liang, J., Jaroniec, M., Qiao, S.Z., 2012. Graphitic carbon nitride materials: controllable synthesis and applications in fuel cells and photocatalysis. Energy Environ. Sci. 5(5), 6717-6731.

[174]Zheng, H. Mathe, M., 2011. Enhanced conductivity and stability of composite membranes based on poly (2,5-benzimidazole) and zirconium oxide nanoparticles for fuel cells. J. Power Sources. 196(3), 894-898.

[175]Zeng, R., Wang, Y., Wang, S., Shen, P.K., 2007. Homogeneous synthesis of PFSI/silica composite membranes for PEMFC operating at low humidity. Electrochim. Acta. 52(12), 3895-3900.

[176]Zimmerman, C.M., Singh, A., Koros, W.J., 1997. Tailoring mixed matrix composite membranes for gas separations. J. Membr. Sci. 137(1-2), 145-154.

# 22

# Key issues in estimating energy and greenhouse gas savings of biofuels: challenges and perspectives

Dheeraj Rathore[1,†], Abdul-Sattar Nizami[2,†], Anoop Singh[3], Deepak Pant[4,*]

[1] *School of Environment and Sustainable Development, Central University of Gujarat, Gandhinagar-382030, Gujarat, India.*

[2] *Solid Waste Research Group, Center of Excellence in Environmental Studies (CEES), King Abdul Aziz University, P.O Box: 80216, Jeddah 21589, Saudi Arabia.*

[3] *Government of India, Ministry of Science and Technology, Department of Scientific and Industrial Research (DSIR), Technology Bhawan, New Mehrauli Road, New Delhi- 110016 India.*

[4] *Separation and Conversion Technologies, VITO-Flemish Institute for Technological Research, Boeretang 200, 2400 Mol, Belgium.*

## HIGHLIGHTS

- Land-use and land-use changes may overestimate the biofuel sustainability.
- Environmental and social acceptability is essential in making biofuel-support policies.
- Different biomass result in different energy balances and GHG savings of biofuels.
- Sustainable biomass or non-food biomass can increase biofuel sustainability ranking.

## GRAPHICAL ABSTRACT

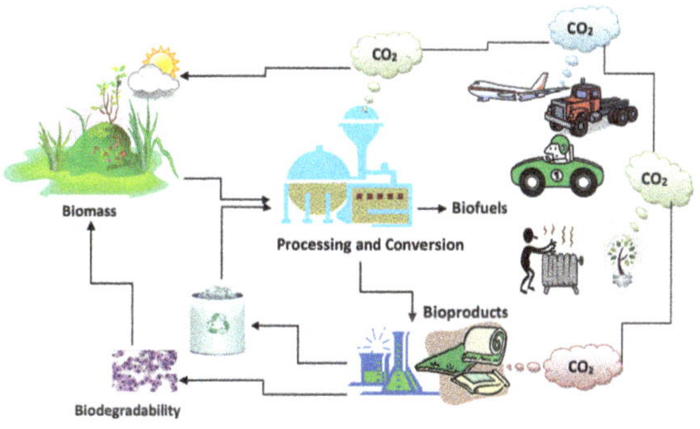


## ABSTRACT

The increasing demand for biofuels has encouraged the researchers and policy makers worldwide to find sustainable biofuel production systems in accordance with the regional conditions and needs. The sustainability of a biofuel production system includes energy and greenhouse gas (GHG) saving along with environmental and social acceptability. Life cycle assessment (LCA) is an internationally recognized tool for determining the sustainability of biofuels. LCA includes goal and scope, life cycle inventory, life cycle impact assessment, and interpretation as major steps. LCA results vary significantly, if there are any variations in performing these steps. For instance, biofuel producing feedstocks have different environmental values that lead to different GHG emission savings and energy balances. Similarly, land-use and land-use changes may overestimate biofuel sustainability. This study aims to examine various biofuel production systems for their GHG savings and energy balances, relative to conventional fossil fuels with an ambition to address the challenges and to offer future directions for LCA based biofuel studies. Environmental and social acceptability of biofuel production is the key factor in developing biofuel support policies. Higher GHG emission saving and energy balance of biofuel can be achieved, if biomass yield is high, and ecologically sustainable biomass or non-food biomass is converted into biofuel and used efficiently.

**Keywords:**
Energy balance
Greenhouse gas (GHG) balance
Biofuels
Sustainability
Life cycle assessment (LCA)

* Corresponding author
E-mail address: deepak.pant@vito.be ; pantonline@gmail.com
† These authors contributed equally.

**Contents**



**Abbreviations**

| | |
|---|---|
| AD | Anaerobic digestion |
| ALT | Atmospheric life time |
| BtL | Biomass to liquid |
| CG | Clean gasoline |
| CHP | Combined heat and power |
| FER | Fossil energy ratio |
| FFV | Flexible fuel vehicles |
| F-T | Fischer-Tropsch |
| GHG | Greenhouse gas |
| GWP | Global warming potential |
| HHV | High heating value |
| LCA | Life cycle assessment |
| MEC | Microbial electrolysis cell |
| NER | Net energy ratio |
| NEY | Net energy yield |
| OFMSW | Organic fraction of municipal solid waste |
| PPO | Pure plant oil |
| TtW | Tank to wheel |
| WtT | Well to tank |
| WtW | Well to wheel |

## 1. Introduction

Biofuels are getting significant attention worldwide due to depletion of fossil fuels and concerns regarding climate change (Popp et al., 2016; Khanna et al., 2010). They are recognized for their potential role in reducing greenhouse gas (GHG) emissions and providing energy security (**Fig. 1**). However, their sustainability is still under an intense debate due to different methodologies, biomass sources, land-use and land-use changes, fuel-blends, and end-use applications (Sims et al., 2010; Glenister and Nunes, 2011; Lankoski et al., 2011). Biofuels exist in solid, liquid, and gas forms and are derived from different biomass sources such as perennial crops, sugarcane, and corn starch as well as agricultural and forestry residues and organic fraction of industrial and municipal wastes (Nizami and Ismail, 2013; Ouda et al., 2016). Liquid biofuels are used as transportation fuel and for electricity generation through turbines and engines (Korres et al., 2011; Singh et al., 2012). The gaseous biofuels are also used as transportation fuel and for electricity generation using specially-designed direct and indirect turbine-equipped plants (Gnansounou et al., 2008a; Sadaf et al., 2016). While, solid biofuels are used in power plants instead of coal as fuel briquettes or pellets (Singh et al., 2010a; Nizami et al., 2015a and b).

Biofuels produced by exploiting fertile lands are criticized due to environmental, social, and economic issues (Sharma et al., 2012). According to Mukherjee et al. (2011), the issues of land-use and high food and animal feed prices are associated with biofuels that are produced using fertile lands. Moreover, negative impacts on forests and grasslands, loss of biodiversity due to large mono-cropped fields, water scarcity and pollution, and air quality degradation are often associated with such biofuels (Doornbosch and Steenblik, 2007; Fargione et al., 2008; Gnansounou et al., 2008b). Therefore, biofuels produced from non-food biomass sources such as corn stover, cereal straw, sugar cane bagasses, perennial grasses, forestry and agricultural wastes, and municipal and industrial organic wastes are receiving preferences (Searchinger et al., 2008). However, such biofuels are not yet produced at a commercial scale, but can influence GHG savings through land-use changes. For instance, biofuels from algae feedstock, if ever produced in an economically-viable manner, can potentially address most of the biofuels-related issues (Sander and Murthy, 2010; Singh et al., 2012).

Biofuels can be beneficial by reducing GHG emissions to keep climate-change impacts within the limits societies could be able to cope with. However, the benefits of biofuels largely depend on the whole life cycle of biofuel production, as the environmental ranking of biofuels based on GHG savings and energy balances vary with measuring methods, system boundary, land-use and land-use changes, functional unit and allocation methods (Kauffman and Hayes, 2013). All these variables and anticipation in results require comprehensive studies on biofuel production systems (Menichetti and Otto, 2009). Therefore, estimating GHG savings and energy balances of biofuels is not only critical from their sustainability point of view, but also is a challenging task (Singh et al., 2010b). Various models or life cycle assessment (LCA) tools are used to explain the results of biofuel studies that are either policy oriented or related to the process or product design or operation (Hong et al., 2013).

LCA is a cradle-to-grave analysis for the energy and environmental impacts of a product, process, or pathway. This is mandatory by the directive of EU (Directive 2009/28/EC) to employ the LCA method to reduce $CO_2$ emissions by 35% until 2017, by 50% until 2018 and by 60% after 2018 (EC-Directive, 2009). The energy efficiency of a biofuel is presented as a ratio of the amount of energy obtained from the fuel to the amount of fossil fuel energy required in its production process (Davis et al., 2008). While, the estimation of energy balances includes both the life cycle energy efficiency of biofuels and the savings from fossil fuels. According to Gnansounou et al. (2009), the inclusion of fossil fuel saving is critical with respect to the replacement efficiency of biofuels with fossil fuels.

The available scientific literature is mainly focused on bioethanol and biodiesel as being the most prominent biofuels. While, the other biofuel resources and systems are more or less ignored in terms of their sustainability (Singh et al., 2010b; Korres et al., 2011; Mukherjee et al., 2011; Sharma et al., 2012). Moreover, there is a strong need to address the key challenges in estimating the GHG savings and energy balances of biofuels along with their possible solutions. Therefore, this review paper aims to examine the various biofuel production systems for their GHG savings and energy balances, relative to conventional fossil fuels. The key issues and future directions for LCA based biofuels studies, especially on estimating GHG savings and energy balances are highlighted.

## 2. Life cycle assessment (LCA) and biofuel studies

LCA is a well-known internationally recognized methodology to evaluate environmental performance of any processes, products or pathways along with their whole or partial life cycle (Gnansounou et al., 2008a). The procedures for LCA are explained in ISO 14040-series (**Fig. 2**). Numerous studies on the LCA of biofuels have been reported by various researchers with different scope, accuracy, consistency level, transparency, and framework. Scope and goal of LCA are the two important steps, on which system definition and system boundary depend. The goal may be based on an operation, design, or policy (Menichetti and

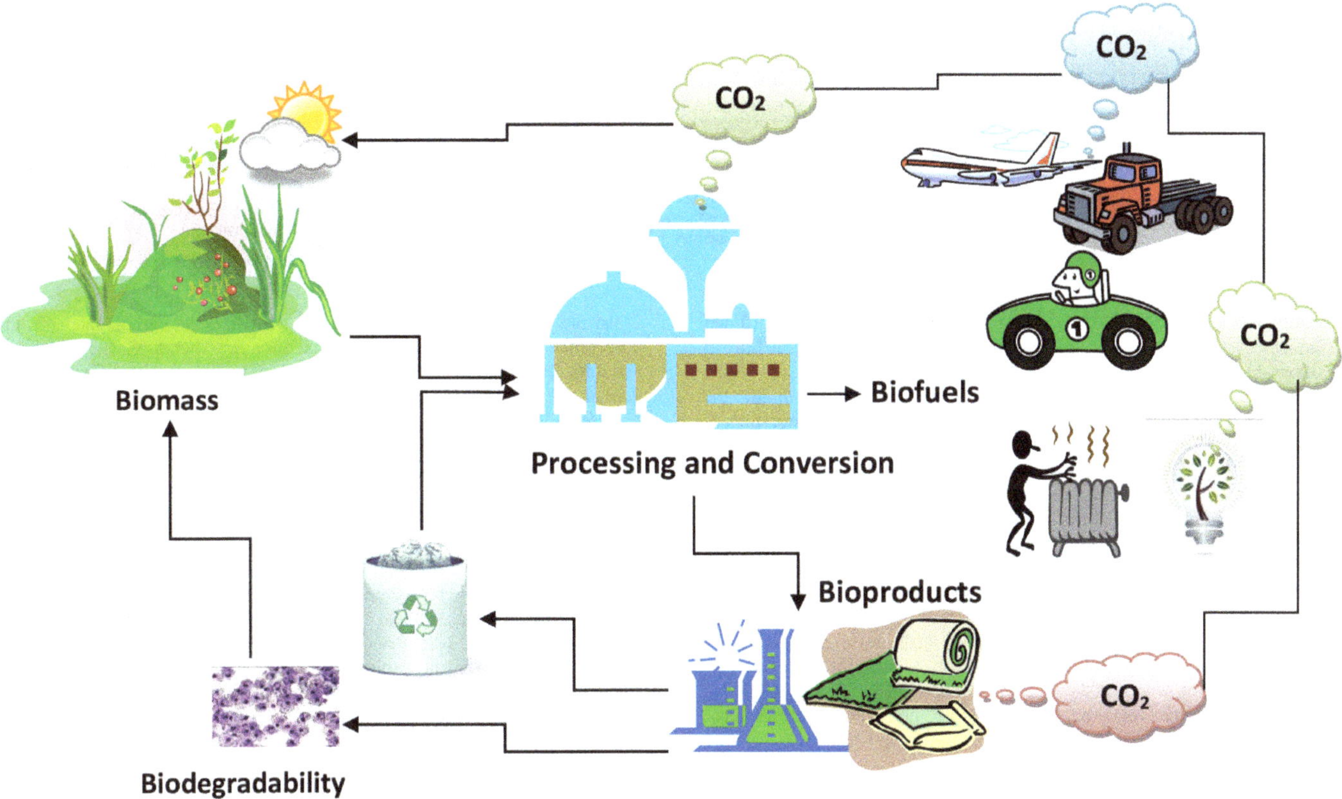


**Fig.1.** The life cycle of biomass to biofuels and bioproducts.

Otto, 2009). In case of operation and design improvement, the definition of the system should be more comprehensive. While, a simple flowchart of biofuel pathways can describe the policy purpose (Gnansounou et al., 2009). In case of policy as an integral part of LCA framework, the boundary of the system should be adopted according to the purpose. For example, if bioethanol is compared using a well to tank (WtT) approach, LCA results have no impacts on fuel combustion in the engine. However, the situation will change, if the comparison is carried out for the same biofuel with various fossil fuels or their blends (Gnansounou et al., 2008a).

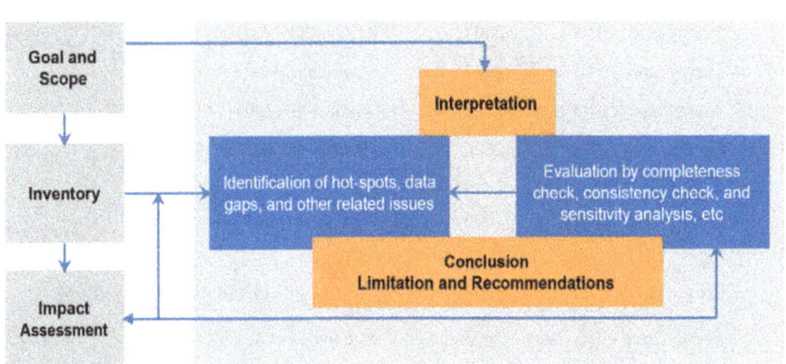


**Fig.2.** The LCA methodological framework.

Carbon dioxide ($CO_2$), water vapours, methane ($CH_4$), sulphur hexafluoride, chlorofluorocarbons, nitrous oxide ($N_2O$), hydrofluorocarbons, and perfluorocarbons are considered as major GHG contributors (Cherubini and Strømman, 2011). However, for GHG savings of biofuel, only anthropogenic sources are considered (Nizami and Ismail, 2013). Moreover, $CO_2$, $CH_4$, and $N_2O$ are taken into account during LCAs, as their origin could be either fossil or biogenic based (IPCC, 2011). Atmospheric life time (ALT) and its potential are the two factors based on which, GHG effect and global warming are described over a defined timescale. The reference for GHG is $CO_2$, and 1 is its value for global warming potential (GWP) that is taken for all time periods, including anthropogenic and radiative forcing. $CH_4$ has an ALT of 9 for 15 years and has a GWP of 84 for 20 years, 28 for 100 years. The other major GHG contributors like $N_2O$ has a GWP of 264 for 20 years, 265 for 100 years (Gnansounou et al., 2008a; Myhre et al., 2013). Most of the LCA studies follow the IPCC guidelines to take reference time of 100 years (Gnansounou et al., 2009; IPCC, 2011).

When comparison is made for a biofuel with fossil fuel, it is critical to select the same applicable service. For instance, in case of mobility applications, researchers use 1 MJ of fuels compared (e.g., E5 and gasoline) as a functional unit or choose 1 kWh of brake power produced by the fuels compared as functional unit (Rathore et al., 2013). Many studies concluded that E5 consumption in litres is different than the consumption of gasoline for the same service (e.g., for 100 km distance travelled). Therefore, less than 1 MJ of E5 is compared with 1 MJ of gasoline (Gnansounou et al., 2008a). The biofuel process comparison is carried out with a certain base line or reference system in order to evaluate the GHG emission savings. For this purpose, most of the LCA based studies use fossil fuels like diesel or gasoline as a reference system. The different results of such studies are due to variations in GHG savings of biofuel co-products that are used to replace the existing fossil fuel products (Singh et al., 2011 and 2012).

In LCA, different allocation methods such as physical and economical allocation are used to divide the environmental burden of a process or product when several functions reflect the same process. Therefore, allocation methods vary by mass (wet or dry), energy content, system expansion, economic value, and carbon content. The recommended

**Table 1.**
Biofuels classification (Pande and Bhaskarwar, 2012).

| Biofuels | First generation biofuels | Second generation biofuels | Third generation biofuels |
|---|---|---|---|
| Features | Fuels produced from raw materials in competition with food and feed industry | Fuels produced from non-food crops (energy crops) or raw materials based on waste residues | Fuels produced using aquatic microorganisms like algae |
| Examples | Bioethanol from sugarcane, sugar beet and starch (corn and wheat), biodiesel from oil based crop like rape seed, sunflower, soybean, palm oil, and waste edible oils- and starch-derived biogas | Biogas from waste and residues, bioethanol and biohydrogen from lignocellulosic materials like residues from agriculture, forestry and industry and fuels from energy crops such as sorghum | Biodiesel produced using algae and algal hydrogen |

allocation method of ISO 14040-series is system expansion, which is difficult to implement as results rely on the reference system (Singh et al., 2010a). If the direct land-use changes especially the carbon storage, is missing in the consideration of previous carbon storage, it may overestimate the performance of a biofuel. The carbon storage is positive when feedstock is produced from a degraded soil (Gnansounou et al., 2008a). There is a large difference between the data provided from existing database and the data obtained from a region, site, or country. Therefore, if the cost of supplementary information is affordable, then the default data should be used with precaution, otherwise data generalization should be avoided (Singh et al., 2010b).

Various sensitivity analysis and scenarios are used to evaluate the sensitivity in LCA studies. Ecoinvent® or SimaPro® are the most used LCA tools for sensitivity analysis. Manual sensitivity analysis can also be carried out in some cases, when results are presented in a range instead of precise values (Gnansounou et al., 2008a; Cherubini et al., 2009). The system input emission factors vary from one database to another. Each database has various inputs, based on which LCA calculates its carbon intensity in accordance with the methodological choices. Therefore, the obtained results for carbon intensities from GREET® model may be different from what achieved by Ecoinvent® database (Gnansounou et al., 2008a and 2009).

### 3. Estimation of GHG savings and energy balances of biofuels

Biofuels are often classified as first, second, and third generation biofuels (**Table 1**). First-generation biofuels utilize food crops as feedstock for biofuel production, while second-generation biofuels utilize non-food biomass. Third-generation biofuels use algae and microbes as fuel source materials (Singh et al., 2012). Various LCA based biofuel studies that have estimated the GHG emission savings and energy balances are grouped in **Table 2**. Following is the detail of GHG savings and energy balances of most prominent biofuel systems, relative to conventional fossil fuels.

**Table 2.**
Literature review on the estimation of energy and GHG balances of biofuels based on LCA studies.

| Feedstock | System adopted | Estimations | Year | References |
|---|---|---|---|---|
| Maize, Switch grass | Various | Energy and GHG | 2006 | Farrell et al. (2008) |
| Rapeseed, Recycled vegetable oil, Wood chip (residues, woodland management and short-rotation coppice), Miscanthus, Straw, Ligno-cell, Beet, Wheat | WtT[1] | Energy and GHG | 2003 | Elsayed et al. (2003) |
| Ethanol: Wheat, Beet, Straw, Wood waste, Sugar cane<br>Methanol: Wood waste, Farmed wood<br>Diesel: Rapeseed, Sunflower | WtT and TtW[2] | Energy and GHG | 2007 | Edwards et al. (2007) |
| Maize, Switchgrass | WtT | Energy and GHG | 2007 | Grood and Heywood (2007) |
| Palm oil | WtW[3] | Energy and GHG | 2007 | Reinhardt et al. (2007) |
| Maize, Sugar cane, Soybean, Palm oil, Waste material | WtW | Energy and GHG | 2007 | Unnasch and Pont (2007) |
| Maize, Switchgrass | WtW | Energy and GHG | 2007 | Wang et al. (2007) |
| Maize, Wheat, Cellulose | WtW | Energy and GHG | 2006 | Menichetti and Otto (2009) |
| Imported soy oil (40%)/ domestic Sunflower oil (10%)/ imported Palm (25%)/domestic and imported rapeseed (25%) | WtW | Energy and GHG | 2006 | Lechón et al. (2007) |
| Wheat straw | WtT and TtW | Energy and GHG | 2006 | Veeraragha van and Riera-Palou (2006) |
| Sugarcane, Maize | WtW | Energy and GHG | 2005 | de Oliveira et al. (2005) |
| Wheat, Barley | WtW | Energy and GHG | 2005 | Lechón et al. (2005) |
| Sugar cane | WtW | Energy and GHG | 2004 | Macedo et al. (2004) |
| Biogas: Woody biomass, Beet, Lignocellulose, Rapeseed | WtT and TtW | Energy and GHG | 2002 | Choudhury et al. (2002) |
| Wheat, Sunflower, Rapeseed | WtT | Energy and GHG | 2002 | Ecobilan PwC (2002) |

[1] WtT : Well to Tank
[2] WtW: Well to Wheel
[3] TtW: Tank to Wheel

*3.1. Bioethanol*

Bioethanol is one of the most widely used biofuels in the world. Globally, bioethanol production grew from 17,000 to 65,614 million litres from 2006 to 2008, respectively (RFA, 2009). In 2015, global bioethanol production reached up to 25576 million gallon with a maximum share of 57% from the United States of America (USA) (RFA, 2016). In the USA, about 13.7 billion gallon of fuel ethanol were added to motor gasoline in 2015. This fuel ethanol accounted for about 10% of the total volume of motor gasoline consumed in the country (US-EIA, 2016). Moreover, it is estimated that bioethanol will provide 7% of total global energy as a transportation fuel by 2030 (Escobar et al., 2009). Bioethanol is produced from starch and sugar crops such as cassava, wheat, barley, corn grain, or sugarcane (Kim and Dale, 2004; Nguyen et al., 2007; Macedo et al., 2008). The non-food biomass sources can also be used for producing bioethanol (Reijnders, 2008; Sassner et al., 2008; Najafi et al., 2009; Gonzalez-Garcıa et al., 2010). The process of bioethanol production is similar to conventional brewing beer process; where starch crops are converted into sugars, then the sugars are fermented into ethanol, and finally the ethanol is distilled into the final product. Bioethanol is blended with gasoline at ratios ranging from 2 to 85% by volume in order to use in flexible fuel vehicles (FFV) (UNICA, 2009; Ansari and Verma, 2012). Moreover, 100% ethanol concentration could also be used in dedicated vehicles (Zhi Fu etal., 2003; Macedo et al., 2008; Hahn-Hägerdal et al., 2009; Gonzalez-Gracia et al., 2012).

An array of feedstocks including sugarcane, maize, sugar beet, cereal crops, cassava, potato, wheat, and cellulosic materials were studied for estimating their energy balance when producing bioethanol (Rosenberger et al., 2001; Kim and Dale, 2005; Malca and Freire, 2006; Renouf et al., 2008; Gracia et al., 2011; Papong and Malakul, 2010). Although bioethanol has a lower energy content than conventional gasoline, but its high-octane value results in higher compression ratios and efficient thermodynamic operation in internal-combustion engines (Nguyen et al., 2007). Energy balance investigation of bioethanol from sugarcane in Mexico showed an energy ratio of 4.8 $GJ_{Ethanol}/GJ_{Fossil}$. However, this value was lower in comparison with the Brazilian sugarcane-ethanol energy ratio of 8.4 $GJ_{Ethanol}/GJ_{Fossil}$ (Gracia et al., 2011). Sugarcane-bioethanol system has high yields of net energy and net potential of reducing GHG emissions than gasoline, as no energy is required to depolymerize carbohydrate into fermentable sugars (Macedo, 1998; Liska and Cassman, 2008). In a comparative study, Koga et al. (2013) reported an energy efficiency of 4.63 MJ $L^{-1}$ energy input in Kon-iku (potato type no. 38) and traditional practice potato (5.68 MJ $L^{-1}$ energy input) in northern Japan for bioethanol production. Nguyen et al. (2007) estimated energy efficiency of cassava-bioethanol and found similar results with corn-grain bioethanol. Liska and Cassman (2008) recalculated metrics reported in an LCA by Malca and Freire (2006) and estimated a net energy ratio (NER) of 1.9 for wheat-bioethanol system. In a different study, Walker et al. (2011) analysed 5 different crops such as miscanthus, willow, winter wheat, rape, and potato for their energy balances.

Gonzalez-Gracia et al. (2010) showed a slight decrease in GHG emissions when shifting from clean gasoline to E10 blends of ethanol regardless of the feedstock type, e.g., alfalfa leaves, Ethiopian mustard, poplar, flax fibres and linseed, hemp fibres and dust. The use of E85 blends seems to be more advantageous than E10 in terms of GHG emission savings. Moreover, Gonzalez-Gracia et al. (2010) suggested that up to 88% of total GHG emission savings can be achieved with lignocellulosic sources such as alfalfa stems. Increased GHG emission savings with high blend of ethanol such as E85 and E100 is because of increased $CO_2$ sequestration during crop cultivation. However, improving the production of crops requires higher use of fertilizer, which in return causes increased emissions of $N_2O$.

GHG savings for bioethanol varies according to the choice of system boundary. Some studies used WtT analysis (Elsayed et al., 2003), while others have considered well to wheel (WtW) analysis (Gnansounou and Dauriat, 2004; Beer and Grant, 2007; Edwards et al., 2007). For instance, WtW analysis is used for calculating complete GHG emission savings of biofuel to a reference system. However, engine performance results for biofuels compared with conventional fuels may influence the final outcomes. Gracia et al. (2011) in their WtT analysis, compared the GHG emission savings of Mexican sugarcane-bioethanol system with a reference fossil fuel. The land-use changes reverted the scenario, especially when the rainforest was converted to sugarcane crop. The conversion of maize-biomass into electrical and thermal energy saved (6.3±0.56) Mg $CO_2$ eq $ha^{-1}$ on average, whereas Miscanthus chips saved (22.3±0.13) Mg $CO_2$ eq $ha^{-1}$ $yr^{-1}$ (Felten et al., 2013). A cross comparison of GHG savings and energy balances of bioethanol studies is presented in **Table 3**.

*3.2. Biodiesel*

Various edible and non-edible crops such as soybean (Balan et al., 2009), oil palm (Yanez et al., 2009; Kamhara et al., 2010), rapeseed (Long et al., 2011; Gardy et al., 2014), Ethiopian mustard (Bouaid et al., 2005 and 2009), sunflower (Rashid and Anwar, 2008; Xin et al., 2009), desert date (Deshmukh and Bhuyar, 2009), castor (Scholz and Silva, 2008), Jatropha (Diwani et al., 2009; Sharma et al., 2012), Pongamia (Das et al., 2009; Kesari and Rangan, 2010), Azadirachta (Nabi et al., 2006) were used as feedstock for biodiesel production. According to Prasad et al. (2007a and b), biodiesel production second and third generation

**Table 3.**
Energy and GHG balance of bioethanol.

| Feedstock | System adopted | Reference system | Functional unit | Energy Balance | GHG Balance | Country | Reference |
|---|---|---|---|---|---|---|---|
| Reed canary grass | Cradle to grave | Coal | $CO_2$ e-C | - | 84% GHG saving | USA | Adler et al. (2007) |
| Switchgrass | Cradle to grave | Coal | $CO_2$ e-C | - | 114% GHG saving | USA | Adler et al. (2007) |
| Hybrid poplar | Cradle to grave | Coal | $CO_2$ eq-C | - | 117% GHG saving | USA | Adler et al. (2007) |
| Corn-soybean | Cradle to grave | Coal | $CO_2$ eq-C | - | 38-41% GHG saving | USA | Adler et al. (2007) |
| Corn stover | Energy product to gate | Gasoline, a hypothetical case of pure ethanol | 1 km driving of mid-size car | Positive | Reduction in GWP | The Netherlands | Luo et al. (2009) |
| Switchgrass and corn stover | Cradle to wheel | Low-sulfur reformulated gasoline | $CO_2$ eq $km^{-1}$ | Positive | up to 70% lower GHG emissions | Canada | Spatari et al. (2005) |
| Household and Biodegradable municipal waste | Cradle to grave | Gasoline | MJ of fuel equivalent | - | Up to 92.5% GHG emission saving | UK | Stichnothe and Azapagic (2009) |
| Corn stover | Cradle to grave | Gasoline | $CO_2$ $km^{-1}$ | - | Reduction of 267 g $CO_2$ $km^{-1}$ | USA | Sheehan et al. (2004) |
| Blue-Green Algae | Cradle to Grave | Gasoline | $CO_2$ eq $MJ_{EtOH}^{-1}$ | Positive | 67% and 87% reductions in the carbon footprint | USA | Luo et al. (2010b) |
| Switchgrass- Corn Stover | Cradle to Grave | Low sulfur reformulated gasoline | $CO_2$ eq $km^{-1}$ | - | Up to 65% lower GHG emissions | Canada | Spatari et al. (2005) |

biofuels pathways has led to promising results (Christi, 2007; Campbell, 2008). However, these pathways are not yet at a commercial scale due to higher production cost and challenges in process and conversion technologies compared with the first generation biodiesel (Khan et al., 2009). Chemical and biological catalysts such as alkali and acidic compounds and lipase are often used in biodiesel production (Kim et al., 2007).

Biodiesel production from soybean comprises soybean production, its transportation to the processing facility, separation of oil and meal and conversion into biodiesel through transesterification process, and finally the distribution of biodiesel (Sheehan et al., 1998). Usually, a multistage stage transesterification process is required to convert crop-oil into biodiesel that can be used in conventional diesel engines. In 2014, global production of biodiesel (most of which as FAME) reached up to 30 billion litres with a maximum share from the USA (16%), followed by Brazil and Germany (both with 11%), Indonesia (10%), and Argentina (9.7%). Europe accounted for 39% of global biodiesel production in 2014 (REN21, 2009). Global biodiesel production is expected to reach up to 39 billion litres by 2024 (OECD-FAO, 2015).

The NER and fossil energy ratio (FER) of biodiesel from soybean were reported higher than those of the corn grain-bioethanol system, while, it was reported to have a 23% smaller net energy yield (NEY) (Hill et al., 2006). This means that soybean-biodiesel requires more land area to yield the same amount of NEY compared with corn grain-bioethanol (Liska and Cassman, 2008). Hill et al. (2006) reported NER of 1.9 for soybean-biodiesel. While, Sheehan et al. (1998), reported FER of 3.215 for soybean-biodiesel using TEAM (Ecobalance, Neuilly-sur-Seine, France) as modelling software. According to Sharma et al. (2012), soybean-biodiesel can generate more energy than what required to grow the crops and convert them into fuel. Such controversy in energy balance of soybean-biodiesel was triggered after a study by Pimentel and Patzek (2005) who reported less energy output from biodiesel in comparison with fossil fuel inputs. They further claimed that the soybean-biodiesel needed 27% more fossil energy than the actual energy of the produced biodiesel. Pradhan et al. (2008) explained this negative value of energy using Pimentel and Patzek model as an arithmetic error and stated that it occurred during calculations related to lime application. Pimentel and Patzek (2005) reported that only 19.3% of the total input energy goes to the soybean meal, however, in reality 82% of the soybean mass goes into meal. Similarly, they assigned 4,800 kg lime $ha^{-1}$ $yr^{-1}$ for the average soybean crop, while lime is used for only acidic soil to correct pH once in several years. Pradhan et al. (2008) reanalysed Pimentel's model with 3 other different models, including Ahmed, GREET, and NREL and concluded that the discrepancy in the energy balance of soybean - biodiesel was due to variation in the allocated energy proportions to biodiesel and its meal co-product.

Palm oil yield per hectare is significant in comparison with soybean oil marking palm oil-biodiesel the most competitive biofuel in terms of gross energy (Liska and Cassman, 2008). Kamhara et al. (2010) reported 3.5 NER for palm oil-biodiesel in Indonesia, while Yanez et al. (2009) and de Souza et al. (2010) reported 4.7 and 2.33 NER, respectively, in Brazil. Similarly, high energy balance was also observed in sunflower oil (Sheehan et al., 1998), canola oil (Fore et al., 2011), rapeseed oil (Janulis, 2004), and microalgal oil (Batan et al., 2010). Zhang et al. (2013) observed a high energy gain for biodiesel from microalgae. However, due to expensive commercial-scale facilities and climate sensitivity of microalgae, the production of microalgae for microalgae-biodiesel is still in the developmental phase. More interesting results were achieved in a study by Zhang et al. (2013) who reported higher energy gain of biodiesel from wastewater sludge while they could also resolve the energy consumption and waste sludge disposal problems.

Leguminous crops such as soybean require less nitrogen fertilizer during crop production, thus have a high potential for GHG emission savings. The cultivation of rapeseed for biodiesel led to GHG emission savings of 3.2±0.38 Mg $CO_2$ eq. $ha^{-1}$ (Felten et al., 2013). GHG emission savings from the rapeseed-biodiesel ranged from 20 to 80% with an average value of 40-60% in comparison with conventional fuel (Menichetti and Otto, 2009). It should be noted that although palm oil-biodiesel is associated with the most promising energy (Liska and Cassman, 2008) and GHG emission savings (Beer et al., 2007; Zah et al., 2007), land-use changes adversely influence the results. Beer et al. (2007) explained that if rainforest and peat forest are converted into crop land for palm oil production, the results of GHG emission savings will revert to negative value ranging from 800 to 2000%, respectively. A WtW analysis showed 1.1 kg $CO_2$ eq. for a 1 km travelling distance with a reference fossil fuel while rapeseed-biodiesel led to 0.48 kg $CO_2$ eq. for a similar distance with a total GHG emission saving of 56% (Finco et al., 2012). However, the study did not include land-use changes that can further reduce GHG emission savings. A study conducted by Finco et al. (2012) for rapeseed-biodiesel showed that agriculture phase was the major contributor to GHG emission (68%) followed by transesterification process (18%) and solvent extraction (8%). Pehnelt and Vietze (2013) also analysed various similar scenarios. They indicated that the GHG emission savings from palm oil-biodiesel ranged from 37.1 to 85%. The GHG savings and energy balances of biodiesel obtained from different sources are shown in **Table 4**.

**Table 4.**
Energy and GHG balance of biodiesel.

| Feedstock | System adopted | Reference system | Functional unit | Energy Balance | GHG Balance | Country | Reference |
|---|---|---|---|---|---|---|---|
| Microalgae | Cradle to grave | Fossil diesel | Kg $CO_2$ eq $ton^{-1}$ biodiesel | Positive in raceway pond and negative in air-lift tubular bioreactors | About 80% lower GWP | UK | Stephenson et al. (2010) |
| Microalgae | Well to pump | Fossil diesel | Kg $CO_2$ (1000 MJ Energy)$^{-1}$ | - | Positive $CO_2$ emissions for the centrifuge process while negative values for the filter press process | USA | Sander and Murthy (2010) |
| Rapeseed | Cradle to grave | Conventional gasoline | 1 PKM | - | Reduced green house gas emission | Argentina | Emmenegger et al. (2011) |
| Microalgae wastewater sludge | Cradle to grave | - | GJ $ton^{-1}$ biodiesel produced | Positive | Sequestered carbon, reduction in GHG emission | USA | Zhang et al. (2013) |
| Rapeseed | Field to wheel | Conventional diesel | 1 km traveled by bus | - | 56% GHG savings | Italy | Finco et al. (2012) |
| *Pongamia pinnata* | Field to wheel | Diesel | 1 MJ energy available in *Pongamia* | - | 1.5 ton + additional 1 ton $CO_2$ sequestration potential by 1 hectare *Pongamia pinnata* | India | Chandrashekar et al. (2012) |
| Jatropha | Well to Tank | Fossil Diesel | 1 MJ of JME | - | 72% GHG savings | Ivory Coast and Mali | Ndong et al. (2009) |
| Microalgae | Cradle to Grave | first generation biodiesel and oil diesel | 1 MJ of fuel in diesel engine | - | Significantly decreased environmental impacts | France | Lardon et al. (2009) |

### 3.3. Pure plant oil (PPO)

Similar to biodiesel, pure plant oil (PPO) is also derived from lipids. Primary process steps such as feedstock production and oil extraction are also similar to those of biodiesel production. However, final production of PPO and its purification procedures are additional steps. Although the name of PPO refers to original vegetable-oil, but it also includes waste oil and oil from animal fats. PPO can be used in diesel engines, but due to its relatively high viscosity (12 times higher than fossil diesel) (WWI, 2006) and combustion properties (Paul and Kemnitz, 2006), engines should be modified and refitted. For PPO, no study has been conducted so far in order to investigate the energy balance and GHG emission savings. The energy consumed for transesterification of biodiesel can be saved in case of PPO while the absence of co-products as is the case for biodiesel, i.e., glycerol, can further save GHG emissions (Dreier and Tzscheutschler, 2001; Quirin et al., 2004).

### 3.4. Biomass to liquid (BtL) biofuels

BtL biofuels are produced by various techniques (Sawayama et al., 1999; Ledford, 2006; Jungbluth et al., 2008; Bensaid et al., 2012). The Fischer-Tropsch (F-T) thermochemical synthesis can utilize a wide range of biomass sources to produce liquid biofuels. F-T synthesis using biomass has been successfully examined at pilot-plant scale and further development for biofuels of high quality is underway (Huber et al., 2006). One such F-T facility is located in Germany (Ledford, 2006). A full life cycle study conducted by Jungbluth et al. (2008), concluded that BtL biofuels from agricultural biomass, particularly short rotation crop did not show a significant reduction in GHG emissions and did not produce sufficient energy, while biomass used from forestry could increase both GHG emission savings and energy balance.

Bensaid et al. (2012) conducted a study on converting biomass wastes into valuable liquid biofuel by choosing a direct liquefaction technology. They reported process power consumption of 0.258 kWh/kg oil corresponding to an output/input energy ratio of 35.8. Moreover, they manage to achieve an output/input energy ratio of 9.7, even without power generation. Oil recovered by thermochemical liquefaction from microalgae such as *Botryococcus braunii* showed 1.6 times more heating value (HHV) (45.9 MJ kg $yr^{-1}$) than coal (28 MJ kg $yr^{-1}$) along with a high energy balance (Sawayama et al., 1999). They also reported high energy balance for liquefaction of sewage as no net energy is used for sewage production. Similarly, Nzihou et al. (2012) concluded that significant benefits in energy and GHG balances could be gained by using solar energy as external heating source for the standard gasification process.

Agrawal et al. (2009) proposed an innovative way of producing biofuels by combining biomass and hydrogen from a carbon-free energy source. The produced biofuels had three times more yield per unit mass of biomass in comparison with the conventional gasification F-T process. The energy contents of the biomass and hydrogen fed into the conversion plant were higher for the hybrid hydrogen–carbon process. It should be noted that hydrogen-cars are one of the examples that can achieve the carbon efficiency of nearly 100% in comparison with 37% for the conventional process. Moreover, the use of second generation biofuels can result in a greater $CO_2$ reduction per biomass unit used (Martinsen et al., 2010). Therefore, changes in biomass source can further improve the potential of GHG emission savings. On the contrary, Monti et al. (2009) found 50 to 60% less impact on GHG emission savings by changing conventional crops to more efficient energy crops.

Life cycle study of BtL biofuels carried out by van Vliet et al. (2009) concluded that GHG emissions from F-T process depend on the efficiency of conversion plants, biomass intermediates, and the use of feedstock. Coal-to-liquid chains without carbon capture and storage were reported to increase transportation-related GHG emissions. While, gas-to-liquid with carbon capture and storage was found to reduce GHG emissions by around 5% in comparison with fossil diesel. Moreover, the net emissions from BtL can be smaller and negative through the application of carbon capture and storage. Therefore, a net climate neutral biofuel can be made by using around 50% BtL with carbon capture and storage and blending it with other fuels. For instance, if biomass gasification and carbon sequestration are operated at industrial scale, and the feedstock is obtained in a sustainable way, the resultant biofuel will be climate neutral. GHG savings and energy balances of BtL biofuels are shown is **Table 5**.

### 3.5. Biomethane

A range of feedstocks such as organic fraction of municipal solid waste (OFMSW), sludge, slaughterhouse waste, biofuel residues, industrial, agricultural and forestry residues, and energy crops can be used in anaerobic digestion (AD) (Prasad et al., 2007; Smith et al., 2009; Sadaf et al., 2016; Tahir et al., 2015). Biogas produced through AD is purified and

**Table 5.**
Energy and GHG balance of BtL.

| Feedstock | System adopted | Reference system | Functional unit | Energy Balance | GHG Balance | Country | Reference |
|---|---|---|---|---|---|---|---|
| Rapeseed<br>Oil palm<br>Jatropha | Cradle to grave | Conventional diesel | 317 GJ | - | GWP reduced by about half | USA | Clarens et al. (2010) |
| Micro algae | Cradle to grave | Coal thermal plant | - | Positive | Reduction in GHG emissions | Japan | Sawayama et al. (1999) |
| Microalgae<br>Canola<br>Switchgrass<br>Corn | Cradle to gate | Comparison | 317 GJ | Positive | GHG savings | USA | Clarens et al. (2010) |
| Sugar cane<br>Sugar crops<br>Jatropha<br>Algae<br>Palm oil<br>Short rotation woody crops<br>Forestry wood<br>Wood residues<br>Agricultural residues<br>Used cooking oil<br>Waste | Cradle to grave | fossil reference system | Variable | Positive | Reduction in GHG emissions | Norway | Cherubini and Strømman (2011) |

upgraded to enriched biomethane (up to 97% $CH_4$) that can be blended with or used as an alternative fuel in natural gas vehicles or as a source of thermal energy (Murphy et al., 2013). For efficient distribution of produced biomethane, the existing network of natural gas grid can be utilized with end-applications of electricity, thermal, and transportation energy generation (Korres et al., 2010). The biomethane generated from grass or grass silage is considered sustainable biofuel by the EU Renewable Energy Directive (EC-Directive, 2009). However, to make grass-biomethane commercially available, an efficient vehicle and carbon sequestration with up to 60% GHG emission savings is required (Singh et al., 2011).

Dressler et al. (2012) reported a low net energy demand of -0.274 to 0.175 kWh/kWh$_{el}$ at Celle region of Germany because of using combined heat and power (CHP) unit in the region. Thyø and Wenzel (2007) concluded that biogas produced from manure has high fossil fuel savings in comparison with conventional manure storage and manure soil application. The GHG emissions correlated with the cultivation of energy maize ranged from 45.4 to 57.7 kg $CO_2$ eq. $t^{-1}$ of fresh maize. While, GHG emissions range from 0.179 to 0.058 kg $CO_2$ eq. kWhe$^{-1}$, when biogas was produced and used from maize (Dressler et al., 2012). They showed more efficiency in term of GHG emission savings in comparison with the study by Vogt (2008a), because part of the mineral fertilizers was substituted by digester output (i.e. digestate). However, the choice of fermenter was not considered by Dressler et al. (2012), which could have further manipulated the GHG emission savings. Vogt (2008b) also showed the influence of open and closed fermenters on GHG emission savings. The direct emissions from the fermenters can account for 25 to 75% of the overall GHG emissions. Depending on the level of carbon sequestration 75 to 150% GHG emission savings were achieved for grass biomethane in comparison to fossil fuel (Korres et al., 2010). Singh and Murphy (2009) reported 82% GHG emission savings for cattle slurry-biomethane in comparison to diesel, while savings of 21 to 53% were achieved for grass-biomethane in comparison to diesel. The GHG savings and energy balances of biomethane studies are shown is **Table 6**.

### *3.6. Biohydrogen*

Biohydrogen is a promising candidate for future energy supplies due to being renewable in nature with no GHG emissions during combustion, as well as easy conversion into electricity through fuel cells (Hallenbeck and Benemann, 2002). It has the largest energy contents per weight in comparison with other fuels. It can be produced by different techniques such as water splitting, coal gasification, and natural gas reforming (Levin et al., 2004). However, these methods for biohydrogen production need high energy inputs using non-renewable resources (Levin et al., 2004). Biological production of hydrogen using bioelectrochemical systems such as microbial electrolysis cells (MEC) can solve this problem by using microorganisms for converting biomass into hydrogen gas ($H_2$) (Das and Veziroglu, 2001; Hallenbeck and Benemann, 2002). Although biohydrogen production rates in MECs are still not high enough, more research is underway to optimize the performance of MECs (Zhu and Beland, 2006). Moreover, Rathore and Singh (2013) discussed in detail the potential role of microalgae for biohydrogen production.

Cheng and Logan (2007) conducted a study on exoelectrogenic bacteria in specially designed reactors to produce $H_2$ from biodegradable organic matters through the electrohydrogenesis process. They observed the overall energy efficiency of 288% for the process that was based on the electricity applied. While, an efficiency of 82% was achieved, when the combustion heat of acetic acid was included in the energy balance, at a biohydrogen production rate of 1.1 $m^3$ $d^{-1}$ per each cubic meter of the reactor. A high yield of biohydrogen was also observed by using glucose, several volatile acids such as acetic, butyric, lactic, propionic, and valeric, and cellulose at maximum stoichiometric yields of 54–91%. The achieved energy efficiencies ranged from 64 to 82%. Djomo and Blumberga (2011) reported 1.08, 1.14, and 1.17 energy ratios for wheat straw, sweet sorghum stalk, and steam potato peels, respectively, without considering the co-products such as protein residues. The energy efficiency was further enhanced by 23–128%, when the co-products were taken into account.

Biohydrogen is considered a clean fuel if does not produce $CO_2$ during combustion. By using biohydrogen instead of natural gas, heavy fuel oil, and coal to produce electricity, 33, 39.5, and 39% $CO_2$ emissions could be avoided, respectively (Ranagnoli et al., 2011). Djomo and Blumberga (2011) reported 55.53, 54.30, and 51.84% GHG emission savings when steam potato peels, sweet sorghum stalk, and wheat strew were used for biohydrogen production, respectively. However, a negative value (3.93%)

**Table 6.**
Energy and GHG balance of biomethane.

| Feedstock | System adopted | Reference system | Functional unit | Energy Balance | GHG Balance | Country | Reference |
|---|---|---|---|---|---|---|---|
| Grass | Cradle to Grave | Fossil diesel | | Positive | 54-75% GHG savings | Ireland | Korres et al. (2010) |
| | | | $CO_2$ eq KWhe$^{-1}$ | | | | |
| Maize | Cradle to grave | Fossil fuel | | | GHG emission 0.179 to 0.058 kg $CO_2$ eq. KWhe$^{-1}$ | Germany | Dressler et al. (2012) |
| | | | $CO_2$ eq KWhe$^{-1}$ | | | | |
| Grass | Cradle to grave | Fossil diesel | | Positive | Up to 85% GHG savings | Ireland | Singh et al. (2010a) |
| Bagasse | Cradle to gate | Landfilling with utilization of landfilled gas | MWh of electricity ton$^{-1}$ of pulp produced | - | Reduction in GHG emissions | Thailand | Kiatkittipong et al. (2009) |
| Maize silage<br>Manure | Cradle to Grave | Gasoline | KM Transport<br>GJ Heat<br>GJ Power | Positive | Reduction in GHG emissions | Germany | Thyø and Wenzel (2007) |
| Biowaste | Cradle to Gate | Incineration | Km$^2$ area ton$^{-1}$ biowaste | - | Reduction in GHG emissions | Spain | Güereca et al. (2006) |
| Silage maize<br>Silage grass<br>Silage rye<br>Forage beet | Cradle to Grave | Grid electricity | 1 Terajoule electricity fed into public electricity system | - | Reduction in GHG emissions | Germany | Hartmann (2006) |
| Energy crops | Cradle to Grave | Natural gas | 1 MJ injected into natural gas grid | Positive | Reduction in GHG emissions | Luxembourg | Jury et al. (2010) |

**Table 7.**
Energy and GHG balance of biohydrogen.

| Feedstock | System adopted | Reference system | Functional Unit | Energy Balance | GHG Balance | Country | Reference |
|---|---|---|---|---|---|---|---|
| Potato steam peels | Cradle to Grave | Defined | 1kg $H_2$ $Kg^{-1}$ Potato peel | - | Reduction of GHG emissions | USA | Djomo et al. (2008) |
| Green algae<br>Cyanobacteria<br>Potatoes peels | Cradle to Well | Not defined | MW $yr^{-1}$<br>GJ $yr^{-1}$ | - | Reduction of GHG emissions | Latvia | Romagnoli et al. (2011) |
| Sugarcane juice | Cradle to Grave | Electricity | Kg $CO_2$<br>MJ | Positive | 57-73% reduction of GHGs | India | Manish and Banerjee (2008) |
| Potato peels | Well to Wheel | Conventional fossil diesel and gasoline | g$CO_2$ $MJH_2^{-1}$ | Positive (45-52% reduction of energy consumption) | 65-69% reduction of $CO_2$ emissions | Portugal | Ferreira et al. (2011) |
| Biomass gasification | Cradle to Grave | Diesel | 1 MJ $s^{-1}$ hydrogen production | Positive | Reduction of GHG emissions | Turkey | Kalinci et al. (2012) |
| Food waste and wheat feed | Cradle to Grave | Diesel | 1 km of Transportation | Positive | Reduction of GHG emissions | UK | Patterson et al. (2013) |
| Microalgae biomass | Cradle to Grave | Electricity | g (1 MJ of $H_2$ produced)$^{-1}$ | - | Reduction of GHG emissions | Portugal | Ferreira et al. (2013) |

was observed for biohydrogen from steam methane. Several other examples for biohydrogen production with their respective GHG savings and energy balances are presented in **Table** 7.

## 4. Challenges and perspectives in GHG savings and energy balances of biofuels

The environmental performance of biofuels based on their GHG savings and energy balances depend on a wide range of factors such as feedstock types, conversion technologies, issues related to land-use and land-use changes along with substituted products like electricity, transportation fuel, and animal feed (Menichetti and Otto, 2009). The distribution of impacts for estimating GHG savings and energy balances vary from study to study (**Table 3**). As long as the GHG emissions are concerned, agricultural activities share a major role due to release of nitrogen gases (i.e., $N_2O$, NOx) and SOx with the use of fertilizers. Moreover, they are also responsible for acidification and eutrophication. The treatment of co-products and the allocation of impacts for co-products also change the LCA results significantly (**Box 1**).

Energy estimations are influenced significantly by technology conversion pathways as well. Moreover, the type of energy input in the form of heat and power from coal, natural gas, petroleum or bagasse, and energy quantity change the results of LCA studies (Wang et al., 2007). For example, when coal is used as a fuel in the corn-ethanol system, the GHG emissions are three times higher than gasoline. However, by using biomass feedstock like wood chips as an energy source, the GHG emission savings were surpassed by 50%. Besides agricultural activities, the fate of co-products, allocation of impacts, life cycle inventory databases and life cycle impact assessment methods could result in different LCA outcomes (**Box 1**). Assumptions on vehicle performance and biofuel transportation distance could also influence the LCA results (Menichetti and Otto, 2009).

Based on the results of a survey, Kim and Dale (2009) presented regional differences in GHG emissions from corn-ethanol and soybean oil production in 40 different counties in the USA. They observed that with the selection of feedstock material for biofuel production, the fertilizer requirement changes, which is another major source of GHG emissions. Dressler et al. (2012) also observed local and regional variations in GHG emissions at three different districts of Germany. They reported 50-56% GHG emissions from the cultivation stage with a minimum emission at the district of Gottingen, because of lower demand of fertilizers, pesticides, and fuel. According to Kim and Dale (2009), $N_2O$ emissions from the soil and nitrogen fertilizers during energy crops cultivation resulted in the highest GHG emissions, accounting for 13 to 57% of all GHG emissions.

Brazil and Indonesia showed a large percentage of GHG emissions from land-use changes that account for 61% of the world $CO_2$ emissions originated from land-use changes (Le Quéré et al., 2009). This means, a biofuel crop with higher-energy productivity may have less land-use change emissions per MJ than a less productive biofuel crop, even from the same field. Therefore, it is possible that a productive biofuel crop can be combined with the optimum management techniques to achieve the EU's target of 35% minimum GHG emission savings for biofuels (Lange, 2010).

The assumptions and data used in most of the LCA based studies on biofuels are taken from the Europe and the USA based technology pathways. Therefore, there is a strong need to focus on the Asian and South American countries as well (**Box 2**). This will increase the representativeness of LCA studies around the globe, including the developing countries. The non-GHG environmental impact factors, including acidification, eutrophication, human health, and toxicity should be considered in future LCA studies. LCA results should be extended to the local and regional needs and conditions by considering other environmental assessment methods. Moreover, LCA results should also take direct land-use change impacts into consideration (**Box 2**).

It is worth quoting that LCA studies do not assess the large-scale development of any technology or product, therefore, the other assessment tools such as agro-economic market models should also be used (**Box 1**). Moreover, there is a strong need to reach national and international consensus on the use and execution of biofuels-related LCAs by considering the GHG emission savings goals. This will lead to a set of approaches and assumptions on significant indicators such as technology status, land-use carbon stock, $N_2O$ emissions, and impact allocation for co-products (Menichetti and Otto, 2009).

## 5. Conclusion

- A wide range of results in terms of net energy balances and net GHG emission savings has been obtained from various biofuel production systems and their biomass sources. The factors causing such variations in results are different feedstock types, conversion technologies, land-use and land-use changes, replaced products like electricity, transportation fuel, and animal feed.

- The net GHG emission savings were expressed in terms of $CO_2$ equivalent in almost all of the presented cases. The agricultural activities, selection of feedstock, and treatment of co-products and the allocation of impacts for co-products are the major contributors to GHG emissions.
- For biomethane, bioethanol, and biodiesel production systems, the choice of reactor is one of the main parameters leading to significant variations in GHG emission savings.
- Land-use and land-use changes may sometimes result in an overestimation of a biofuel efficiency. Environmental and social sustainability of biofuels production are the key factors for the development of biofuel support policies.
- Higher GHG emission savings and energy balances with biofuels can be achieved, when biomass yields are high, particularly when ecologically sustainable biomass or non-food biomass are converted into biofuel and used efficiently.

**Box 1.**
Key issues in LCA bases studies on biofuels (Cherubini and Strømman, 2011; IPCC, 2011; Cherubini et al., 2009; Menichetti and Otto, 2009).

- LCA studies are being carried out continuously, but their number is still small in comparison with other processes.
- Most of the LCA based studies were carried out by following the US or European conditions and adopting their agriculture processes and conversion technologies.
- Most common LCA based studies have referred to the first-generation biofuels, while a few studies investigated the second-generation biofuels.
- Most of the researchers have considered the traditional feedstock such as rapeseed wheat, sugar corn, corn, *etc.* for LCA studies. However, a few have tried to assess the LCA of recent biofuel crops such as sweet sorghum and Jatropha.
- In most of the LCA based studies, emphasis has been placed on the energy consumption either only from non-renewable sources or total energy sources. However, a few studies included all the other potential impacts of the process such as eutrophication, acidification, toxicity potential, and ozone depletion. Moreover, water impacts have been included in very limited number of studies.
- There is no LCA based study accounting the impacts on biodiversity. The methods for development of biodiversity indicators are still under discussion.
- Impacts on direct land-use and land-use changes due to crop production have less been studied. Limited studies described alternative land-use as reference system and calculated the carbon stock. Moreover, potential impacts on land due to its indirect use for high bioenergy production demand are not measured.
- The complexity level of LCA studies is quite high due to variable nature of assumption and hypothesis, emission factor, yield, heating value, and other background methodological choices. However, limited studies included a quality data review required for LCA in accordance with ISO standards.
- Variations have been observed in managing the co-products, and impact allocation methods.
- Social issues are sometime ignored during the LCA process. This shows that the pure focus is on the environment aspects of the LCA process.
- Different databases are used for LCA studies. However, old databases have been used more frequently, which could affect the results quality regardless of the quality of primary data collection.

**Box 2.**
Future directions for LCA based studies on biofuel (Cherubini and Strømman, 2011; IPCC, 2011; Cherubini et al., 2009; Menichetti and Otto, 2009).

- The future biofuel systems have to satisfy all aspects of environmental, economica, and social factors, especially the impacts on biodiversity, water resources, human health, and toxicity, and food security.
- The assumptions and data used in LCA based studies on biofuels should be based on regions rather than European and North American conditions as well, such as like Brazil, China, and Southeast Asia.
- More research and development is required to commercialize the second-generation biofuels that are made from non-food biomass sources to solve the food, feed, and fibre production issues. There is a possibility that such LCA studies will be based on uncertainties, as there is no such biofuel system commercially established yet. Therefore, uncertainties and parameter sensitivities should be handled carefully. arametric type LCA will be useful tool in this regard.
- The future LCA studies need to consider the integrated multi-fuel and multi-product systems like biorefineries.
- There is also a need to properly define the system boundaries in connection with land-use and land-use change.
- The GHG emission savings can be increased for a biofuel based on increased carbon sequestration when using perennial grasses established in set-aside and annual crop land.
- LCA findings of biofuels from dedicated crops should be expressed in per hectare basis.
- LCA results of biofuels from biomass residues should be expressed in per unit output basis.
- For transportation biofuels, the LCA results should be expressed in per km basis.
- There is a need to solve the issues related to liquid biofuels that require more fossil-based energy than the generation of heat and electricity from biomass.
- The emphasis should be given to the residues and organic wastes that can be employed for biofuels production since they have been shown to lead to maximum GHG emission savings due to direct reduction of emissions related to waste disposal.
- High GHG emission savings and energy balances can be achieved if agricultural co-products and process residues are used for biofuel production. However, the effect of residues removal from the soil and the GHG implications should also be considered.
- Higher GHG emission savings can be achieved if biomass is utilized in making value-added products.

## References

[1] Adler, P.R., Grosso, S.J.D., Parton, W.J., 2007. Life-cycle assessment of net greenhouse-gas flux for bioenergy cropping systems. Ecol. Appl. 17(3), 675-691.

[2] Agrawal, R., Singh, N.R., Ribeiro, F.H., Delgass, W.N., Perkis, D.F., Tynerb, W.E., 2009. Synergy in the hybrid thermochemical-biological processes for liquid fuel production. Comput. Chem. Eng. 33(12), 2012-2017.

[3] Ansari, F.T., Verma, A.P., 2012. Experimental determination of suitable ethanol-gasoline blend for spark ignition engine. Int. J. Eng. 1(5), 10 pages.

[4] Atilgan, B., Azapagic, A., 2015. Life cycle environmental impacts of electricity from fossil fuels in Turkey. J. Cleaner Prod. 106, 555-564.

[5] Arvidsson, R., Persson, S., Froling, M., Svanstromb, M., 2011. Life cycle assessment of hydrotreated vegetable oil from rape, oil palm, and Jatropha. J. Cleaner Prod. 19, 129-137.

[6] Balan, V., Rogers, C.A., Chundawat, S.P.S., Sousa, L.D.C., Slininger, P.J., Gupta, R., Dale, B.E., 2009. Conversion of extracted oil cake fibers into bioethanol including DDGS, canola, sunflower, sesame, soy, and peanut for integrated biodiesel processing. J. Am. Oil Chem. Soc. 86(2), 157-165.

[7] Batan, L., Quinn, J., Willson, B., Bradley, T., 2010. Net energy and greenhouse gas emission evaluation of biodiesel derived from microalgae. Environ. Sci. Technol. 44(20), 7975-7980.

[8] Beer, T., Grant, T., 2007. Life-cycle analysis of emissions from fuel ethanol and blends in Australian heavy and light vehicles. J. Cleaner Prod. 15 (8–9), 833–837.

[9] Beer, T., Grant, T., Campbell, P.K., 2007. The greenhouse and air quality emissions of biodiesel blends in Australia. CSIRO Mar. Atmos. Res. (CMAR), Australia.

[10] Bensaid, S., Conti, R., Fino, D., 2012. Direct liquefaction of ligno-cellulosic residues for liquid fuel production. Fuel. 94, 324-332.

[11] Bouaid, A., Diaz, Y., Martinez, M., Aracil, J., 2005. Pilot plant studies of biodiesel production using *Brassica carinata* as raw material. Catal. Today. 106(1-4), 193-196.

[12] Bouaid, A., Martinez, M., Aracil, J., 2009. Production of biodiesel from bioethanol and Brassica carinata oil: oxidation stability study. Bioresour. Technol. 100(7), 2234-2239.

[13] Campbell, M.N., 2008. Biodiesel: algae as a renewable source for liquid fuel. Guelph. Eng. J. 1, 2-7.

[14] Chandrashekar, L.A., Mahesh, N.S., Gowda, B., Hall, W., 2012. Life cycle assessment of biodiesel production from pongamia oil in rural Karnataka. Agric. Eng. Int.: CIGR J. 14(3), 67-77.

[15] Cheng, S., Logan, B.E., 2007. Sustainable and efficient biohydrogen production via electrohydrogenesis. PNAS. 104(47), 18871-18873.

[16] Cherubini, F., Strømman, A.H., 2011. Life cycle assessment of bioenergy systems: State of the art and future challenges. Bioresour. Technol. 102(2), 437-451

[17] Cherubini, F., Bird, N.D., Cowie, A., Jungmeier, G., Schlamadinger, B., Woess-Gallasch, S., 2009. Energy-and greenhouse gas-based LCA of biofuel and bioenergy systems: key issues, ranges and recommendations. Resour. Conserv. Recycl. 53, 434-447.

[18] Choudhury, R., Weber, T., Schindler, J., Weindorf, W., Wurster, R., 2002. GM well to wheel analysis of energy use and greenhouse gas emissions of advanced fuel/ vehicle systems- a European study. Ottobrunn, Germany.

[19] Christi, Y., 2007. Biodiesel from microalgae. Biotechnol. Adv. 25(3), 294-306.

[20] Clarens, A.F., Resurreccion, E.P., White, M.A., Colosi, L.M., 2010. Environmental life cycle comparison of algae to other bioenergy feedstocks. Environ. Sci. Technol. 44(5), 1813-1819.

[21] Das, D., Veziroglu, T.N., 2001. Hydrogen production by biological processes: a survey of literature. Int. J. Hydrogen Energy. 26(1), 13-28.

[22] Das, L.M., Bora, D.K., Pradhan, S., Naik, M.K., Naik, S.N., 2009. Long-term storage stability of biodiesel produced from Karanja oil. Fuel. 88(11), 2315-2318.

[23] Davis, S.C., Anderson-Teixeira, K.J., De Lucia, E.H., 2008. Life-cycle analysis and the ecology of Biofuels. Trends Plant Sci. 14(3), 140-146.

[24] de Souza, S.P., Pacca, S., de Ávila, M.T., Borges, J.L.B., 2010. Greenhouse gas emissions and energy balance of palm oil biofuel. Renew. Energ. 35(11), 2552-2561.

[25] Deshmukh, S.J., Bhuyar, L.B., 2009. Transesterified Hingan (Balanites) oil as a fuel for compression ignition engines. Biomass Bioenergy. 33(1), 108-112.

[26] de Oliveira, M.E.D., Vaughan, B.E., Rykeil, E.J., 2005. Ethanol as fuel: energy, carbon dioxide balances, and ecological footprint. BioScience. 55(7), 593-602.

[27] Diwani, G.E., Attia, N.K., Hawash, S.I., 2009. Development and evaluation of biodiesel fuel and byproducts from Jatropha oil. Int. J. Environ. Sci. Technol. 6(2), 219-224.

[28] Djomo, S.N., Humbert, S., Blumberga, D., 2008. Life cycle assessment of hydrogen produced from potato steam peels. Int. J. Hydrogen Energy. 33(12), 3067-3072.

[29] Djomo, S.N., Blumberga, D., 2011. Comparative life cycle assessment of three biohydrogen pathways. Bioresour. Technol. 102(3), 2684-2694.

[30] Doornbosch, R., Steenblik, R., 2007. Biofuels. Is the cure worse than the disease?. OECD SG/SD/RT 2007-3, Paris.

[31] Dreier, T., Tzscheutschler, P., 2001. Ganzheitliche Systemanalyse für die Erzeugung und Anwendung von Biodiesel und Naturdiesel im Verkehrssektor. Bayerisches Staatsministerium für Landwirtschaft und Forsten, Gelbes Heft 72, München.

[32] Dressler, D., Loewen, A., Nelles, M., 2012. Life cycle assessment of the supply and use of bioenergy: impact of regional factors on biogas production. Int. J. Life Cycle Assess. 17, 1104-1115.

[33] EC-Directive, 2009. Directive 2009/28/EC of The European Parliament and of The Council of 23 April 2009 on the promotion of the use of energy from renewable sources and amending and subsequently repealing Directives 2001/77/EC and 2003/30/EC. Official Journal of the European Union.

[34] Ecobilan PwC, 2002. Bilans énergétiques et gaz à effet de serre des filières de production de biocarburants. Technical report, Final version, November 2002. Prepared for ADEME/ DIREM, by Ecobilan PwC, Neuilly-sur-Seine, France.

[35] Edwards, R., Larive, J-F., Mahieu, V., Rouveirolles, P., 2007. Well-to-wheels analysis of future automotive fuels and powertrains in the European context, version 2c. European Commission Joint Research Centre, Concawe and EUCAR.

[36] Elsayed, M.A., Matthews, R., Mortimer, N.D., 2003. Carbon and Energy Balances for a Range of Biofuels Options. Project No. B/B6/00784/REP, URN 03/836, Sheffield Hallam University, Resources Research Unit.

[37] Emmenegger, M.F., Pfister, S., Koehler, A., de Giovanetti, L., Arena, A.P., Zah, R., 2011. Taking into account water use impacts in the LCA of biofuels: an Argentinean case study. Int. J. Life Cycle Assess. 16, 869-877.

[38] Escobar, J.C., Lora, E.S., Venturini, O.J., Yanez, E.E., Castillo, E.F., Almazan, O., 2009. Biofuels: Environment, technology and food Security. Renew. Sust. Energy Rev. 13(6-7), 1275- 1287.

[39] Fargione, J., Hill, J., Tilman, D., Polasky, S., Hawthorne, P., 2008. Land clearing and the biofuel carbon debt. Science. 319, 1235-1238.

[40] Farrell, A.E., 2008. Biofuels and greenhouse gases: a view from California. GBEP 2nd TF Meeting on GHG Methodologies, Washington.

[41] Felten, D., Fröba, N., Fries, J., Emmerling, C., 2013. Energy balances and greenhouse gas mitigation potentials of bioenergy cropping systems (*Miscanthus*, rapeseed, and maize) based on farming conditions in Western Germany. Renew. Energ. 55(13), 160-174

[42] Ferreira, A.F., Ortigueira, J., Alves, L., Gouveia, L., Moura, P., Silva, C.M., 2013. Energy requirement and $CO_2$ emissions of bio$H_2$ production from microalgal biomass. Biomass Bioenergy. 49, 249-259.

[43] Ferreira, A.F., Ribau, J.P., Silva, C.M., 2011. Energy consumption and $CO_2$ emissions of potato peel and sugarcane biohydrogen production pathways, applied to Portuguese road transportation. Int. J. Hydrogen Energy. 36(21), 13547-13558.

[44] Finco, A., Bentivoglio, D., Rasetti, M., Padella, M., Cortesi, D., Polla, P., 2012. Sustainability of rapeseed biodiesel using life cycle assessment. Presentation at the international association of agricultural economists (IAAE) triennial conference, Foz do Iguaçu, Brazil.

[45] Fore, S.R., Porter, P., Lazarus, W., 2011. Net energy balance of small-scale on-farm biodiesel production from canola and soybean. Biomass Bioenergy. 35(5), 2234-2244.

[46] Gardy, J., Hassanpour, A., Lai, X., Rehan, M., 2014. The influence of blending process on the quality of rapeseed oil-used cooking oil biodiesels. Int. Sci. J. Environ. Sci. 3, 233-240.

[47] Glenister, D., Nunes, V., 2011. Understanding sustainable biofuels production, the EU renewable energy directive and international initiatives to verify sustainability - A discussion about the global importance of ensuring biofuels are produced sustainably and the

international initiatives to drive the market in a sociality accettable and environmentally friendly well-managed direction", White Paper, Systems and Services Certification, SGS.

[48] Gnansounou, E., Dauriat, A., 2004. Etude Comparative de Carburants par Analyse de Leur Cycle de Vie. Swiss Federal Institute of Technology of Lausanne. Laboratory of Energy Systems (LASEN), Reports. 415.105:2002 and 420.100:2004, prepared for Alcosuisse.

[49] Gnansounou, E., Dauriat, A., Panichelli, L., Villegas, J.D., 2009. Estimating energy and green house gas balance of various biofuel: concept and Methodologies; working paper. EPFL-ENAC-LASEN.

[50] Gnansounou, E., Dauriat, A., Villegas, J., Panichelli, L., 2009. Life cycle assessment of biofuels: Energy and greenhouse gas balances. Bioresour. Technol. 100(21), 4919-4930.

[51] González-García, S., Moreira, M.T., Feijoo, G., 2010. Comparative environmental performance of lignocellulosic ethanol from different feedstocks. Renew. Sust. Energy Rev. 14(7), 2077-2085.

[52] González-García, S., Moreira, M.T., Feijoo, G., 2012. Environmental aspects of eucalyptus based ethanol production and use. Sci. Total Environ. 438, 1-8.

[53] Gracia, C.A., Fuentes, A., Hennecke, A., Riegelhaupt, E., Manzini, F., Masera, O., 2011. Life Cycle greenhouse gas emission and energy balances of sugarcane ethanol production in Mexico. Appl. Energy. 88(6), 2088-2097.

[54] Groode, T.A., Heywood, J.B., 2007. Ethanol: a look ahead. MIT publication, Cambridge, MA, USA.

[55] Güereca, L.P., Gassó, S., Baldasano, J.M., Jiménez-Guerrero, P., 2006. Life cycle assessment of two biowaste management systems for Barcelona, Spain. Resour. Conserv. Recycl. 49(1), 32-48.

[56] Hahn-Hägerdal, B., Galbe, M., Gorwa-Grauslund, M.F., Lidén, G., Zacchi, G., 2009. Bioethanol-the fuel of tomorrow from the residues of today. Trends Biotechnol. 24, 549-556.

[57] Hallenbeck, P.C., Benemann, J.R., 2002. Biological hydrogen production; fundamentals and limiting processes. Int. J. Hydrogen Energy. 27(11-12), 1185-1193.

[58] Hartmann, J.K., 2006. Life-cycle-assessment of industrial scale biogas plants. Dissertation, zur Erlangung des Doktorgrades, der Fakultät für Agrarwissenschaften, der Georg-August-Universität Göttingen.

[59] Hill, J., Nelson, E., Tilman, D., Polasky, S., Tiffany, D., 2006. Environmental, economic, and energetic costs and benefits of biodiesel and ethanol biofuels. Proc. Natl. Acad. Sci. 103, 11206-11210.

[60] Hong, Y., Nizami, A.S., Pourbafrani, M., Saville, B.A., MacLean, H.L., 2013. Impact of cellulase production on environmental and financial metrics for lignocellulosic ethanol. Biofuels, Bioprod. Biorefin. 7(3), 303-313.

[61] Huber, G.W., Iborra, S., Corma, A., 2006. Synthesis of transportation fuels from biomass: chemistry, catalysts, and engineering. Chem. Rev. 106(9), 4044-4098.

[62] IPCC, 2011. Summary for Policymakers, in: Edenhofer, O., Pichs-Madruga, R., Sokona, Y., Seyboth, K., Matschoss, P., Kadner, S., Zwickel, T., Eickemeier, P., Hansen, G., Schlömer, S., von Stechow C. (Eds.), IPCC special report on renewable energy sources and climate change mitigation. Cambridge University Press, Cambridge, United Kingdom and New York, NY, USA.

[63] Janulis, P., 2004. Reduction of energy consumption in biodiesel fuel life cycle. Renew. Energ. 29(6), 861-871.

[64] Jungbluth, N., Büsser, S., Frischknecht, R., Tuchschmid, M., 2008. Life cycle assessment of biomass-to-liquid fuels. Draft Report. ESU Services GmbH, Uster, Germany.

[65] Jury, C., Benetto, E., Koster, D., Schmitt, B., Welfring, J., 2010. Life Cycle Assessment of biogas production by monofermentation of energy crops and injection into the natural gas grid. Biomass Bioenergy. 34(1), 54-66.

[66] Kalinci, Y., Hebasli, A., Dincer, I., 2012. Life cycle assessment of hydrogen production from biomass gasification systems. Int. J. Hydrogen Energy. 37(19), 14026-14039.

[67] Kamahara, H., Hasanudin, U., Widiyanto, A., Tachibana, R., Atsuta, Y., Goto, N., Daimon, H., Fujie, K., 2010. Improvement potential for net energy balance of biodiesel derived from palm oil: a case study from Indonesian practice. Biomass Bioenergy. 34(12), 1818-1824.

[68] Kauffman, N.S., Hayes, D.J., 2013. The trade-off between bioenergy and emissions with land constraints. Energy Policy. 54, 300-310.

[69] Keller, D., Wahnschaffe, U., Rosner, G., Mangelsdorf, I., 1998. Considering human toxicity as an impact category in Life Cycle Assessment. Int. J. Life Cycle Assess. 3(2), 80-85.

[70] Kesari, V., Rangan, L., 2010. Development of *Pongamia pinnata* as an alternative biofuel. Crop - current status and scope of plantations in India. J. Crop Sci. Biotechnol. 13(3), 127-137.

[71] Khan, S.A., Rashmi, H.M.Z., Prasad, S., Banerjee, U.C., 2009. Prospects of biodiesel production from microalgae in India. Renew. Sust. Energy Rev. 13(9), 2361-2372.

[72] Khanna, M., Scheffran, J., Ziberman, D., 2010. Handbook of bioenergy economics and policy. Springer Science and Business Media.

[73] Kiatkittipong, W., Wongsuchoto, P., Pavasant, P., 2009. Life cycle assessment of bagasse waste management options. Waste Manage. 29(5), 1628-1633.

[74] Kim, S., Dale, B.E., 2005. Environmental aspects of ethanol derived from no tilled corn grain: non-renewable energy consumption and greenhouse gas emissions. Biomass Bioenergy. 28(5), 475-489.

[75] Kim, S., Dale, B.E., 2009. Regional variations in greenhouse gas emissions of biobased products in the United States-corn based ethanol and soybean oil. Int. J. Life Cycle Assess. 14(6), 540-546.

[76] Kim, S., Dale, B.E., 2004. Global potential bioethanol production from wasted crops and crop residues. Biomass Bioenergy. 26(4), 361-375.

[77] Kim, S.J., Jung, S.M., Park, Y.C., Park, K., 2007. Lipase catalyzed transesterification of soybean oil using ethyl acetate, an alternative acyl acceptor. Biotechnol. Bioprocess Eng. 12(4), 441-445.

[78] Koga, N., Kajiyama, T., Senda, K., Iketani, S., Tsuda, S., 2013. Energy efficiency of potato production practices for bioethanol feedstock in northern Japan. Eur. J. Agron. 44, 1-8.

[79] Korres, N.E., Singh, A., Nizami, A.S., Murphy, J.D., 2010. Is grass biomethane a sustainable transport biofuel. Biofuels, Bioprod. Biorefin. 4(3), 310-325.

[80] Korres, N.E., Thamsiriroj, T., Smyth, B.M., Nizami, A.S., Singh. A., Murphy, J.D., 2011. Grass biomethane for agriculture and energy. in: Lichtfouse, E. (Ed.), Genetics, biofuels and local farming systems, sustainable agriculture reviews. Springer Science and Business Media B.V., pp. 5-49.

[81] Lange, M., 2011. The GHG balance of Biofuels taking into account land use changes. Energy Policy. 39(5), 2373-2385.

[82] Lankoski, J., Ollikainen, M., 2011. Biofuel policies and the environment: Do climate benefits warrant increased production from biofuel feedstocks. Ecol. Econ. 70(4), 676-687.

[83] Lardon, L., Helias, A., Sialve, B., Steyer, J.P., Bernard, O., 2009. Life-cycle assessment of biodiesel production from microalgae. Environ. Sci. Technol. 43(17), 6475-6481.

[84] Le Quéré, C., Raupach, M., Canadell, J., Marland, G., Bopp, L., Ciais, P., Conway, T., Doney, S., Feely, R., Foster, P., Friedlingstein, P., Gurney, K., Houghton, R., House, J., Huntingford, C., Levy, P., Lomas, M., Majkut, J., Metzl, N., Ometto, J.P., Peters, I., Prentice, C., Randerson, J., Running, S., Sarmiento, J., Schuster, U., Sitch, S., Takahashi, T., Viovy, N., van der Werf, G., Woodward, F., 2009. Trends in the sources and sinks of carbon dioxide. Nat. Geosci. 2, 831-836.

[85] Lechón, Y., Cabal, H., Lago, C., de la Rua, C., Sáez, R., Fernández., M., 2005. Análisis del ciclo de vida de combustibles alternativos para el transporte. Fase I. Análisis de ciclo de vida comparativo del etanol de cereales y de la gasolina. Energía y Cambio Climático. Ed. Centro de Publicaciones Secretaría General Técnica Ministerio de Medio Ambiente, Madrid.

[86] Lechón, Y., Cabal, H., de la Rúa, C., Lago, C., Izquierdo, L., Sáez. R., 2010. Life cycle environmental aspects of biofuel goals in Spain. Scenarios. 15th European Biomass Conference and Exhibition-From Research to Market Deployment. 7-11. May 2007, Berlin.

[87] Ledford, H., 2006. Liquid fuel synthesis: Making it up as you go along. Nature. 444, 677-678.

[88] Levin, D.B., Pitt, L., Love, M., 2004. Biohydrogen production: prospects and limitations to practical application. Int. J. Hydrogen Energy. 29(2), 173-185.
[89] Liska, A.J., Cassman, K.G., 2008. Towards standardization of life-cycle metrics for biofuels: Greenhouse gas emissions mitigation and net energy yield. J. Biobased Mater. Bioenergy. 2, 187-203.
[90] Long, Y.D., Guo, F., Fang, Z., Tian, X.F., Jiang, L.Q., Zhang, F., 2011. Production of biodiesel and lactic acid from oil using sodium silicate as catalyst. Bioresour. Technol. 102, 6884-6886.
[91] Luo, L., van der Voet, E., Huppes, G., Udo de Haes, H.A., 2010. Allocation issues in LCA methodology: A case study of corn stover-based fuel ethanol. Int. J. Life Cycle Assess. 14, 529-539.
[92] Luo, D., Hu, Z., Choi, D.G., Thomas, V.M., Realff, M.J., Chance, R.R., 2010. Life cycle energy and greenhouse gas emissions for an ethanol production process based on blue-green algae. Environ. Sci. Technol. 44(22), 8670-8677.
[93] Macedo, I.C., Seabra, J.E.A., Silva, J.E.A.R., 2008. Greenhouse gases emissions in the production and use of ethanol from sugarcane in Brazil: the 2005/2006 averages and a prediction for 2020. Biomass Bioenergy. 32(7), 582-595.
[94] Macedo, I.D.C., 1998. Greenhouse gas balance and energy balance in bioethanol production and utilization in Brazil. Biomass Bioenergy. 14, 77-81.
[95] Malca, J., Freire, F., 2006. Renewability and life cycle energy efficiency of bioethanol and bio-ethyl tertiary butyl ether (bio ETBE): assessing the implication of allocation. Energy. 31, 3362-3380.
[96] Manish, S., Banerjee, R., 2008. Comparison of biohydrogen production processes. Int. J. Hydrogen Energy. 33, 279-286.
[97] Martinsen, D., Funk, C., Linssen, J., 2010. Biomass for transportation fuels-A cost effective option for the German energy supply?. Energy Policy. 38(1), 128-140.
[98] Macedo, I.M.R., Lima, V., Leal, J.E., da Silva, A.R., 2004. Assessment of greenhouse gas emissions in the production and use of fuel ethanol in Brazil. Government of the State of São Paulo, São Paulo, Brazil.
[99] Menichetti, E., Otto, M., 2009. Energy balance and greenhouse gas emissions of biofuels from a life-cycle perspective. In: Howarth R.W., Bringezu, S. (Eds.), Biofuels: Environmental Consequences and Interactions with Changing Land Use. Proceedings of the Scientific Committee on Problems of the Environment (SCOPE) International Biofuels Project Rapid Assessment, Ithaca NY, Cornell University, USA, pp. 81-109.
[100] Monti, A., Fazio, S., Venturi, G., 2009. Cradle-to-farm gate life cycle assessment in perennial energy crops. Eur. J. Agron. 31(2), 77-84.
[101] Mukherjee, P., Varshney, A., Johnson, S., Jha, T.B., 2011. *Jatropha curcas*: a review on biotechnological status and challenges. Plant Biotechnol. Rep. 5(3), 197-215.
[102] Murphy, J.D., Browne, J., Allen, E., Gallagher, C., 2013. The resource of biomethane, produced via biological, thermal and electrical routes, as a transport biofuel. Renew. Energ. 55, 474-479.
[103] Myhre, G., Shindell, D., Breon, F., Collins, W., Fuglestvedt, J., Huang, J., Koch, D., Lamarque, J.F., Lee, D., Mendoza, B., Nakajima, T., 2013. Anthropogenic and natural radiative forcing. Clim. Change. 423.
[104] Nabi, M.N., Akhter, M.S., Shahadat, M.M.Z., 2006. Improvement of engine emissions with conventional diesel fuel and diesel-biodiesel blends. Bioresour. Technol. 97(3), 372-378.
[105] Najafi, G., Ghobadian, B., Tavakoli, T., Yusaf, T., 2009. Potential of bioethanol production from agricultural wastes in Iran. Renew. Sust. Energy Rev. 13(6-7), 1418-1427.
[106] Ndong, R., Montrejaud-Vignoles, M., Girons, O.S., Gabrielle, B., Pirot, R., Domergue, M., Sablayrolles, C., 2009. Life cycle assessment of biofuels from *Jatropha curcas* in West Africa: a field study. GCB Bioenergy. 1(3), 197-210.
[107] Nguyen, T.L.T., Gheewala, S.H., Garivait, S., 2007. Fossil energy savings and GHG mitigation potentials of ethanol as a gasoline substitute in Thailand. Energy Policy. 35(10), 5195-5205.
[108] Nizami, A.S., Ismail, I.M.I., 2013. Life cycle assessment of biomethane from lignocellulosic biomass, in: Life cycle assessment of renewable energy sources. Green Energy and Technology book series, Springer-Verlag London Ltd. pp. 79-94.
[109] Nizami, A.S., Ouda, O.K.M., Rehan, M., El-Maghraby, A.M.O., Gardy, J., Hassanpour, A., Kumar, S., Ismail, I.M.I., 2015a. The potential of Saudi Arabian natural zeolites in energy recovery technologies. Energy. 1-10. doi:10.1016/j.energy.2015.07.030
[110] Nizami, A.S., Rehan, M., Ouda, O.K.M., Shahzad, K., Sadef, Y., Iqbal, T., Ismail, I.M.I., 2015b. An argument for developing waste-to-energy technologies in Saudi Arabia. Chem. Eng. 45, 337-342.
[111] OECD-FAO., 2015. AgriculturaL Outlook., 2015-2024.
[112] Nizami, A.S., Shahzad, K., Rehan, M., Ouda, O.K.M., Khan, M.Z., Ismail, I.M.I., Almeelbi, T., Basahi, J.M., Demirbas, A. 2016. Developing waste biorefinery in Makkah: a way forward to convert urban waste into renewable energy. Appl. Energy. DOI: 10.1016/j.apenergy.2016.04.116
[113] Ouda, O.K.M., Raza, S.A., Nizami, A.S., Rehan, M., Al-Waked, R., Korres, N.E., 2016. Waste to energy potential: A case study of Saudi Arabia. Renew. Sust. Energy Rev. 61, 328-340.
[114] Nzihou, A., Flamant, G., Stanmore, B., 2012. Synthetic fuels from biomass using concentrated solar energy. Energy. 42(1), 121-131.
[115] Pande, M., Bhaskarwar, A.N., 2012. Biomass conversion to energy, in: Baskar, C., Baskar, S., Dhillon, R.S. (Eds.), Biomass Conversion. Springer-Verlag Berlin Heidelberg.
[116] Papong, S., Malakul, P., 2010. Life cycle energy and environmental analysis of bioethanol production from cassava in Thailand. Bioresour. Technol. 101(1), S112-S118.
[117] Patterson, T., Esteves, S., Dinsdale, R., Maddy, J., 2013. Life cycle assessment of biohydrogen and biomethane productionand utilisation as a vehicle fuel. Bioresour. Technol. 131, 235-245.
[118] Paul, N., Kemnitz, D., 2006. Biofuels- Plant raw materials, Products. Fachagentur, Nachwachsende Rohstoffe eV (FNR), WPR communications, Berlin, 43.
[119] Pehnelt, G., Vietze, C., 2013. Recalculating GHG emissions saving of palm oil biodiesel. Environ. Dev. Sust. 15, 429-479.
[120] Pimentel, D., Patzek, T.W., 2005. Ethanol production using corn, switchgrass, and wood, biodiesel production using Soybean and Sunflower. Nat. Resour. Res. 14(1), 65-76.
[121] Popp, J., Lakner, Z., Harangi-Rákos, M., Fári, M., 2014. The effect of bioenergy expansion: Food, energy, and environment. Renew. Sust. Energy Rev. 32, 559-578.
[122] Pradhan, A., Shrestha, D.S., Van Gerpen, J., Duffield, J., 2008. The energy balance of soybean oil biodiesel production: a review of past studies. Trance. ASABE. 51(1), 185-194.
[123] Prasad, S., Singh, A., Jain, N., Joshi, H.C., 2007a. Ethanol production from sweet sorghum syrup for utilization as automotive fuel in India. Energy Fuels. 21(4), 2415-2420.
[124] Prasad, S., Singh, A., Joshi, H.C., 2007b. Ethanol as an alternative fuel from agricultural, industrial and urban residues. Resour. Conserv. Recycl. 50(1), 1-39.
[125] Quirin, M., Gärtner, S., Pehnt, M., Reinhardt, G., 2004. CO2 mitigation through biofuels in the transport sector. Main report, IFEU, Heidelberg.
[126] Rehan, M., Nizami, A.S., Shahzad, K., Ouda, O.K.M., Ismail, I.M.I., Almeelbi, T., Iqbal T., Demirbas, A., 2016. Pyrolytic liquid fuel: a source of renewable energy in Makkah. Energy Sources, Part A: Recovery, Utilization, and Environmental Effects. DOI:10.1080/15567036.2016.1153753
[127] Rashid, U., Anwar, F., 2008. Production of biodiesel through base-catalyzed transesterification of sunflower oil using an optimized protocol. Energy Fuels. 22(2), 1306-1312.
[128] Rathore, D., Singh, A., 2013. Biohydrogen Production from Microalgae, in: Gupta, V.K., Tuohy, M.G. (Eds.), Biofuel Technologies. Springer-Verlag Berlin Heidelberg.
[129] Rathore, D., Pant, D., Singh, A., 2013. A comparison of life cycle assessment studies of different biofuels, in: Singh, A., Pant, D., Olsen, S.I. (Eds.), Life Cycle Assessment of Renewable Energy Sources. Green Energy and Technology Series, Springer-Verlag London, pp. 269-289.
[130] Reijnders, L., 2008. Ethanol production from crop residues and soil organic carbon. Resour Conserv. Recycl. 52(4), 653-658.

[131] Reinhardt, G.A., Gärtner, S., Rettenmaier, N., Münch, J., Falkenstein., E.V., 2007. Screening life cycle assessment of Jatropha, Germany.

[132] REN21. 2009. Renewables Global Status Report: 2009 Update (Paris: REN21 Secretariat).

[133] Renouf, M.A., Wegener, M.K., Nielsen, L.K., 2008. An environmental life cycle assessment comparing Australian sugarcane with US corn and UK sugar beet as producers of sugars for fermentation. Biomass Bioenergy. 32(12), 1144-1155.

[134] RFA. 2009. Ethanol Industry Outlook, Renewable Fuels Association.

[135] RFA. 2016. Fueling a high octane future. Ethanol Industry Outlook. Renewable Fuels Association.

[136] Romagnoli, F., Blumberga, D., Pilicka, I., 2011. Life cycle assessment of biohydrogen production in photosynthetic processes. Int. J. Hydrogen Energy. 36(13), 7866-7871.

[137] Rosenberger, A., Kaul, H.P., Senn, T., Aufhammer, W., 2001. Improving the energy balance of bioethanol production from winter cereals: the effect of crop production intensity. Appl. Energy. 68(1), 51-67.

[138] Sadaf, Y., Nizami, A.S., Batool, S.A., Chaudhary, M.N., Ouda, O.K.M., Asam, Z.Z., Habib, K., Rehan, M., Demibras, A., 2016. Waste-to-energy and recycling value for developing integrated solid waste management plan in Lahore. Energy Sources. Part B: Economics, Planning, and Policy.

[139] Sander, K., Murthy, G.S., 2010. Life cycle analysis of algae biodiesel. Int. J. Life Cycle Assess. 15, 704-714.

[140] Sassner, P., Galbe, M., Zacchi, G., 2008. Techno-economic evaluation of bioethanol production from three different lignocellulosic materials. Biomass Bioenergy. 32(5), 422-430.

[141] Sawayama, S., Minowa, T., Yokoyama, S-Y., 1999. Possibility of renewable energy production and $CO_2$ mitigation by thermochemical liquefaction of microalgae. Biomass Bioenergy. 17(1), 33-39.

[142] Scholz, V., da Silva, J.N., 2008. Prospects and risks of the use of castor oil as a fuel. Biomass Bioenergy. 32(2), 95-100.

[143] Searchinger, T., Heimlich, R., Houghton, R.A., 2008. Use of US croplands for biofuels increases greenhouse gases through emissions from land-use change. Science. 319, 1238-1140.

[144] Sharma, V., Ramawat, K.G., Choudhary, B.L., 2012. Biodiesel Production for Sustainable Agriculture, in: Lichtfouse, E. (Ed.), Sustainable Agriculture Reviews. Sust. Agric. Rev. pp, 133-160.

[145] Sheehan, J., Aden, A., Paustian, K., Killian, K., Brenner, J., Walsh, M., Nelson, R., 2004. Energy and environmental aspects of using corn stover for fuel ethanol. J. Ind. Ecol. 7, 117-146.

[146] Sheehan, J., Camobreco, V., Duffield, J., Graboski, M., Shapouri, H., 1998. Life cycle inventory of biodiesel and petroleum diesel for use in an urban bus. NREL/SR-580-24089. Golden, Colo. National Renewable Energy Laboratory.

[147] Sims, R.E.H., Mabee, W., Saddler, J.N., Taylor, M., 2010. An overview of second generation biofuel technologies. Bioresour. Technol. 101(6), 1570-1580.

[148] Singh, A., Murphy, J.D., 2009. Biomethane from animal waste and grass for clean vehicular biofuel in Ireland. Twelfth International Waste Management and Landfi ll Symposium, CISA, Environmental Santiary Engineering Centre, Italy Cagliari, Sardinia, Italy.

[149] Singh, A., Nizami, A.S., Korres, N.E., Murphy, J.D., 2011. The effect of reactor design on the sustainability of grass biomethane. Renew. Sust. Energy Rev. 15(3), 1567-1574.

[150] Singh, A., Pant, D., Olsen, S.I., Nigam, P.S., 2012. Key issues to consider in microalgae based biodiesel production. Energy Educ. Sci. Technol. Part A Energy Sci. Res. 29(1), 563-576.

[151] Singh, A., Smyth, B.M., Murphy, J.D., 2010a. A biofuel strategy for Ireland with an emphasis on production of biomethane and minimization of land take. Renew. Sust. Energy Rev. 14(1), 277-288.

[152] Singh, A., Pant, D., Korres, N.E., Nizami, A.S, Prasad, S., Murphy, J.D., 2010b. Key issues in life cycle assessment of ethanol production from lignocellulosic biomass: Challenges and perspectives. Bioresour. Technol. 101(13),5003-5012.

[153] Smyth, B.M., Murphy, J.D., O'Brien, C., 2009. What is the energy balance of grass biomethane in Ireland and other temperate northern European climates?. Renew. Sust. Energy Rev. 13(9), 2349-2360.

[154] Spatari, S., Zhang, Y., MacLean, H.L., 2005. Life cycle assessment of switchgrass- and corn stover-derived ethanol-fueled automobiles. Environ. Sci. Technol. 39 (24), 9750-9758.

[155] Stephenson, A.L., Kazamia, E., Dennis, J.S., Howe, C.J., Scott, S.A., Smith, A.G., 2010. Life-cycle assessment of potential algal biodiesel pro-duction in the United Kingdom: a comparison of raceways and air-lift tubular bioreactors. Energy Fuels. 24(7), 4062-4077.

[156] Stichnothe, H., Azapagic, A., 2009. Bioethanol from waste: life cycle estimation of the greenhouse gas saving potential. Resour. Conserv. Recycl. 53 (11), 624-630.

[157] Tahir, M.S., Shahzad, K., Shahid, Z., Sagir, M., Rehan, M., Nizami, A.S., 2015. Producing methane enriched biogas using solvent absorption method. Chem. Eng. Trans. 45,1309-1314.

[158] Thyø, K.A., Wenzel, H., 2007. Life Cycle Assessment of biogas from maize silage and from manure - for transport and for heat and power production under displacement of natural gas based heat works and marginal electricity in northern Germany. 2nd draft report for Xergi A/S Sofiendalsvej 7, 9200 Aalborg SV.

[159] Unnasch, S., Pont, J., Hooks, M., Chan, M., Waterland, L., Rutherford, D., 2007. Full fuel cycle assessment: well to wheels energy inputs, emissions, and water impacts. Consultant Report prepared by TIAX LLC for California Energy Commission, Cupertino CA, USA.

[160] United States - Energy Information Administration (US-EIA), 2016.

[161] UNICA; Brazilian Sugarcane Industry Association, 2009. Mexico requests unica to lead technical mission to discuss ethanol.

[162] van Vliet, O.P., Faaij, A.P., Turkenburg, W.C., 2009. Fischer–Tropsch diesel production in a well-to-wheel perspective: a carbon, energy flow and cost analysis. Energy Convers. Manage. 50, 855-876.

[163] Veeraraghavan, S., Riera-Palou, X., 2006. Well-to wheel performance of Iogen lignocellulosic ethanol. Shell Global Solutions International.

[164] Vogt, R., 2008a. Basisdaten, zu THG-Bilanzen für Biogas-Prozessketten und Erstellung neuer THG Bilanzen. Institut für Energie- und Umweltforschung Heidelberg GmbH.

[165] Vogt, R., 2008b. Optimierungen für einen nachhaltigen Ausbau der Biogaserzeugung und nutzung in Deutschland.

[166] World Watch Institute (WWI), 2006. Biofuels for transportation, global potential and implications for sustainable agriculture and energy in the 21st century. Submited report prepared for BMELV in cooperation with GTZ and FNR.

[167] Xin, J., Imahara, H., Saka, S., 2009. Kinetics on the oxidation of biodiesel stabilized with antioxidant. Fuel. 88(2), 282-286.

[168] Yáñez Angarita, E.E., Silva Lora, E.E., da Costa, R.E., Torres, E.A., 2009. The energy balance in the palm oil-derived methyl ester (PME) life cycle for the cases in Brazil and Colombia. Renew. Energ. 34(12), 2905-2913.

[169] Zah, R., Böni, H., Gauch, M., Hischier, R., Lehmann, M., Wäger, P., 2007. Ökobilanz von Energieprodukten: Ökologische Bewertung von Biotreibstoffen. EMPA, St. Gallen, Switzerland.

[170] Zhang, X., Yan, S., Tyagi, R.D., Surampalli, R.Y., 2013. Energy balance and greenhouse gas emissions of biodiesel production from oil derived from wastewater and wastewater sludge. Renew. Energ. 55, 392-403.

[171] Zhi Fu, G., Chan, A.W., Minns, D.E., 2003. Life cycle assessment of bio-ethanol derived from cellulose. Int. J. Life Cycle Assess. 8(3), 137-141.

[172] Zhu, H., Beland, M., 2006. Evaluation of alternative methods of preparing hydrogen producing seeds from digested wastewater sludge. Int. J. Hydrogen Energy. 31(14), 1980-1988.

# 23

# Development and evaluation of a novel low power, high frequency piezoelectric-based ultrasonic reactor for intensifying the transesterification reaction

Mortaza Aghbashlo[1,†,*], Meisam Tabatabaei[2,3,†,*], Soleiman Hosseinpour[1,*], Seyed Sina Hosseini[1], Akram Ghaffari[2], Zahra Khounani[2,3], Pouya Mohammadi[3]

[1] *Department of Mechanical Engineering of Agricultural machinery, Faculty of Agricultural Engineering and Technology, College of Agriculture and Natural Resources, University of Tehran, Karaj, Iran.*

[2] *Agricultural Biotechnology Research Institute of Iran (ABRII), Agricultural Research, Education, and Extension Organization (AREEO), Karaj, Iran.*

[3] *Biofuel Research Team (BRTeam), Karaj, Iran.*

## HIGHLIGHTS

- A novel low power, high frequency piezoelectric-based ultrasonic reactor was investigated for the first time for rapid biodiesel production.
- 6:1 alcohol/oil molar ratio, 10 min sonication time, and 60 °C temperature as the optimal conditions.
- Conversion efficiency of 97.12% at a specific energy consumption of as low as 378 kJ/kg was achieved.
- Promising replacement for high power, low frequency ultrasonic systems was introduced.

## GRAPHICAL ABSTRACT

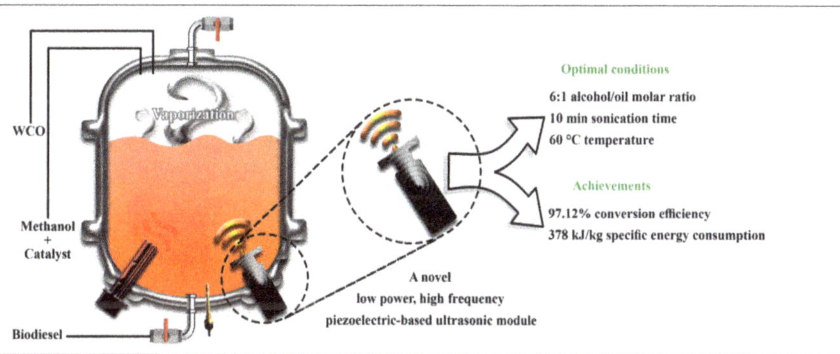


## ABSTRACT

In this study, a novel low power, high frequency piezoelectric-based ultrasonic reactor was developed and evaluated for intensifying the transesterification process. The reactor was equipped with an automatic temperature control system, a heating element, a precise temperature sensor, and a piezoelectric-based ultrasonic module. The conversion efficiency and specific energy consumption of the reactor were examined under different operational conditions, i.e., reactor temperature (40–60 °C), ultrasonication time (6–10 min), and alcohol/oil molar ratio (4:1–8:1). Transesterification of waste cooking oil (WCO) was performed in the presence of a base-catalyst (potassium hydroxide) using methanol. According to the obtained results, alcohol/oil molar ratio of 6:1, ultrasonication time of 10 min, and reactor temperature of 60 °C were found as the best operational conditions. Under these conditions, the reactor converted WCO to biodiesel with a conversion efficiency of 97.12%, meeting the ASTM standard satisfactorily, while the lowest specific energy consumption of 378 kJ/kg was also recorded. It should be noted that the highest conversion efficiency of 99.3 %, achieved at reactor temperature of 60 °C, ultrasonication time of 10 min, and alcohol/oil molar ratio of 8:1, was not favorable as the associated specific energy consumption was higher at 395 kJ/kg. Overall, the low power, high frequency piezoelectric-based ultrasonic module could be regarded as an efficient and reliable technology for intensifying the transesterification process in terms of energy consumption, conversion efficiency, and processing time, in comparison with high power, low frequency ultrasonic system reported previously. Finally, this technology could also be considered for designing, developing, and retrofitting chemical reactors being employed for non-biofuel applications as well.

**Keywords:**
Biodiesel
Transesterification process
Process intensification
Piezoelectric-based ultrasonic reactor
Low power, high frequency ultrasonic system
Specific energy consumption

* Corresponding author E-mail address: maghbashlo@ut.ac.ir (M. Aghbashlo) E-mail address: meisam_tab@yahoo.com & meisam_tabatabaei@abrii.ac.ir (M. Tabatabaei) E-mail address: shosseinpour@ut.ac.ir (S. Hosseinpour)

† These authors contributed equally.

## 1. Introduction

The increasing energy demands along with the exhaustion of fossil fuel resources are steadily pushing researchers to explore environmentally-clean and economically-viable alternative energy carriers in order to fill the gap between the energy demand and supply in a sustainable manner. Among alternative energy sources developed to date, biodiesel has gained a growing attention as a good substitute to mineral diesel because of providing eco-friendly combustion, having similar physicochemical properties to those of mineral diesel, and being renewable, biodegradable, non-toxic, and lubricant (Aghbashlo et al., 2015). However, there are many damning reports published in the last decade where biodiesel production from food- and feed-grade feedstocks is portrayed as the main culprit for the food price hikes and shortages because of a controversial competition over the use of land and water (Collins, 2008; Mitchell, 2008). In order to address this issue, the use of waste cooking oil (WCO) as a promising alternative source to the first generation feedstocks has gained increasing popularity due to its environmentally-acceptable and economically-viable features (Aghbashlo and Demirbas, 2016).

Transesterification of triglycerides or fatty acids with an alcohol in the presence of a homogeneous or heterogeneous catalyst using a mechanically-stirred reactor is one of the most popular procedures applied commercially for synthesizing biodiesel from various feedstocks. However, mechanical stirring method require a long processing time and a huge amount of energy to boost the interfacial region between alcohol and oil due to their low immiscibility as the main factor affecting the yield of the transesterification process. In order to meet the increasing biodiesel demands worldwide and to accelerate the transesterification reaction, various techniques have been developed using heterogeneous catalysts including alcohol supercritical temperature, transesterification *via* radio frequency microwave, alcohol reflex temperature, and ultrasonication (Ramachandran et al., 2013). Amongst the technologies developed to date, ultrasound technique is taken into account as one of the most promising process-intensification technologies due to its low cost and high safety. Ultrasonic irradiations can efficiently reduce the required process time and temperature compared with the mechanically-stirred transesterification process through strengthening the mass transfer of the liquid–liquid systems (Yin et al., 2012). The irradiated ultrasound waves create micro fine bubbles in the phase boundary of the two immiscible liquids which in turns facilitate and intensify the emulsification process due to a high amount of energy released by the asymmetric collapse of the generated microbubbles.

In this sense, numerous research attempts have been conducted to apply high power, low frequency ultrasound technology for biodiesel production from various feedstocks in recent years. For instance, Deshmane et al. (2008) investigated the methylation of palm fatty acid distillate in the presence of concentrated $H_2SO_4$ using ultrasonic irradiations at a frequency of 22 kHz and a power of 120 W. In another study, Santos et al. (2010) produced methyl esters from *Oreochromis niloticus* oil using a low-frequency (40 kHz), high-intensity ultrasound system. Furthermore, Salamatinia et al. (2012) intensified the biodiesel production process using ultrasonic irradiations at 20 kHz and 200 W in the presence of SrO catalyst. Later, Maddikeri et al. (2013) applied an ultrasonic horn at a frequency of 22 kHz and power of 750 W to intensify the methylation of WCO with methyl acetate in the presence of potassium hydroxide as catalyst. In addition, Michelin et al. (2015) explored the fatty acid ethyl esters production process from *Macauba* coconut oil using solvent-free enzymatic (immobilized lipase) transesterification reaction under ultrasound irradiations (40 kHz and 132 W). Recently, Subhedar and Gogate (2016) synthesized biodiesel from the WCO using methyl acetate and immobilized lipase by means of ultrasound irradiations (20 kHz and 120 W). More comprehensive information could be found in a review paper published by Ramachandran et al. (2013) on the application of ultrasound technology for intensifying the transesterification process from various feedstocks.

Although the above-mentioned studies revealed the potential of ultrasound technology for making biodiesel from oils, the required processing time and energy can still be discounted by developing innovative and emerging technologies. Therefore, the objective of the current study was to develop and evaluate a low power, high frequency piezoelectric-based ultrasonic reactor for enhanced transesterification process. Moreover, the effects of experimental variables, i.e., ultrasonication time (6–10 min), reaction temperature (40–60 °C), and alcohol/oil molar ratio (4:1–8:1) on the conversion efficiency and specific energy consumption of the WCO to methyl esters were also studied. To the best of our knowledge, this is the first study reporting the use of a low power (31 W) and high frequency (1.7 MHz) piezoelectric-based ultrasonic module for the intensification of the transesterification reaction. In fact, most research attempts were focused on using high power (100–2400 W) and low frequency (20–40 kHz) ultrasound systems to convert triglycerides into fatty acids alkyl esters. Accordingly, the ultrasound system introduced herein can be a more efficient and reliable technology over those reported in the literature from the energy consumption and processing time perspectives. The findings of the current survey would be helpful to designers and engineers in applying the proposed low power, high frequency ultrasound technology for biodiesel production in an environmentally-sustainable and economically-viable manner as well as in retrofitting the available systems.

## 2. Materials and Methods

### *2.1. Materials*

The WCO containing linoleic acid (50.57%), oleic acid (33.94%), and palmitic acid (15.48%) was obtained from a local restaurant (Karaj, Iran). The FFA content was determined at about 1% using the standard titrimetry method (Tiwari et al., 2007). Filtration using normal sieves was carried out twice to remove food debris from the WCO. Moreover, the water present was also eliminated. Methanol (99%) and potassium hydroxide used throughout this study were obtained from Merck (Germany).

### *2.2. Piezoelectric-based ultrasonic reactor, biodiesel production methodology, and analysis*

In this study, a novel low power, high frequency piezoelectric-based ultrasonic reactor was constructed and evaluated for biodiesel production from WCO. A schematic representation of the developed cylindrically-shaped piezoelectric-based ultrasonic reactor is manifested in **Figure 1**. The stainless steel (S316) chamber used had a total volume of 5 L with a wall thickness of 2 mm. The reactor was 20 cm in diameter and 16 cm in height. The heating of the liquid media inside the reactor was carried out by means of a spiral-shaped 500 W heating element. In order to control and adjust the temperature of the liquid media, a digital waterproof

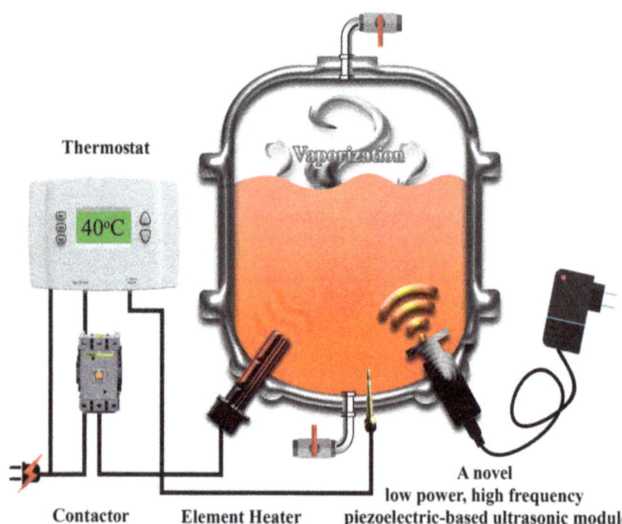


**Fig.1.** A schematic representation of the high-frequency, low-power piezoelectric-based ultrasonic reactor developed in this study for the transesterification of WCO.

temperature sensor with a precision of 0.1 °C was located within the chamber. Moreover, automatic temperature controlling of the reactor was performed using a 5 W digital thermostat. A low-power (31 W), high-frequency (1.7 MHz) piezoelectric-based ultrasonic module was installed within the chamber in order to accelerate the methylation process of WCO. The specifications of the equipment and instruments used in the construction of the reactor are summarized in **Table 1**.

**Table 1.**
Specifications of the equipment and instruments used in the construction of the reactor.

| Equipment and instruments | Specifications |
|---|---|
| Ultrasonic module | Size: ϕ 46×22.9 (Piezo: ϕ 20×1.2) mm<br>Resonant Frequency: 1.70 MHz<br>Resonant Impedance: < 2 Ω<br>Coupling factor:> 52%<br>Electrostatic capacity: 1800 pF<br>Spray volume: <380 mL/h |
| Electrical heater | Power: 500 w<br>Type: Tubular heating element<br>Material: Stainless steel |
| Thermocouple | Accuracy: 0.1 degree of Celsius<br>Material: Stainless steel |
| Controller and displayer | Model: SU-105 IP<br>Temp. range: -50.0 °C~150 °C<br>Output: Main/Aux (DC12V)<br>Main/Aux(Relay) |

In the current survey, the effects of ultrasonication time (6, 8, and 10 min), reaction temperature (40, 50, and 60 °C), and alcohol/oil molar ratio (4:1, 6:1, and 8:1) were investigated on the specific energy consumption and conversion efficiency of WCO to methyl esters. An initial volume of 1500 mL of the liquid medium was considered for all experiments in order to avoid any volume effects on the studied parameters. A balance with a precision of 0.001 g was used to weigh all materials used in the experiment. A schematic representation of the experimental methodology employed in the present study for the production and analysis of biodiesel from WCO is provided in **Figure 2**.

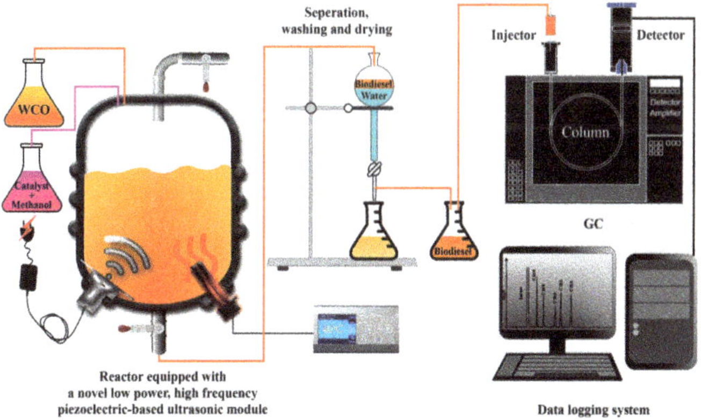


**Fig.2.** A schematic illustration of the experimental methodology applied in the present survey for production and analysis of biodiesel from WCO.

To synthesis biodiesel, first potassium hydroxide (1 wt.%) and methanol were mixed for 2 min by a magnetic stirrer. The resultant potassium methoxide was then gradually transferred to the reactor together with the WCO. After heating the liquid to the desired temperature, the ultrasound module was turned on to intensify the transesterification process. The electrical power utilized by the system was recorded during each experiment in order to compare the energy consumption of the performed trials. Afterwards, the reaction products were immediately drained from the reactor and transferred into an ice-bath for 30 min to cease the transesterification process. In order to segregate crude glycerol from biodiesel, the reaction products were left for 2 h. The unreacted methanol, formed soap, and potassium hydroxide were then separated from the produced biodiesel by thrice washing the crude biodiesel with 500 mL hot water (60 °C). Finally, the purified biodiesel was dehydrated in an oven at 80 °C for 4 h in order to vaporize the absorbed water.

The conversion efficiency of the biodiesel samples was determined by means of a Varian CP-3800 gas chromatograph (GC) (Varian, Inc., Palo Alto, CA) equipped with a CP-Sill 88 fused silica column (100 m, 0.25 mm I.D., film thickness 0.25μm) and a flame ionization detector (FID) detector. More specifically, 0.5. mL of *n*-hexane solution containing lauric acid methyl ester as internal standard (4 mg/mL) was added into each pre-weighed biodiesel sample and 1 μL of the mixture was injected into the GC (split mode with a split ratio of 80:1). Nitrogen was used as the carrier gas. The purity of the produced biodiesel was the computed using the following equation (Wang et al., 2006):

$$\text{Purity (\%)} = \left(\frac{\text{area of FAME/area of reference} \times \text{weight of reference}}{\text{weight of biodiesel}}\right) \times 100 \qquad \text{Eq. 1}$$

The computed conversion efficiency was used to calculate the molar fraction of methyl esters, triglyceride, diglyceride, and monoglyceride, and glycerin. In order to simplify the mass balance calculations, it was postulated that the FFA (linoleic acid) of the WCO was completely reacted with potassium hydroxide generating soap and water at the end of experiment. Moreover, the molar fractions of triglyceride, diglyceride, and monoglyceride were assumed to be equal at the end of the reaction.

## 3. Results and discussions

**Table 2** tabulates the full mass balance for the low power, high frequency piezoelectric-based ultrasonic biodiesel synthesis from WCO at various alcohol/oil molar ratios, ultrasonication times, and reactor temperatures. It is obvious from the data presented in the table that the reactor could satisfactorily methylate the WCO to biodiesel at the highest level of alcohol/oil molar ratio, ultrasonication time, and reactor temperature, while the transesterification reaction was sluggish at the lowest levels of these parameters. The maximum amount of fatty acids methyl esters were synthesized at alcohol/oil ratio of 8:1, ultrasonication time of 10 min, and reactor temperature of 60 °C. However, mass balance or even conversion efficiency could lead to misleading conclusions when analyzing a reactor employed for biodiesel production. This could be ascribed to the fact that these parameters recognize the optimum operational conditions on the basis of the process yield alone, while such conditions might have the highest energy requirement.

**Figure 3** shows the effects of experimental variables on the conversion efficiency of the WCO to biodiesel. Conversion efficiency varied between 9.73% and 99.3% and increased as the alcohol/oil molar ratio elevated. Even though the required stoichiometric alcohol/oil molar ratio is 3:1 for the complete conversion of triglyceride to fatty acids esters, higher ratios (excess alcohol) should be applied to drive the reaction towards achieving higher yields. It should also be noted that further increase of alcohol/oil molar ratio beyond an optimal value could lead to only a marginal increase in the conversion efficiency and could also reverse the transesterification reaction due to an excessive dilution of oil. Gupta et al. (2015) noted that the access of triglycerides molecules to active site of the catalyst used was diminished at higher alcohol/oil molar ratios due to oil dilution. Moreover, further increasing the alcohol/oil ratio could result in difficulties during biodiesel purification and separation process.

In the present study, a steady increase in the conversion efficiency was observed from 55.53% to 97.12% by increasing the alcohol/oil molar ratio from 4:1 to 6:1 at the constant temperature of 60 °C

**Table 2.**
Full mass balance (g) for the low power, high frequency piezoelectric-based ultrasonic biodiesel processor.

| Operational conditions | | | Inputs (g) | | | | Outputs (g) | | | | | | | | | | |
|---|---|---|---|---|---|---|---|---|---|---|---|---|---|---|---|---|---|
| A/O[1] Molar ratio | Ultrasonication time (min) | Reactor T (°C) | Total TG[2] | FFA | MeOH | KOH | TG | DG[3] | MG[4] | $C_{19}H_{34}O_2$ | $C_{19}H_{36}O_2$ | $C_{17}H_{34}O_2$ | Glycerin | Soap | Water | MeOH | KOH |
| 4 | 6 | 40 | 1159.9 | 3.7 | 171.0 | 13.3 | 517.8 | 347.8 | 181.4 | 57.35 | 38.49 | 17.56 | 12.0 | 4.3 | 0.2 | 158.6 | 12.6 |
| 4 | 8 | 40 | 1159.9 | 3.7 | 171.0 | 13.3 | 469.3 | 315.2 | 164.4 | 107.19 | 71.94 | 32.81 | 22.4 | 4.3 | 0.2 | 147.7 | 12.6 |
| 4 | 10 | 40 | 1159.9 | 3.7 | 171.0 | 13.3 | 439.0 | 294.9 | 153.7 | 138.33 | 92.85 | 42.35 | 28.9 | 4.3 | 0.2 | 140.9 | 12.6 |
| 4 | 6 | 50 | 1159.9 | 3.7 | 171.0 | 13.3 | 451.0 | 302.9 | 158.0 | 126.01 | 84.58 | 38.57 | 26.3 | 4.3 | 0.2 | 143.6 | 12.6 |
| 4 | 8 | 50 | 1159.9 | 3.7 | 171.0 | 13.3 | 440.4 | 295.8 | 154.2 | 136.94 | 91.91 | 41.92 | 28.6 | 4.3 | 0.2 | 141.2 | 12.6 |
| 4 | 10 | 50 | 1159.9 | 3.7 | 171.0 | 13.3 | 413.6 | 277.8 | 144.9 | 164.42 | 110.36 | 50.33 | 34.3 | 4.3 | 0.2 | 135.3 | 12.6 |
| 4 | 6 | 60 | 1159.9 | 3.7 | 171.0 | 13.3 | 386.7 | 259.7 | 135.4 | 192.05 | 128.90 | 58.79 | 40.1 | 4.3 | 0.2 | 129.2 | 12.6 |
| 4 | 8 | 60 | 1159.9 | 3.7 | 171.0 | 13.3 | 269.8 | 181.2 | 94.5 | 312.19 | 209.54 | 95.57 | 65.1 | 4.3 | 0.2 | 103.1 | 12.6 |
| 4 | 10 | 60 | 1159.9 | 3.7 | 171.0 | 13.3 | 255.1 | 171.3 | 89.3 | 327.23 | 219.64 | 100.17 | 68.3 | 4.3 | 0.2 | 99.8 | 12.6 |
| 6 | 6 | 40 | 1082.3 | 3.5 | 239.4 | 13.3 | 176.8 | 118.8 | 61.9 | 368.26 | 247.17 | 112.73 | 76.8 | 4.0 | 0.2 | 159.2 | 12.6 |
| 6 | 8 | 40 | 1082.3 | 3.5 | 239.4 | 13.3 | 123.7 | 83.1 | 43.3 | 422.82 | 283.80 | 129.43 | 88.2 | 4.0 | 0.2 | 147.4 | 12.6 |
| 6 | 10 | 40 | 1082.3 | 3.5 | 239.4 | 13.3 | 96.0 | 64.5 | 33.6 | 451.26 | 302.89 | 138.14 | 94.1 | 4.0 | 0.2 | 141.2 | 12.6 |
| 6 | 6 | 50 | 1082.3 | 3.5 | 239.4 | 13.3 | 116.9 | 78.5 | 41.0 | 429.77 | 288.46 | 131.56 | 89.7 | 4.0 | 0.2 | 145.8 | 12.6 |
| 6 | 8 | 50 | 1082.3 | 3.5 | 239.4 | 13.3 | 92.3 | 62.0 | 32.3 | 455.03 | 305.42 | 139.30 | 94.9 | 4.0 | 0.2 | 140.4 | 12.6 |
| 6 | 10 | 50 | 1082.3 | 3.5 | 239.4 | 13.3 | 71.2 | 47.8 | 24.9 | 476.79 | 320.02 | 145.96 | 99.5 | 4.0 | 0.2 | 135.6 | 12.6 |
| 6 | 6 | 60 | 1082.3 | 3.5 | 239.4 | 13.3 | 105.5 | 70.9 | 36.9 | 441.52 | 296.35 | 135.16 | 92.1 | 4.0 | 0.2 | 143.3 | 12.6 |
| 6 | 8 | 60 | 1082.3 | 3.5 | 239.4 | 13.3 | 45.3 | 30.4 | 15.9 | 503.33 | 337.84 | 154.08 | 105.0 | 4.0 | 0.2 | 129.8 | 12.6 |
| 6 | 10 | 60 | 1082.3 | 3.5 | 239.4 | 13.3 | 15.4 | 10.4 | 5.4 | 534.04 | 358.45 | 163.48 | 111.4 | 4.0 | 0.2 | 123.2 | 12.6 |
| 8 | 6 | 40 | 1014.7 | 3.3 | 299.3 | 13.3 | 145.9 | 98.0 | 51.1 | 365.61 | 245.40 | 111.92 | 76.3 | 3.7 | 0.2 | 219.7 | 12.7 |
| 8 | 8 | 40 | 1014.7 | 3.3 | 299.3 | 13.3 | 91.6 | 61.5 | 32.1 | 421.45 | 282.88 | 129.02 | 87.9 | 3.7 | 0.2 | 207.5 | 12.7 |
| 8 | 10 | 40 | 1014.7 | 3.3 | 299.3 | 13.3 | 51.2 | 34.4 | 17.9 | 462.89 | 310.69 | 141.70 | 96.6 | 3.7 | 0.2 | 198.5 | 12.7 |
| 8 | 6 | 50 | 1014.7 | 3.3 | 299.3 | 13.3 | 128.6 | 86.4 | 45.0 | 383.41 | 257.35 | 117.37 | 80.0 | 3.7 | 0.2 | 215.8 | 12.7 |
| 8 | 8 | 50 | 1014.7 | 3.3 | 299.3 | 13.3 | 71.0 | 47.7 | 24.9 | 442.53 | 297.03 | 135.47 | 92.3 | 3.7 | 0.2 | 202.9 | 12.7 |
| 8 | 10 | 50 | 1014.7 | 3.3 | 299.3 | 13.3 | 53.8 | 36.1 | 18.8 | 460.27 | 308.93 | 140.90 | 96.0 | 3.7 | 0.2 | 199.1 | 12.7 |
| 8 | 6 | 60 | 1014.7 | 3.3 | 299.3 | 13.3 | 39.4 | 26.5 | 13.8 | 475.05 | 318.85 | 145.42 | 99.1 | 3.7 | 0.2 | 195.8 | 12.7 |
| 8 | 8 | 60 | 1014.7 | 3.3 | 299.3 | 13.3 | 12.8 | 8.6 | 4.5 | 502.38 | 337.20 | 153.79 | 104.8 | 3.7 | 0.2 | 189.9 | 12.7 |
| 8 | 10 | 60 | 1014.7 | 3.3 | 299.3 | 13.3 | 3.5 | 2.3 | 1.2 | 511.97 | 343.63 | 156.72 | 106.8 | 3.7 | 0.2 | 187.8 | 12.7 |

[1] A/O molar ratio: alcohol to oil molar ratio [2] TG: triglyceride [3] DG: diglyceride [4] MG: monoglyceride

and ultrasonication time of 10 min. The conversion efficiency marginally increased to 99.31% with increasing the alcohol/oil molar ratio to 8:1. It should be quoted that the optimum alcohol/oil molar ratio is profoundly affected by various factors including system configuration, reactor volume, catalyst properties, operating temperature, etc. (Mootabadi et al., 2010).

Increasing the ultrasonication time and liquid media temperature increased the conversion efficiency (**Fig. 3**). The longer exposure time to ultrasonic irradiations apparently enhanced the emulsification of the immiscible reactants (alcohol–oil phases) due to an increase in the number of cavitation microbubbles formed, leading to the higher yield of fatty acid methyl esters. In addition, increasing the liquid temperature enhanced the reaction rate since reactant molecules could attain a sufficient amount of energy to prevail the energy barrier. Furthermore, increasing the reactor temperature increased the miscibility and solubility of methanol in the WCO, leading to an increase in the kinetic energy of the reactants and their better contact. This could also be attributed to the fact that the diffusion resistance between the different phases of the mixture decreased owing to a drop in viscosity at higher temperatures (Gupta et al., 2015). However, further increase of the liquid media temperature beyond 60 °C might have negatively impacted the cavitational effects of the ultrasound irradiations, lowering the conversion efficiency unfavorably (Maran and Priya, 2015). In fact, this occurred because of the low boiling temperature of methanol, i.e., 64.7 °C (Lee et al., 2011). In better words, elevating the liquid media temperature beyond an optimal value could lead to evaporation of a portion of the involved alcohol, decreasing the possibility of successful collision of triglyceride molecules with the alcohol. This could also lead to the supersaturation of cavitational bubbles with methanol vapor, resulting in their implosion with lower intensity (Gupta et al., 2015). Furthermore, increasing the liquid medium temperature could have also reduced the cavitational effect by facilitating the propagation and dissipation of the ultrasound waves (Korkut and Bayramoglu, 2016). These clearly indicate why optimum operational variables should be exclusively applied for different transesterification systems.

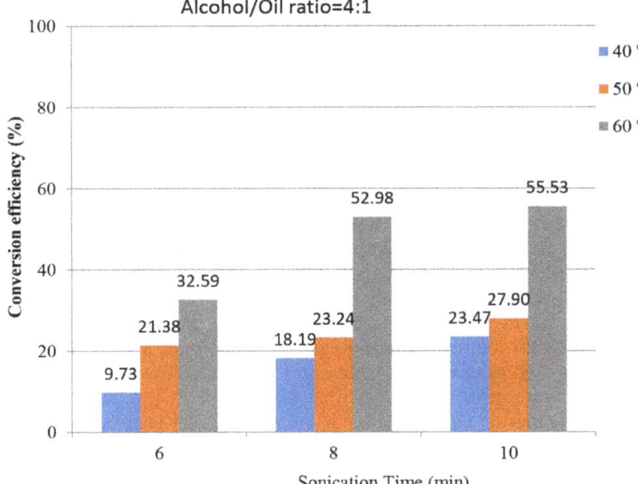


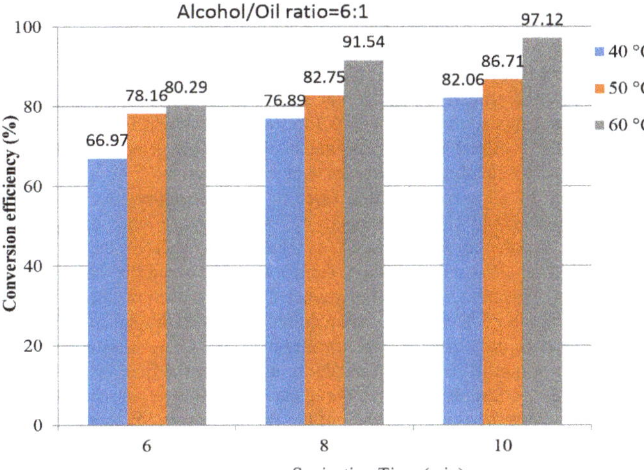


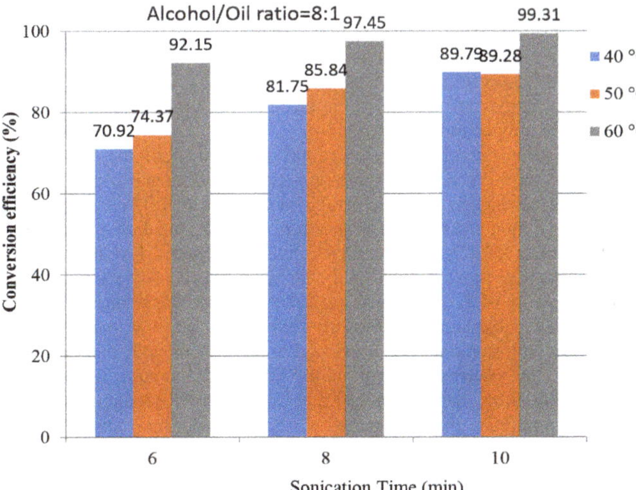


**Fig.3.** The effects of experimental variables on the conversion efficiency of the WCO to biodiesel.

For better understanding of the effects of various operational conditions on the conversion efficiency of the rector, a linear regression model was developed. First, the experimental variables were normalized between 1 and 2 using the following equation (**Eq. 2**):

$$X' = \frac{X - X_{min}}{X_{ax} - X_{min}} + 1 \qquad \text{Eq. 2}$$

The developed model for the conversion efficiency was as follows (**Eq. 3**):

$$CE\,(\%) = -70.37 + 57.32 \times M + 13.85 \times U + 19.91 \times T \qquad \text{Eq. 3}$$

where CE is the conversion efficiency while M, U, and T are the normalized alcohol/oil molar ratio, normalized ultrasonication time, and normalized reactor temperature, respectively.

Although the developed model fitted acceptably to the conversion efficiency with a multiple R value of about 0.90, advanced modeling systems allowing accurate prediction of this important parameter should be examined in future studies. According to the developed model, the main contributors to the conversion efficiency were (in a descending order of importance): alcohol/oil molar ratio, reactor temperature, and ultrasonication time. However, the interaction of these factors on the conversion efficiency should be scrutinized using advanced statistical tools.

**Table 3.**
The electrical energy utilized by various components of the reactor at different alcohol/oil molar ratios, ultrasonication times, and reactor temperatures.

| **Operational conditions** | | | **Energy consumption (kJ)** | | | |
|---|---|---|---|---|---|---|
| Alcohol/oil molar ratio | Sonication time (min) | Reactor T (°C) | Module | Heater | Thermostat | Total |
| 4:1 | 6 | 40 | 11.16 | 162.50 | 6.85 | 180.51 |
| 4:1 | 8 | 40 | 14.88 | 162.50 | 8.05 | 185.43 |
| 4:1 | 10 | 40 | 18.60 | 162.50 | 9.25 | 190.35 |
| 4:1 | 6 | 50 | 11.16 | 269.00 | 8.98 | 289.14 |
| 4:1 | 8 | 50 | 14.88 | 269.00 | 10.18 | 294.06 |
| 4:1 | 10 | 50 | 18.60 | 269.00 | 11.38 | 298.98 |
| 4:1 | 6 | 60 | 11.16 | 367.50 | 10.95 | 389.61 |
| 4:1 | 8 | 60 | 14.88 | 367.50 | 12.15 | 394.53 |
| 4:1 | 10 | 60 | 18.60 | 367.50 | 13.35 | 399.45 |
| 6:1 | 6 | 40 | 11.16 | 162.50 | 6.85 | 180.51 |
| 6:1 | 8 | 40 | 14.88 | 162.50 | 8.05 | 185.43 |
| 6:1 | 10 | 40 | 18.60 | 162.50 | 9.25 | 190.35 |
| 6:1 | 6 | 50 | 11.16 | 269.00 | 8.98 | 289.14 |
| 6:1 | 8 | 50 | 14.88 | 269.00 | 10.18 | 294.06 |
| 6:1 | 10 | 50 | 18.60 | 269.00 | 11.38 | 298.98 |
| 6:1 | 6 | 60 | 11.16 | 367.50 | 10.95 | 389.61 |
| 6:1 | 8 | 60 | 14.88 | 367.50 | 12.15 | 394.53 |
| 6:1 | 10 | 60 | 18.60 | 367.50 | 13.35 | 399.45 |
| 8:1 | 6 | 40 | 11.16 | 162.50 | 6.85 | 180.51 |
| 8:1 | 8 | 40 | 14.88 | 162.50 | 8.05 | 185.43 |
| 8:1 | 10 | 40 | 18.60 | 162.50 | 9.25 | 190.35 |
| 8:1 | 6 | 50 | 11.16 | 269.00 | 8.98 | 289.14 |
| 8:1 | 8 | 50 | 14.88 | 269.00 | 10.18 | 294.06 |
| 8:1 | 10 | 50 | 18.60 | 269.00 | 11.38 | 298.98 |
| 8:1 | 6 | 60 | 11.16 | 367.50 | 10.95 | 389.61 |
| 8:1 | 8 | 60 | 14.88 | 367.50 | 12.15 | 394.53 |
| 8:1 | 10 | 60 | 18.60 | 367.50 | 13.35 | 399.45 |

**Table 3** summarizes the electrical energy utilized by various components of the reactor under different experimental conditions. **Figure 4** also manifests the effects of experimental variables on the specific energy consumption during the conversion of WCO to biodiesel using the

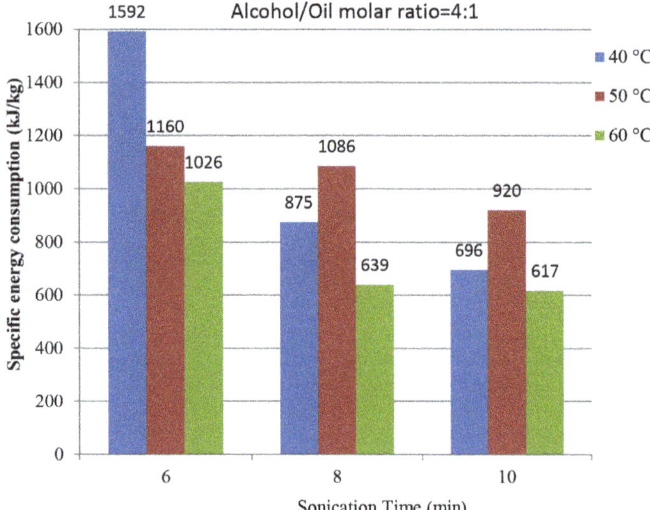


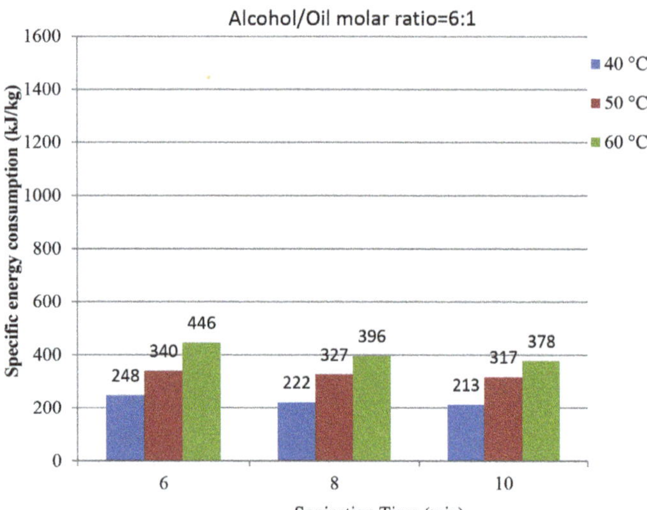


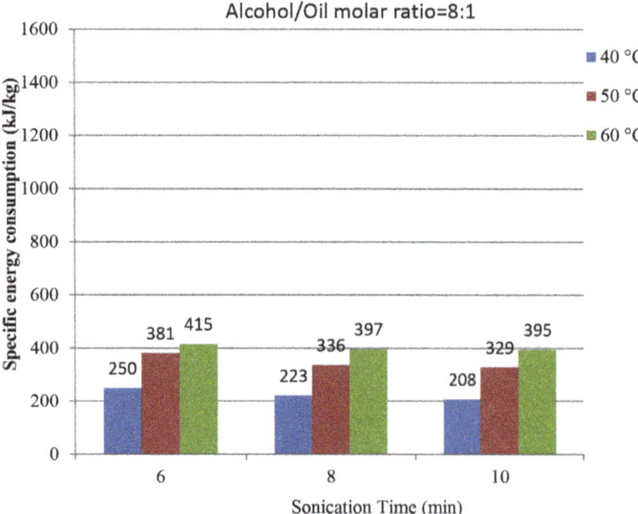


**Fig.4.** The effects of experimental variables on the specific energy consumption during piezoelectric-based ultrasonic conversion of WCO to biodiesel.

piezoelectric-based ultrasonic reactor. The total energy utilized for the transesterification of WCO by the developed piezoelectric-based ultrasonic reactor ranged from a minimum value of 180.51 kJ to a maximum value of 399.45 kJ. Increasing the ultrasonication time and reactor temperature increased the total energy utilization of the transesterification process. While, the total energy utilized by the reactor was not affected by alcohol/oil molar ratio. The total energy consumption cannot be a perfect metric in order to make a decision on the experimental variables. To address this issue, the specific energy consumption was also defined for decision making on the operational variables of the reactor by considering the total energy utilized and the amount of biodiesel produced during the transesterification process.

The specific energy consumption values varied between 208 kJ/kg and 1592 kJ/kg. In general, the specific energy consumption decreased by increasing the ultrasonication time and alcohol/oil molar ratio, while it was increased by enhancing the reactor temperature. The lowest specific energy consumption (378 kJ/kg) while meeting the ASTM standard, i.e., 96.5% conversion efficiency, was found at the reactor temperature of 60 °C, ultrasonication time of 10 min, and alcohol/oil molar ratio of 6:1. The second lowest specific energy consumption while meeting the ASTM standard was determined at 395 kJ/kg for reactor temperature of 60 °C, ultrasonication time of 10 min, and alcohol/oil molar ratio of 8:1. Between these conditions, the first one could be suggested as optimal conditions for biodiesel production from WCO using the developed reactor due to the lower alcohol utilization and the trivial difference in the conversion efficiency.

Nevertheless, it is worth highlighting that comprehensive techno-economic assessments should still be performed using advanced tools like exergy and its extension as well as life cycle assessment in order to find the most cost-effective and environment-friendly arrangements/conditions of the proposed ultrasonic reactor technology. In fact, the conversion efficiency of ultrasound-assisted reactors being developed for biodiesel production could be optimized by manipulating the operational parameters of the reactor such as alcohol/oil molar ratio, reaction temperature, ultrasonication time, etc. However, it should be emphasized that multi-objective optimization studies using advanced evolutionary- or knowledge-based techniques are required to lower the specific energy consumption and increase the conversion efficiency, simultaneously, while ensuring that the ASTM standards are met for the final fuel as well.

In order to further scrutinize the effect of operational variables on the specific energy consumption (SE) of the WCO transesterification process, a linear mathematical model (R=0.77) was also developed as follows (**Eq. 4**):

$$\mathrm{SE\ (kJ/kg)} = 1.75 - 0.63 \times \mathrm{M} - 0.20 \times \mathrm{U} + 0.02 \times \mathrm{T} \qquad \text{Eq. 4}$$

According to the developed model, the order based on which the operational variables impacted the specific energy consumption was: alcohol/oil molar ratio > ultrasonication time > reactor temperature. In fact, both alcohol/oil molar ratio and ultrasonication time decreased the specific energy consumption, while this criterion was negatively affected by reactor temperature.

Taking in consideration the results obtained from the conversion efficiency and specific energy consumption, the optimum conditions for biodiesel production were alcohol/oil molar ratio of 6:1, ultrasonication time of 10 min, and reaction temperature of 60 °C, yielding a conversion efficiency of 97.12%. The optimal conditions obtained herein were also compared with those reported in the literature in order to further evaluate the suitability of the developed piezoelectric-based ultrasonic reactor for synthesizing biodiesel (**Table 4**). Obviously, the low power, high frequency ultrasonic reactor developed in the present study outperformed most of the systems reported in the published literature in terms of processing time, conversion efficiency, and power requirement. It is interesting to note that the optimal alcohol/oil molar ratio found through this study was markedly lower than those previously reported in this domain. This could be ascribed to the intensified reaction resulted from high frequency ultrasonication which in turns lowered the amount of alcohol required to maintain the rate of transesterification process in an acceptable level.

**Table 4.**
The optimal conditions reported in the literature for biodiesel production from various feedstocks using low frequency, high power ultrasound systems compared with the low power, high frequency piezoelectric-based ultrasonic system developed herein.

| Feedstock | Catalyst | Optimum reaction condition | Conversion efficiency (%) | Reference |
|---|---|---|---|---|
| Palm oil | SrO | C=3 wt%, M=15:1, T=65 °C, t=60 min, F=20 kHz, , $P_{max}$=200 W | 95.2 | Mootabadi et al. (2010) |
| Palm oil | BrO and SrO | C=2.8 wt%, M=9:1, T=65 °C, t=50 min, F=20 kHz, $P_{max}$=200 W with amplitudes 70% and 80% for BaO and SrO | 95 | Salamatinia et al. (2010) |
| Jatropha oil | $Ca(OMe)_2$ | C=5.5 wt%, M=11:1, T=64 °C, t=60 min, F=35 kHz, P=35 W | 96 | Choudhury et al. (2014) |
| WCO | Calcium diglyceroxide | C=1 wt%, M=9:1, T=60 °C, t=30 min, F=22 kHz, P=120 with 50% duty cycle | 93.5 | Gupta et al. (2015) |
| Canola oil | Dolomite and CaO | C=5 wt%, M=9:1, T=60 °C, t=90 min, F=20 kHz, P=45 W for dolomite<br>C=3 wt%, M=9:1, T=60 °C, t=75 min, F=20 kHz, P=45 W for CaO | 97.4 for dolomite<br>95.5 for CaO | Korkut and Bayramoglu (2016) |
| WCO | Immobilized novozym 435 | C=3 w/v%, M=9:1, t= 180 min, F=20 kHz, P=80 W with 60% duty cycle | 96.1 | Subhedar and Gogate (2016) |
| WCO | KOH | C=1 w/v%, M=6:1, t= 10 min, F=1700 kHz, P=31 W | 97.1 | Present study |

C: Catalyst, t: reaction time, F: frequency, P: power

Moreover, a quick glance at the published literature revealed that the amount of catalyst used by most research studies was often higher than what applied in the current survey (**Table 4**). This finding could be attributed to the adequate contact among the reactants owing to the uniform mixing patterns or dispersions resulted from the cavitational events, reducing the quantity of catalyst required to achieve a high conversion efficiency. Therefore, by using the piezoelectric-based ultrasound technology, an excess use of catalyst as well as the consequent down-stream difficulties can be avoided. Overall, the transesterification of WCO under low power, high frequency ultrasonic irradiation could be an efficient, time-saving, and cost-effective method for synthesizing biodiesel. This could be ascribed to the severe turbulence and liquid circulation currents generated owing to the propagation of high frequency ultrasonic waves within liquid medium, intensifying the emulsification process more intensely compared with the low frequency ultrasonic systems.

## 4. Conclusions

This study successfully demonstrated the reliability and productivity of a novel low power, high frequency piezoelectric-based ultrasonic reactor for intensification of the transesterification process. The developed reactor converted over 97% of the triglycerides to methyl esters at alcohol/oil molar ratio of 6:1 and reaction temperature of 60 °C within 10 min with an specific energy consumption of 378 kJ/kg. Although the transesterification time was comparatively lower than those of high power, low frequency ultrasound systems reported in the published literatures, the obtained conversion efficiency herein satisfactorily met the ASTM standard. Therefore, the high frequency ultrasound system outperformed those previously reported in terms of conversion efficiency, energy requirement, and processing time. This revealed that the piezoelectric-based ultrasonic reactor could assuredly replace the low frequency ultrasonic horn and bath systems reported in the literature for biodiesel production.

Nevertheless, future studies should scrutinize the cost-effectiveness and eco-friendly features of ultrasound-assisted biodiesel production systems though advanced engineering tools like exergy analysis and life cycle assessment approach. Moreover, future research works should also be directed towards the techno-economic assessment of large-scale industrial biodiesel production plants using the proposed low power, high frequency ultrasound technology.

## Acknowledgements

The author would like to acknowledge the supports provided by SGP/GEF/UNDP (Project No.: IRA/SGP/OP5/Y3/STAR/CC/13/13(189)) and Biofuel Research Team (BRTeam) for constructing the ultrasonic reactor. We would also like to extend our appreciation to University of Tehran and Iranian Biofuel Society (IBS) for supporting this study. The authors are also thankful to Dr. Majid Mohadesi (Kermanshah University of Technology, Iran) and Eng. Ali Dadak for providing the GC internal standard and assisting with the graphical designs, respectively.

## References

[1] Aghbashlo, M., Demirbas, A., 2016. Biodiesel: hopes and dreads. Biofuel Res. J. 3(2), 379.

[2] Aghbashlo, M., Tabatabaei, M., Mohammadi, P., Pourvosoughi, N., Nikbakht, A.M., Goli, S.A.H., 2015. Improving exergetic and sustainability parameters of a DI diesel engine using polymer waste dissolved in biodiesel as a novel diesel additive. Energy Convers. Manage. 105, 328-337.

[3] Choudhury, H.A., Goswami, P.P., Malani, R.S., Moholkar, V.S., 2014. Ultrasonic biodiesel synthesis from crude *Jatropha curcas* oil with heterogeneous base catalyst: mechanistic insight and statistical optimization. Ultrason. Sonochem. 21(3), 1050-1064.

[4] Collins, K., 2008. The role of biofuels and other factors in increasing farm and food prices-a review of recent developments with a focus on feed grain markets and market prospects. Supporting material for a review conducted by Kraft Foods Global, Inc.

[5] Deshmane, V.G., Gogate, P.R., Pandit, A.B., 2008. Ultrasound-assisted synthesis of biodiesel from palm fatty acid distillate. Ind. Eng. Chem. Res. 48(17), 7923-7927.

[6] Gupta, A.R., Yadav, S.V., Rathod, V.K., 2015. Enhancement in biodiesel production using waste cooking oil and calcium diglyceroxide as a heterogeneous catalyst in presence of ultrasound. Fuel. 158, 800-806.

[7] Korkut, I., Bayramoglu, M., 2016. Ultrasound assisted biodiesel production in presence of dolomite catalyst. Fuel. 180, 624-629.

[8] Lee, S.B., Lee, J.D., Hong, I.K., 2011. Ultrasonic energy effect on vegetable oil based biodiesel synthetic process. J. Ind. Eng. Chem. 17(1), 138-143.

[9] Maddikeri, G.L., Pandit, A.B., Gogate, P.R., 2013. Ultrasound assisted interesterification of waste cooking oil and methyl acetate for biodiesel and triacetin production. Fuel Process. Technol. 116, 241-249.

[10] Maran, J.P., Priya, B., 2015. Comparison of response surface methodology and artificial neural network approach towards efficient ultrasound-assisted biodiesel production from muskmelon oil. Ultrason. Sonochem. 23, 192-200.

[11] Michelin, S., Penha, F.M., Sychoski, M.M., Scherer, R.P., Treichel, H., Valério, A., Di Luccio, M., de Oliveira, D. Oliveira, J.V., 2015.

Kinetics of ultrasound-assisted enzymatic biodiesel production from Macauba coconut oil. Renew. Energy. 76, 388-393.

[12] Mitchell, D., 2008. A note on rising food prices. world bank policy research working paper series. Washington DC: the World Bank. Development Prospects Group.

[13] Mootabadi, H., Salamatinia, B., Bhatia, S., Abdullah, A.Z., 2010. Ultrasonic-assisted biodiesel production process from palm oil using alkaline earth metal oxides as the heterogeneous catalysts. Fuel. 89(8), 1818-1825.

[14] Ramachandran, K., Suganya, T., Gandhi, N.N., Renganathan, S., 2013. Recent developments for biodiesel production by ultrasonic assist transesterification using different heterogeneous catalyst: a review. Renew. Sust. Energy Rev. 22, 410-418.

[15] Salamatinia, B., Abdullah, A.Z., Bhatia, S., 2012. Quality evaluation of biodiesel produced through ultrasound-assisted heterogeneous catalytic system. Fuel Process. Technol. 97, 1-8.

[16] Santos, F.F., Malveira, J.Q., Cruz, M.G., Fernandes, F.A., 2010. Production of biodiesel by ultrasound assisted esterification of *Oreochromis niloticus* oil. Fuel. 89(2), 275-279.

[17] Subhedar, P.B., Gogate, P.R., 2016. Ultrasound assisted intensification of biodiesel production using enzymatic interesterification. Ultrason. Sonochem. 29, 67-75.

[18] Tiwari, A.K., Kumar, A., Raheman, H., 2007. Biodiesel production from jatropha oil (*Jatropha curcas*) with high free fatty acids: an optimized process. Biomass Bioenergy. 31(8), 569-575.

[19] Wang, Y., Ou, S., Liu, P., Xue, F., Tang, S., 2006. Comparison of two different processes to synthesize biodiesel by waste cooking oil. J. Mol. Catal. A: Chem. 252(1-2), 107-112.

[20] Yin, X., Ma, H., You, Q., Wang, Z., Chang, J., 2012. Comparison of four different enhancing methods for preparing biodiesel through transesterification of sunflower oil. Appl. Energ. 91(1), 320-325.

# PERMISSIONS

All chapters in this book were first published in Biofuel Research Journal, by Green Wave Publishing of Canada; hereby published with permission under the Creative Commons Attribution License or equivalent. Every chapter published in this book has been scrutinized by our experts. Their significance has been extensively debated. The topics covered herein carry significant findings which will fuel the growth of the discipline. They may even be implemented as practical applications or may be referred to as a beginning point for another development.

The contributors of this book come from diverse backgrounds, making this book a truly international effort. This book will bring forth new frontiers with its revolutionizing research information and detailed analysis of the nascent developments around the world.

We would like to thank all the contributing authors for lending their expertise to make the book truly unique. They have played a crucial role in the development of this book. Without their invaluable contributions this book wouldn't have been possible. They have made vital efforts to compile up to date information on the varied aspects of this subject to make this book a valuable addition to the collection of many professionals and students.

This book was conceptualized with the vision of imparting up-to-date information and advanced data in this field. To ensure the same, a matchless editorial board was set up. Every individual on the board went through rigorous rounds of assessment to prove their worth. After which they invested a large part of their time researching and compiling the most relevant data for our readers.

The editorial board has been involved in producing this book since its inception. They have spent rigorous hours researching and exploring the diverse topics which have resulted in the successful publishing of this book. They have passed on their knowledge of decades through this book. To expedite this challenging task, the publisher supported the team at every step. A small team of assistant editors was also appointed to further simplify the editing procedure and attain best results for the readers.

Apart from the editorial board, the designing team has also invested a significant amount of their time in understanding the subject and creating the most relevant covers. They scrutinized every image to scout for the most suitable representation of the subject and create an appropriate cover for the book.

The publishing team has been an ardent support to the editorial, designing and production team. Their endless efforts to recruit the best for this project, has resulted in the accomplishment of this book. They are a veteran in the field of academics and their pool of knowledge is as vast as their experience in printing. Their expertise and guidance has proved useful at every step. Their uncompromising quality standards have made this book an exceptional effort. Their encouragement from time to time has been an inspiration for everyone.

The publisher and the editorial board hope that this book will prove to be a valuable piece of knowledge for researchers, students, practitioners and scholars across the globe.

# LIST OF CONTRIBUTORS

**Mostafa Rahimnejad, Gholamreza Bakeri and Ghasem Najafpour**
Biotechnology Research Lab., Faculty of Chemical Engineering, Babol University of Technology, Babol, Iran

**Mostafa Ghasemi**
Fuel Cell Institute, Universiti Kebangsaan Malaysia, 43600 UKM Bangi, Selangor Darul Ehsan, Malaysia

**Sang-Eun Oh**
Department of Biological Environment, Kangwon National University, Chuncheon, Kangwon-do, Republic of Korea

**Meisam Tabatabaei**
Microbial Biotechnology and Biosafety Department, Agricultural Biotechnology Research Institute of Iran (ABRII), AREEO, Karaj, Iran
Biofuel Research Team (BRTeam), Karaj, Iran

**Keikhosro Karimi**
Department of Chemical Engineering, Isfahan University of Technology, 84156-83111 Isfahan, Iran

**Ilona Sárvári Horváth**
Swedish Centre for Resource Recovery, University of Borås, 501 90 Borås, Sweden

**Rajeev Kumar**
Center for Environmental Research and Technology (CE-CERT), Bourns College of Engineering, University of California, Riverside, California, USA

**Temitope E. Odetoye**
Department of Chemical Engineering, University of Ilorin, PMB1515, Ilorin, Nigeria
European Bioenergy Research Institute, CEAC, Aston University, Birmingham, United Kingdom

**Kolawole R. Onifade**
Department of Chemical Engineering, LadokeAkintola University of Technology, Ogbomoso, Nigeria

**Muhammad S. AbuBakar**
European Bioenergy Research Institute, CEAC, Aston University, Birmingham, United Kingdom

**James O. Titiloye**
European Bioenergy Research Institute, CEAC, Aston University, Birmingham, United Kingdom
College of Engineering, Swansea University, Swansea, SA2 8PP, United Kingdom

**C. Pothiraj**
Department of Botany, Alagappa Government Arts College, Karaikudi 630003, Tamilnadu, India

**A. Arun**
Department of Energy Science, Alagappa University, Karaikudi 630004, Tamilnadu, India

**M. Eyini**
Centre in Botany, Thiagarajar College, Madurai 625009, Tamilnadu, India

**Roman A. Voloshin**
Controlled Photobiosynthesis Laboratory, Institute of Plant Physiology, Russian Academy of Sciences, Botanicheskaya Street 35, Moscow 127276, Russia

**Vladimir D. Kreslavski**
Controlled Photobiosynthesis Laboratory, Institute of Plant Physiology, Russian Academy of Sciences, Botanicheskaya Street 35, Moscow 127276, Russia
Institute of Basic Biological Problems, Russian Academy of Sciences, Pushchino, Moscow Region 142290, Russia

**Sergey K. Zharmukhamedov**
Institute of Basic Biological Problems, Russian Academy of Sciences, Pushchino, Moscow Region 142290, Russia

**Vladimir S. Bedbenov**
Controlled Photobiosynthesis Laboratory, Institute of Plant Physiology, Russian Academy of Sciences, Botanicheskaya Street 35, Moscow 127276, Russia

**Seeram Ramakrishna**
Department Center for Nanofibers and Nanotechnology, Department of Mechanical Engineering, National University of Singapore, 117576, Singapore

**Suleyman I. Allakhverdiev**
Controlled Photobiosynthesis Laboratory, Institute of Plant Physiology, Russian Academy of Sciences, Botanicheskaya Street 35, Moscow 127276, Russia
Institute of Basic Biological Problems, Russian Academy of Sciences, Pushchino, Moscow Region 142290, Russia
Department of Plant Physiology, Faculty of Biology, M.V. Lomonosov Moscow State University, Leninskie Gory 1-12, Moscow 119991, Russia

**Keshini Beetul and Daneshwar Puchooa**
Faculty of Agriculture, University of Mauritius, Réduit, Mauritius

**Shamimtaz Bibi Sadally, Nawsheen Taleb-Hossenkhan and Ranjeet Bhagooli**
Faculty of Science, University of Mauritius, Réduit, Mauritius

**Mamatha Devarapalli and Hasan K. Atiyeh**
Department of Biosystems and Agricultural Engineering, Oklahoma State University, Stillwater, OK 74078, USA

**Erick Heredia-Olea, Esther Pérez-Carrillo and Sergio O. Serna-Saldívar**
Centro de Biotecnología FEMSA, Escuela de Ingeniería y Ciencias, Tecnológico de Monterrey, Avenida Eugenio Garza Sada 2501 Sur, CP 64849, Monterrey, Nuevo León, México

**Mehrdad Mashkour and Mostafa Rahimnejad**
Biofuel and Renewable Energy Research Center, Department of Chemical Engineering, Babol Noshirvani University of Technology, Babol, Iran

**Esra Uçkun Kiran and Antoine P. Trzcinski**
Advanced Environmental Biotechnology Centre, Nanyang Environment & Water Research Institute, Nanyang Technological University, 1 Cleantech Loop, Singapore 637141, Singapore

**Yu Liu**
Advanced Environmental Biotechnology Centre, Nanyang Environment & Water Research Institute, Nanyang Technological University, 1 Cleantech Loop, Singapore 637141, Singapore
Division of Environmental and Water Resources Engineering, School of Civil and Environmental Engineering, Nanyang Technological University, 50 Nanyang Avenue, Singapore 639798, Singapore

**Paniz Izadi and Mostafa Rahimnejad**
Biotechnology Research Lab., Faculty of Chemical Engineering, Babol University of Technology, Babol, Iran

**Yasuo Kojima and Yoshiaki Kato**
Department of Applied Biological Chemistry, Faculty of Agriculture, Niigata University, 2-8050 Ikarashi, Nishi-ku, Niigata, 950-2181, Japan

**Minami Akazawa**
Graduate School of Science and Technology, Niigata University, Niigata, 950-2181, Japan

**Seung-Lak Yoon**
Department of Interior Materials Engineering, Gyeongnam National University of Science and Technology, 150 Chiram-Dong, Jinju, Gyeongnam, 660-758, Korea

**Myong-Ku Lee**
Department of Paper Science & Engineering, Kangwon National University, 192-1 hyoja 2-dong, Chuncheon, 200-701, Korea

**Victoria Rosalía Durán-Padilla, Norma Angélica Chávez-Vela and Juan Jáuregui-Rincón**
Departamento de Ingeniería Bioquímica, Universidad Autónoma de Aguascalientes (UAA), Av. Universidad 940 Ciudad Universitaria, Aguascalientes, Aguascalientes, México

**Gustavo Davila-Vazquez**
Tecnología Ambiental, Centro de Investigación y Asistencia en Tecnología y Diseño del Estado de Jalisco, A.C.(CIATEJ), Normalistas 800, Colinas de La Normal, Guadalajara, Jalisco, México

**José Raunel Tinoco-Valencia**
Escalamiento y Planta piloto, Instituto de Biotecnología, Universidad Nacional Autónoma de México (UNAM), Av. Universidad 2001, Chamilpa, Cuernavaca, Morelos, México

**Vipan K Sohpal**
Department of Chemical Engineering, Beant College of Engineering & Technology, Post Box No 13, Gurdaspur Punjab, India

**Amarpal Singh**
Department of Electronics & Communication Engineering Beant College of Engineering & Technology, Post Box No 13, Gurdaspur Punjab, India

**Aline S. Simões, Lucas Ramos, Larissa Freitas, Julio C. Santos and Heizir F. de Castro**
Engineering School of Lorena, University of São Paulo Estrada Municipal do Campinho s/n, 12602-810 Lorena, São Paulo, Brazil

**Gisella M. Zanin**
State University of Maringa, Department of Chemical Engineering, Av. Colombo 5790, E-46, 87020-900, Maringa - PR, Brazil

**Raju S. Thombal and Vrushali H. Jadhav**
Department of Organic Chemistry, National Chemical Laboratory (CSIR-NCL), Pune-411008, India

**Somayeh FazeliNejad, Jorge A. Ferreira, Tomas Brandberg, Patrik R. Lennartsson and Mohammad J. Taherzadeh**
Swedish Centre for Resource Recovery, University of Borås, SE 501 90, Borås, Sweden

**Juciana Clarice Cazarolli, Patrícia Dörr de Quadros, Francielle Bücker and Fátima Menezes Bento**
Department of Microbiology, Federal University of Rio Grande do Sul. Rua Sarmento Leite, N° 500, 90050-170, Porto Alegre, RS, Brazil

**Ruiz Frazão Santiago and Eduardo Homem de Siqueira Cavalcanti**
Corrosion and Degradation Division, National Institute of Technology, Av. Venezuela, N°82, Sala 608, 200081-312, Rio de Janeiro, RJ, Brazil

**Clarisse Maria Sartori Piatnicki and Maria do Carmo Ruaro Peralba**
Department of Inorganic Chemistry, Federal University of Rio Grande do Sul. Av. Bento Gonçalves, N°9500, 91501-970, Porto Alegre, RS, Brazil

**Ilona Sárvári Horváth**
Swedish Centre for Resource Recovery, University of Borås, 501 90 Borås, Sweden

**Meisam Tabatabaei**
Microbial Biotechnology Department, Agricultural Biotechnology Research Institute of Iran (ABRII), AREEO, Karaj, Iran
Biofuel Research Team (BRTeam), Karaj, Iran

**Keikhosro Karimi**
Department of Chemical Engineering, Isfahan University of Technology, Isfahan 84156-83111, Iran
Microbial Industrial Biotechnology Group, Institute of Biotechnology and Bioengineering, Isfahan University of Technology, Isfahan 84156-83111, Iran

**Rajeev Kumar**
Center for Environmental Research and Technology (CE-CERT), Bourns College of Engineering, University of California, Riverside, California, USA

**Somayeh Farzad, Mohsen Ali Mandegari and Johann F. Görgens**
Department of Process Engineering, University of Stellenbosch, Private Bag X1, Matieland, 7602, South Africa

**Kolsoum Pourzare and Saeed Farhadi**
Membrane Research Laboratory, Lorestan University, Khorramabad, P.O. Box 68137-17133, Iran

**Yaghoub Mansourpanah**
Membrane Research Laboratory, Lorestan University, Khorramabad, P.O. Box 68137-17133, Iran
Membrane Separation Technology (MST) Group, Biofuel Research Team (BRTeam), Karaj, Iran

**Dheeraj Rathore**
School of Environment and Sustainable Development, Central University of Gujarat, Gandhinagar-382030, Gujarat, India

**Abdul-Sattar Nizami**
Solid Waste Research Group, Center of Excellence in Environmental Studies (CEES), King Abdul Aziz University, P.O Box: 80216, Jeddah 21589, Saudi Arabia

**Anoop Singh**
Government of India, Ministry of Science and Technology, Department of Scientific and Industrial Research (DSIR), Technology Bhawan, New Mehrauli Road, New Delhi- 110016 India

**Deepak Pant**
Separation and Conversion Technologies, VITO-Flemish Institute for Technological Research, Boeretang 200, 2400 Mol, Belgium

**Mortaza Aghbashlo, Soleiman Hosseinpour and Seyed Sina Hosseini**
Department of Mechanical Engineering of Agricultural machinery, Faculty of Agricultural Engineering and Technology, College of Agriculture and Natural Resources, University of Tehran, Karaj, Iran

**Meisam Tabatabaei and Zahra Khounani**
Agricultural Biotechnology Research Institute of Iran (ABRII), Agricultural Research, Education, and Extension Organization (AREEO), Karaj, Iran
Biofuel Research Team (BRTeam), Karaj, Iran

**Akram Ghaffari**
Agricultural Biotechnology Research Institute of Iran (ABRII), Agricultural Research, Education, and Extension Organization (AREEO), Karaj, Iran

**Pouya Mohammadi**
Biofuel Research Team (BRTeam), Karaj, Iran

# Index

www.ingramcontent.com/pod-product-compliance
Lightning Source LLC
Chambersburg PA
CBHW050448200326
41458CB00014B/5106

* 9 7 8 1 6 8 2 8 6 6 0 9 2 *